DAVID HORNER is Professor of Australian Defence History in the Strategic and Defence Studies Centre, Australian National University. A graduate of the Royal Military College, Duntroon, he served as an infantry platoon commander in South Vietnam and has had many years of regimental and staff experience. In 1998, as an Army Reserve colonel, he became the first head of the Army's Land Warfare Studies Centre. He has written numerous books on military command, operations, defence policy and strategy, including *Defence Supremo*, *Blamey: The Commander-in-Chief* and *Breaking the Codes* (co-authored with Desmond Ball).

SAS: Phantoms of War

A history of the Australian Special Air Service

Updated edition of *SAS: Phantoms of the Jungle*

David Horner

ALLEN&UNWIN

First published in 1989
This edition published 2002

Allen & Unwin
83 Alexander Street
Crows Nest NSW 2065
Australia
Phone: (61 2) 8425 0100
Fax: (61 2) 9906 2218
Email: info@allenandunwin.com
Web: www.allenandunwin.com

National Library of Australia
Cataloguing-in-Publication entry:

Horner, D.M. (David Murray), 1948– .
   SAS: phantoms of war: a history of the Australian Special Air Service.

   2002 ed.
   Bibliography.
   Includes index.
   ISBN 1 86508 647 9

   1. Australia. Army. Special Air Service Regiment—History.
   2. Australia Army—Commando troops. I. Title.

356.1670994

Set in 10.5/12 pt Times by Excel Imaging Pty Ltd, Australia,
and Midland Typesetters, Maryborough, Vic.
Printed by South Wind Production Services, Singapore

10 9 8 7 6 5 4 3 2 1

Contents

Maps

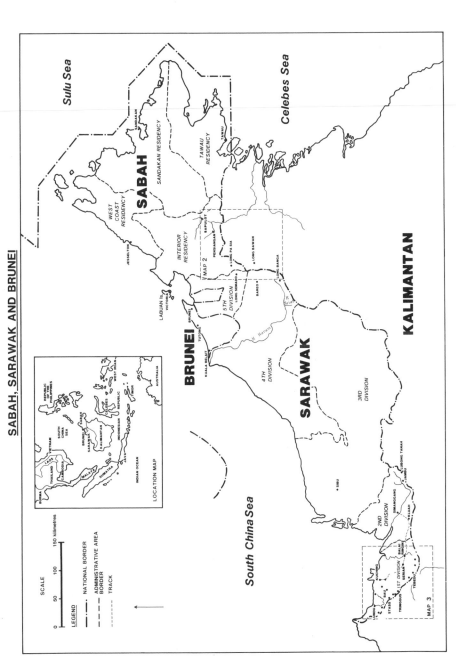

Map 1   This map shows the two areas where the SAS served in Borneo. 1 SAS Squadron generally operated in the area marked as Map 2 in 1965, while 2 SAS Squadron operated in the area marked as Map 3 in 1966.

BORDER AREA: SABAH, SARAWAK, KALIMANTAN

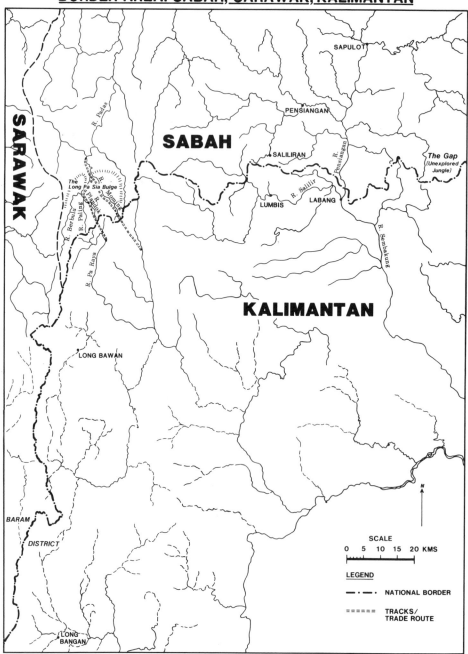

Map 2   This map shows the area where 1 SAS Squadron generally operated in 1965, as
described in chapters 5 to 8. The area along the border between Kalimantan and
Sarawak and Sabah consisted of a tangle of steep mountains separated by numerous
rivers, and was covered by thick jungle. Maps were unreliable and it was difficult to know
when the border had been crossed.

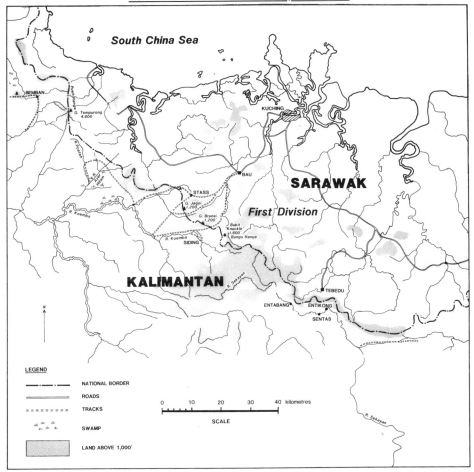

Map 3   This map shows the area where 2 SAS Squadron operated in 1965, as described in chapters 9 and 10. The border was usually along a jungle-covered ridge leading down to rivers, which were the main means of communication in this area of Kalimantan. The Sarawak side consisted of scattered patches of jungle with small cultivated areas around the kampongs.

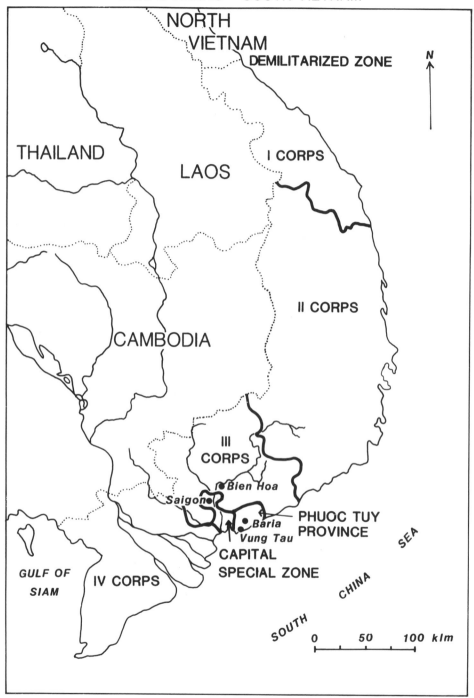

Map 4 This map shows the location of Phuoc Tuy Province in relation to the remainder of South Vietnam.

PHUOC TUY PROVINCE

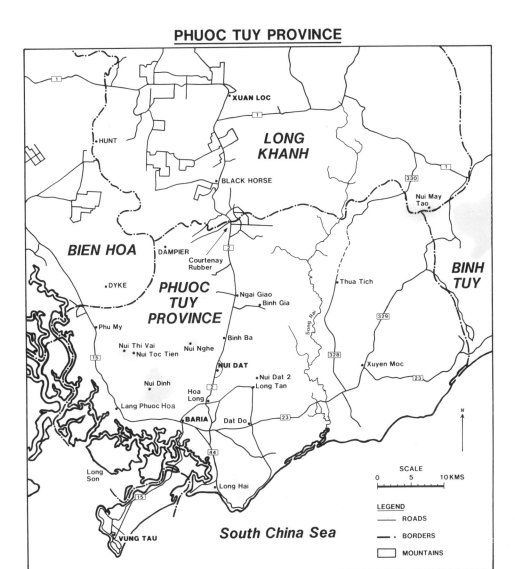

Map 5 From 1966 to 1971 the Australian SAS Squadron was based at Nui Dat and operated in the area covered by this map. Missions were conducted throughout Phuoc Tuy Province and across the border into Bien Hoa, Long Khanh and Binh Tuy provinces. The border and mountain areas were generally covered with extensive areas of jungle, while the areas around the villages and towns were usually cultivated, either with rice or rubber plantations.

EAST TIMOR

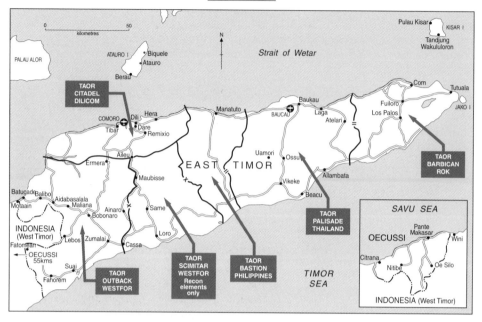

Map 6   In September 1999 the 3 SAS Squadron group was deployed to East Timor, and was based at Dili. It conducted operations throughout the territory, but mainly in the western areas, until replaced by 1 SAS Squadron in December 1999.

Preface
to the 2002 edition

The thirteen years since the initial publication of my history of the Australian SAS has seen momentous changes, with the development of new capabilities within the regiment and its deployment on numerous operations. As a result, it has been decided to republish the original book with two new chapters bringing up to date the story to the end of 2000. This includes deployments on peace operations during the 1990s, culminating in the operation in East Timor in 1999–2000. There is some irony here—the regiment's first operations in 1963 were in the Indonesian territory of Borneo (Kalimantan), its most recent overseas operations in the Indonesian territory of East Timor, now an independent country. The wide range of activities carried out by the regiment over the past decade has made it impossible to describe them in the same depth as in the original account, nor can details of some recent activities be published at this stage. Also, it is not possible to mention the names of as many individual soldiers as was the case in the earlier account. The story that emerges, however, shows how the regiment has outgrown the Vietnam legacy and has secured its place as the Australian Defence Force's force of choice. The book remains, therefore, as a comprehensive account of the regiment's service during a period of more than 40 years. I have made no attempt to rewrite the original chapters, merely deleting the old final chapter and replacing it with two new chapters. I have made a very few minor corrections to the rest of the material.

As with the earlier edition, I could not have written the new chapters without the support of the commanding officer—during 1999–2000, Lieutenant Colonel Tim McOwan. During a research visit to Swanbourne he gave me access to unit files and facilitated interviews with serving unit members. I am grateful to him, the many SAS officers and soldiers who shared their stories with me, and to successive Commanders Special Forces, Brigadiers Philip McNamara and Duncan Lewis, who cleared the way for the project. As before, the additional chapters have no official status, although they have been reviewed to protect SAS techniques and other matters concerning national security.

David Horner
Canberra
August 2001

Preface
to the 1989 edition

The purpose of this book is to describe the operations of the Australian Special Air Service from its establishment in 1957 to the present time. It is not intended to be an organisational history of the regiment or a year book of events and personalities. Rather it seeks to show how the SAS has been employed in peace and war.

The book also seeks to put to rest some misconceptions about the role and purpose of the SAS. The first concerns secrecy. It is true that SAS operational techniques are best kept secret, and obviously while on operations the SAS is careful about security, particularly concerning the details of future missions. But in many respects the SAS has little to hide. The regiment is as much under military discipline and control as any other unit in the Australian Army. Its roles are clearly stated and are in accordance with government policy. No particular effort is made to hide the identity of SAS soldiers and, except for specific exercises, they wear uniform while on duty.

Another misconception concerns the nature of the men in the SAS. They are not Rambo types, super-soldiers, or wild men itching for a fight. There is no doubt that they are tough, but they are intelligent, highly disciplined and professional in their approach.

It is hoped that through describing the development of the SAS and its experiences on operations, this book will promote a better understanding of the nature of the SAS. It is also hoped that it will give recognition to the relatively small group of men who have served Australia with courage and devotion, but with little public acknowledgement, over a period of 30 years.

When the Director of Special Action Forces (DSAF), Colonel Rod Curtis, first approached me to write this book I was excited at the prospect, even though I was not an SAS soldier. Under normal circumstances I would not have been able to undertake the work. It could not easily be undertaken as a part-time project and the Army is not able to employ its personnel to write unit histories as their primary duty. I was, however, at the time a visiting fellow in the Department of History of the University College, Australian Defence Force Academy, and a history of the SAS seemed to be an appropriate research topic for me to pursue in my capacity as a full-time academic. I had not anticipated that after a year of work, and with the book about half written, I would be posted to the staff of the Australian Joint Services Staff College. Fortunately the Commandant,

Brigadier Steve Gower, was sympathetic to my plight. While the book was not my main work at the college, his forbearance enabled me to complete the task, even though much of it was undertaken in my own time.

The book could not have been written without the assistance of a large number of serving and former members of the SAS. I am grateful for their generosity with their time, and their willingness to share their experiences with an outsider to the SAS family. The names of those with whom I had significant communication are listed in the bibliography. The staff of the Directorate of Special Action Forces, and its two directors during the period when the book was being written, Colonels Rod Curtis and Chris Roberts, proviced excellent support and smoothed the way past many obstacles. The two commanding officers of the SAS Regiment, Lieutenant-Colonels Terry Nolan and Jim Wallace, placed the resources of the regiment at my disposal. Along with their second-in-command, Major Andy Leahy, they and the remainder of the regiment made me very welcome during my three research visits to Campbell Barracks and on my visit to a major counter-terrorist (CT) exercise in 1987. Corporal Steve Danaher, the curator of the regiment's historical collection, painstakingly compiled the active service nominal roll, and also arranged the copying of many of the photographs in the book. The maps were drawn by Corporal Mal Sharp of the regiment's training support wing.

The primary sources for the book were twofold. The first source was the records of the SAS Regiment (SASR). With respect to peacetime activities these records were patchy, and needed to be supplemented by research into Army Headquarters, Army Office and Defence records. However, the regiment's operational activities were well documented and almost 1400 patrol reports were available. The security classification of these reports had to be downgraded so that they could be used in the book.

The second source was the participants themselves. I was aware that the patrol reports might not necessarily tell the full story, or even, at times, the correct story, so in cases where I decided to describe a mission in more than just a few sentences, I made every effort to obtain the views of the participants. The SAS is composed of strong individuals, all with their own strongly held points of view. The only solution was to give as many people as possible the opportunity to comment on my drafts; in all, a total of some 40 serving and former SAS soldiers read and commented on draft chapters.

Before undertaking the work, I sought and received agreement that as author it would be my decision as to what was published. The only caveat was that the book had to be cleared by SAS staff to safeguard SAS techniques. The small number of Army and Defence files from the closed period also had to be cleared, and I am grateful to Major Louise Davidson

for assisting with that process. I need hardly add that I accept full responsibility for any errors of fact or interpretation.

While I had access to Australian SAS records, I obtained only limited
access to other Australian official files. The views expressed in the book
do not necessarily reflect those of the Australian Department of Defence,
nor are they endorsed by it. Thus while the book received the support of
the SASR, it has no official status.

To save space and so as not to unduly distract the reader I have
deleted about 80 per cent of the notes. The deleted notes refer mainly to
patrol reports, interviews or private letters. Copies of the typescript with
complete notes will be lodged with the SASR, DSAF and the Australian
War Memorial where it will be available to researchers.

I have tried to minimise the use of military jargon in the book, but it might
be helpful to keep in mind a number of definitions currently in use by the
Australian Defence Force. These are:

Reconnaissance

A mission undertaken to obtain, by visual observation or other detection
methods, information about the activities and resources of an enemy or
potential enemy, or to secure data concerning the meteorological, hydro-
graphic, or geographic characteristics of a particular area.

Special Action Forces

Army units which perform operational roles that are not normal for
conventional forces. They include Special Air Service, Commando and
special forces signals units.

Special Forces

Military personnel with cross-training in basic and specialised military
skills, organised into small multiple purpose detachments with the
mission to train, organise, supply, direct, and control indigenous forces
in guerilla warfare and counter-insurgency operations, and to conduct
unconventional warfare operations.

Special Operations

Secondary or supporting operations which may be adjuncts to various
other operations and for which no one service is assigned primary
responsibility.

Preface to the 1989 edition

Special Warfare

Embraces all the military and para-military measures and activities related to unconventional warfare, assistance to indigenous forces and other operations outside the scope of conventional forces.

Surveillance

The systematic observation of aerospace, surface or sub-surface areas, places, persons, or things, by visual, aural, electronic, photographic, or other means.

Unconventional Warfare

General term used to describe operations conducted for military, political or economic purposes within an area occupied by the enemy and making use of the local inhabitants and resources[1].

Writing a book is like undertaking a long distance running race. Ultimately the writer must rely on his own skill, judgement and endurance. But it helps immeasurably to have the support of an understanding and loving family. I am deeply grateful to my wife, Sigrid, and my three children for their patience and encouragement over the two years it has taken to complete the book.

David Horner
Canberra
February 1989

Abbreviations

2ic	Second-in-command
AATTV	Australian Army Training Team Vietnam
ADC	Aide-De-Camp
ADF	Australian Defence Force
AFV	Australian Force Vietnam
AIB	Allied Intelligence Bureau
AIF	Australian Imperial Force
ALSG	(1st) Australian Logistic Support Group
AO	Area of Operations
APC	Armoured Personnel Carrier
ASC	Australian Services Contingent (Somalia)
ASLAV	Australian Light Armoured Vehicle
ATF	(1st) Australian Task Force
ARVN	Army of the Republic of Vietnam
CDF	Chief of the Defence Force
CIA	Central Intelligence Agency
CMF	Citizen Military Forces
CNRT	National Council of Timorese Resistance
CSAR	Combat Search and Rescue
CSM	Company Sergeant Major
CSOTF	Combined Special Operations Task Force
CT	Counter-Terrorist (capability or force)
DAAG	Deputy Assistant Adjutant General
DAQMG	Deputy Assistant Quartermaster General
DJFHQ	Deployable Joint Force Headquarters
DOBOPS	Director of Borneo Operations
DZ	Drop Zone
FAC	Forward Air Controller
FALINTIL	Armed Forces for the National Liberation of East Timor
FARELF	Far East Land Forces
GPMG	General Purpose Machine Gun
GOC	General Officer Commanding
GSO1	General Staff Officer grade one
IDP	Internally Displaced Person
INTERFET	International Force East Timor
ISD	Inter-Allied Services Department

JSOTF	Joint Special Operations Task Force
JTF	Joint Task Force
KIA	Killed in action
LCM	Landing Craft Medium
LCT	Liaison and Communication Team
LCVP	Landing Craft Vehicle and Personnel
LFT	Light Fire Team
LMG	Light Machine Gun
LofC	Line(s) of Communication
LP	Landing Place
LRRP	Long Range Reconnaissance Patrol
LUP	Lying Up Place or Position
LZ	Landing Zone
MACV	Military Assistance Command Vietnam
NATO	North Atlantic Treaty Organisation
NCO	Non-Commissioned Officer
NEFIS	Netherlands East Indies Regional Section
NGO	Non-government Organisation
NVA	North Vietnamese Army
OAG	Offshore Assault Group
OAT	Offshore Assault Team
OC	Officer Commanding (a squadron or company)
OP	Observation Post
OTLP	One Time Letter Pad
PF	Popular Force
PIR	Pacific Islands Regiment
PNGDF	Papua New Guinea Defence Force
RAAF	Royal Australian Air Force
RAN	Royal Australian Navy
RAR	Royal Australian Regiment
R and C	Rest and Recuperation
RSM	Regimental Sergeant Major
RV	Rendezvous
SAC-PAV	Standing Advisory Committee on Commonwealth State Cooperation for Protection Against Violence
SAF	Special Action Forces
SAS	Special Air Service
SASR	Special Air Service Regiment

Abbreviations

SBS	Special Boat Squadron or Service (Royal Marines)
SEAL	Sea Air and Land (team—US Navy)
SEATO	South East Asia Treaty Organisation
SLR	Self Loading Rifle
SMG	Sub Machine-Gun
SO2	Staff Officer grade two
SOC	Special Operations Component
SOCCE	Special Operations Command and Control Element
SOCCENT	Special Operations Command Central Command
SOE	Special Operations Executive
SOP	Standing Operating Procedure
SQMS	Squadron Quartermaster Sergeant
SRD	Services Reconnaissance Department
SSM	Squadron Sergeant-Major
SUR	Ship Underway Recovery
TAG	Tactical Assault Group
TAOR	Tactical Area of Responsibility
TF	Task Force
TNI	Indonesian Armed Forces
UN	United Nations
UNAMET	United Nations Assistance Mission East Timor
UNAMIR	United Nations Assistance Mission in Rwanda
UNOSOM	United Nations Operation in Somalia
UNSCOM	United Nations Special Commission
UNTAET	United Nations Transitional Administration East Timor
US	United States
USAF	United States Air Force
WIA	Wounded in action
VC	Viet Cong

1
The tractor job:
2 Squadron, March 1968

It was March 1968 and Vietnam was at the height of the dry season. In the jungles, villages and paddy fields of the Republic of South Vietnam the Americans and their allies were fighting countless battles with the Viet Cong and North Vietnamese Army (NVA), while for many South Vietnamese villagers the routine of tending their crops went on as it had for decades. From the passenger seat of the Army Cessna light aircraft Sergeant Frank Cashmore looked down at the dry, yellow-brown paddy fields of Phuoc Tuy Province passing slowly 1000 metres below. To his right he could see a haze of smoke where farmers were burning off the rice stubble in preparation for the growing season. To his left and also ahead was the expanse of green jungle, known as the Hat Dich area, that covered the north west sector of the province. This was the traditional home of the 274th VC Regiment. But Cashmore had little thought for the wider issues of the war. On his knees was spread his contact-covered map with possible Landing Zones (LZ) circled in black, for the selection of an LZ was one of the crucial decisions facing an SAS patrol commander. Behind him sat Corporal Danny Wright, second-in-command of his patrol, and together they were planning their first operational mission in Vietnam.

Frank Cashmore knew that an SAS patrol was most vulnerable at the moment of its insertion into enemy-held territory. Would a large VC reception party be waiting on the edge of the LZ? Would the aircraft activity alert a nearby VC force so that it could move quickly to intercept and perhaps ambush the newly arrived SAS? Had the VC planted anti-personnel mines or booby traps across the LZ? After all, the SAS had been operating in Phuoc Tuy for the past twenty months and the VC had a reasonable understanding of SAS insertion techniques. And there were other considerations; for this operation Cashmore's men would be carrying heavy loads and it was important that the LZ be relatively close to the target area.

To his left Cashmore could now see the straight red-brown scar of the Firestone Trail through the green vegetation. Hacked out of the jungle by the great Rome ploughs of the US Army engineers, it facilitated the movement of armoured vehicles and enabled the reconnaissance flights to detect whether anyone had walked or driven on it during the night. As the aircraft flew north Cashmore saw the trail swing east, cutting across their flight path a kilometre ahead. It crossed the grassy open patch known as LZ Dampier, and stretched away towards the Courtenay rubber plantation. Cashmore became more alert. LZ Dampier was the target area. They turned east, following the Firestone Trail, and shortly before reaching LZ Dampier they spotted a large bomb crater that might provide suitable cover during the operation. Not wanting to risk another flight over that area they turned south. Then Cashmore saw what he was looking for; about 600 metres south of Dampier was a small clearing that had been occupied by elements of the 2nd Battalion, The Royal Australian Regiment (2 RAR) during Operation DUNTROON in January 1968. This would be a suitable LZ. Cashmore tried to take in as much as he could as the Cessna kept on its course without slackening speed.

It was by sheer chance that Frank Cashmore had been given the tractor job.[1] Along with 30 other members of the advance party of 2 SAS Squadron, he had arrived at Nui Dat on 7 February 1968 and had begun the process of in-country training and familiarisation before the rest of the squadron arrived on 27 February. There was a lot for a young SAS patrol commander to learn from the experienced hands of 1 Squadron who were preparing to return to Australia: the rules of engagement; the orders for the observation posts manned by the SAS on Nui Dat hill; the layout of the sprawling Task Force base half hidden beneath the trees of the rubber plantation below the SAS enclave on Nui Dat hill; and the procedures for calling for support from the helicopters of No 9 Squadron RAAF. Cashmore knew the SAS Standing Operating Procedures backwards, but there were many local factors to be absorbed.

One important factor was the enemy situation. Barely a week before the SAS advance party arrived the VC had begun their celebrated Tet offensive. Two Australian infantry battalions, 2 and 7 RAR, were already operating on the border of Bien Hoa and Long Khanh Provinces. The remaining battalion, 3 RAR, which had just completed its acclimatisation and training, was soon in action clearing the enemy from the provincial capital of Baria and the nearby town of Long

Sergeant Frank Cashmore, commander of the patrol that successfully conducted the tractor ambush in March 1968. Aged 25 at the time, it was his first operational patrol in Vietnam.

Dien, seven kilometres south west and south of the Task Force base respectively. Both Long Dien and Hoa Long, only three kilometres south of Nui Dat, had to be cleared again, and in each case 1 SAS Squadron was asked to assist with the cordon and search. The new arrivals from 2 Squadron accompanied 1 Squadron and gained valuable experience, although it was not the sort of task for which the SAS was trained. Before 1 Squadron departed Cashmore was able to join only one SAS patrol, a 21 man ambush patrol delivered by armoured personnel carriers six kilometres west of Nui Dat. The country was mainly flat, open paddy fields with high grass and the ambush was in a thicket of bamboo. As it was the dry season, water was scarce, the sun was hot, and movement through the dry grass was noisy. No enemy were seen on the patrol, which lasted from 9 to 12 February, and it was scarcely an ideal introduction to SAS patrolling in Vietnam.

As a well-trained SAS patrol commander, however, Cashmore learned from these and other experiences. Raised in Collie, Western Australia, he had enlisted in the Army with the express purpose of becoming an SAS soldier. At that time new soldiers were required to serve for eighteen months in another Army posting before applying for SAS. Cashmore went to infantry, joining 2 RAR at Terendak in Malaya. In August 1962 the battalion was deployed to the Thai–Malay border region for anti-terrorist operations, and Cashmore spent two months on these operations as a Bren-gunner—good training for later patrols in a more hostile environment.

While Cashmore was in Malaya, Captain Mike Jeffery from the SAS visited the battalion on a recruiting tour. Cashmore applied, returned to Australia, completed the cadre course and joined the SAS in mid 1963. By early 1966 when 2 Squadron was deployed to Borneo he had reached the rank of lance corporal. The fourteen day patrols into Indonesian territory were arduous and testing, but as in Malaya, Cashmore did not fire a shot although he saw a few Indonesian soldiers. Nevertheless, it was excellent preparation for Vietnam.

On return from Borneo in August 1966 Cashmore was promoted to corporal and towards the end of 1967, when the squadron was on exercise in New Guinea, he was promoted to sergeant to replace a patrol commander who had not performed adequately. Cashmore, now 25 years old, arrived in Vietnam as a recently promoted patrol commander in February 1968. Emotional and highly strung, Cashmore was under considerable pressure as he prepared for his first patrol. Fortunately Corporal Danny Wright was an experienced SAS soldier. Aged 28, he had joined the SAS in 1960 and served as second-in-command of a patrol in Borneo, although like Cashmore, he had seen no action. Returning from Borneo, Wright had left the army to run a parachute school until as he put it, the 'bugle called when the SAS became involved in Vietnam'. But the remainder of Cashmore's patrol who joined him at Nui Dat on 27 February—Privates Kim McAlear (aged 19), Adrian Blacker (21) and David Elliott (22)—had only recently joined the SAS and had no operational experience. Blacker recalled that he and McAlear were still 'quite naive': to them it merely seemed as though they were preparing for an exercise, just as they had in Western Australia several weeks earlier.

It was probably because of this lack of experience that Cashmore's patrol was one of the last to be given a mission. By 9 March, when they were warned that they were to undertake a reconnaissance of the area around LZ Dampier between 15 and 20 March, the squadron had already deployed eleven patrols. It was sheer chance that Cashmore had already been tasked to go to the very area where an ambush mission was now required, and suddenly he found himself joining his squadron commander, Major Brian Wade, for the short Land Rover journey to Task Force headquarters in the rubber plantation east of Nui Dat hill. The Task Force was engaged in Operation PINNAROO clearing the VC from the notorious Long Hai hills south of Nui Dat,

4

Corporal Danny Wright, second-in-command of the patrol and commander of the demolitions group for the tractor ambush. (PHOTO, T. BOWDEN)

but the Task Force commander, Brigadier Ron Hughes, had a special task for the SAS. Cashmore and Wade were met by Hughes' SO2 Operations, Major Ian MacLean, who described their mission.

Each morning an aircraft from 161 (Independent) Recce Flight flew around the borders of the province looking for sign of enemy activity, and an observer had reported what appeared to be the tracks of a tractor and trailer across LZ Dampier. It was not known where the tractor had come from, but the French owner of the Courtenay rubber plantation had reported that his Fordson Major tractor had been stolen. It was thought that the tractor was being used to transport stores and ammunition from the more heavily populated area of the rubber plantations to the VC bases in the Hat Dich.

In the aftermath of the VC Tet offensive it was important to maintain pressure on the enemy in Phuoc Tuy Province as two Australian infantry battalions had been deployed outside the province since late January. The destruction of the tractor would seriously interfere with VC resupply activities and keep them on the defensive with a salutary reminder that they could not move with impunity in Phuoc Tuy Province. That the VC were sensitive to the security of the tractor was shown by information that the tractor might be escorted by 60 VC, 30 in the front and 30 in the rear.

Cashmore's task was to destroy the tractor. Both Cashmore and the other patrol members later asserted that they were told that since the tractor was owned by a French firm, for political reasons they had to conduct a deniable operation; that is, there should be no indication that the Australians were involved. Wade and Hughes both denied that this was required. But Wade did instruct his men that they should make it appear as though the tractor and its cargo had exploded spontaneously; there should be no indication that the SAS was in the area. [2]

The task presented considerable problems. First, they needed better information on the area. Cashmore obtained from 2 RAR good, recent, oblique, aerial photographs of the Firestone Trail near to LZ Dampier.

The second problem was the technique to be used to destroy the tractor. Wade had become familiar with the technique of 'demolition ambush' when he had attended the US Special Warfare Course at Fort Bragg in 1965. Given that the contents of the trailer and the size of the VC escort party were unknown, he saw demolition ambush as the only practicable solution and briefed Cashmore accordingly. Cashmore selected Corporal Danny Wright to head the demolition team, and Corporal Dave Scheele joined the patrol as his assistant. A quiet, 28 year old Dutchman, Scheele had completed his National Service with the Dutch commandos, served with 2 RAR and been on operations with 2 SAS Squadron in Borneo.

Once the technique had been decided and the team selected, Wade, Cashmore, Wright and Scheele visited 1 Field Engineer Squadron to investigate using anti-tank mines. None were available in Vietnam and the engineers recommended that they use Beehive explosive charges. Beehives were shaped charges that stood on short legs and were designed to blow a hole into the ground. The engineer squadron sergeant-major (SSM), Warrant Officer Turner, suggested that they remove the legs of the Beehive charges and set them upside down. Once initiated the charges would punch a hole right through the tractor and trailer into any explosives or weapons stored on the trailer. The engineers supplied four fifteen pound (6.8 kilogram) Beehive charges. To counter the enemy force accompanying the tractor it was decided to set up four Claymore mines.

The next problem was that the planners did not know what was on the trailer, and if it was packed with explosives they would not want to be too close when the demolition was initiated. Wade ordered that the SAS party had to be at least 100 metres from the explosion, so the

ELECTRICAL PRESSURE SWITCH
TRACTOR AMBUSH C/S 1/3 2 SQN /69

DESCRIPTION
IMPROVISED PRESSURE SWITCH
WITH SELF DESTRUCT DEVICE
ELEVATION

MANUFACTURED BY
1 FIELD SQN ENGINEERS
AT SAS REQUEST

NOTE
OPERATIONAL REQUIREMENTS DEMANDED
THE SWITCH ONLY FUNCTION ABOVE A
SPECIFIC WEIGHT THEREFORE THE SPRINGS
WERE CUT TO RESIST ANY LESSER WEIGHT.

LEGEND
BD – BAKING DISH
PP – PRESSURE PLATE
POS BSL – POS BATTERY STRAP & LEAD
NEG BSL – NEG BATTERY STRAP & LEAD
ZZ – CUT CLUTCH PLATE SPRINGS
SDD – SELF DESTRUCT DEVICE

Sketch showing the construction of the pressure switch to detonate the beehive charges using a baking dish, battery straps and clutch plate springs.

explosion would have to be detonated by the weight of the tractor. But what if the tractor did not appear on the first night? The patrol would have to be prepared to remain in position for up to seven nights, and the charges would have to be expertly camouflaged to survive the scrutiny of any VC who might pass on foot during this period. There was also a chance that an innocent Vietnamese Lambro or motor scooter might drive along the track, so the demolition device would have to be adjusted so that it could only be detonated by the heavy weight of the tractor.

Wright and Scheele tackled the problem energetically and worked out a wiring diagram to detonate electrically the four Beehives and the four Claymores which would be sited to catch the flank protection escorts. The only thing missing was a suitable pressure plate to close the circuit when the target's heaviest wheels were over it. The SSM of 1 Field Squadron designed a switch made from a baking dish, four Land Rover clutch springs (cut to compress with the weight of the main wheel of a Fordson Major tractor), an eighth of an inch (3mm) steel plate, twelve inches (30cm) by eighteen inches (46cm), and battery straps. The terminals were fixed to blocks of wood, one on the baking dish and the other on the steel plate. Two pounds (900 grams) of C4 explosive were taped underneath the baking dish to ensure the destruction of the switch itself. Cells from a disassembled 64 radio set battery were then placed in a beer can and an on/off switch, a self-destruct switch and two cable-connecting terminals were soldered onto the top. The device was constructed by members of the detachment of 152 Signals Squadron attached to the SAS Squadron.

By now Brigadier Hughes was becoming agitated by the apparent delay in undertaking the mission, but Wade explained to him that proper preparation was necessary. Rehearsals were conducted day and night for three days until every move was perfected. McAlear and

Blacker were to provide security to each flank while Wright and Scheele set up the demolitions. Communications between the sentries and the command group would be by URC 10 radios on 241 frequency; if enemy troops arrived unexpectedly the sentries would provide sufficient warning for the demolition team to hide until they had passed. Cashmore would keep overall control while Elliott removed the equipment from the packs in the correct order and handed it to the demolition men, ensuring that everything taken from the packs and not actually used was returned and not left on the track. It was soon apparent that they would have to carry a considerable weight and Wright decided to use only a single Don 10 wire as it would have to stretch up to 100 metres. Claymore leads of only 33 metres would require too many joins, creating more electrical resistance and thus a need for a larger and heavier power source.

A full dress rehearsal was carried out on the track that ran up Nui Dat hill past the squadron picture theatre known as Ocker's Opry House. It took three and a half hours to dig the explosives with dummy charges into the track and to set up the ambush. When all was ready Cashmore walked up to the squadron officers' and sergeants' mess and invited Wade to drive his Land Rover down the track. He did so but to their dismay the switch failed to work. The Land Rover's tyre had missed the steel plate. But once the tyre connected with the steel plate the switch worked perfectly.

The next day Wright and Scheele again checked all the charges and inspected the wires inch by inch. Meanwhile, Cashmore discussed the insertion of the patrol with the RAAF helicopter crews. Just as he was ready for his final briefing, Warrant Officer Turner arrived with one more suggestion: he was carrying an auger which he thought would be useful for digging the holes for the Beehive charges.

It took some time to organise the load to be carried by the patrol. Since they had to remain in position for up to seven days they had to carry fourteen water bottles for each man, plus seven days' rations. Then there were the demolitions: four Beehive charges, four Claymores, the pressure switch with its baking dish and steel plate, the batteries and the 100 metres of wire. In addition to their normal radio sets there were the three URC 10s. Each man also carried his usual weapon and ammunition. Cashmore had the additional burden of the steel auger—1.5 metres long with a metre wide handle. It was one of the heaviest loads ever carried by an SAS patrol in Vietnam.

Soon after 9 am on 17 March the six members of the patrol staggered across Kangaroo Pad (the main helicopter landing place at Niu Dat) and heaved themselves into the waiting Iroquois helicopter. Perspiration was already running down the brown and green face-cream they had just applied and staining their mottled camouflage uniforms as they strained to lift their heavy packs. The rush of air as the Iroquois gained height and gathered speed brought cool relief. But their respite was brief; it took barely fifteen minutes to reach the LZ.

An SAS insertion was nerve-racking at the best of times. By 9.30 am Cashmore's heart was pumping faster as he prepared to leap first from the helicopter descending towards the LZ. As the skids came to within half a metre of the ground he jumped. The weight of his pack drove him face down into the rock-hard surface. Gasping for breath, he struggled to his feet in time to see the other men suffering the same fate, and the helicopter lifting slightly as it was relieved of the weight of each man. Within seconds they had reached the treeline, but realised that Dave Elliott was still struggling across the LZ. He had been the last man to jump, and by then the helicopter was almost a metre off the ground.

Quickly they helped Elliott to the edge of the trees and then discovered that he had torn the ligaments in his ankle and could barely walk. Cashmore spoke on his radio to the helicopters—instructed to circle some distance away for 20 minutes—and asked them to return. Meanwhile the team members carried out a mental check to ensure that Elliott would not be evacuated with any equipment that might be vital to the mission. They decided to sacrifice some of the food and water. Unfortunately no one remembered that he was carrying one of the URC 10 241 radios that were part of the warning group.

Cashmore later described the approach march as particularly difficult: 'We were young and inexperienced with regard to Vietnam. This was Charlie's backyard and we were in it, plus all this weight, and it would be interesting to know how much weight we lost through sweat'. Shielded from even the slightest breeze, and with the sun beating down on their heads they pushed through stifling kunai grass, thickets of spiky bamboo and into the tangled vegetation of the secondary jungle. 'We looked like Christmas trees', observed Cashmore, 'it was noisier than the average patrol'. Wright and Scheele were carrying all of the explosives and were finding the going particularly difficult. Adrian Blacker recalled that the hot weather and dry grass reminded him of their training at Bindoon, Western Australia, in the heat of February.

It took the patrol about one hour to move 200 metres north of the LZ, where they stopped and went into a Lying Up Position (LUP). While resting one of them suddenly realised that Elliott had been carrying one of the three URC 10s. They discussed in whispers whether to continue with the mission. Wright wondered whether the two remaining radios on the 241 frequency could communicate with the URC 10 on 243 (International Distress Emergency only) frequency that the patrol carried to facilitate emergency extractions. In the close confines of the LUP this appeared to work so it would be possible to continue with the plan to deploy the two sentries to provide early warning.

The patrol moved off again and soon came across a clear area 2 metres by 2 metres with a tied clump of grass in the centre. Suspecting that it had been mined by the VC, they turned north west. They had a further 500 metres to walk, but it was not until mid afternoon that they reached the Firestone Trail. Cashmore recalled that by this time he had a thumping headache and he set the patrol down in another LUP. After a short rest Cashmore and Wright crept forward to observe the track, striking it about 200 metres west of LZ Dampier. Cashmore received his second shock for the day when he discovered that the tractor had not driven down the centre of the trail with one wheel in each rut, but had gone west with one wheel on the track and the other in the grass, and on its return journey had one wheel in the other rut. The Firestone Trail was 30 to 40 metres wide, the tractor was relatively small, and Cashmore believed that he could not risk placing the pressure plate in only one rut and the explosives between two of the three wheel tracks. Covered by Wright, Cashmore stepped out onto the track and walked along it towards LZ Dampier as if he owned it, feeling as he said later, 'like a spare prick at a wedding'. With one man on the track and the other covering from the jungle they moved along the track for about 100 metres until they came across a slight bend and discovered that in changing direction the tractor had cut the corner and for a short distance the wheels had been in both ruts. This was the spot for the ambush, and looking around, 50 metres to the north across the Firestone Trail, they found the large bomb crater they had earlier identified as possible cover for the patrol during the laying of the explosives.

That night the patrol moved down to the track to observe the night's activities. At 11.20 pm the bright red tractor and trailer appeared, moving west along the Firestone Trail, and remarkably, travelling with a single, centrally placed light blazing through the jungle. From a distance through the grass, it became clear that there was another advantage in planting the demolitions on a bend; the tractor's light would sweep past any sign of the demolitions that had escaped camouflage. At 1.45 am the tractor returned, travelling towards the east, and at last, after a long day the patrol could crawl back into their LUP and sleep.

Having timed the tractor's movements Cashmore decided that they had sufficient information to mount the ambush on the following night, and they spent the next day resting and checking their equipment. It was a hot, uncomfortable day. There were only scattered trees

The tractor job

CHARGE PROFILE
TRACTOR AMBUSH C/S 1/3 2 SQN /69

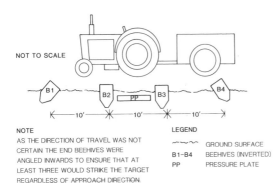

NOT TO SCALE

NOTE
AS THE DIRECTION OF TRAVEL WAS NOT
CERTAIN THE END BEEHIVES WERE
ANGLED INWARDS TO ENSURE THAT AT
LEAST THREE WOULD STRIKE THE TARGET
REGARDLESS OF APPROACH DIRECTION.

LEGEND
~~~~ GROUND SURFACE
B1–B4   BEEHIVES (INVERTED)
PP   PRESSURE PLATE

Sketch drawn by Danny Wright showing how the beehive charges used in the tractor ambush were set into the track. The pressure plate can be seen in the centre.

and it seemed to the men that they had to keep moving constantly to remain under the shade.

Last light came at about 7 pm and after waiting a further 20 minutes they moved down to the ambush site, knowing from the previous night's experience that they had until 11.20 pm to lay, camouflage and activate the system. They crossed the Firestone Trail from south to north, carefully obliterating any trace that they might have left. The two sentries prepared to move into position and before leaving the patrol they checked their radios; they were working, but only just. McAlear went west and Blacker, armed with a silenced Sterling sub-machine gun, moved east to a position from where he could observe across LZ Dampier.

Once the sentries were in position the demolition team moved out onto the track and, in the absence of Elliott, Cashmore unpacked the demolition kits and passed them to the two demolition men. Immediately there were problems. The auger that had been carried to dig in the Beehive would not penetrate the hard-caked mud of the track. They tried to dig the holes with their machetes, but again with little success. To make the task easier they cut about 25 milimetres of compressed paper from the top of the Beehives, but still they would not fit into the holes. Eventually they decided to use the central ridge of dried mud between the ruts to conceal the Beehives, but they still could not dig down far enough, and eventually they had to slightly build up the ridge.

It was exhausting work and towards the end, while they were camou-flaging the charges, Danny Wright realised that he had misplaced his British issue machete. It would be a dead give away, and for the next ten minutes Wright, Scheele and Cashmore carried out a frantic 'search by feel' before it was located. With the demolitions laid and camouflaged it was now time to recall the sentries. But all attempts to

11

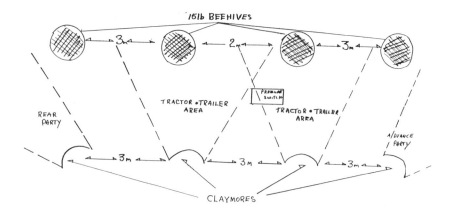

Original sketch from the patrol report showing a plan view of the setting distance and arcs of fire of the Claymore mines. The four Claymores were set up to cover the enemy rear and advance parties, plus the area of the tractor and trailer.

raise them by radio failed. As they subsequently discovered, the 243 radio would not work to the 241 radio at a distance beyond 100 metres. Wright later described the outcome: 'the two demo-team men had the hair-raising experience of moving out individually to locate a lone sentry, each on his first operational patrol. Accusations of slackness were soon disproved when each sentry showed an URC 10 still emitting "hash" and a test call made back to the central point with no answer forthcoming. The demo-team had been working exposed without protection for approximately three and a half hours'.

The sentries had barely returned to the central point when at 10.20 pm they heard the tractor start up some distance away to the east. It was twenty minutes earlier than the previous night. The patrol did not panic but it was clear that time was becoming short. They connected the Claymores and began to lay the final 50 metres of cable through the grass to the bomb crater. With the tractor approaching they realised that they would not have time to connect the wires and withdraw the 100 metres buffer distance, so they decided to connect the power source and then remain in the bomb crater.

The chug chug of the tractor was becoming louder by the second. The power source constructed by the signallers from a beer can included two clips on the top, each with a hole through which the wires had to be connected with one hand while depressing a spring clip with the other. Danny Wright had to perform a delicate operation as contact with the wrong terminal would create a premature explosion. Because of the proximity of the enemy he could not use a torch and had to hold the power source up to the moonlight to ensure that he

The tractor job

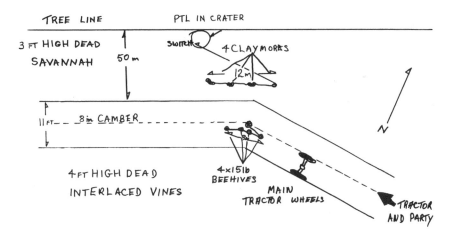

Original sketch from the patrol report of the tractor ambush showing the location of the beehive charges, the Claymore mines and the bomb crater.

placed the wire in the correct terminal. To the other members of the patrol Wright appeared to be taking his time. The chug chug became even louder and the normally phlegmatic Dutchman, Dave Scheele, kept saying: 'For fuck's zake, Danny, get ze vire in ze hole'. After what seemed to be an eternity, but was in fact only seconds there was a click and Danny said, 'its in—stand by'. Dave leaned close to Danny and whispered: 'If this goes off, I kiss you'.

While the patrol had been running the cable from the demolitions to the bomb crater Frank Cashmore had taken a compass fix on a branch sticking up above the skyline. Now, as the rest of the patrol huddled in the bomb crater, he listened to the tractor, trying to determine when it was in line with the compass bearing. After the frantic effort to connect the power source over a minute passed. By now it was 11.10 pm and it seemed that the tractor should have reached the pressure plate. Cursing to himself Cashmore stood up. Immediately above the kunai he could see the single light of the tractor advancing along the trail. It was impossible to see if anyone was walking nearby as the grass was over waist height and he was standing in the crater. Just as he was convinced that they had failed there was a huge explosion. 'It was the most horrific explosion I have ever witnessed in my military career', said Cashmore. 'It blew me arse over head backwards. Four giant orange flames went up into the sky, plus four Claymores...It was just

13

unbelievable.' Adrian Blacker was also watching the track. To him the tractor with its single light looked like a train approaching. 'When the explosion went off it was almost like an old movie of a train wreck. For an instant the light flashed skywards before going out.'

The men flattened themselves against the bottom of the crater while great clumps of earth, tin and steel rained down. Danny and Dave were ecstatic. 'We've done it—magic', thought Danny. After the tension and hard work of the last five days they had achieved their aim. Desperately Frank Cashmore tried to quieten the two demolition men. They lay there, weapons at the ready, not daring to move while from the trail there was the sound of moaning and four voices speaking excitedly in Vietnamese. Wright, who had attended a Vietnamese language course, thought that he heard one voice saying that they ought to go for help. A little later they heard what sounded like someone collecting weapons and clearing the breeches in quite a professional manner.

Kim McAlear, at the age of 19 years, the youngest member of 2 Squadron, recalled that

> It was beyond my wildest dreams of what war was all about. To say that I was scared shitless after the explosion is the understatement of all time. The enormity of our actions hit home, with wounded and dying screaming and moaning, but more worrying was the presence of a group of well organized, coherent enemy who commenced to assess the situation and organize evacuation of their comrades. Had we been detected at that time, caught in a bomb crater at least 50 metres from the treeline in reasonably open country, I feel, the patrol would have been up the proverbial creek without the paddle. This is where good leadership and the thorough training of the preceding 18 months as a team came to the fore. Even though Frank Cashmore was the junior patrol commander in 2 Sqn, his positive leadership and subsequent actions for the remainder of the patrol until extraction got us through what could only be described as a very nerve-racking situation.

Blacker's thoughts were similar. Until then he had thought of the operation as just another training exercise; now, listening to the wounded VC drowning in their own blood, he realised that this was the real thing.

After about 45 minutes Cashmore decided to have a look, and taking Adrian Blacker with him he eased himself out of the hole and began to inch forward. According to Cashmore, 'we were moving like two baby elephants, on hands and knees through the brittle kunai'. They had barely travelled a metre when they heard someone clear his throat and cock a rifle. Cashmore and Blacker slid back into the hole. For all they knew the VC might be preparing to sweep through the area. The patrol crawled out of the hole and withdrew about 20 metres to the treeline on the northern edge of the clearing. Until then there had been a light cloud cover, but now as they crawled towards the jungle the moon appeared and brightly illuminated the figures against the yellow grass. To Blacker this was the most nerve-racking part of the whole operation.

The jungle was too dense for silent night movement so they moved only about 50 metres north east away from the contact area, where, fatigued and drained, they crawled under a large bush in a semi-open patch. Cashmore felt like he had 'been through a washing machine'. Some of them were too tired to remove their packs. They spent the night listening for the sounds of organised reaction; there was none.

At dawn they crawled out from beneath their protective bush, looked around for signs of enemy, then moved stealthily towards LZ Dampier, barely 150 metres away. As they approached the LZ, Blacker, who was leading, saw a lone VC, heavily bandaged, staggering across the open area. Mindful of their orders to avoid contact, the patrol went to ground, waited for a while, and when there was no more activity radioed Nui Dat that the mission had been successful and that they wanted an immediate extraction.

Back at Nui Dat the SAS squadron operations room telephoned No 9 Squadron RAAF. Within a few minutes the helicopters were in the air, and by 9.10 am the SAS patrol had been lifted out of LZ Dampier. As the helicopter gained height they crossed the ambush site, only 200 hundred metres away. All Cashmore could see was 'a hell of a mess with a big black hole'.

The patrol was met by Major Wade and after a short debrief Cashmore and Wade drove down to Task Force headquarters. At this stage neither Wade nor Cashmore knew the extent of the success, but although Wade was keen to insert more SAS patrols into the area, permission was denied. He then asked that an infantry company be inserted but one was not available at short notice. Wade was trying to persuade Brigadier Hughes that they should at least be credited with killing the driver of the tractor when over the radio came the voice of JADE 6, an American Forward Air Controller circling over LZ Dampier. He started counting the bodies out loud, and according to Cashmore, 'Wade appeared to grow about two feet'. Clearly the success was greater than they had imagined and Hughes personally congratulated Cashmore on the performance of his patrol.

Meanwhile, a 161 Recce Flight Sioux helicopter flown by Lieutenant Ross Hutchinson with SAS photographer Private Dick Meisenhelter aboard had been despatched to photograph the wreckage. On the first photographic run the helicopter took fire from a wounded survivor

The ambush site, looking west along the Firestone Trial. The lying up place (LUP) was in the jungle to the left of the picture, while the bomb crater was in the long grass to the right. The wreck of the tractor can be seen scattered over the track.

and consequently did not get good photographs. The helicopter returned to Nui Dat where Meisenhelter armed himself with an M60 machine-gun, returned to the site, quelled the opposition and took the photographs which enabled an assessment of the mission to be made. Seeking more information, Captain Phillip Perrin, the Task Force artillery intelligence officer, flew to the area. The aircraft made an unauthorised touch-down and Perrin salvaged a new, unfired Chicom 75 mm recoilless rifle from the wreckage.[3] From an examination of the wreckage it appeared that there had been six to eight cases carrying 75 mm rockets. The wreckage was napalmed by Skyraiders of the South Vietnamese airforce late that afternoon.

In accordance with the normal procedure, the 75 mm recoilless rifle was handed in to the Task Force intelligence section for examination. It is not known what intelligence was gained from the weapon, but it was retained in the Task Force armoury. Both the SAS and 161 Recce Flight were furious that this trophy had been denied to them, and some nights later a group of SAS NCOs and officers from 161 Recce Flight carried out a night raid and recovered the weapon. Under direct orders from Brigadier Hughes, the weapon was returned and is now in the Infantry Museum, Singleton.

The patrol had achieved the mission of stopping the supply vehicle and as a result of air-photographic interpretation they were credited with fifteen kills, but it must have been many more. A VC defector connected with the unit operating the tractor gave a figure of 21. The ambush probably came as a considerable blow to the VC because at about the same time a Thai unit attacked the VC in the Courtenay rubber plantation. Some weeks later the US Senior Provincial Adviser, Lieutenant-Colonel John Jessup, led an ambush party of Americans and Australians (mainly ex-SAS) from the Long Range Reconaissance Patrol Wing at Van Kiep, near Baria, into a rubber plantation near Baria. They ambushed a VC tractor and trailer causing substantial casualties.

An analysis of Cashmore's patrol reveals many of the reasons for the success of the SAS over years of operations. Despite their lack of operational experience the soldiers were highly trained and their planning was marked by attention to detail and comprehensive rehearsals. They refused to be hurried in their preparations and their plans were flexible enough to allow for unexpected incidents such as the injury to Private Elliott, the loss of the radio and the early arrival of the tractor. Bound together by a close comradeship, the men had confidence in each other and the operation was conducted with professionalism and discipline. They were, of course, lucky that the VC did not approach while the ambush was being laid.

The mission also showed the extent to which the SAS relied on the support of other units. 161 Recce Flight was involved at various stages: in locating the original tracks, conducting the reconnaisance for the patrol commander, photographing the aftermath and recovering the 75 mm recoilless rifle. The engineers designed and built the pressure switch and provided the Beehives. No 9 Squadron RAAF, as always, inserted the patrol, returned to evacuate the injured soldier, and rapidly extracted the patrol following the ambush. It was a graphic demonstration of the fact that while the SAS demanded from each member of its small operational patrols a high degree of individual courage and initiative, those highly trained men still needed substantial outside support.

The tractor job was one of the SAS's outstanding patrols. But it was only one of almost 1400 operational patrols conducted by the Australian SAS in seven years of active service. At least a dozen other patrol missions could have been used to demonstrate the wide-ranging skills, the detailed planning and the cool execution which was commonplace in the SAS. By the time of the tractor job the Australian SAS had been in existence for over ten years and the success of the operation was the result of years of preparation and training. What seemed standard and

Taken from a helicopter the morning after the tractor ambush, this photograph
shows the devastation caused by the blast. The remains of the trailer, parts of the
tractor and some of the bodies can be seen clearly. Other bodies were located
outside the area of this photograph. The 75 mm recoilless rifle can be seen in the
centre of the picture.

routine in 1968 had not always been so. Frank Cashmore and his men
can rightly claim the credit for their success.[4] But a large slice of the
credit is also due to the entire SASR, its training system, the high
standards it demands and its uncompromising search for excellence.
And knowing the ethos of the SAS, Frank Cashmore's patrol, like a
hundred other patrols, would have it no other way.

2
Towards special forces
Australia: 1940–1957

The Australian Army has always been ambivalent about special operations and special forces. Australians in general have been suspicious of anything which includes the word 'special'. In a society where 'Jack is as good as his master' and where 'tall poppies' are ruthlessly cut down, any organisation or group of individuals which thinks of itself as special or elite is under suspicion. In some quarters of the Australian community 'secret armies' have been seen as contrary to democracy or civil liberties. [1]

It is not just an Australian phenomenon. Regular forces in many armies have often had good reason to resent special forces. Some of the 'private armies' of the Second World War were wasteful of resources, concerned with the glorification of individuals, and provided little benefit to the general prosecution of the war.

Yet just as Australians have taken pride in the performances of their elite, such as sportsmen or scientists, the Australian services have taken pride in the performances of their special forces. There has been a feeling that Australians are uniquely suited to special operations.

Special forces as we know them are a relatively modern development. But the idea of unconventional warfare, irregular warfare or guerilla war, is almost as old as history. In more recent times, such as in the American War of Independence, irregular troops played a significant role. In the wars between the British and the French in North America, both sides used Indians as well as white colonists in an irregular capacity. And the Peninsular War of 1808–1814 saw the first use of the term 'guerilla'. By the latter half of the nineteenth century guerilla warfare was widespread in the colonies of the great empires—the British, French, American, Russian, Dutch and

19

Spanish—but seemed to have less relevance to wars between industrial powers. Nonetheless, irregular warfare was pursued in the US Civil War, 1861–1865, and the Franco-Prussian War, 1870–1871. The Boer War, 1899–1902, introduced the word 'commando' and demonstrated that a relatively small, highly motivated and well-armed guerilla force could pin down a large modern conventional army. In the First World War T. E. Lawrence raised and commanded an effective Arab force that harassed the Turks and contributed to their defeat in Arabia.

Despite this long history of unconventional warfare, special forces as they are understood in western countries today, grew out of the Second World War. It was the British who led the way and the Australian special forces owe their origin to British developments. By the beginning of 1940 three separate organisations had been established in Britain: a propaganda organisation (Electra House); an organisation within the Foreign Office to conduct clandestine operations such as sabotage (Section D); and a War Office section, known as GS(R), to conduct guerilla warfare. The latter section led to the formation of the Independent Companies, the forerunner of the Commandos.

The appointment of Winston Churchill as British Prime Minister and the fall of France in June 1940 caused the British to seek ways to carry the war to German-occupied Europe, and in July 1940 a new body, the Special Operations Executive (SOE) was formed 'to co-ordinate all action, by way of subversion and sabotage, against the enemy overseas'. Churchill exhorted the Minister for Economic Warfare, who was given responsibility for SOE, to 'go and set Europe ablaze'.[2] Initially SOE absorbed the three organisations, Electra House, Section D, and GS(R), but in due course the propaganda wing was given to a new organisation, the Political Warfare Executive. Part of GS(R) remained with the War Office and grew into the Directorate of Combined Operations and eventually Combined Operations Headquarters.

Thereafter the Commandos and SOE tended to go their own ways. The Commandos, as part of Combined Operations, mounted raids against Hitler's European fortress, demonstrating the capabilities of Combined Operations, but as the war progressed were often used to supplement regular formations in conventional operations. The SOE undertook numerous operations behind enemy lines in many theatres of war. They raised and armed guerillas, conducted sabotage and gathered intelligence. Their first operation began in Ethiopia in November 1940. By the latter months of 1940 the general fields of responsibility of the SOE and the Commandos had been established.

Australia's involvement with these new developments was confused and hesitant. Even before the outbreak of war the Royal Australian Navy had begun to establish a coastwatcher organisation—based on civil servants, plantation managers and missionaries—in the islands to the north of Australia. It was anticipated that if the islands were occupied by the enemy the coastwatchers would remain behind to continue operating from behind enemy lines. In early 1941 naval intelligence officers were appointed to supervise the mainly civilian coastwatchers and to expedite the flow of intelligence. Once the war with Japan began in December 1941 the civilian coastwatchers were given naval rank or rating in the RANVR.

The Australian Army did not begin to train any special forces until late June 1940 when it received a cable from the Chief of the Imperial General Staff in Britain. The cable outlined a proposal that special branches be established in the dominions to conduct para-military activities against the enemy, including raids, demolitions and organising civil resistance and sabotage in enemy or enemy-occupied territories. Britain proposed sending a Military Mission to assist in the setting up of Australian forces to conduct these activities. As Lieutenant-General Vernon Sturdee—who became Chief of the General Staff when his predecessor was killed in an air crash in August 1940—wrote some years later, negotiations 'were somewhat protracted owing largely to the need for British secrecy and partly to our uncertainty as to what it was all about and whether such units would be of value in the Australian Army'.[3]

When Britain had first sent its proposals to Australia the SOE had not been formed, so when the Military Mission, headed by Lieutenant-Colonel J. C. Mawhood, arrived in Australia in November 1940 there was some confusion about the purpose of the Mission. In January 1941 the British government cabled Australia to clarify the proposals initially forwarded the previous June. The proposals included: 'steps to counter possible Fifth Column activities on the part of the enemy in Australia'; the 'constitution of independent companies which would receive special training to fit them to take part in combined operations'; the training of personnel in sabotage; and the 'formation of military missions which would be available to organise guerilla operations in enemy territory in which [a] large proportion of the population was hostile to enemy and friendly to us'. In Britain, the first of these activities was conducted by MI5, under the control of the Home Office. The second group of activities was the responsibility of the Commandos, and the last two groups of activities were being undertaken by the SOE. Mawhood dabbled secretly in all of these disparate activities and raised the suspicion of the Australian army chiefs.

Meanwhile, acting on British advice, the Australian Army had begun to train Independent Companies whose role would include raids, demolitions, sabotage, subversion and organising civil resistance; in other words, the activities carried out by the SOE in Britain. The first

Australian course, under British instructors who had trained with the British Independent Companies, began at Wilson's Promontory in Victoria on 24 March 1941. One of the British officers, Captain Frederick Spencer Chapman, recalled that 'We talked vaguely of guerilla and irregular warfare, of special and para-military operations, stay-behind parties, resistance movements, sabotage and incendiarism, and, darkly and still more vaguely, of "agents"; but the exact role of the...Independent Companies had never been made very clear'.[4] Another instructor was Captain Mike Calvert, who had already fought in Finland before joining the British Independent Companies. He was to be instrumental in reforming the British SAS after the Second World War.

In May 1941 an SOE mission arrived at Singapore and Mawhood visited Singapore to exchange ideas. He reported back to Australia that the Independent Companies could be used for raids on enemy lines of communication but that they would not be required to engage in the sort of activities then being undertaken by the SOE. He suggested that an SOE-type organisation be established in Australia. But Mawhood had completely lost the confidence of the Australian Chief of the General Staff and nothing more was done about establishing an SOE organisation in Australia before the outbreak of war with Japan in December 1941.

By mid 1941, perhaps anticipating a Japanese advance, the Independent Companies were training for a slightly different task. As Spencer Chapman explained, the role would 'be to stay behind, live off the country or be provisioned by air, and be a thorn in the flesh of the occupying enemy, emerging in true guerilla style to attack vital points and then disappear again into the jungle. We also visualised long-range penetration of the enemy lines by parties so highly skilled in fieldcraft and in living off the country that they could attack their targets and get back again without getting detected'.[5]

By the second half of 1941 three Independent Companies had been formed, each with seventeen officers and 256 men. The fourth company was still at Wilson's Promontory, but training was suspended pending a firm decision about the role of the companies. The Australian official historian, Dudley McCarthy, commented on the ambivalence towards special forces: 'From the beginning the army was not single-minded in its attitude towards Independent Companies. There was a feeling among some officers that well-trained infantry could do all that was expected of the commandos, and that the formation of these special units represented a drain on infantry strength that was out of proportion to the results likely to be achieved. The supporters of the commando idea replied that the new companies would relieve the infantry of the task of providing detachments for special tasks.' The units were to be sent to the Middle East, but as the threat from Japan developed it was decided to use them in the string of islands to the north and north east of Australia where it was necessary to establish Army and RAAF outposts to warn of the approach of Japanese forces. Their mission was to remain behind and harass the invaders.

In July 1941 the 2/1st Independent Company was deployed with sections spread from Manus Island and New Ireland to the Solomons and the New Hebrides. When war came the bulk of the company in New Ireland was overrun and captured.

The 2/2nd Independent Company spent some time in the Northern Territory before being deployed to Portuguese Timor in December 1941. Assisted by the local population, they successfully harassed the Japanese, causing them to divert troops from other areas for operations in Timor. In September the 2/4th Independent Company, which had resumed training on the outbreak of the Pacific war, reinforced the 2/2nd Independent Company. The force was withdrawn in January 1943, but it had been an outstanding example of the value of highly trained and motivated men operating behind enemy lines. Their success was a model for later SAS training. 'At no time', wrote Otto Heilbrunn, 'had a guerilla force or guerilla-type force in any theatre of World War II withstood for so long such a superior force with such telling results.'[6]

In January 1942 the 2/3rd Independent Company was despatched to New Caledonia as a gesture to the Free French and to carry out demolitions if required. They returned to Australia the following August.

As well as these units, a 48-man contingent from the 8th Australian Division in Malaya travelled to Burma to train for guerilla warfare. As Mission 204 they set out for China in January 1942 and the 24 survivors did not return to Australia until November that year, having suffered from sickness, starvation and Chinese indifference and inefficiency, but not having been involved in any real action.

The Independent Companies therefore soon found themselves in the front line of the war against Japan while the majority of the Australian Imperial Force was still in the Middle East. The Australians in the Middle East had only a passing acquaintance with special forces. In June 1940 the British Commander-in-Chief Middle East, General Sir Archibald Wavell, agreed to set up a long range patrol, later known as the Long Range Desert Group. The commander of the new group thought that the best recruits might be men from the Australian outback. However, the General Officer Commanding the AIF in the Middle East, Lieutenant-General Sir Thomas Blamey, pointed out that the Australian government was opposed to its men serving outside Australian formations. New Zealand troops were allowed to join the

Long Range Desert Group and eventually the New Zealand contribution was seven officers and 86 other ranks. But the New Zealand commander in the Middle East was soon anxious to get his men back.

While in the Middle East the Australians saw the British Commandos in operation in Crete in May 1941 where they covered the withdrawal, in Lebanon in June when they assisted with the advance of the 7th Australian Division along the coast, and with the 9th Australian Division in the defence of Tobruk. Also, in April 1941 Blamey was appointed Deputy Commander-in-Chief in the Middle East and in that capacity would have been acquainted with the SOE operation conducted by Brigadier Orde Wingate in Ethiopia. It is less certain that Blamey would have been aware of the formation, towards the end of 1941, of the SAS. By September he was in deep argument with the new Commander-in-Chief, General Sir Claude Auchinleck, over the relief of Tobruk, and in November he visited Australia for consultations, arriving back early in December.

The story of the formation of the SAS by Lieutenant David Stirling in the latter months of 1941 has been told in a number of books and is now well established folklore in the SAS. The new unit's first action in November 1941 was a disastrous failure, but its next raid the following month was successful and by the end of 1942 the SAS was established well enough to survive the capture of its founder in January 1943. Stirling's idea was, as he wrote later, 'based on the principle of the fullest exploitation of surprise and of making the minimum demands on manpower and equipment'.[7] Initially he envisaged parties of five men but later reduced this to four. The parties were to be infiltrated behind enemy lines by parachute drop or with the assistance of the Long Range Desert Group, and had to be highly trained to use any means of insertion available. The unit continued to expand and operated with outstanding success in North Africa, Italy, Greece and North West Europe.

Well before these events, in March 1942, General Blamey had been recalled to Australia. It is not known exactly how much he had learned of special forces during his period of almost two years in the Middle East, but some of his later correspondence indicates that he had gained some appreciation of the nature of special operations. When he returned to Australia as Commander-in-Chief of the Australian Military Forces he was quick to seize upon the potential of special operations.

Just as the fall of France stimulated activity in Britain, the outbreak of war against Japan, and more particularly, the surrender of Singapore in February 1942 accelerated defence preparations in Australia. On 9 December 1941 an adviser urged the Minister for the Army, F. M. Forde, to consider the formation of a guerilla army within Australia, but at that stage the Chief of the General Staff,

General Sturdee, thought that the proposal was defeatist.[8] In March 1942, however, two SOE officers, Majors Egerton Mott and A. E. B. Trappes-Lomax, arrived in Australia after service in Singapore and Java to assist in setting up an SOE organisation.

A few days after their arrival, Blamey took up his new appointment, and General Douglas MacArthur became Allied Commander-in-Chief of the South West Pacific Area. Both commanders were enthusiastic about the possibilities expounded by the British officers. The proposal to form an SOE organisation was put to the Prime Minister, John Curtin, who agreed, and in May 1942 a new special operations organisation, the Inter-allied Services Department (ISD), was set up in Melbourne under Blamey's control.

By June 1942 there were a number of special operations and intelligence-gathering organisations in Australia, including the ISD, a branch of the British Secret Intelligence Service known as Secret Intelligence Australia, the coastwatchers, and a propaganda organisation, the Far East Liaison Office. On 6 July 1942 the Allied Intelligence Bureau (AIB) was formed to assume responsibility for all these organisations. Answering directly to MacArthur's headquarters, the mission of the AIB was 'to obtain and report information of the enemy in the South West Pacific Area, exclusive of the continent of Australia and Tasmania, and in addition, where practicable, to weaken the enemy by sabotage and destruction of morale and to lend aid and assistance to local efforts to the same end in occupied teritories'.[9]

In September 1942 the Far East Liaison Office was separated from the AIB, and in April 1943 the AIB was re-organised into five sections. The Philippines Regional Section was responsible for operations in the Philippines; almost immediately it assumed a semi-idependent status and as the war drew closer to the Philippines it became a completely autonomous, US operation. The Netherlands East Indies Regional Section (NEFIS III) conducted operations in the Netherlands East Indies and was mainly a Dutch organisation. The North East Regional Section took over many of the operations of the coastwatchers and was responsible for obtaining intelligence in the South West Pacific Area east of the Dutch New Guinea border. The next section was the Services Reconnaissance Department (SRD), the new name for ISD; it conducted SOE-type operations, including obtaining intelligence, and the 'execution of subversive and highly specialised sabotage chiefly by means of undercover methods'.[10] The final section was Secret Intelligence Australia; with connections to the Secret Intelligence Service in London it conducted espionage, especially in the Netherlands East Indies.

It took some time for SRD operations to build up steam, and in the meantime the Commandos continued their operations behind enemy lines. Although the 2/2nd Independent Company was still in Timor, most action was in New Guinea. In May 1942 the 2/5th Independent Company was deployed to the Wau area of New Guinea to harass the Japanese around Lae and Salamaua. They conducted their operations with enthusiasm and imagination, and in October were reinforced by the 2/7th Independent Company. At about the same time, the 2/6th Independent Company flew over the Owen Stanley Range to operate on the flanks of the American division advancing towards Buna. In January 1943 the 2/3rd Independent Company, led by the legendary George Warfe, arrived at Wau. Some weeks later he was leading a small patrol, following hard on the enemy along a track. They ran out of rations but continued for several days raiding small parties of Japanese. On one of these occasions they came silently upon twelve Japanese who were preparing a meal. Warfe and his men lay concealed in the jungle and patiently waited for the meal to be cooked. This took nearly an hour, then they attacked, disposed of the enemy and proceeded to eat the ready-cooked meal.

The Independent Companies did sterling work in New Guinea, but as the war progressed tended to operate as flanking units for larger formations. In mid 1943 their name was changed to Cavalry (Commando) Squadrons and eventually in 1944 to Commando Squadrons. By this time they had been brigaded to form the 2/6th, 2/7th and 2/9th Australian Cavalry (Commando) Regiments, the headquarters of which had originally been the Cavalry Regiments of the 6th, 7th and 9th Divisions. In the last year of the war the eleven Commando Squadrons fought in Borneo, New Guinea and Bougainville. Although often used for tasks which could just as easily have been undertaken by normal infantry, they nonetheless retained their special forces ethos. They were tough, highly trained and able to operate for long periods in isolated areas.

Many of the men from the Commandos eventually transferred to either M or Z Special Units. These two units were the administrative holding units for Army personnel serving with the AIB. M Special Unit was the North East Regional Section operating in New Guinea and the Solomons, while Z Special Unit was the Services Reconnaissance Department operating in Borneo and the Netherlands East Indies. Two well known SRD missions were Operation JAYWICK, the successful raid on Japanese shipping at Singapore in 1943, and Operation RIMAU, the unsuccessful attempt in 1944.[11]

Because of inter-allied jealousy the AIB did not realise its full potential until 1945. By that time the Philippines Regional Section had been separated completely from the AIB and the remaining AIB organisations included 1659 Australian and British personnel, 1100 natives, 268 Dutch and 19 Americans. The AIB possessed its own flight of aircraft and its own surface craft. General Sturdee, commanding the First Australian Army operating in New Guinea, New Britain and Bougainville in 1945, claimed that about half of his operational intelligence was collected by AIB parties. At the end of the war the Controller of the AIB, Brigadier Kenneth Wills, estimated that his organisation had killed 7061 Japanese, had taken 141 prisoners and rescued 1054 servicemen and civilians from enemy-occupied territory. The AIB had conducted 364 operations; in 35 operations they had suffered personnel killed or captured, and a further fifteen operations were classified as unsuccessful.[12] Because of racial differences, the AIB in the South West Pacific and South East Asia did not usually operate in civilian clothes in populated areas as the SOE did in Europe. Thus with its emphasis on small groups operating deep behind enemy lines the AIB provided the closest parallel to the SAS.

Australian special forces owe their beginning to a number of organisations, the first of which were the coastwatchers. The Independent Companies were formed to undertake SAS-type operations, but their role changed as the war progressed. The 2/2nd Independent Company operations in Timor in 1942 still remain a classic example of how to conduct guerilla warfare behind enemy lines. With the formation of the AIB in 1942, SAS-type operations were conducted by both the SRD (Z Special Unit) and the successor of the coastwatchers, M Special Unit.

Neither the Commandos nor the AIB survived the general rush towards demobilisation at the end of the Second World War. Indeed the only reason full-time Australian Army infantry units were retained was to provide the Australian contribution to the British Commonwealth Occupation Force in Japan. Despite the experience of the Second World War when Australia was directly threatened, after the war the focus of Australian defence planning returned quickly to the old concept of supplying troops as part of a British Commonwealth commitment, with the Middle East seen as the likely area for future operations. When the new Defence policy was announced in July 1947 the Regular Army was to consist of the three infantry battalions then serving in Japan and an armoured regiment. The Citizen Military Force (CMF), which was to be formed in 1948, was ambitiously planned to expand to two infantry divisions and an armoured brigade, but no special forces were to be raised.

Special forces had more success in Britain, where, at the end of the war the commander of the SAS Brigade, Brigadier Mike Calvert, wrote to the Chief of the Imperial General Staff suggesting an investigation into the results achieved by the SAS and other similar formations. The Directorate of Tactical Investigations at the War Office, headed by

Major-General Sydney Rowell, was ordered to undertake the investigation. Rowell was an Australian officer on loan to the War Office and he returned to Australia in January 1946 to take up the appointment of Vice Chief of the General Staff. Undoubtedly, before he left London he would have been aware of the preliminary findings of the study.

Brigadier Calvert fired the first shot of the enquiry with a letter to the SAS commanding officers refuting some of the criticism that had been made of special forces. He pointed out that these forces had tended to be called 'private armies' because there had 'been no normal formations in existence to fulfil this function, a role which [had] been found by all commanders to be a most vital adjunct of their plans'. To the criticism that special forces skimmed the regular units of their best officers and men, Calvert replied that in a regular unit there was far less opportunity for men to display their initiative, resourcefulness, 'independence of spirit' and self-confidence. To the argument that a normal battalion could do the same job, Calvert claimed that his experience showed that they definitely could not: 'in my opinion, no normal battalion I have seen could carry out an SAS role without 80 per cent reorganisation'.[13]

Calvert followed this letter with a detailed submission to the War Office, and in due course, after receiving other submissions, the War Office reached the following conclusions:

1 There was unlikely ever again to be war with static front lines, except perhaps for short periods.

2 Small parties of well trained and thoroughly disciplined troops operating behind enemy lines achieve results out of all proportion to the numbers involved.

3 Their operations are, and should be quite distinct from non-regular groups such as Special Operations Executive, or Secret Service.

4 The full potential of such units is not yet fully known but there is clearly scope for tremendous development.

5 The role of SAS troops should never be confused with the normal role of the infantry. The SAS task is more specialised. The SAS does not necessarily drain the infantry of its best men but will often take a person who is no better than average in his ordinary tasks and transform him into a specialist. A man of great individuality may not fit into an orthodox unit as well as he does to a specialist force. In wartime the best leaders were independent, well-travelled men who were often good linguists; university men, who had made full use of their brains at and after university and were mature, were often successful.[14]

As a result of this study, in 1947 a territorial unit, 21 SAS (Artists) was born. But it took the outbreak of war in Korea in 1950 to initiate moves to form a full-time SAS unit. As it happened, the SAS squadron formed to serve in Korea was not required there, but it linked up with a unit known as the Malayan Scouts that Calvert had formed to carry the war to the communist terrorists then operating in Malaya. Along with a Rhodesian squadron, the squadron from 21 SAS and the Malayan Scouts were brought together to form 22 SAS Regiment as a regular army unit in 1952 and it served effectively in Malaya until 1959.

Although Australia committed forces to Korea in 1950, the only Australian commitment to Malaya during the early 1950s was from the RAAF. Thus while visiting officers might have been told about the SAS, few Australian Army officers had the opportunity to observe how successfully the SAS was operating in Malaya until the first Australian Army units began to arrive towards the end of 1955. In the meantime, a number of people in Australia had begun agitating for the formation of an SAS unit in Australia. One of these agitators was Stewart Harris, a journalist with the Brisbane *Courier Mail*, who had served in 21 SAS for a year before migrating to Australia in 1951.

Observing the potential for SAS in Australia, he wrote to Headquarters 21 SAS Regiment for support, and was advised to seek help from Major-General A. J. H. Cassells, who had recently commanded the Commonwealth Division in Korea, and was head of the British Army staff in Australia. Cassells discussed the proposal with the Australian Chief of the General Staff, Lieutenant-General Sir Sydney Rowell, who replied that the matter was 'outside his scope' and suggested that Harris should write to the Minister for the Army, Mr Josiah Francis. In due course Francis informed Harris that 'training in the Australian Army on the lines you describe is receiving attention. In this regard, Australia is in constant touch with the United Kingdom'. It is not clear whether Francis meant that training was underway or was merely being considered. There is no evidence that training was actually being conducted.[15]

Meanwhile, Harris continued to receive encouragement from London; as Major A. Greville-Bell, DSO, of 21 SAS wrote to him in December 1952, 'I feel that there is a tremendous case for the SAS in an army as small as the Australian one, particularly since by temperament and environment and quick wittedness, I have always imagined that the Australian soldier would provide the ideal SAS material'. He suggested that Harris try to enlist the support of the new Australian Governor-General, Field Marshal Slim, when he arrived in Australia in February 1953. Harris followed up the suggestion and the Governor-General was 'most interested' in the possibility of forming an SAS Regiment in Australia. Slim discussed the proposal with both Mr Francis and General Rowell, but was told that it did 'not seem very likely that the formation of a Special Air Service Regiment in Australia [was] feasible in the immediate future'. Harris also approached Lieutenant-General Sir Lesie Morshead, a well-respected commander from the Second World War, but he replied that he could not 'envisage our Army authorities adopting such a scheme here, certainly not at this stage'.[16]

At this stage the efforts to form an SAS Regiment in Australia appeared to have been defeated, but in 1954 Captain John Slim, the Field Marshal's son, was posted to Australia to join the United Kingdom Services Liaison Staff in Melbourne. John Slim had served with the SAS in Malaya and before leaving London discussed with David Stirling the possibility of forming an SAS unit in Australia. On arrival in Melbourne he got in touch with Stewart Harris but also began agitating himself. He wrote a paper which he gave to Rowell and Francis, and he also spoke to the Prime Minister, Robert Menzies. Slim recalled that he also had the opportunity of putting his views to Rowell's successor, Lieutenant-General Sir Henry Wells, who became Chief of the General Staff in December 1954.

The period from 1954 to 1960, described by Professor T. B. Millar as one of 'posture without preparedness', was a rather thin time for Australian defence. As mentioned earlier, following the end of the Second World War, for the first time the Australian government had raised a regular infantry force of three battalions, primarily for service as part of the British Commonwealth Occupation Force in Japan. Despite the requirements of the Korean War from 1950 to 1953, the Australian Regular Army was not expanded and there was some difficulty maintaining the force in Korea, which eventually grew to two battalions. The Citizen Military Force was still seen as the backbone of the Australian ground forces. But the conclusion of the Korean war, the defeat of France in Indo-China in 1954 and the signing of the South-East Asia Collective Defence Treaty, soon to be known as SEATO, later that year, changed the strategic environment. The following year Australia agreed to send an infantry battalion to Malaya as part of the British Commonwealth Far East Strategic Reserve.[17]

The Australian government was loath to react quickly to the new strategic situation if it involved the outlay of more money. The infantry battalion that was sent to Malaya was available only because the Australian commitment to Korea had declined substantially in 1954. New Zealand, however, had an even smaller regular force than Australia, and when the New Zealand Prime Minister asked the British government what assistance New Zealand might offer as part of its commitment to the Strategic Reserve, he was told that New Zealand might supply a replacement unit for the Rhodesian SAS squadron, then operating as part of 22 SAS Regiment. In March 1955 it was announced that New Zealand was to raise an SAS squadron for service in Malaya. The British request perhaps reflected a belief that New Zealand might have difficulty sending an infantry battalion to Malaya, but the chief motive reflected the oustanding success that had already been achieved by the SAS in Malaya.

It is difficult to determine what effect these developments had on thinking in Australia, or who was the main advocate for an SAS capability among the higher echelons of Australian defence. Lieutenant-General Sir Thomas Daly, who was Brigadier General Staff (A) at Army Headquarters in 1955 and 1957, cannot recall that he played a major role in advocating an SAS capability, and thought that the initiative might have come from the Chief of the General Staff, Lieutenant-General Sir Henry Wells. The BGS (B) in 1956, Brigadier, later Lieutenant-General Sir Mervyn Brogan, was not sure that Wells would have been the initiator as he was too conventional a soldier. Nonetheless, Wells had been Commander-in-Chief of the British Commonwealth Forces in Korea in 1953 and 1954, and may have been aware that as a result of the experiences in Korea the US Army had decided to set up their own special forces. On 20 June 1952 the 10th Special Forces Group (Airborne) was formed, although this group was deployed to Europe rather than to Korea.

It is possible that the BGS (A) in 1956, Brigadier John Wilton, had been alerted to the role of SAS when he had visited Malaya in 1951 and had attended the Imperial Defence College in 1952. But the most likely initiator of the idea was the Director of Infantry, Colonel Maurice (Bunny) Austin, who was always full of progressive ideas. It is known that towards the end of 1954, when Rowell was still Chief of the General Staff, Austin was promoting the idea of an independent parachute company. Towards the end of the Second World War the Australian Army had raised a parachute battalion, and there were a number of advocates for raising a similar capability in the post war period. Eventually it was decided to raise an airborne platoon of the Royal Australian Regiment.

Formed on 23 October 1951, the platoon had the roles of: land/air warfare tactical research and development; demonstrations to assist land/air warfare training and Army recruiting; airborne firefighting; airborne search and rescue; and aid to the civil power in 'national catastrophes'. Initially the platoon was formed by detaching to the School of Land/Air Warfare at RAAF base Williamtown a normal rifle platoon from the regular infantry battalion stationed in Sydney. But on 1 January 1953 the platoon became a separate unit on the Army order of battle, and it began to take on an elite aspect. All members undertook parachute training, and selected members completed courses in air despatch, airportability, advanced map-reading, advanced photo-reading, demolitions, medical, driving, small boat handling, bush firefighting, survival, rifle platoon tactics, NCO promotion, and special operations. Considering the fact that the platoon numbered less than 40 personnel, this training programme was quite ambitious. An effort was made to ensure that volunteers for the platoon had seen active service, preferably in Korea, but there does not appear to have been an operational role for the platoon.[18]

While the idea of a parachute unit was gaining strength within the Australian Army, liaison visits to Malaya had revealed the success of the SAS in its operations against the communist terrorists there. In January 1955 Lieutenant-General Wells visited Malaya to investigate how Australia could best contribute to the Commonwealth Strategic Reserve and he would have been informed of the role of the SAS. It was obvious that there was a role for special forces in modern warfare. Financial restraints dictated, however, that if Australia were to raise true special forces they would have to be part of the Citizen Military Forces, and the experience of the Second World War inclined the Australian Army to raise Commando units rather than SAS units. Consequently, in January 1955 it was announced that two CMF Commando companies were to be raised, one in Sydney and the other in Melbourne. These new units drew their heritage from the Second World War Commando companies, and the first commander of 1 Commando Company, Major W. H. (Mac) Grant, had served with the 2/12th Commando Squadron in 1944 and 1945.[19] Grant was a regular officer, and unlike other CMF units the Commandos have always been commanded by regular officers. The first commander of 2 Commando Company (Melbourne) was Major Peter Seddon, an artilleryman who had completed parachute training at Williamtown.

The changing strategic environment was reflected in the 1956 Strategic Basis document which put greater emphasis on possible military activity in the South East Asian region. The document stated that 'in view of the present assessment of the international outlook, preparations to enable Australia to participate effectively in cold war activities, and to increase her preparedness to participate in limited wars should take priority, in that order, over measures directed solely for preparedness for global war'. As a consquence, Australia had to 'have adequate forces immediately available to deter or defeat any hostile action which might prejudice vital Australian interests' in South East Asia.[20]

If the Army was to have forces available for deployment to South East Asia at short notice it was necessary to review the structure of the Army's operational forces, and staff studies during 1956 concluded that the minimum requirement would be a 'mobile, hard-hitting' regular brigade group in addition to the battalion group with the Strategic Reserve in Malaya. To free personnel for the regular brigade group there would be a requirement to re-assign some of the Regular Army manpower then devoted to the Citizen Military Forces and the National Service scheme, and amendments would therefore be needed to be made both to the CMF and the National Service scheme. On 31 October 1956 the Military Board presented this plan to the Minister for the Army, Mr J. O. Cramer. It was proposed that a new headquarters for the brigade be raised and that the brigade group would consist of two infantry battalions, one armoured regiment, one field artillery regiment, one field engineer squadron, one special air service squadron, and elements of signals, supply, transport, medical, ordnance and workshop units. The complete strength was to be 252 officers and 4331 other ranks.

Many of the major units for this new brigade were already in existence, although very much under strength, but the SAS squadron was a completely new unit and the Military Board proposal spelt out the reasons for including it in the new brigade:

> This is a unit new to the Australian Army. The squadron consists of specially trained men capable of being parachuted behind or infiltrating enemy lines and of operating in small detachments for comparatively lengthy periods. They are trained in sabotage, the collection of intelligence, and similar commando-like roles.
>
> Exercises in NATO armies have shown that these squadrons are suited for the difficult task of locating and advising to a commander targets most suitable for engagement by either tactical atomic weapons, ground or air launched, such as guided missiles and cannon, or by rocket and bomb attack by the Tactical Air Force.
>
> In Malaya, Special Air Service units are proving of great value in operations calling for deep jungle penetration. Such troops can be dropped into areas virtually inaccessible to other ground forces. Here they establish liaison with local tribesmen, create confidence in remote villages, gain information of Communist activities and take rapid punitive action. In general they dominate the area in which they operate. It is clear how vital to the operations of our forces such units could be in the event of war in South East Asia.
>
> The nucleus for a Special Air Service squadron is available in the existing airborne platoon and in the number of other men who have been trained as parachutists.[21]

The comment about NATO was interesting. Since they were fully occupied in Malaya, no regular SAS units were deployed in Europe,

SAS: Phantoms of War

but US Special Forces were, and had learnt much from the French 2nd Demi-Brigade de Parachutistes de Choc.

The government took some time to consider these proposals, and was criticised in the press for this delay, but on 4 April 1957 the Prime Minister, Robert Menzies, announced that there had been a review of defence policy. He stated that owing to the nuclear deterrent the threat of global war was now considered unlikely, but in view of communist efforts at expansion, limited war could occur in the neighbouring region at any time, and Australia shared a responsibility to prevent the occurrence of such outbreaks. 'In the upshot, speed and a capacity to hit [would] determine victory', and it was therefore necessary to have available for immediate employment highly trained, effective and compact units. The National Service scheme was to be modified and the Army's annual intake would be reduced from 39 000 to 12 000, selection being by ballot. This would enable the Army to raise a regular field force of a brigade group, in addition to the battalion in Malaya. A few days later the Minister for the Army, Mr J. O. Cramer, stated that the Army's future role would be in South East Asia, and that 'a complete reshuffle was now going on in the Army'. An Army spokesman added that the new brigade would include an SAS Company which was 'a form of Commando group'.[22]

Thus although there were many matters to be resolved, the decision was taken to raise the SAS Company. 1 Australian Infantry Brigade was to be located at Holsworthy, near Sydney, but the location for the SAS Company was the subject of considerable discussion. The Directorate of Infantry had always envisaged that the SAS Company would absorb the Airborne Platoon and would be based near Williamtown. Under a joint service arrangement, the Army was required to provide a parachute training capability at the School of Land/Air Warfare. Possible locations for the SAS Company were Gan Gan, near Port Stephens, NSW, and RAAF Rathmines, near Lake Macquarie, NSW. But there was strong political pressure to locate the SAS Company in Western Australia.

The problem created by the requirement to locate the SAS in Western Australia was analysed by the Director of Military Training in June 1957:

1. The first problem is to clear our minds on precisely what functions the SAS Sqn is designed to carry out. The abbreviation SAS stands for Special *Air* Services [sic] Squadron. If the title is in fact correct then the unit must be located close to airborne training establishments...

If other considerations dictate that this unit should be sited elsewhere then we should title it differently. The primary purpose in having the airborne platoon at Williamtown is to keep in being the complex organisation necessary for training in airborne operations...

It is understood that consideration is being given to siting the SAS Sqn in WA. If this is so it will be extremely difficult for satisfactory paratroop training to be carried out except at considerable cost.[23]

For a while it seemed that the SAS Company would be located at Rottnest Island, just off the Western Australian coast near Fremantle, but eventually Campbell Barracks at Swanbourne, a western suburb of Perth looking out across the Indian Ocean, was chosen. The barracks were already used for National Service training, but that training was to be cut back.

Some consideration was given to maintaining the airborne platoon as a detachment of the SAS Company, but that idea was rejected, and it was decided that the platoon would remain at Williamtown. It was also suggested that both the SAS Company and the airborne platoon be allotted to the Royal Australian Regiment to give them a 'spiritual home'. This proposal was supported by the Brigadier General Staff (B), but rejected by the Deputy Chief of the General Staff, Major-General H. G. Edgar, in a statement which was to guide the raising of the SAS:

> The SAS Coy is the first component of a new type of infantry regiment, viz the SAS Regt. In the not distant future it may be necessary and possible to raise an SAS Regt less one coy. It is an additional type of infantry regiment and its role and organization differ considerably from those of the battalions of the RAR. Therefore, while appreciating the question of a 'spiritual home' which will be well cared for by the GOC of the Command concerned, I am firmly of the opinion that the new regiment which is being created by the raising of one coy should have its own identity and build its own traditions outside the RAR.[24]

General Edgar's view was clear; the SAS was not to be just another infantry unit. For the first time Australia was to have a Regular Army special forces unit. The decision to introduce special forces into a conventional army marked a fundamental change in attitude to land warfare and would have a profound influence on strategic concepts. But there is no evidence that the impact on warfare and strategy was considered in any depth in 1957. Nor is there any evidence that the Army had a clear understanding of what the role of the SAS should have been or how it should have been trained. To a large extent the role of the SAS in Australia's defence would depend on the SAS itself.

3
The 1st SAS Company
Swanbourne, 1957 – 1962

The 1st Special Air Service Company (Royal Australian Infantry) was born not with a detailed instruction outlining its role and tasks, or even with a philosophical statement of why it was needed. Rather, on 1 April 1957 the Deputy Chief of the General Staff simply issued an instruction authorising the raising of 1 Independent Brigade Group. The date for the completion of the reorganisation and assembly of the formation was to be 1 September 1957. The only mention of the SAS Company was in the attached staff table which showed that the SAS Company was to have an authorised strength of 16 officers and 144 other ranks and the equipment of one standard rifle company less one platoon. Since the SAS company was to be substantially larger than an infantry company it is difficult to see why it was given a reduced equipment table. The staff table indicated that the company was to be located in Eastern Command (New South Wales).[1]

Two months later, on 10 June 1957 the Director of Staff Duties at Army Headquarters issued an instruction that approval had been given to raise 1 SAS Company in Western Command and that it was to be located at Swanbourne. Before the end of the month the establishment of the SAS Company had been changed to 11 officers and 168 other ranks, and action was underway to man the company. The new company relied to a certain extent on men who had already completed parachute training at Williamtown, but volunteers were called for from all corps. By this time it had been decided that personnel posted to the unit were to be 'trained parachutists or volunteers for such training', 'of appropriate age and undoubted medical fitness', possessing 'a reasonable standard of education', with a slightly higher level of aptitude than that applying for ordinary soldiers, between 18 and 35

36

Major W.W. Gook, the
first Officer Commanding
1 SAS Company, 25 July
1957–31 January 1958,
had the task of setting in
place the administrative
arrangements for the new
unit.

years of age, with a good disciplinary record, between 5 feet 3 inches
(160 centimetres) and 6 feet 1 inch (185 centimetres) in height, and
weighing no more than 13 stone (82.6 kilograms).[2]

But the first task was to select a commander, and the officer chosen
was Major Len Eyles, the Senior Army Instructor and second-in-
command of the Parachute Training Wing at Williamtown. It was a
logical appointment, considering the current thinking that the SAS
Company was little more than an extension of the Army's limited
parachute capability of the Airborne Platoon, and the Director of
Infantry, Colonel Austin, travelled from Melbourne to Williamtown to
inform Eyles of his appointment. It was clear that the raising of the
company would place an additional burden on the Parachute Training
Wing, and since the wing was short of instructors, it was decided that
Eyles should remain there until the new year. In any case, he would
have the opportunity of training the men who were to form the
majority of his new unit.

In the meantime, Major William (Wally) Gook, a 40 year old infantry officer then serving at Swanbourne with 17 National Service Battalion, was appointed to administer command of the company until
Eyles arrived. Gook had been captured in Crete while serving with the
2/11th Battalion in 1941, and after the Second World War had filled
various administrative positions in Western Command. One of his
young officers described him as a 'lovely old chap, kindly, sincere and
honest'. At about the same time, posting orders were issued for nine
officers to join the unit. These were Captains R. L. Burnard,

The first officers of 1 SAS Company, 1958. *Rear, left to right:* Captain K. H. Kirkland, Lieutenants T. R. Phillips, W. G. Woods, B. Wade, and K. J. Bladen. *Front, left to right:* Captains H. M. Lander and R. L. Burnard, Major L. A. Eyles, Captains I. G. McNeill and R. R. Hannigan. Kirkland retired as a major-general, Burnard and Wade as brigadiers, and Lander and Eyles as colonels. Wade is still serving as a brigadier. McNeill wrote the official history of the Australian Army Training Team, Vietnam.

H. M. Lander, K. H. Kirkland, and R. R. Hannigan, Lieutenant I. G. McNeill, who was to be promoted to temporary captain, and Lieutenants I. B. Mackay, B. Wade, W. J. Brydon and W. G. Woods. While no obvious effort was made to select officers with particular qualifications for special forces, it was nonetheless a high quality group. Three officers—Wade, Brydon and Woods—were platoon commanders in 17 National Service Battalion at Swanbourne.

The public announcements surrounding the decision to raise the SAS Company indicated high expectations for the new unit. For example, the Minister for the Army, Mr J. O. Cramer, said that specially selected troops, chosen for their 'rugged individuality', were under training. The SAS was to be used as the 'eyes and ears' of the vanguard in any action taken to defend Australia, and as a tough spearhead to penetrate enemy territory. On long range reconnaissance the company could be dropped into enemy country to carry out specified tasks

The 1st SAS Company was located in these old huts at Campbell Barracks at Swanbourne, a seaside suburb of Perth W.A.

before withdrawing or consolidating in an area until relieved by a larger force. 'They will be Australia's most highly-trained fighting men', he said. 'Among the men will be specialists in demolitions, cliff climbing and small craft handling, especially two-man canoes'. Mr Cramer added that every man would complete at least one parachute jump every three months, and hard training manouevres would be the order of the day.

Brigadier J. S. Andersen, the commander of 1 Infantry Brigade Group, matched the rhetoric of the Minister, commenting that he hoped 'to make each man wear out two pairs of boots in ten days'. He described the troops as 'the cream of the army', and warned that they would have to conduct themselves well in public as 'there were plenty of keen young men waiting to take the place of any who displeased'. The type of man wanted was one who could act on his own with a minimum of direction, endure any physical task given him, take his place as an ordinary soldier, confidently carry out clandestine operations, and be a credit to Australia in his conduct. While visiting Western Australia Andersen explored Rottnest Island looking for 'formidable cliffs' for the SAS to scale. He said that 'he found none that he could not scale in his dinner jacket and would look further afield'.

1 SAS Company was formed officially on 25 July 1957 when Major Gook, Captain Kirkland, Lieutenants Brydon, Woods and Wade, and a small group of soldiers marched into the unit, located in old wooden buildings at Campbell Barracks, Swanbourne. Gook and his officers were faced with a challenging task. Although informed that the role of the company was medium reconnaissance, they had no training directives, store or equipment.[3]

Remarkably, the unit was soon placed on fourteen days' operational notice for service in Malaya, but there was no indication what role it might be expected to play. The closest model was that of the Australian CMF Commando companies, still only two years old, and the organisation of the company reflected the Commando approach. For example, the headquarters comprised the company commander, the second-in-command (Captain Burnard), the adjutant/quartermaster (Captain Kirkland), and the company sergeant-major (Warrant Officer Class One C. M. Veness). The headquarters platoon was commanded by a lieutenant and included three sections: administration (clerical, stores, medical, catering, and electrical and mechanical engineer personnel), mortars and assault pioneers. The four SAS platoons were each commanded by a captain with a reconnaissance officer (a lieutenant) and a platoon sergeant. The platoons comprised three sections, each commanded by a sergeant and consisting of nine men. Attached to each platoon were three signallers and a medic, while attached to the company was a signals detachment of a lieutenant, two sergeants and six signallers.

The reconnaissance officers never understood what role they actually fulfilled. They had already commanded infantry platoons, and one recce officer thought that his main task was to make cups of coffee for his platoon commander.

It was soon clear that the SAS Company would require quite different stores and equipment to that which the initial equipment table had entitled it, and in October 1957 the Directorate of Infantry presented a detailed request for climbing and parachute equipment, particularly Karabiners, ropes, canoes, outboard motors, life jackets, explosives for destroying aircraft, Commando knives, parachute helmets and handcuffs. Approval was granted reluctantly in May 1958, with the Director of Military Operations commenting that it was 'not envisaged that SAS units carrying out their main role of medium reconnaissance will normally require the special items of equipment...However, there may well be special circumstances when issue of the items may be required to meet operational needs and so it is considered that there is a peace training requirement'.[4]

By 11 November 1957 the company had a posted strength of 10 officers and 43 other ranks, with a further 64 soldiers who had completed the parachute course at Williamtown, on the way to Swanbourne. An additional 61 soldiers had been provisionally selected to begin training at Williamtown. It was realised that not all of this group would qualify, but the Parachute Training Wing could not accept any more. While there were certainly many fine soldiers in these intakes, there were also unsuitable soldiers. The selection procedure was carried out purely on paper. There were few regular units in Perth, and some soldiers were allocated to the SAS simply on compassionate grounds to enable them to return to their home state.

With the increasing numbers of soldiers available, in the last weeks of 1957 training began in earnest for the 1 Brigade exercise (described at the time as a concentration) planned for February 1958. A report in the *West Australian* newspaper provides a graphic picture of 1 Platoon training near Collie, south of Perth. 'For a day and a night I did everything the red-beret, jungle-style commandos did', wrote journalist Philip Bodeker. 'I swung on wires high above streams, scaled ropes with explosives bursting under me, charged up a hill firing from the hip while Bren gun fire sheared leaves, twigs and bark from the trees overhead...This was no ordinary training camp...The theory, according to platoon commander Captain Dick Hannigan, was this:

> Perth had been captured by enemy forces.
> All defences had been destroyed and the enemy was now in a
> 100-mile radius of Perth.
> The two SAS platoons, one at Geraldton [north of Perth], the other
> at Collie had been parachuted into the state from the eastern states
> for guerilla work behind enemy lines.

Bodeker wrote:

> The SAS commando's reveille is three plugs of gelignite exploded behind the row of tents. They blast the camp to life at 5.30 am. Then it's a quick cup of tea and off for a 20 minute muscle-toughening course...Then followed a swim in the river—it was only 6 am—breakfast, and we were ready for the confidence course. The confidence course about half a mile long, is covered at a run under Bren and simulated mortar fire...Then it was 10 am—time for the assault course. This course is done with live ammunition, again at the double. The commandos ran across the Collie River on narrow fallen tree trunks while gelignite and Bren gun bullets whipped up the water around them into drenching waterspouts. Six abreast, section leader in the centre, we advanced up a heavily timbered hill rifles at the hip. Half a dozen figures appeared in front of us. 'Fire!' shouted the section leader, and our rifles bucked at our hips. We shot down two more ambushes of tin

enemies, captured the hill...and it still wasn't lunchtime. There was still the river-crossing course—crossing by pulley on a wire—and the shooting gallery. 'There's only one thing wrong with this place,' a young commando told me later. 'We have to go home in three day's time'.

As the newspaper report indicated, already the SAS was wearing the red berets, while their badge was the crossed rifles of the Royal Australian Infantry Corps.

By the time Eyles assumed command on 1 February 1958 his new company was at reasonable, although not full strength, and had already undertaken hard training. In addition to the original group of officers, Lieutenants K. J. Bladen and T. R. Phillips had joined the unit, the latter as company signals officer. Apart from McNeill, all of the captains had seen active service in Korea, as had a high proportion of the NCOs. Many NCOs, such as Ray Simpson, Roy Weir and Curly Lamb, had experience from the Second World War. The signals sergeant, Sergeant Arthur Evans, had served in the British Army in the Long Range Desert Group in North Africa and Italy. Lieutenant Neville Smethurst, who joined the company from 17 National Service Battalion in 1958, recalled 30 years later when he was a major-general, that the NCOs were some of the finest he had seen in the Army. They certainly taught him and the other young officers who had already commanded infantry platoons, much about soldiering.

As mentioned, Eyles had come from the Parachute Training Wing. Born in 1926, he had graduated from the Royal Military College, Duntroon, in 1947 and had served with the British Commonwealth Occupation Force in Japan. He was a platoon commander, second-in-command of a company, adjutant of the 3rd Battalion, The Royal Australian Regiment during operations in Korea, and was mentioned in despatches. After a number of other infantry postings, he had then been posted to the School of Land/Air Warfare at Williamtown before training as a Parachute Jumping Instructor in Britain. He returned to Williamtown as Senior Army Instructor in April 1956. Eyles brought an aggressive, go-ahead style to his new command, and immediately set about the task of formulating a training policy based on the general role of medium reconnaissance.[5]

But he did not have time to develop a coherent training programme, because on 10 February 1958 the company left Perth for the 1 Brigade concentration at Holsworthy and its first major exercise with the brigade at Kangaroo Valley, south of Sydney. The SAS's role was to act as controlled enemy, engaging in infiltration and harassment. They were highly successful, already showing the benefit of their intensive training. At the end of the exercise, training in parachuting, roping, canoeing and minor tactics was conducted in the Gan Gan, Williamtown

Colonel L. A. Eyles, Officer Commanding 1 SAS Company, 1 February 1958–11 September 1960; Commanding Officer SAS Regiment, 29 March 1967–12 December 1969. He had a major influence in establishing the role and esprit de corps of the new unit in 1958.

area, under instructors (both regular and CMF) from 1 Commando Company, Sydney, and using Commando equipment. That the SAS was taking some time to become established was demonstrated by the fact that when the company returned to Perth, all the soldiers' steel trunks went astray, eventually being located at Industrial Sales and Service (ISAS) rather than being directed to 1 SAS.

In October 1958 Eyles spent two weeks in Malaya on operations with the British SAS. While this was valuable experience, it was realised that the Australian SAS could not merely copy the British SAS; it had to develop its own methods of operation. For example, the Australian SAS continued with its infantry–commando style organisation, and did not introduce the four-man patrols that had originally been advocated by David Stirling in 1941, and were still being used in Malaya. Apart from the visit to Malaya there was not, at this stage, a close liaison between the SAS units of the two countries. When the company wrote to the British SAS requesting a copy of its history or a description of its activities, it received a copy of the biography of David Stirling, *The Phantom Major*, by Virginia Cowles. This book was read avidly by members of the company, and it was obvious that

1 SAS Company Ball, 6 June 1958, *Front, left to right:* T. Wilkinson, F. Neilson, G. Scott, J. Thorburn, F. Thurston; *Centre:* J. Sheather, A. Brooks, J. Southern, Captain M. Lander, Sergeant C. Lamb, M. Hawksworth, M. Kelly.
Rear: M. Sorenson, E. Livock (obscured), R. King, B. Sutherland, a store dummy in parachute rig, M. Grovenor.

there was a close parallel between the terrain of North Africa and northern Western Australia.

Already Eyles was becoming concerned at the problem of maintaining interest among soldiers with above average intelligence and ability, and he developed a number of long range exercises in northern Western Australia. From 16 June to 22 July 1958 he led a vehicle-mounted reconnaissance from Perth to Darwin and return. The group of 40 consisted mainly of 3 Platoon under the command of Captain Mal Lander. The main aims of the exercise, known as PERDAR, were to assess the feasibility of future long range vehicle patrols in the Kimberleys, test long range communications, undertake repair and maintenance of vehicles remote from base facilities, and practise navigation in areas of poor roads and tracks.

The party followed the north west coastal highway to Port Headland; the Great Northern Highway to Wyndham via Broome, Derby, Fitzroy Crossing and Halls's Creek; and then to Katherine and Darwin. The party returned inland from Katherine via Wavehill, Hall's Creek, Marble Bar, Port Hedland and Meekatharra. As Eyles

The SAS Company's expedition to Darwin in June and July 1958 began the SAS's long-standing interest in operating in northwest Australia. Private E. C. Livock, back to camera, is plucking a bush turkey, during a stop on the trip to Darwin.

wrote later, 'it detracts from realities of the day to write of following "highways" to Darwin. While some areas of the road were quite good, others were atrocious. The pot holes, bulldust which permeated everything, and the infamous Pardoo Sands between Port Hedland and Broome, had to be experienced to be believed. Between Wyndham and Katherine much of the track was the result of driving a grader or dozer through scrub, one blade width. As the newly created surface deteriorated to bulldust the sliced off stumps of young scrub protruded and played havoc with tyres'. The exercise was featured in an article in *People* magazine which claimed that 'the cherry-berets...can kill enemies silently with piano wire and bare hands'. The article explained that 'Australia's regiment of terror troops' was 'a mixture of commando, paratroop, "Z" Force and long range desert groups'.[6]

In August 1958 the company second-in-command, Captain Ray Burnard, commanded Exercise DRYSDALE, which lasted for about two months. The main body came from 1 Platoon, commanded by Captain Dick Hannigan, and the exercise took place in the North Central Kimberleys, north of the Drysdale River.

45

An SAS convoy returning to Campbell Barracks, Swanbourne, after an exercise in northwest Australia, 1958. The lead vehicle is driven by Private Jack Bailey with Captain Ray Burnard as passenger. (PHOTO, J. BAILEY)

Exercise GRAND SLAM, held for the 1st Infantry Brigade Group in the Mackay area of central Queensland, was one of the Army's largest exercises for many years. In April 1959 the SAS Company boarded the train for the eastern states, and Lieutenant Neville Smethurst recalled the company sergeant-major, that grand soldier, Warrant Officer Class One Paddy Brennan, moving down the train relieving soldiers of elicit beer. Only a skilled man-manager could induce soldiers to hand it over as they did with a laugh and no malice.

The SAS concentrated at Williamtown for parachute refresher training, and then 140 soldiers emplaned in new RAAF C130 Hercules transport aircraft for the flight to Mackay.[7] The operational phase of the exercise lasted from 21 to 27 May 1959, but during the training phase the Chief of the General Staff, Lieutenant-General Sir Ragnar Garrett, visited the unit, and was dismayed to hear the colourful language used by one of the company's experienced sergeants, Ray Simpson. He complained to the GOC Northern Command, Major-General Tom Daly, who was controlling the exercise, that the SAS Company appeared to be lacking in discipline and perhaps should be disbanded. Daly directed the Commandant of the Jungle Training

Members of 3 Platoon, 1 SAS Company after a parachute jump into the Safety Bay area north of Perth. *Left to right:* J. Sheather, G. Brand (back to camera), J. Southern, M. Grovenor, and E. C. Livock (obscured).

Centre, Colonel Arthur MacDonald, to spend some time with the company; he discovered that it was a well disciplined and competent unit, and there was no more talk of disbanding it. As with the previous brigade exercise, the SAS Company formed part of the enemy force known as 'The Phantom Army'. It performed well, but after almost two years of existence it had still not been exercised in its primary role—medium reconnaissance for the Brigade.

From 9 to 12 August 1959 the company underwent its first parachute jumps in Western Australia. The jumps were made from Dakota aircraft into a DZ in the Safety Bay area south east of Rockingham. The public took great interest and several hundred spectators gathered to watch. That same month another exercise (RAVEN) was held in the Kimberley area. This expedition built a string of helicopter landing pads in preparation for survey activities for the geophysical year of 1960.

In early 1960, following the annual parachute descents at Pearce, the company set off on a ten-day exercise covering about 150 kilometres with the intention of being extracted by sea at Safety Bay. At the last moment the extraction was cancelled when the SAS were advised of the accident at the mouth of Port Phillip Bay in which a number of members of the Melbourne Commando Company were drowned.

In February, 1 Platoon, commanded by Captain Ian McNeill, returned to the Kimberley area for Exercise ANDERSON, and Sergeant H. J. A. Haley was awarded the British Empire Medal for outstanding performance of his duties under trying conditions during heavy flooding. An extract from a public relations article written by Lieutenant Graeme Belleville, later killed while serving with the Training Team in Vietnam, captures the flavour of the exercise:

> On the most recent expedition to the Kimberley district...it was not all driving either. At one stage, two of the SAS men walked 10 miles [16 kilometres] into Mount Elizabeth cattle station in the West Kimberleys when their vehicle could not get through the impassable country due to rains. Then with a mule carrying their load of rations and wireless gear, they walked 90 miles to a helicopter pad near Mount Hann. Their task was to receive a parachute drop of aviation fuel from a RAAF Dakota flying over the helicopter pad. The pad was later used by Army surveyors in their task of mapping the Kimberleys.
>
> Two other members of the company [Lieutenant Mike Jeffery and Private John Sexton] had to abandon a 12 foot rubber dinghy when a crocodile climbed aboard in Montague Sound on the WA coast. Both men hastily abandoned the dinghy and the crocodile stayed aboard. Sharks were in the area, but the two men gingerly pulled up the anchor and towed the dinghy ashore with the crocodile as passenger. Luckily, when they reached the shore the crocodile scurried off over the stern.

By mid 1960 the company had built up to its full establishment of 12 officers and 182 other ranks, plus the signals attachment of one officer and 38 other ranks. Specialist training in small craft handling, shallow water diving, roping and small scale raids was now an established part of the company's training activities. These, together with constant emphasis on platoon tactics, physical fitness and individual skills had welded the company into a solid fighting unit with a very high esprit de corps. But by 1960 it was evident that the emphasis of SAS training should switch from vehicle-mounted reconnaissance to specialised infantry patrolling in tropical conditions.

The emphasis on operations in South East Asia was confirmed by a government announcement in March 1960 that the Army would adopt a 'pentropic' organisation. Indeed the February 1960 edition of the *Australian Army Journal*, which explained the working of the new organisation, claimed that the new structure would 'reduce the vulnerability of our field forces to nuclear attack, and...improve their capacity to meet the particular requirements of a war in South-East Asia'. In essence, the Pentropic Division consisted of five battle groups, two regular and three from the Citizen Military Forces. The new organisation had little direct effect on the SAS, but the effort required to provide guidance to the Army on the operation of the Pentropic Division meant that for the first time the role of the SAS Company was spelt out in considerable detail.

Under the new Army organisation, SAS companies would be provided on the scale of one per Pentropic Division, with the primary role of long range and medium reconnaissance and battlefield surveillance. The SAS was neither organised nor equipped to hold ground for protracted periods, nor to undertake normal infantry offensive operations, but the company was capable of limited defensive and offensive action. To achieve this capability only minor changes were made to the organisation, which was now to consist of 13 officers and 218 other ranks. There were to be four platoons, each of three nine-man sections, but the platoon headquarters was to include one light mortar and one rocket launcher. The idea was that platoons and even sections could operate independently. The headquarters platoon included administration, mortar, assault pioneer and signals sections.

In outlining the capabilities of the company, Army Headquarters stated that:

> When the going is suitable and the situation demands, the SAS company can operate on a jeep-mounted basis, but will more usually operate on foot in jungle terrain...Even if the initial move to the start point of a patrol is made by sea or air, the subsequent operations, if on foot, will be slow, and the duration of patrols where local inhabitants are unfriendly will depend on the amount which can be man-packed and the practicability of resupply.

The missions which might possibly be allotted to the SAS Company were:

1 Medium reconnaissance in advance of the forward troops when contact has not been made or has been lost.

2 Maintaining a watch on an exposed flank and providing limited flank protection.

3 Long-range reconnaissance behind enemy lines, including the location or confirmation of possible nuclear targets.

4 Small-scale harassing operations behind the enemy lines, such as—
    a Minor tactical demolitions.
    b Destruction of equipment.
    c Disruption of communications.
    d Destruction of headquarters.
    e Battlefield surveillance over areas inaccessible by other means.

f Internal security tasks.

g Traffic control when not required for other tasks.[8]

The general review of the Army that had preceded the introduction of the Pentropic Division had confirmed that the SAS and Commando Companies were to remain as independent sub-units for the 'forseseeable future', and as a result, in January 1960 the Director of Infantry recommended that the SAS Company be shown as a unit of the Royal Australian Regiment (RAR). He argued that: 'The infantry (and major) component of 1 SAS Coy relies wholly for its personnel on the RAR and all ranks normally transfer to a battalion of RAR when moving out of this unit for promotion or any other reason'. Major Eyles strongly opposed the suggestion for fear that the SAS would lose its identity, but on 14 November the Deputy Chief of the General Staff, Major-General I. T. Murdoch, approved the designation of the unit as '1st Special Air Service Company, The Royal Australian Regiment'. The company therefore changed its beret badge from Infantry Corps to that of the Royal Australian Regiment. At the same time the company became part of the Combat Support Group of the 1st Pentropic Division.

Both Major Eyles, and his successor, Major Clark, continued to oppose the new designation, and on 28 February 1961 the Commander of Western Command wrote to Army Headquarters requesting that the decision be reconsidered. The Deputy Chief of the General Staff, by now Major-General T. S. Taylor, replied that the objections had been based on the assumption by the two SAS commanders that the company would expand to a much larger unit in the event of war. 'Such an assumption, however, is not valid'. It will be recalled that it had been for that very reason that General Edgar had rejected the idea of including the SAS in the RAR in 1957. Furthermore, under the new organisation, one SAS company was allocated to each Pentropic division. In time of war it was envisaged that there would be more than one division and therefore more SAS companies.

The redefining of the role of the SAS was followed, on 12 September 1960, by the arrival of Major Lawrie ('LG') Clark as the new company commander. One month short of his 33rd birthday, Clark had graduated from Duntroon in 1947 and had served with 3 RAR in Korea where he had earned the Military Cross. He had been an instructor at Duntroon, adjutant of a CMF Battalion, and in 1958 had attended the Canadian Staff College at Kingston. On completion of the course he had joined the Australian Army Staff in Washington before attending a US Special Forces course at Fort Bragg and the US Ranger and Airborne courses at Fort Benning, USA.

Colonel L. G. Clark, MC, Officer Commanding 1 SAS Company (RAR) 12 September 1960–23 June 1963; Commanding Officer SAS Regiment, 13 December 1969–25 January 1972. His major innovation was the introduction of the Recondo course in 1960 which set the standards for the SAS for many years.

Clark took the opportunity to absorb many ideas that could be applied to the SAS. As he said later, he found the US 'concept of pushing people nicely beyond their perceived limits was very good'. One concept, described in the US *Army* magazine in February 1960, was that of the Recondo (reconnaissance–commando) patrol of opportunity. Recondo was 'dedicated to the domination of certain areas of the battlefield by small, aggressive, roving patrols of opportunity which have not been assigned a definite reconnaissance or combat mission'.[9]

Soon after assuming command of the SAS Company Clark re-organised training and established a training platoon to conduct specialist courses in roping, driving, small scale amphibious raids, small craft handling, and physical efficiency. Until this time the selection of SAS personnel had been based on interviews, parachute qualifications and performance in the unit. Clark inaugurated SAS selection

51

tours and cadre courses to determine a soldier's suitability before joining the unit. The first selection tour was conducted by Lieutenant Mike Jeffery. All Regular Army infantry units, and the artillery and armoured regiments were visited, and to save travelling expenses, the selection courses were run at the Infantry Centre at Ingleburn, near Sydney. Jeffery recalled that this was a considerable responsibility to place on a lieutenant, and he was tremendously impressed when after his course had conducted a forced march of over 160 kilometres in 40 hours from Bathurst to Ingleburn, Colonel Gerry O'Day, the Infantry Centre commandant, called out all soldiers in the centre to welcome them when they marched into Ingleburn in pouring rain at 3 am one morning. The course, which lasted for six weeks, introduced the soldiers to the unit's specialist techniques as well as providing a complete revision of weapon handling, navigation and patrolling. It continued to be run at the Infantry Centre until February 1964.

SAS selection procedures were reviewed at a conference at Army Headquarters in October 1960. The conditions of eligibility remained little changed from those determined in 1957, but the required characteristics are worth quoting in full.

> SAS training is arduous and exacting. The role of the unit demands a particular type of soldier of outstanding personal qualities. These personal qualities and make up count more than technical efficiency in his own arm or service. Technical efficiency can be taught, the personal qualities required for long range, long-term operations in enemy territory however are part of a man's character and, although they may be developed over a period, they must be learnt in childhood. These qualities are: initiative, self-discipline, independence of mind, ability to work without supervision, stamina, patience and a sense of humour. The aim is to find the individualist with a sense of self-discipline rather than the man who is a good member of a team. The self disciplined individualist will always fit well into a team when team work is required but a man selected for team work is by no means always suitable for work outside the team.

Lieutenant Reg Beesley, who joined the company in January 1961, found 'LG' Clark to be an innovative trainer who introduced many new ideas. He was 'a hard man who suffered fools poorly' and allowed his officers to get on with their tasks without interference. He was 'somewhat of an academic', and Beesley thought that he was shy and did not always relate easily to the soldiers.

Building on his US experience, in November 1960 Clark introduced the first Recondo course of nineteen day duration, designed to teach and assess patrol and raiding techniques. Students, both officers and soldiers, were all given the title of 'Ranger', and were assessed as patrol commanders. They were required to pass the course to stay in the unit. Failure meant repeating the course, which was conducted under the most gruelling conditions and required the utmost endurance and self-discipline. Lack of sleep and constant activity often caused hallucinations. The course, considered to be the most important one in the unit, was in three phases, with the final phase based at Collie, where raids and exercises were conducted over a wide area and involved parachute insertions, amphibious raids and river crossings. Activities covered a broad range of terrain and vegetation. For convenience, students were divided into five-man patrols, despite the fact that at the time the standard SAS patrol organisation was nine men. Having conducted nine Recondo courses, Colonel Beesley considered the course to be the most realistic peacetime training he experienced in his career. Major-General Jeffery recalled, almost 30 years later, that it was the toughest and best course he had undertaken in the Army. It showed many soldiers that they possessed a greater limit of endurance and ability to operate under arduous conditions than they had imagined. The best soldier on each course was awarded a chrome-plated Commando dagger.

Major Clark later described the course in an article in the *Australian Army Journal*:

The Recondo course is realistic, tough, and to a degree hazardous. It is the closest approach to combat conditions that can be achieved in peacetime. The number and variety of situations faced by a student equal those which a soldier would gain in two or three campaigns. Fatigue, thirst, hunger, the necessity for quick, sound decisions and the requirement for demonstrating calm, forceful leadership under conditions of stress are all encountered. Here an individual student, selected with no notice and at any time, must impose his personality and will on others to achieve the patrol mission. He operates mostly at night, every night, under adverse conditions of weather and ground when the physical condition of his men at times approaches exhaustion. They are constantly harassed by an active 'enemy' which forces them into unexpected situations calling for prompt valid decisions'.[10]

The pressure of the course was described by one student:

The patrol was required early that night to hold for several hours a feature overlooking a road junction. We had been on the move for four days, and I had had no more than a total of five hours sleep. I finished my reserve water bottle that morning and I was limping badly from a large raw blister on my left heel. The enemy attacked our feature as we were ready to withdraw over the eight miles [13 kilometres] of sand dunes to the sea, for exfiltration at 0300 hours by DUKW [amphibious vehicle]. During the attack one of the patrol 'broke' his leg, and the fourteen of us made a stretcher from bush timber, and, four or six at a time, started to carry him back. My own rucksack and rifle weighed over sixty pounds, and I doubted I would ever reach the DUKW on my own, let alone sharing the carrying of a casualty—and still remaining alert for an enemy. My first turn at carrying was agony, and I doubted I would ever last until 0300. By 0200 it was obvious we would not reach the DUKWs in time, and we would be faced with a day of harassment by the enemy, and a twenty-mile [32 kilometre] walk back through the FDLs [forward defensive localities] next night. The pain in my foot, the thirst and the agony of the heavy loads made me decide that by 0300 I would be able to go no farther.

At 0400 I realised that my problems were probably no worse than those of my fellows, and we had developed a camaraderie to see the thing through. At 0630 dawn broke, and we could see ourselves one mile [1.6 kilometre] from the beach-head, and the DUKWs had gone. But I knew I could go on carrying that casualty forever, for I had already passed through all the physical and mental barriers in my mind I could forsee. I realised that obstacles are there to be negotiated, and, having crossed one, got less and less in stature...[11]

On 1 August 1961 a conference was held at Army Headquarters in Canberra to discuss the role and training requirements of the SAS Company. It was agreed that the primary role of the SAS was re-connaissance, both medium and long range. The SAS was not organised for area defence nor for airborne assault operations. Its secondary role was to undertake small scale harassing operations in the nature of raids and to assist in internal security operations. In contrast, the primary role of the Commandos was to undertake small scale raids. The conference agreed that one platoon of the SAS should be nominated as a 'special' parachute platoon, trained in free fall and other advanced techniques. The SAS was not to assume responsibility for the Airborne Platoon at Williamtown.

To take account of the newly confirmed roles, Clark gave each platoon the task of specialising in one particular mode of operations: 1 Platoon in ground operations, including roping and cliff climbing; 2 Platoon, watercraft operations and diving; 3 Platoon, parachute oper-ations including free falling; and 4 Platoon, vehicle operations.

Platoon training was carried out in the Avon Valley training area, and on 5 August 1960, during a platoon patrol exercise, Private Anthony Smith was drowned while crossing the Avon River in the Lower Chittering area about 60 kilometres from Perth. Private John Coleman, a stocky Englishman who had served in Cyprus and Suez, recalled swimming the flood-swollen river: 'I had just reached the far bank when I heard [Private] Vince [Manning] shout "Grab Tony". I turned and saw Tony's face racing past. I dived in and as I grabbed him we were swept away from the bank and down stream by the current. He was still limp but his eyes were open. I talked to him all the time. But he didn't answer. We went under twice in a sort of undertow where a big stream came into the side of the river. Tony was like a dead weight. The third time we went down I lost him. I just couldn't hold him any longer. I'd had it'. In the subsequent enquiry Clark fought vigorously to prevent the conclusion being reached that such crossings needed boats and ropes; if training was to be realistic accidents were sometimes going to happen.[12]

While reconnaissance remained the main role of the SAS, Clark gave considerable attention to raids and harassment. His belief that this was an important role was emphasised in an article he wrote for the *Australian Army Journal* which was published in May 1961. He suggested that the task of organising guerilla warfare should be included in the roles of either the SAS or the Commandos. For operations in South-East Asia he proposed that teams of a 'Special Army Force' should be deployed to organise and train indigenous forces in areas where it was not appropriate for conventional forces to operate.[13]

During June 1961 the company began the first of two three-week exercises, called SHARK BAY 1 and 2, in the Hamelin Pool area near Shark Bay. Designed to practise air and amphibious landings, and long range reconnaissance and raiding, they were the biggest combined Regular–Citizen Military Force exercises conducted in Western Australia to that time. Participants included the 10th Light Horse Regiment, 22 Construction Squadron, an electrical and mechanical engineer workshop, an artillery regiment and a signals unit. One patrol from 3 Platoon deployed for an immediate contact drill and found itself confronted by a wild camel, which was duly despatched.

In October 1961 the GSO1 1st Division, Lieutenant-Colonel C. H. A. East, visited the SAS Company to evaluate its level of training. He reported that morale was high, keenness and enthusiasm were apparent throughout the unit, the standard of the SAS soldier was high and their individual performances were impressive. East was concerned, however, that while the unit morale was good, it might not be possible to maintain if they continued the 'currently heavily loaded training activities programme'. Many members of the company were married and if soldiers were absent from home too frequently there could well be domestic unrest. There was a further concern. The company had not trained jointly with other units of the 1st Division, and they had not had the opportunity to exercise their role of reconnaissance for the division. 'The temptation', wrote East, 'to employ 1 SAS Coy RAR as enemy in any of these exercises must be resisted'.

During November 1961 and January 1962 two exercises were held with 42 Commando Royal Marines, which was embarked on board the carrier, HMS *Bulwark*. In one exercise the Commandos were airlanded by helicopter on a deserted part of the Western Australian coast near Lancelin while the SAS, operating as an infantry company, acted as enemy. The company had its first opportunity to work with the Sioux and Whirlwind helicopters from the *Bulwark*. Parachuting was conducted from the Whirlwinds, one of which crashed soon after the last parachutist jumped.

One morning in January 1962 Clark arrived for an early parade at Swanbourne to find the State Governor, General Sir Charles Gairdner, was there and had issued orders that the company was to be deployed to fight bushfires that were out of control south of Perth. Clark represented that Gairdner had bypassed the normal chain of command, but the Governor stated that he could not locate the Commander, Western Command. In any case, Clark realised the importance of the task, and within half an hour 90 soldiers were on the way. Many soldiers were recalled from leave by press, radio and television. The company was deployed again, later in the month, to fight the disastrous fires which wiped out the town of Dwellingup. The company was responsible for saving considerable property, and provided extensive communications throughout the striken areas.

In May 1962 the company took part in its first SEATO exercise, Exercise AIR COBRA, held in Thailand. 3 Platoon, commanded by Captain Don Anstey, flew to Bangkok, and from there to Korat, a Thai Airforce recruit training depot about 300 kilometres north east of Bangkok. The platoon was attached to the US 4th Marine Light Reconnaissance Company with Lieutenant Reg Beesley acting as liaison officer. The exercise involved a 65 kilometre patrol in four-wheel drive vehicles, one patrol per vehicle, with the tasks of reconnaissance and calling in airstrikes. The platoon spent six weeks in Thailand; however as it was primarily an air force exercise it did not prove to be particularly realistic.

Exercise AIR COBRA was a salutary reminder that the strategic situation in South East Asia was deteriorating. Throughout 1961 there had been concern about a Communist take-over in Laos, and the SEATO exercise was an effective means of raising the confidence of the Thai government. Indeed, on 23 May 1962, while the SAS was still in Thailand, the Minister for External Afairs announced that Australia had agreed to help 'maintain the territorial integrity' of Thailand.[18] Five days later he revealed that a detachment of RAAF Sabres would be stationed at Ubon.

Concern over Laos was matched by increasing worry over the ability of the South Vietnamese government to resist VC insurgency. On 24

Emplaning for a demonstration jump on Air Force Day, RAAF Base, Pearce, W.A. 1961. The last man is Corporal T. Thorne. The second man, Lieutenant G. R. Belleville, was killed on 12 February 1966 while serving with the Training Team in Vietnam.

May the Minister for Defence, Athol Townley, announced that Australia would provide a team of 30 Army instructors to assist the South Vietnamese Army with jungle warfare techniques, village defence, engineering and signals. The first contingent of the Australian Army Training Team, Vietnam, was split equally between officers and senior NCOs. Two SAS officers, Major Lawrie Clark and Lieutenant Ian Gollings, were selected, while another officer, Captain Brian Wade, had previously spent almost two years in the SAS. There were four NCOs from the SAS—Sergeants Joe Flannery, Des Kennedy, Ray Simpson and Roy Weir—all with previous operational service.[14] When Clark departed on 14 July 1962 Captain Geoff Cohen, the second-in-command, assumed command of the SAS Company. He had been filling that position since 24 June when Clark had started attending pre-embarkation courses.

Under Cohen's command the company took part in Exercise NUT-CRACKER in the Colo–Putty area of New South Wales during October and November 1962. The company moved by train in 'dog boxes' (sitting room only) to Holsworthy, New South Wales, and for nearly

57

five days sat shoulder to shoulder. NUTCRACKER was a conventional war exercise involving nearly 8000 troops with both 1 and 3 RAR advancing on separate axes. For the first time since it had been raised over five years before, the company was exercised in its main role of providing medium and long range reconnaissance for a large conventional force.

The period of five years from 1957 to 1962 was stimulating for the soldiers and young officers of the SAS Company. They had been given the freedom to develop a wide range of important skills, and they had built a reputation for toughness, discipline and effectiveness. The company had certainly made an impression on the public in Perth, as demonstrated by an article in the *West Australian*: 'Five years ago this month the Australian Army called on its toughest, fittest, and best trained soldiers to trade in their traditional slouch hats for the exclusive red beret', wrote journalist Brian Pash. 'Since then the adventurous and specialised activities of the SAS have captured the imagination of the West Australian public. Especially the youth who regard the red beret and shoulder patch wings as symbols of physical and moral achievement. For training had bred in each SAS man initiative, self-discipline, ability to work without supervision, stamina, patience, and a typical Australian sense of humour'.

Reflecting on these years, Colonel Reg Beesley thought that this sort of publicity was not useful. Some soldiers believed the press reports that they were 'super-soldiers', but in his view, 'super-soldiers' thought that they had nothing more to learn and they therefore became a danger to their own patrol. Fortunately, most SAS soldiers did not succumb to this publicity.

Clearly, the men of the SAS Company could be proud of their achievements. But there had also been considerable frustration. There had been battles to obtain suitable equipment, and financial restrictions had meant that the company had been unable to train regularly in the Eastern States, particularly at the Jungle Training Centre at Canungra in south Queensland, or with other regular units. For the more senior officers and NCOs of the company it was clear that many high ranking officers in the remainder of the Army had no idea how the SAS could be employed. The deployment of the Training Team to Vietnam offered the chance for overseas service for officers and senior NCOs of the company, but there seemed little possibility of the SAS being deployed as a company or even as a platoon, in an operational situation. Unlike the infantry battalions, or even artillery and engineer sub-units, there was no likelihood of a rotation to the Commonwealth Far East Strategic Reserve in Malaya. But as 1962 drew to a close, events in and near Malaya were developing in such a way that would have a fundamental influence on the role and organisation of the SAS and would demonstrate that special forces were a vital capability for any nation in modern limited war.

4
To Borneo
Formation of the regiment:
1963–1964

Although most SAS soldiers did not know it at the time, the training and development of the SAS Company during 1963 and 1964 was undertaken against the background of a likely deployment to Borneo. The outbreak of hostilities in Borneo in 1963 between the British and Malaysian security forces on the one hand and the Indonesian irregular, and later regular, troops on the other hand, offered the first possibility for the deployment of the SAS in a role for which they were being specifically trained.

Like anybody else in the community, the members of the SAS could read in the newspapers what had been happening in South-East Asia. During 1962 the Indonesian President, Sukarno, had complained that the proposed formation of the new Federation of Malaysia by the joining of Malaya, Singapore, Sarawak, Brunei and British North Borneo (Sabah) was neocolonialist—a claim he eventually used to justify the use of armed forces against it in alleged support of the newly emerging forces opposing colonialism and imperialism. Furthermore, Indonesia was moving to secure control of Dutch New Guinea (West Irian), and this would bring Indonesian troops to a common border with the Australian territories of Papua and New Guinea. When these events were coupled with developments in Laos and Vietnam, it was obvious that the strategic situation in the region was deteriorating.

It is unlikely that the Indonesians directly instigated the revolt in Brunei which broke out in December 1962 and was quickly put down by British troops flown in from Singapore. However the rebels had been trained and supported by Indonesian forces in northern

Kalimantan (Indonesian Borneo). Brunei did not join the new Federation, but Sukarno raised doubts as to whether the people of Sarawak and Sabah were joining the Federation of their own free will, and he backed this rhetoric with a number of indiscriminate raids into Sarawak and Sabah by so-called volunteers, trying to foment opposition to the formation of Malaysia.

In this delicate political climate the Australian SAS Company carried out its first overseas exercise. (Only a platoon had exercised in Thailand in May 1962.) During September 1962 the British Commander of the Far East Land Forces in Singapore had mentioned to a visiting Australian officer that he would like the Australian SAS to train in Malaya. The Australians declined the offer, but when the Chief of the General Staff, Lieutenant-General Sir Reginald Pollard, visited the SAS Company in October he was impressed with their keenness to have an overseas tour. He therefore proposed an exercise in Papua New Guinea in the first half on 1963. The RAAF wished to provide aircraft in April or May 1963, but Pollard thought that it 'would be politically inadvisable' to move the company to New Guinea during this period 'as it might seem to be connected with the handing over of the administration of Western New Guinea from the United Nations control to Indonesia on 1 May 1963'. [1]

The exercise was therefore brought forward to begin on 14 February 1963, but it was not without opposition from the Minister for Defence, Athol Townley, who feared offending the Indonesians. The Chairman of the Chiefs of Staff Committee, Air Chief Marshal Sir Frederick Scherger, argued that if Australia was reluctant to exercise on its own territory before the Indonesians took over in West New Guinea, it would be even more reluctant to do so afterwards. Australia could not tolerate exercises in its own territory being determined by a foreign country. The Minister for External Affairs, Garfield Barwick, supported Scherger, and Townley 'reluctantly' agreed to the exercise proceeding. [2]

Exercise LONG HOP was the first complete move of the SAS Company outside Australia, and the company became the first Australian unit to exercise in Papua New Guinea since the Second World War. The exercise consisted of a period of acclimatisation training from 14 to 22 February, the actual exercise from 22 February to 1 March and specialist training until 25 March. The idea of the exercise was to replay the advance of the Japanese over the Kokoda Track in 1942, with two companies of the 1st Battalion, the Pacific Island Regiment (1 PIR) acting as the Japanese. Initially the platoons were deployed independently into the Popondetta area with the task of locating,

Crossing the Kumusi River at Wairopi, Papua New Guinea, during Exercise LONG HOP, March 1963. This was the first time that the complete SAS Company had exercised overseas. The SAS were opposed by the Pacific Island Regiment in the exercise that followed the path of the Japanese advance from Buna over the Kokoda Track.

harassing and delaying the advance of the PIR. The Australians received an enthusiastic welcome and were showered with fruit by natives lining the routes around the villages as they moved into the exercise area. Perhaps this was part of the enemy's plot, because the sudden change in diet resulted in multiple cases of diarrhoea.

Once the exercise began the PIR proved an able adversary. Their knowledge of the jungle and the intelligence they received from the villagers, many of whom had relatives in the PIR, resulted in the capture of a number of SAS members. The company was not outdone and had its own collaborators in some of the plantation owners who transported troops concealed in their vehicles. Once the PIR reached the Kumusi River at Wairopi the SAS operated as an independent company attempting to delay the PIR as it advanced along the Kokoda Track over the Owen Stanley Ranges. The exercise finished at Macdonald's Plantation, about 30 kilometres from Port Moresby.

Captain Geoff Cohen presenting the Officer Commanding's chrome helment to Major Alf Garland, in March 1963 after their return to Perth from Exercise LONG HOP in Papua New Guinea. Cohen had commanded the company from 24 June 1962 to 14 March 1963.

This exercise in Papua New Guinea and subsequent exercises overseas proved of inestimable value in preparing the unit for future operational deployment. In particular, the soldiers learned much about jungle fieldcraft from the PIR soldiers. Furthermore, as the final report noted, 'the problems imposed by an unsympathetic population were a new training experience in the maintenance of security'.[3] Major Garland, who had observed the company during the exercise, and who assumed command on its conclusion, saw that the exercise had revealed a weakness of the SAS in the area of communications. As a result, the signals officer, Lieutenant Daryl Slade, began training his operators to use sky wave rather than ground wave transmissions, and eventually this training was extended to all ranks of the unit. The training of all ranks in the receiving and sending of morse and the use of one-time code pads was to pay off in Borneo.

Major Alf Garland assumed command of the company in Papua New Guinea on 15 March 1963, four days short of his 31st birthday. Stocky, possessed of a strong and determined personality, with dark hair and a dark moustache, Garland had graduated from Duntroon in 1953 and had served with both the 3rd and 1st Battalions of The Royal Australian Regiment in Korea in the period after the armistice. While serving with 3 Cadet Brigade in Melbourne in 1956 and 1957 he had completed a parachute course, and had made known his interest in the SAS and the Commandos. This interest bore fruit in December 1961 when he was posted from the Infantry Centre to undertake training in the United States. During 1962 he attended a special forces course at Fort Bragg, a ranger course at Fort Benning, and was attached to the 187th Airborne Battlegroup of the 101st US Airborne Division and to the US Fleet Marine Force Reconnaissance Unit. It was intended that he would complete his training with attachments to the British SAS and the Royal Marines, but in August 1962, soon after Major Clark was posted from the SAS Company to the Training Team in Vietnam, Garland was ordered to cut short his training programme and to return to take command of the company early in 1963. In the eyes of some observers, Garland was unduly influenced by United States methods, but he had read about the work of the British SAS in the Second World War and he was determined that the Australian SAS would give less emphasis to raiding and operating as an independent airborne company and more emphasis to its reconnaissance and intelligence gathering role.

The need to concentrate on reconnaissance was given extra impetus by the visit of the Commanding Officer of 22 SAS Regiment, Lieutenant-Colonel John Woodhouse, to Perth in July 1963. Woodhouse brought more definite information on the role of the SAS in Borneo and indicated that he was keen for the Australians to be involved.[4]

Throughout 1963 the British SAS had proved its value in Borneo. After putting down the rebellion in Brunei in December 1962, the British commander in Borneo, Major-General Walter Walker, had realised that the 1100 kilometres of mountainous and jungle border separating the Indonesian and British sections of the island offered Sukarno a ready opportunity to embarrass the efforts to form the Federation of Malaysia, and he sought to build up his defences to thwart the Indonesian plans. The vast distances meant that he would need to rely heavily on intelligence and early warnings of when and where the Indonesians were crossing the border. To gain this intelligence, he planned to use the British SAS in three general roles: firstly, reconnaissance by small SAS patrols; secondly, by training and leading local units called Border Scouts; and thirdly, by conducting 'hearts and minds' operations in the distant border villages and by encouraging the local people to report any Indonesian military activity. A Squadron 22 SAS Regiment, which was deployed to Borneo in January 1963,

Major Alf Garland and Lieutenant Colonel John Woodhouse, Commanding Officer of the British 22 SAS Regiment, during the latter's visit to Perth in July 1963. Woodhouse described the role of the British SAS in Borneo to Garland and indicated his desire for the Australian SAS to become involved. (PHOTO A. GARLAND)

did the initial work in developing these tasks, and it was replaced by D Squadron in April, with A Squadron returning in August.

Brunei declined to join the Federation of Malaysia and Singapore, which came into being in September 1963, and by this time the scale of Indonesian incursions had increased. By September the British had deployed seven battalions to Sarawak and in the following months the Indonesians began to send regular troops on substantial incursions into Malaysian territory. In December 1963 the Malay Regiment suffered considerable casualties in an Indonesian raid in Sabah.

The SAS operations during 1963 were tiring and exacting but involved few contacts with the enemy. Nevertheless, 22 SAS Regiment was stretched to the limit. Its two squadrons were being rotated through Borneo and the Commanding Officer, Lieutenant-Colonel Woodhouse, was making every effort to obtain permission to double his sabre squadrons to four. His dilemma is shown in a letter to his squadron commander in Borneo on 31 October 1963: 'You can be sure I appreciate all you are doing and in event of a sudden flare up of [operations] I will support necessary increases in your [strength] as fast as possible. Meanwhile it is in the Regiment's interest to scrape by in Borneo—this involving a calculated risk that large scale [operations] will not break out before Christmas'. To help solve his problems he added that he 'would like to see the Australian SAS work with us in Borneo'.

During his visits to Borneo Woodhouse mentioned to both General Walker and to visiting British government ministers the desirability of obtaining help from the Australian SAS. These approaches soon gained a result, with the British Commander-in-Chief in the Far East passing on to the Australian Chiefs of Staff his view that it would be of great help if Australia and New Zealand could provide in order of priority an SAS Squadron, additional RAAF personnel for duty at Butterworth in northern Malaya, an RAAF Iroquois squadron, an infantry battalion and additional ships for troop lift.[5]

Early in December 1963 the British Prime Minister, Sir Alec Douglas-Home, formalised the request in a letter to the Australian Prime Minister. Menzies replied that there was 'no clear military justification' for any 'Australian military contribution' for the time being. He agreed to the deployment of a detachment of engineers stationed in Malaya if the British wanted them, but he 'feared that the Australian loss of influence in Djakarta if their troops were released for operations, far outweighed the military value of a contribution of a few hundred troops ahead of need'. Yet despite Menzies' refusal, early in 1964 Garland was summoned to Canberra to be informed by the Deputy Director of Military Operations, Colonel S. P. Weir, that the Prime Minister had decided on a graduated response to the British request. If the situation in Borneo did not improve, eventually the SAS would be deployed and Garland was to prepare his unit for operations in Borneo.

This information came as no surprise to Garland. In the latter half of 1963 he had received a number of letters from Woodhouse informing him of the efforts to get the Australian SAS into Borneo. Then, in December 1963 he had received a letter from Lieutenant-Colonel J. W. Norrie, in the Directorate of Infantry, proposing a reorganisation of the SAS to enable it to relieve a British SAS company 'in South East Asia'. Norrie did not have a clear appreciation of the role of SAS in Borneo:

> The proposed role is basically infantry in nature and full use would not be made unless the occasion arose of the highly specialised SAS skills. However the SAS would gain valuable operations in a theatre where they could practise their secondary roles of internal security and harassing raids as well as train members of another force in skills peculiar to SAS.

Norrie suggested that the force to be sent to South East Asia should consist of 96 SAS soldiers organised as a squadron of two troops. He added that the Director of Military Operations and Plans, Brigadier Ken Mackay, wanted to 'capitalise on publicity' and intended using the terms 'squadron' and 'troops' rather than 'company' and 'platoons'. The SAS squadron overseas would rotate on a six monthly basis with the company in Australia. The company remaining in Australia would consist of two full platoons and would continue its main role of medium reconnaissance with the battle group.[6]

Garland discussed these proposals with the Commander of Western Command, Brigadier Hunt, who directed him to begin preliminary planning to prepare a mounting instruction. In the meantime, Garland replied to Norrie that there were a number of practical questions which need to be resolved. These included: would his new squadron be deployed as an independent unit or would it be organic to a British unit; how much of his own stores would he have to take; and would the squadron travel out of Australia by civil or military aircraft, as this would determine the amount of stores which could be taken? Clearly much planning was necessary, and in particular the likely role of the SAS had to be resolved.

The idea of the SAS operating in an infantry role was certainly contrary to the views of Colonel Woodhouse. For example on 2 January 1964 he wrote to Garland:

> Your Brig Frank Hassett was here on an 'unofficial' visit yesterday and spent an hour with us. There is plenty of room for you as well as us, and the Kiwis for that matter! I have the impression that he thinks your SAS should become independent of the infantry, and said we found this worked best.

Hassett was returning to Australia, after attending the Imperial Defence College, to take up his appointment as Deputy Chief of the General Staff, and he would be influential in separating the SAS from the Royal Australian Regiment.

Woodhouse advised Garland that 'the most valuable preparatory training in the event you come here would be Malay language and medical training'. He added: 'things are getting hot here but the bastards have so far not come through areas where we have patrols deployed—another good reason to get more SAS here. We should have the third British Squadron operational by 1 August 1964...we are as keen to see you here as I know you are to come'.[7]

Although Garland was considering the possibility of operations in Borneo, his company continued its specialist training in Australia. During October the Parachute Training School conducted the first free fall course for SAS soldiers and a team led by the adjutant, Captain

Officers of 1 SAS Company (RAR), towards the end of 1963. *Front, left to right:* Captains H. W. Irwin and R. P. Beesley, Colonel J. G. Ochiltree, OBE, Director of Infantry, Major A. B. Garland, Captains I. D. McFarlane, D. L. Hill, and P. N. Greenhalgh. *Rear:* Lieutenant T. R. Kelly, Captain O. J. O'Brien, Lieutenants P. M. McDougall, G. C. Skardon, and D. J. Slade, Captains G. E. Williams and D. G. Robertson.

Reg Beesley, became the first in Australia to free fall at night with equipment. Exercise SKY HIGH conducted by the 1st Division in the Colo–Putty area of New South Wales from 8 to 29 November 1963 indicated that many senior commanders still had little or no appreciation of the correct role of the SAS. The company simulated an air drop off the back of lorries with the independent parachute company task of securing the airfield, and then was used as a controlled enemy as well as undertaking some reconnaissance tasks in support of the friendly forces. As Garland put it: 'it was a long hard fight to try and convince people that that was not what the SAS was all about'.

Specialist training continued and Garland arranged for members of the SAS to go to the School of Army Health in Victoria for training in the duties of medical assistant, and subsequently for these soldiers to obtain on-the-job training by attachment to the casualty departments of hospitals in Perth and Fremantle. Language training of sorts was also started in the unit. One of Garland's innovations, which did not survive his period as commander, was the wearing of blue cravats.

In April 1964 the Minister for the Army, Jim Forbes, visited the company and Garland took the opportunity of asking him when they would be deployed to Borneo. Forbes replied that maybe the next time he saw them would be in Borneo. With this encouragement, Garland continued to develop an organisation which would be suitable to conduct operations in Borneo. At this time the unit still consisted of four combat platoons each of three sections of nine and a headquarters platoon comprising a signals platoon, administrative elements, medical elements and pioneer and mortar sections. The unit had the capability of providing in the field at any one time twelve fighting patrols each of at least nine men or 24 reconnaissance patrols each of at least four men. In practice patrols could be larger since a signaller would be attached to each section. Each platoon was commanded by a captain with a lieutenant as his second-in-command. The platoon included four sergeants and numbered 34 personnel. In all, the company consisted of about 230 men.[8]

The prospect of a SEATO exercise with American, British and Philippine forces in the Philippines in May gave Garland the opportunity to try out a new organisation. Each platoon was reduced to form a troop of 21 personnel commanded by a lieutenant and patrols were to number six men. This was essentially the organisation which Garland used in Exercise LIGTAS, the most significant SEATO exercise since the organisation had been formed ten years earlier.

On 11 May half of the company (the experimental squadron with 70 soldiers) deployed to Okinawa where it took part in an orientation exercise. On 13 May, 69 members of the company participated in a parachute drop with members of the US Army Special Forces stationed on Okinawa. After the parachute drop the Commanding Officer of the US Army Special Forces on Okinawa presented each soldier who had participated with a US Army Parachute Badge along with a special order authorising each member to wear the badge. The remainder of the company married up with the experimental squadron at Subic Bay on 24 May, just before the initial deployment of troops into the main exercise area. The company now included 183 soldiers on exercise.

The main exercise involved an assault onto the island of Mindoro similar to the wartime Japanese invasion, and in addition to the Australian SAS it included elements of the 3rd US Marine Division, 503 US Airborne Battlegroup, 40 Commando Royal Marines, the US Air Force and the Fleet Air Arm. The SEATO fleet consisted of 75 ships carrying about 20 000 troops in the assaulting force, and they were supported by about 300 aircraft. On 1 June three SAS platoons plus the company headquarters spearheaded the assault by parachuting into the exercise area to secure the DZ and LZ for use by the US airborne forces and the Royal Marine air landed forces. This was the first time an Australian unit had jumped as a unit outside Australia since the jump into Markham Valley in New Guinea 1943. The other platoon, under Lieutenant McDougall, operated in conjunction with guerilla forces against the invasion force. Exchanges with US Marine reconnaissance units and with US and Philippine Special Forces took place involving about twenty soldiers.

After the landing the company, except for the platoon with the enemy, resumed its reconnaissance role until the conclusion of the exercise on 8 June. Meanwhile the enemy platoon mounted a cunning attack on the exercise headquarters. A Philippines brigadier-general had stated that since he was well guarded by his marines he could not be captured. The Australians took up a collection from among their numbers and gathered sufficient pesos to pay two local women to divert the marine guards. The brigadier-general was duly captured.

One interesting aspect of the exercise was that Headquarters Western Command in Perth believed that Exercise LIGTAS was a cover for the the operational deployment of the Company to Borneo. As a result, the equipment which the company had been anxious to obtain for up to a year was suddenly provided, much of it purchased locally. The brigadier and his staff were quite annoyed when they learned that the exercise was exactly that.

Exercise LIGTAS had shown quite clearly that a squadron headquarters should not be deployed into the field under the circumstances of the exercise. It was unable to provide the necessary communications, control and support to deployed patrols when it was itself deployed. However, the exercise had shown the viability of Garland's new organisation with its reduction in the size of troops, and had demonstrated that there were sufficient personnel to form almost two squadrons from the old company organisation. Furthermore, it was clear that if a squadron were to be deployed to Borneo, another squadron would have to be formed at Swanbourne to replace it once it completed its overseas tour. It came as no real surprise therefore when on 20 August Army Headquarters issued instructions for the raising of a new unit, the Special Air Service Regiment.

Under the new establishment, the regiment was to number fifteen officers and 209 other ranks. The 1st SAS Company Royal Australian Regiment was to be disbanded and the link with the RAR was to cease. The new regiment was to consist of a headquarters of three officers and 37 other ranks, and two SAS squadrons each numbering six officers

US Brigadier General R. G. Davis, Commander of the SEATO Expeditionary Brigade, accompanied by Major Garland, inspecting the SAS Company at Subic Bay in the Philippines, during Exercise LIGTAS, June 1964. On the left are Sergeant Tony Tonna, Private Jack Gebhardt, Lance Corporal Eddie Chenoweth and Private Sonny Edwards.

and 86 other ranks. The SAS squadrons were to be restricted by one troop and would comprise three troops each of 21 plus appropriate specialists on squadron headquarters.

An Army Headquarters instruction issued on 2 September indicated that the regiment had been raised to enable the SAS to undertake 'specialized tasks in the existing cold war conditions as well as maintaining a medium reconnaissance capacity for limited war'. The new organisation was designed specifically for the types of operations required of such a unit in South East Asia and New Guinea. Both SAS squadrons were placed on seven days notice to move from 10 September with priority being given to 1 Squadron. 2 Squadron was to be brought to operational readiness as quickly as possible, at the expense of regimental headquarters if necessary.

Although the Army Headquarters instruction made no mention of Borneo, the roles outlined for the squadrons were clearly directed toward operations in that area. The first role was described as reconnaissance, including border surveillance, based on the employment of small self-contained patrols. These patrols were expected to operate without resupply for up to five days. The second role was the collection of intelligence on the location and movement of enemy forces. This task was expected to be carried out by the indigenous population or border scouts. The SAS patrols would work with the indigenous population and would operate for lengthy periods in or near enemy territory. The third task was the organisation, training and control of indigenous irregular forces. The training of irregular units in the highlands of Vietnam by the Australian Army Training Team was given as an example of this role.

The fourth role was ambushing and harassing strong enemy forces. This included delaying an enemy in time for the concentration and movement of larger bodies of friendly forces to attack in force, and harassing enemy bases in enemy controlled territory, including ambushing their resupply parties. Apart from the fleeting opportunities to ambush and harass which might be presented to a patrol, it was visualised that this role would be undertaken generally at troop level, and on some occasions at squadron level.

The final role was limited civic action projects. It was also noted that if the need arose for commando-type operations the SAS squadrons could be called upon until the Citizen Military Force Commando units became fully operational. Thus a capacity to undertake squadron operations had to be maintained.

On 4 September 1964 Major-General J. S. Andersen, the Commander of the 1st Division, reviewed a regimental parade at Campbell Barracks inaugurating the formation of the Special Air Service Regiment. Plans to call the unit the 1st SAS Regiment were dropped as it was realised that only one SAS regiment would ever be raised in Australia and the entire regiment would never actually be deployed on operations. The date chosen as the birth date of the new unit was the twenty first anniversary of the Lae–Nadzab operation in New Guinea in 1943, the first Australian combined land, sea and airborne operation.

Major Garland took command of 1 SAS Squadron, formed largely from the hand-picked experimental squadron he had taken to Okinawa during the lead up to Exercise LIGTAS, and Captain Reg Beesley, who had been adjutant of the SAS Company, became commander of the embryonic 2 SAS Squadron. One curious aspect of the organisation was the appointment of Major G. R. S. (Bill) Brathwaite as Officer Commanding Headquarters SAS Regiment. The establishment showed that he was also to act as commanding officer of the two SAS squadrons.

Warrant Officer Class One Larry (Bat) Moon being presented with the Long Service and Good Conduct Medal by Brigadier-General Davis in the Philippines. Moon was CSM and then RSM from 24 January 1964 to 21 March 1965, and since the Regiment was formed on 4 September 1964 was therefore the first RSM of the Regiment.

This unusual organisation continued until the beginning of December 1964, but in the meantime the government had been considering the suitability of the size and shape of the Army for possible operations. Throughout 1964 the government had become increasingly concerned at the deteriorating strategic situation in South East Asia. In June the Training Team in Vietnam had been increased to eighty men and had been given permission to accompany the South Vietnamese forces on operations. Some AATTV soldiers, such as Captain Ian Gollings who became second-in-command of the SAS Company on 31 January 1964, had unofficially accompanied South Vietnamese forces on operations a year earlier. In the middle of 1964 an RAAF Caribou flight had been deployed to Vietnam. In August there was the incident in the Gulf of Tonkin and the US President was given authority to conduct the war in South Vietnam.

While Vietnam was worrying, Indonesia appeared even more of a threat. In April the Malaysian government formally sought the Australian government's permission to use Australian forces in Malaya to deal with Indonesian incursions on the Malaysian mainland if the need arose. The Malaysian government also asked that the 3rd Battalion, The Royal Australian Regiment (3 RAR), then based at Malacca, continue to be rotated to the Thai border, that an engineer unit be made available for service in Malaya or Borneo, that two to four minesweepers and four helicopters be provided, and that a number of air transport sorties be made available to assist operations on both the Thai border and in Borneo.[9]

While the Australian government was considering these requests, Sir Alec Douglas-Home cabled Menzies that the British and Malaysian security forces in Borneo were to be permitted to cross the Indonesian border in hot pursuit for a distance of up to 3000 yards. He raised the possibility of an Australian battalion or SAS troops being made available for operations in Borneo. The Defence Committee advised the government that while 3 RAR was 'a good, well-trained and well-equipped unit', and the SAS Company was 'highly trained and permanently on seven days notice to move', a decision to provide the SAS or a battalion should be taken only after it was considered 'that there was no further possibility of deterring the Indonesians by other means'. The Defence Committee added that if a policy of hot pursuit were to be instituted it should be done with non-European troops to avoid Indonesian claims that Malaysia was 'neo-colonialist'. The government therefore decided not to provide a battalion or the SAS at this stage, but did agree to send an engineer construction squadron to Borneo, and to provide two minesweepers, four helicopters and other air support to Malaysia. 3 RAR could be used against Indonesian incursions into mainland Malaya if the need arose.[10]

This eventuality was not long in coming. In August and September Indonesian guerillas parachuted into Johore State, and when more 'guerillas' landed by sea near Malacca in October, 3 RAR was used to round them up. The Australian argument that their troops should not be employed in Borneo was now starting to wear thin. The incursions into mainland Malaysia in the latter half of 1964 brought Indonesia and Malaysia (with its British and Australian allies) close to war, when in September some British planners were reported to have considered conducting sea and air strikes against Indonesian bases.[11] The threat of war with Indonesia presented Australian defence planners with a dangerous situation, as they were concerned at the possibility that Indonesian troops might infiltrate across the border into Papua New Guinea. The Australian government therefore faced the possibility of having simultaneously to deploy troops to three areas, Vietnam, Borneo and New Guinea.

On 10 November 1964 the Prime Minister made a major defence statement in the House of Representatives which was aimed squarely at meeting these threats. The most controversial aspect was the decision to introduce selective National Service from mid 1965. The Regular Army was to be increased from 22 750 to 37 500 men and the Citizen Military Forces were to be expanded. The Pentropic division was to be replaced by a lighter, air portable division consisting of nine battalions. The Pacific Islands Regiment was to be expanded with a second battalion being raised early in 1965. The SAS Regiment was to be expanded to provide a base squadron and four SAS squadrons. In addition new weapons and equipment were to be purchased for the Army. Both the Navy and the Air Force were to be increased in size and provided with new equipment over the following three years.

Considering the fact that the government decided to send a battalion to Vietnam in April 1965 as the first step in Australia's combat commitment, many members of the public later saw the Defence expansion of November 1964 as laying the groundwork for that commitment. However, while Vietnam was probably a consideration, the expansion was directed primarily towards Indonesia. In his speech the Prime Minister stated: 'It is clear that Indonesia is carrying on an active and entirely unjustified armed aggression against Malaysia'. The *Sydney Morning Herald* of 11 November headlined its editorial: 'Preparing Against War with Indonesia'. The editorial stated that the 'basic message to the nation of the Prime Minister's historic defence review last night was that there was "a real risk of war" with Indonesia. It is evident that the Government considers that this risk has suddenly and steeply increased'.

The changing strategic situation had led directly to the expansion of the SAS. When it had become likely in mid 1964 that an SAS squadron would be sent to Borneo it had been necessary to form a second squadron as a reserve and a replacement. But when the possibility of having to face Indonesian infiltration in New Guinea had also eventuated, the Army had realised that this task would require the deployment of the second SAS squadron. With the likelihood of both SAS squadrons being deployed simultaneously it would be necessary to create further squadrons in Australia as a reserve.

In response to the Prime Minister's announcement, in early December Army Headquarters issued a revised establishment removing the restriction of one troop for each SAS squadron, and forming a base squadron, separate from regimental headquarters.[12] Captain F. W. Holding, a Second World War soldier who had recently been commissioned, took over Base Squadron from Brathwaite as well as performing his duties as quartermaster. The commanding officer was now to become a lieutenant-colonel. Brathwaite retained this position although he was not actually promoted until April 1965. He was informed that he was to raise a third SAS squadron by December 1965 and to form a fourth squadron by June 1966.

From the time when the SAS Regiment was formed Garland stepped up his preparations for the deployment of his squadron to Borneo. He continued to be briefed directly by Army headquarters and Major Brathwaite, his nominal commanding officer, took no part, nor was he involved, in the training or preparation for the move. Garland based his preparations on information forwarded by staff channels from the Australian Army component of Far East Land Forces, including six-monthly summaries of 22 SAS Regiment operations. In August he had received a copy of the 22 SAS Standing Operating Procedures for 'Counter Guerrilla Operations in Jungle'. He was also able to benefit from the direct experience of an Australian SAS officer, Lieutenant Geoff Skardon, whom Garland had arranged to be attached in March 1964 to the British 42nd Commando operating in the First Division of Sarawak.

Aged 25, Skardon had graduated from the Officer Cadet School at Portsea in June 1960 and had been an SAS officer since January 1962. At the beginning of July 1964 he was detached from 42 Commando, which had since returned to Malaya, to join A Squadron 22 SAS for six weeks during its third operational tour in Borneo.[13] Skardon's experience on patrol highlighted one of the chief dilemmas faced by SAS soldiers on operations.

On 6 August Skardon and three SAS soldiers were patrolling about 1000 metres on the Malaysian side of the border in Sabah in the area known as the Long Pa Sia Bulge, near the Sarawak border. In thick jungle the forward scout, Trooper White, came unexpectedly upon an Indonesian soldier and quickly shot him. Immediately the patrol commander, Lance-Corporal Blackman, ordered the patrol to 'Get Out', in accordance with the SAS drill of 'shoot-and-scoot'.

The Australian SAS did not operate on a 'shoot-and-scoot' basis and Skardon failed to act on this command. Rather, dropping his bergen pack he moved forward to cover White's withdrawal. Then White was hit by fire from the remainder of the Indonesian patrol and Skardon continued to advance under heavy enemy fire to pull White to safety. Skardon tried to carry White to the cover of a sunken creek, but it was soon apparent that White was dead. With the enemy firing heavily and starting to surround him Skardon had to leave White and make his escape. As well as having discarded his bergen Skardon was also forced to release his belt when he was caught by a tenacious 'wait-a-while' vine.

Skardon was now convinced that he was going to die. The enemy were closing in on his only line of escape, the creek, and once they reached the bank he would be caught like a rat in a barrel. Inexplicably, the Indonesians stopped a few feet from the bank of the creek and Skardon was able to walk down the creek between the enemy patrols to make his escape. Without map or compass he made his way back to the original helicopter landing point, arriving there about midday the next day. Dickens wrote: 'that [Skardon] would succeed was predictable after his display of courage and resolution in action, but it was hard and tortuous going, fraught with anxiety just the same'.[14] A subsequent SAS patrol found White and the dead Indonesian still undisturbed and the bergens were recovered. Spent cartridges indicated that about thirty Indonesians had been in ambush.

Had Skardon been right in disregarding the shoot-and-scoot drill? If White had been less seriously wounded what would have happened if the remainder of the patrol had left him behind? But in going forward Skardon had endangered himself. As Dickens put it, the 'policy had been instituted to save life, both for its own sake and to preserve regimental morale; but now the question had squarely to be faced: whether the latter would be worse affected by leaving a possibly wounded man to die or be captured by a savage foe, or by losing more lives trying to save him'. In his report on the action the commander of A Squadron, Major Peter de la Billière, commented: 'Brave though White and Skardon's actions were, no casualties need have been incurred at all if the patrol had fired one shot and scooted'. But Colonel Woodhouse saw the complexity of the situation and replied to de la Billière: 'I believe troops will welcome, and morale demands, an order that if a man is known to have fallen the patrol will remain in the close vicinity until either they see for certain that he is dead or they recover him alive. I think we should expect to fight to the death for this'. In the latter months of 1964 a satisfactory procedure for such a situation had still not been resolved, but this incident played a key role in causing a rethink in the British attitude towards action on contact.[15]

In January 1965 Garland received a fairly sketchy document from Army Headquarters outlining the command relationships and role of 22 SAS Regiment in Borneo. The document indicated that British squadrons were operating with sixteen patrols each of four men. Its roles were to report on enemy infiltration, to gather intelligence by winning the hearts and minds of the border tribes, to construct LZs near cross-border routes to facilitate the swift deployment of infantry in the 'cut-off' role, to collect topographical information, and to harass the enemy returning to Indonesia. As for tactics, the document noted:

SAS are trained to avoid standing and fighting and will not attempt to hold defensive positions...On the frontier SAS will aim to give early warning of enemy crossings without disclosing their own presence to the enemy. Subsequently they may ambush any attempt by the enemy to establish a porter L of C, be prepared to construct LZs and brief infantry reinforcements into their area, or follow up the enemy as ordered by SAS HQ or Force HQ.[16]

The roles of the British SAS in Borneo were remarkably similar to those outlined by Army Headquarters for the Australian SAS in September 1964.

In preparing for the deployment Garland realised that if his squadron were to fly to Singapore it would have to fly over Indonesian territory and therefore all information concerning the deployment would have to be kept secret. Consequently, throughout November and December 1964, 1 SAS Squadron conducted a series of mobilisation exercises designed to improve the efficiency with which the squadron could be mustered for deployment. Following quick administrative checks, the squadron would mount vehicles with all its mobilisation equipment and move out as if deploying on operations. These exercises also accustomed the citizens of Perth and Fremantle to seeing and hearing announcements on television and radio and at drive-in picture theatres ordering members of the SAS to report to Campbell Barracks.

One difficult decision to be faced was an order from Army Headquarters that the fourth troop in 1 Squadron was to be left behind at Swanbourne. The Australian squadron would be substantially larger than the British squadron it was replacing and there was a shortage of trained manpower in the remainder of the regiment. There would therefore be an unacceptable delay in bringing 2 Squadron to operational readiness. D Troop, commanded by Lieutenant Anatoly (Tony) Danilenko, was bitterly disappointed by this decision. Although made purely on manpower grounds, it became a precedent in the future deployment of all squadrons. In subsequent deployments the four troops in each squadron were put in competition with each other, with the weakest troop being left behind. Often the troop commander involved was so disappointed at this rejection that he left the SAS and sometimes even the Army. Danilenko was discouraged but not beaten. He later served with 2 Squadron in Borneo, and in 1968 was killed while serving with the Training Team in Vietnam.

While Garland knew that 1 Squadron was to be deployed to Borneo, the question remained as to when. Meanwhile, the war in Borneo had heated up. General Walker had been authorised to conduct operations up to 5000 yards (4500 metres) across the Indonesian border, and in the strictest secrecy Gurkha patrols began ambushing Indonesian troops and supply parties as they moved towards the border. Knowledge of these 'Claret' operations, as they were called, was restricted to only a very few senior commanders and key staff officers and the men involved. The task of the SAS was to conduct reconnaissance across the border in the area of the greatest threat, which in the latter months of 1964 was the Fifth Division of Sarawak (north eastern Sarawak) and the neighbouring Interior Residency of Sabah. Occasionally there were minor contacts, but because of the small size of the SAS patrols, they were not usually given ambush tasks.

Then, in December 1964 the focus of operations changed from the north eastern to the south western end of the border. Intelligence reported that Indonesians were massing up to a division of their best troops in the First Division area opposite the Sarawak capital of Kuching. Sukarno had promised that 'before the cock crows in 1965' he would crush Malaysia. B Squadron 22 SAS, then beginning its first tour in Borneo, was redeployed from its base at Brunei to Kuching to meet this developing threat and additional infantry battalions were sent to the area. By the end of the year Walker had eighteen British battalions (including eight Gurkha and two Royal Marine Commandos) and three Malay battalions in Borneo. Also, at the end of the year Walker was given permission to extend his operations up to 10 000 yards (9000 metres) across the border, and he intended that his infantry patrols would take more offensive action to keep the enemy commander off balance. They could now attack specific targets such as camps, forming-up-places and supply dumps in patrols of up to company strength.

On 15 January 1965 the British Prime Minister, Harold Wilson, cabled the Acting Prime Minister, John McEwen, that he had given permission for operations to be extended to 10 000 yards, and in a round-about fashion mentioned the need for additional SAS troops and another battalion. McEwen's reply was non-committal: 'It seems to us that more and more, the two conflicts in Vietnam and against Indonesia are coming to form part of a common pattern and a common threat'.[17] Then, on 21 January the Malaysian government made a direct request for SAS troops, for 3 RAR and for another battalion to be sent to Borneo. The Australian government did not have the resources to send two battalions as they were then considering sending one battalion to Vietnam, but at the end of January they approved the sending of 1 SAS Squadron and 3 RAR to Borneo. The SAS was to be permitted to operate across the Indonesian border.

Staff at Army Headquarters and in Perth now had to move quickly. On 4 February a warning order was received that Operation TRUDGE was under way. 1 SAS Squadron plus elements of Base Squadron totalling 100 men were to move to Malaysia for an operational tour of six months. While the soldiers completed training, administration and pre-embarkation leave, Major Garland flew to Canberra for a briefing from the Deputy Chief of the General Staff, Major-General Hassett. He returned to Perth on 10 February.

At 3.20 pm on Saturday 13 February, the advance party consisting of Major Garland, the operations officer, Lieutenant Peter McDougall, the quartermaster sergeant, Staff Sergeant Marsden (Taffy) Davis, the intelligence sergeant, Sergeant George Gridley and Garland's driver, Lance-Corporal Eddie Chenoweth, all in civilian clothes, left Perth for Singapore by a Qantas flight. While at the Perth airport Garland had a meeting with his second-in-command, Captain Ian Gollings, who had just flown in from a briefing in Canberra. Gollings was able to give Garland a draft of his directive from the Chief of the General Staff.

Meanwhile, although they might have guessed at the reason for some of the activities, the majority of the soldiers were still in the dark about their impending departure. At 4.45 pm on 16 February another mobilisation exercise was called. This time it was the real thing. The Squadron moved to RAAF Base Pearce under a veil of secrecy, and at 7 am on 17 February a Qantas aircraft took off for Singapore following a route that would take it around the north of Sumatra. Only one member of the squadron failed to make the rendezvous and was left behind. Before they stepped onto the plane a message from the Minister for the Army was read to the troops:

> Before you leave for service in Borneo I would like you to know that you will carry with you the best wishes of all your fellow Australians. You are going overseas at a difficult time but I know your high degree of training will stand you in good stead when you join in Australia's part in the defence of Malaysia and therefore of Australia itself. I know also that you will acquit yourselves in accordance with the excellent reputation you enjoy. Good luck and Godspeed.[18]

5
Hearts and minds
1 Squadron: February–May
1965

When Major Alf Garland and his advance party flew into Brunei Town on the evening of 16 February 1965 the setting tropical sun was already silhouetting the town's magnificent new golden-domed mosque. Constructed of Shanghai granite and white Italian marble, the mosque dominated any view of the town, particularly during festivals when it was illuminated at night, and it reminded Garland and his men of two salient facts about Brunei. The first was that this small British protectorate, ruled by the 49 year old Sultan, His Highness Sir Omar Ali Saifuddin, was a strictly Muslim state. There would be little night-life to attract the Australians when they returned from patrol. The second was that because of its reserves of oil, Brunei was, per head of population, the wealthiest nation in South-East Asia.

However, apart from the mosque, opened in 1958, there was little of the opulence or abject poverty that marked some Persian Gulf oil kingdoms at that time . The people were neatly dressed, healthy, and appeared to all ride bicycles. Largely destroyed in the Second World War, the town, with a population of about 11 000, had been rebuilt and included a royal palace, administrative and religious offices, a sports stadium and an Olympic pool. Nearby, and surrounding the mosque, was the historic water village of Kampong Ayer with its squalid collection of old tin and wooden huts perching on stilts in the Brunei River. Government efforts to persuade its 20 000 Malay inhabitants to move to better homes on dry land had met with little success.

In ordinary times there were few more tranquil places than Brunei Town, but in February 1965 the undeclared war along the Indonesian frontier was at its height. To a certain extent the Indonesians had

81

Brunei as the soldiers of 1 SAS Squadron saw it when they arrived in February 1965 on their first tour of duty overseas. The new gold-domed Mosque dominated the historic water village of Kampong Ayer.

abandoned guerilla tactics and were now seeking to launch their regulars in concerted attacks on the Security Forces' border positions. The infantry battalions of the Security Forces vigorously patrolled the frontier, but they were spread thinly, and while their success was partly the result of their energy and skill, it also had to be attributed to the intelligence provided by the SAS, the Border Scouts and from the tribes along the borders. The continuing campaign to win the hearts and minds of the native people of Sarawak and Sabah was therefore a vital aspect of the campaign.

Brunei Town was one of the main bases for the war, and the next day Garland and his operations officer, McDougall, began receiving their briefings on the current situation. They learned that the overall commander of the operations, the Director of Borneo Operations, Major-General Walter Walker, had his joint headquarters on the island of Labuan, a 50 kilometre helicopter ride across Brunei Bay. Also at Labuan was the Commander, Land Forces, Major-General Peter Hunt, but he was very shortly to move his headquarters to Brunei Town. Hunt's staff were concerned mainly with administration because operational control of the four infantry brigades along the frontier was retained by Walker. West Brigade, with five battalions and a front almost 300 kilometres, was responsible for the defence of the First and Second Divisions of Sarawak. With his headquarters at Kuching, the brigade commander, Brigadier W. W. (Bill) Cheyne, faced up to a division of well-trained Indonesian troops barely 50 kilometres away. Mid West Brigade with two battalions covered over 700 kilometres of the Third Division of Sarawak. Brigadier Harry Tuzo, the commander of the Central Brigade with its headquarters in Brunei Town, had two battalions and was responsible for the Fourth and Fifth Divisions of Sarawak plus the Interior Residency of Sabah, a border area of almost 430 kilometres. East Brigade, a Malaysian brigade, had three battalions (two Malaysian and one British) and covered the 130 kilometres of eastern Sabah in the Tawau Residency.

B Squadron 22 SAS had been operating in the West Brigade area since December, and during February 1965 it was replaced by D Squadron 22 SAS. In December the Guards Independent Parachute Company had been deployed across the front of the Central Brigade area, and it was planned that they would be relieved by 1 SAS Squadron. In addition, a half squadron of New Zealand SAS had arrived in February, to be deployed further west in Sarawak.[1] Thus the Director of Borneo Operations would soon have a considerable number of SAS patrols at his disposal, and to control these forces and to advise Walker on their use a small SAS headquarters was established on Labuan, near Walker's headquarters. Command was exercised by either the commanding officer of 22 SAS, Lieutenant-Colonel Mike Wingate Gray, or his second-in-command, Major John Slim. The existence of Headquarters SAS Far East was a recognition of the fact that the SAS was a strategic force over which control needed to be maintained at the highest level. High level control was also necessary because of the sensitivity of the Claret operations. However, in view of the increased number of SAS units it was decided that when SAS squadrons were operating in a particular brigade area they would be placed under operational control of the brigade commander.

Throughout his tour Major Garland enjoyed good relations with Brigadier Tuzo, the able gunner commanding the 51st Gurkha Brigade, operating as Central Brigade. A patient, educated soldier, well schooled in the requirements of the NATO battlefield, Tuzo suffered from the disadvantage that he lacked a detailed understanding of low-level infantry operations in what was essentially an infantry war. At times Garland thought that Tuzo tended to act as though he owned 1 SAS Squadron, but Garland never forgot that he was under the command of the Director of Borneo Operations. Brigadier Tuzo had a high regard for Garland, describing him as 'dedicated, effective and single-minded'. Like other Australian independent commanders in earlier wars, Garland had his directive from the Chief of the General Staff which spelt out his command responsibilities and included the key statement:

> Should you, at any time, consider that the task which you are given or a situation develops which you consider would endanger the national interest of Australia or be likely to imperil unduly your command or any part of it, you are to report the situation at once to the Commander, Australian Force, FARELF [Far East Land Forces]. At the same time, you are to inform your formation commander of your action and the reasons for doing so.[2]

A few days after arriving in Brunei Garland flew across to Walker's headquarters on Labuan to be briefed on cross-border operations. He was told that each operation had to be personally approved by the Director of Operations and that knowledge of the operations was limited to a select few. For example, in Central Brigade knowledge was limited to the commander, his brigade major, the DAAG, the DAQMG and the intelligence staff. The initial 'Golden Rules' for Claret operations were as follows:

> Every operation to be authorised by DOBOPS [Walker]...Only trained and tested troops to be used...
> Depth of penetration must be limited and the attacks must only be made to thwart offensive action by the enemy...
> No operation which required air support—except in an extreme emergency—must be undertaken.
> Every operation must be planned with the aid of a sand-table and thoroughly rehearsed for at least two weeks. Each operation to be planned and executed with maximum security. Every man taking part must be sworn to secrecy, full cover plans must be made and the operations to be given code-names and never discussed in detail on telephone or radio. Identity discs must be left behind before departure and no traces—such as cartridge cases, paper ration packs, etc—must be left in Kalimantan.
> On no account must any soldier taking part be captured by the enemy—alive or dead.[3]

These rules, introduced in 1964, were eased in 1965, but the operations always retained a high level of secrecy.

While Garland was being briefed the main body of the squadron was making its way from Australia. Although the squadron had left Perth in civilian clothes, by the time they arrived in Singapore on the evening of 17 February they had changed into uniform. Several months earlier the regiment had been given permission to change from the red to the sand-coloured beret, but stocks of the new beret had not arrived in the unit. Thus on arrival in Singapore the members of the squadron were still wearing their red berets. The next day the Indonesian radio announced that a Parachute Regiment had arrived in Singapore for service in Borneo, red berets being the headdress of the British paratroops.

On 22 February the squadron embarked on HMT *Auby*, arriving at Brunei on 26 February. The following day squadron headquarters was set up in the 'Haunted House', a large brick house lent by the Sultan of Brunei. Used by the Japanese secret police, the Kempei Tai, during the Second World War, the house was said to contain the ghost of a young girl who had been killed there. The house included space for an operations room and communications centre on the top floor, as well as accommodation with showers for patrols returning from the jungle. Perhaps the most important aspect of the house was that it overlooked the Sultan's residence, the 'Istana', and the SAS in the Haunted House had the ready-reaction task of rescuing the Sultan and moving him to safety if there was any repetition of the 1962 rebellion in the Sultan's own security forces.

Although it had only recently been formed, when 1 Squadron arrived in Brunei its organisation was typical of Australian SAS squadrons for the next eight years. The second-in-command, Captain Ian Gollings, was a 29 year old Portsea graduate who had first joined the SAS in 1962 and had served with the Training Team in Vietnam later in 1962 and 1963. He was concerned mainly with the administration of the squadron. Lieutenant Peter McDougall, the operations officer, had the task of briefing and debriefing the patrols and supervising the operations room. Aged 29, he was a former soldier and had graduated from Portsea in 1959. All three troop commanders, Second Lieutenants Trevor Roderick (24) and Peter Schuman (22) and Lieutenant Tom Marshall (28) had been soldiers and although the British SAS, who had captains as troop commanders, might look askance at the Australian practice of using lieutenants, they failed to realise that all the Australian troop commanders had practical experience of soldiering. Marshall had served as an NCO in Malaya during the Emergency. The squadron sergeant-major, Warrant Officer Alex (Blue) Thompson was a hard, experienced soldier who had served in Korea and Malaya; his task was to man the operations room, but more often he would be found on patrol. The three troop sergeants were Joe Flannery, Roy Weir and Chris Pope, and within each troop there were three sergeant patrol commanders. Squadron headquarters included the quartermaster sergeant (staff sergeant), an intelligence clerk, a radio mechanic, a small arms fitter, a pay clerk and a technical storeman, all sergeants. Including an increment from base squadron, the squadron totalled 100 personnel.

On the morning of 28 February the main body of the squadron moved by vehicle to the Tutong base camp, about 50 kilometres south east along the coast, to begin four weeks of acclimatisation and familiarisation training. In Singapore the Australians had been issued with a substantial amount of British equipment, Claymore mines (obtained by the British from USA), M26 grenades, Sarbe radio beacons and C128 suitcase radios. Indeed even British webbing and jungle green uniforms were issued. A little later they were issued with some AR15 (Armalite) rifles. Ten members of the squadron stayed behind in Singapore to attend a Malay language course beginning on 27 February.

While training continued, Major Garland discussed with John Slim the most effective way of tackling his tasks. On arrival from Australia his squadron was organised to allow for the deployment of six-man patrols, but when faced with a border of over 400 kilometres, including some of the highest mountains and thickest jungle in Borneo, Garland realised that he would have to follow the British example and deploy four-man patrols. Fortunately he had sufficient NCOs to provide the necessary patrol commanders. Indeed Garland saw this as one of the advantages of four-man patrols in that all officers and senior NCOs could be given command of a patrol.[4]

During the previous year Garland had changed the emphasis of the role of SAS to reconnaissance and had reduced the size of the patrols from eight to six men, but the squadron had never really developed tactics specifically for small group operations. Rather, training still relied on infantry minor tactics in which the basic unit was a platoon and fire and movement was paramount. Now, with the introduction of even smaller four-man patrols the squadron needed to learn new tactics, and while at Tutong they were trained by Sergeant Smith and his patrol from the British SAS in British techniques, particularly those applying to operations in Borneo. The training syllabus included physical training, code of conduct briefings, contact and ambush drills, jungle navigation, tropical hygiene, range practices, unarmed combat, helicopter procedures, and specialist training. This was a vital training period for the Australians, and the techniques learned at Tutong were to remain essentially the same in the Australian SAS for the next twenty years.

In accordance with the Director of Operations' rules for cross-border operations, new units were required to operate on the Malaysian side of the border for four weeks before they could begin Claret operations. While the British SAS units, which had now been operating in Borneo on rotation for over two years, were not subject to this rule, it was a wise precaution for the Australians who would begin with a series of familiarisation patrols. On 12 March Garland issued the squadron's first operational instruction, warning that the first patrols would be deployed on or about 28 March. It was intended that

Major Garland, Staff Sergeant Taffy Davis and Major-General J. S. Andersen, the Australian Adjutant General, who was visiting the SAS Company headquarters in Brunei, 1965.

three or four patrols would be deployed on Operation KEEN EDGE into the valley of the upper Batang Baram River in the south east of the Fourth Division. Once KEEN EDGE was underway patrols would be deployed to the Pensiangan area in Sabah on Operation HARD STAB to support the 2/7th Gurkha Rifles. A third series of patrols, known as Operation SHARP LOOK, would be deployed to the wild 'Gap' area between the Interior Residency of Sabah and the more settled coastal area which was the responsibility of East Brigade.

Broadly, the tasks allotted to the patrols were to collect topographical information on tracks, rivers and kampongs in the patrol areas, to win the hearts and minds of the indigenous people, to conduct surveillance of known border crossing points, and to shadow any Indonesion incursions, reporting strengths, direction and identification. Garland reminded his men that the role of the SAS was reconnaissance. They were not to take offensive action unless he ordered them to do so. If a chance encounter took place, patrols were to apply the principle of 'shoot-and-scoot'.[5] In essence, most patrols were to be involved with winning hearts and minds.

Lying astride the equator, Borneo is hot and humid, and is one of the largest expanses of tropical rain forest outside South America. Generally the island is mountainous with its main spine running from Mt Kinabalu (4100 metres) in the north along the Indonesian border to the south. The Central Brigade area included the Kelabit uplands, with heights ranging from 900 metres on the plateau to Mt Murud at 2422 metres. A blanket is useful here at night. Borneo is one of the most difficult countries to move around in on foot and the quickest route is almost invariably the indirect route. Great winding valleys, broad, swiftly flowing rivers which drain away the torrential rain, steep ridges, landslides and thick vegetation presented a physical challenge to the fittest of troops. And the maps were often wrong and of small scale.

Yet it was a fascinating experience to move silently through jungle observing the abundant wildlife and the array of vegetation. In the towering trees could be seen varieties of monkeys, and if one was lucky, the giant orange orang-utan. On the valley floor were deer, honey bears, and many smaller animals. The largest predator was the beautiful spotted leopard, but it was notoriously shy of man. The honey bears could be dangerous but were rarely seen.

The largest mammals were the wild ox, the two-horned rhinoceros and importantly, in view of later experiences of the SAS, the Indian elephant. It has generally been stated that the Borneo elephants were feral descendants of introduced stock, and the members of 1 Squadron were told that they were introduced to assist with logging at the beginning of the century. However, some authorities believed that the elephants were given to the Sultan of Sulu by the East India Company in 1750 and subsequently were liberated in North Borneo. Zoologists are still in doubt about the status of the species in Borneo, but in the early 1960s it was thought that there were about 2000 elephants in the south central area of Sabah. Elephants are unpredictable creatures and the Borneo elephant was far less shy of man than his Malayan counterpart. The naturalist, John MacKinnon, wrote that sometimes 'they flee at the first sniff of man, sometimes they pay no attention. But one night I had to move my sleeping quarters several times in an attempt to throw off a huge beast that seemed to be pursuing me. Eventually I found sanctuary among the buttrees roots of one of the forest giants. On two other occasions I had to scurry up trees to get out of the paths of some angry, charging bulls'.[6]

Crocodiles were found in the rivers, and there were abundant lizards and snakes of all varieties, including pythons and the deadly king cobra. The wild pigs, which moved in scattered hordes following supplies of fallen fruit, were hunted by the nomadic Punan people with blowpipe and poisoned dart. But this fascinating world was spoiled by the biting bugs—leeches, sand flies, mosquitoes, fire ants, hornets, bees and wasps. While the danger from ferocious animals was perhaps not quite as great in Borneo as in some other jungle areas, nonetheless, when linked with the rugged terrain and the undeveloped nature of the country, it was clearly a perilous land even if there were no enemy present.

The first patrols were deployed on 28 March, and by early April twelve patrols had been deployed. In the KEEN EDGE area patrols were commanded by Second Lieutenant Roderick, and Sergeants Weir, Jarvis, Sheehan and Foxley; in the HARD STAB area, patrols were commanded by Lieutenant Marshall, Warrant Officer Thompson, and Sergeant Flannery; and in the SHARP LOOK area, patrols were commanded by Second Lieutenant Schuman and Sergeant Pope. Sergeant O'Keefe, located at Pengsiangan, provided liaison with the local Gurkhas for both HARD STAB and SHARP LOOK. About this time Brigadier Tuzo noted that the Australian squadron had just deployed to the right of his brigade area. 'As expected, they appeared to be extremely keen and tough and are to operate on the same lines as the British SAS.'

The usual system during the hearts and minds operations was for the patrols to live near but not in the native villages. By providing medical and other assistance they attempted to gain the confidence of the local villagers and they encouraged the natives to pass on any information they might have received about the activities of the Indonesians across the border. For the small four-man patrols it could be boring and lonely work, spending weeks by themselves, living out of their packs with their only outside contact being by radio and the scheduled resupply drop.

The patrols in the KEEN EDGE area were required to cover a large area populated by Dayaks. But the area also included the more primitive Punans. Traditionally a nomadic people, most had begun to rely on agriculture. But many were still nomadic and they lived in small groups of about 35 individuals hunting game and living off wild sago palm. They lived in flimsy huts, had few belongings, wore only a loincloth, and were almost devoid of ceremony.

The patrol reports forwarded by Lieutenant Trevor Roderick reveal some typical experiences. On 6 April he reported that on the second night they had slept about 200 to 300 metres from three longhouses (with a population of about 300), but at sick parade time, 7.30 am, they had been found by the locals. 'Frightened Christ out of us.' He therefore went further into the jungle. The locals did not follow them but were waiting on the edge of the jungle for them to come out in the morning. 'It was a trifle embarrassing.'

Roderick tried to make some assessment of the Border Scouts and reported that Sergeant John Iban Ngerong was very intelligent, strong

Helicopter insertion along the Malaysian–Indonesion border, Borneo, 1965. Small landing pads were constructed in rugged, jungle-covered terrain to allow patrols to be inserted and extracted.

and healthy. However, he found one Border Scout to be a 'lazy con man but quite a character. Useful for small jobs and interpreter, speaks limited English and Malay. Bachelor and lecherous bloke, likes his sex and grog and worships Dollars. A swift kick now and then might help him. I keep scaring Christ out of him, when I tell him I am going to take him to the border.'

The Punans impressed Roderick as a proud and shy people and he thought that their health was very good. But within a few days he had changed his mind. 'Remember I said that the people around here are healthy...Christ what a statement to make. We visited the Longhouse yesterday and came across some people from Long Selungo (same general area as Punans). The babies were in one hell of a mess, they had sores all over their bodies and were covered in warts. I suspect this is from malnutrition and the parents not keeping them clean'. Furthermore, Roderick had been told that there were quite a few people who had not come in for treatment because they were too sick. Nevertheless, there was much to learn. 'Had an eye opening exhibition yesterday arvo. The Punans now come to us for treatment instead of having to be chased but are still a little wary of us. Anyway one of them gave us a demo with his blowgun. He hit a 1 gal kero tin at 100 yards. Man what a shot.'

On 14 April Roderick reported that he had no news of any enemy in his area, but as there was little movement across the border either way, the ability of the local people to help was limited. While the locals could be relied upon to inform him if anything happened, they would do so only if they thought it was important, and since they did not care about the Indonesians one way or the other, events could take place and they would not bother to mention them. The only way Roderick thought he would find out was if a Border Scout was told or overheard something. 'The Border Scouts know the score as I have spent hours talking with them and they can be relied on to pass info on...The sergeant Border Scout seems very reliable and sensible. His English is OK but of the yes when you want a no answer...type.'

Roderick and his patrol spent most of his time just winning the confidence of the local people. 'Have made a cricket bat for local school and given first instruction on game...The locals reckon we are number 1 because we will go just about anywhere to look at their sick. They appreciate our sitting down and having a nag, although it gets bloody boring for us at times.' An important item to be supplied by air to assist the natives was kerosene, and sometimes the Australians would throw plastic explosives into the rivers and catch fish to give to the villages. 'We have been visited by dozens [of locals]', wrote Roderick, 'all wanting to thank us for kero, fish etc. The odd sly one of course suggesting that we stay at his house, to get some fish etc for him...Whether or not they would still be friendly and co-operative if all aid stopped for some reason, I don't know. When we knock someone back for something he is so poker faced I don't know if he is shirty or what his feelings are. I never met a more poker faced mob in all my life. I'm sure that if I gave one of them a 44 gal drum of kero, 2 shotguns and an outboard motor, his expression would stay the same.'

The main support provided by the patrols was medical assistance, and Roderick reported that there were a 'lot of sick and twice as many who think they are sick...We look like having a bit of a flu epidemic. Chicken pox is rife at the moment, 50 percent of the kids in the longhouses have it'. Private Bruce English, the patrol medic, was kept busy sewing up injuries, dispensing medicine and even assisting at childbirth. 'The women have queer ideas on having kids', reported Roderick. 'When they are ready to drop the kid, they sit as close to the fire as possible, drink rice water as hot as they can stand, and massage the stomach with a hot stone or beer bottle type hot water bottle. They reckon this keeps the blood warm and that warm blood is bloody good stuff for child birth. Brilliant idea ey. They lose on the average 1 out of 8 mothers this way.'

Roderick discussed with the local headman the need to eradicate the sick dogs in the area. 'Dogs are getting knocked off by the dozen', he wrote. 'We clobbered one the other day and there was a heap of worms in his middle that would disgrace a double serve of spagetti.'

Apart from the dangers to health, the patrols could also be in danger from local hospitality. Sergeant Roy Weir spent a week building a viaduct around a waterfall with explosives. It took him a further two weeks before he could get away from the village. On another occasion he was moving up to the border to make a series of helicopter pads when he stopped at a village. The local chief had just died and people had come for miles around for the wake. Weir and his patrol were not allowed to leave and had to drink the local highly alcoholic brew made from tapioca root. After several days Weir and his men decided that they would have to make a break and they crept out of the village in the middle of the night, moving as far and as fast as they could through the jungle, stopping to sleep just before dawn. They awoke to find the new chief standing over them with hunting dogs. Weir gave him a few presents and tins of linseed oil and was allowed to continue on his journey.

Weir and his patrol were located at the village of Long Banga with the task of providing liaison with B Company, 2/6th Gurkha Rifles. But every second or third day they set out early to visit a village, often involving a walk of up to four hours. On arriving in a village they would first talk to the head man, using the smattering of Malay that Corporal Littler had learned during the language course in Singapore during the previous month, and then they would hold a sick parade. Everyone was sick, especially if it looked as though pills were to be distributed. No matter how often or clearly the men and women, who said that they had diarrhoea or a migraine, were told that they must take one pill each morning and evening, they invariably believed that by taking the whole lot at once they would be cured a lot quicker. Taking five days' tablets at once must have led to some awful cases of constipation.

Conditions in the HARD STAB area were slightly different. Here the jungle-dwelling Muruts built their longhouses with a remarkable springing dance floor mounted on curved saplings. They planted hill rice, sweet potatoes, manioc and maize in the jungle clearings, and they also gathered jungle fruits, fished the rivers and hunted wild pig and deer. During 1964 the British SAS had painstakingly built up the confidence of the local people, particularly in the village of Saliliran, which was only 3000 metres from the border and initially had been passively hostile. The report from Warrant Officer Thompson on 11 April is typical. He had heard that 150 enemy had arrived in the town of Lumbis across the border, but his agents had not returned from that area at the time he was writing. Meanwhile he was getting to know the local Murut people, who he reported were 'not a very energetic race. The men work about two days a week...Their main entertainment is drinking "Tapoi". They brew this in a large urn, it takes about a week to mature, then they attack it. As they drink it they add water, it takes about 3 days to render [the] brew useless, then they make a new brew. While waiting for it to mature they may do some work or hunt'. The authorities overlooked the making of tapoi as it was thought to be a rich source of vitamin B.

Medical support was again important and Thompson reported that he had made a good start: 'at Malalia a 3 year old boy had a badly infected penis due to having forced a bamboo splinter up the eye, cured him with penicillin etc. Next case was at Saliliran, a pregnant women had a badly infected face, cured her with penicillin etc. A woman had a baby on [7 April] however umbilical cord was not tied off, so had to fix. A distribution of vit pills, ointment, elastoplast goes over well. Our medic George Garvin is doing a very good job. They are bludgers as far as cigs and tobacco are concerned. They have supplied us with a couple of meals of curried squirrel (very good), curried chicken and rice.'

The village of Saliliran had a population of about 100 with many refugees from across the border and Thompson gave a detailed picture of Likinan, the village headman.

Age approx 40, speaks fluent Malay. Has wife Sumoi and one daughter. Seems quite friendly and appears ambitious. Probably reliable. He has a skull of Indo soldier he personally chopped off, hidden away, also a uniform of Indo Cpl. I feel he has been spoilt by arrival of army personnel in this area over past 18 months. He has come to expect things as a matter of course rather than a privilege. However he is the most important man this side of the border in this area and therefore worth wasting money on. He has a DEW [defence early warning] system over the border by which he can get early warning of any enemy proposed incursions. If so he intends to move all women and children out of Saliliran back to Talimbakus. Recently a chap name of Anlai had his right arm cut off for giving information. This has upset Likinan to some degree. He has put the acid on me to try and get him a radio. When he spoke about this I more or less told him "bullshit"; however when I thought about it later, I came to the conclusion that to spread the word of Malaysia it would be a good means. These people get no news from outside world at all and therefore a radio would be of great morale assistance. Likinan has told me he wants it for the kampong not for himself. If such thing is feasible I would suggest a good size set in the 200 [Malaysian] dollars or more bracket...As a point of interest when they hit the Tapoi they sing the Malay National Anthem. I rather like this chap, but I think he is out for all he can get from the white man in general.

Thompson believed that with the expected influx of more refugees Saliliran would become the largest kampong in the area, if it was not so already. The people had confidence in white soldiers and it was worth spending a little money to reinforce this attitude.

While the patrols in the KEEN EDGE and HARD STAB area were concerned with hearts and minds, the patrols in the SHARP LOOK area, the 'Gap' area to the east of HARD STAB, were involved more in reconnaissance, trying to understand the tangled country, constructing helicopter LZs and determining if the Indonesians were using the area. It was extremely difficult country and there was no disgrace in becoming temporarily lost. On 17 April Sergeant Chris Pope fell more than 10 metres from a rocky ridge into the Tutomolo River and the only thing that saved him was that he landed on his bergen pack. He landed unconscious in about 2.4 metres of water. Despite the fact that there were dangerous rocks in the river, Private Les Murrell dived fully clothed into the water and supported Pope until he had come around. Pope reported that he was 'knocked queer', had wrenched his leg and his patrol had rested up for the remainder of the day. Nevertheless, he had no wish to leave the area until he had completed all his LZs.

Towards the end of April Major Garland began redeploying his patrols in preparation for cross-border operations early in May. However, the experience of the hearts and minds campaign had been invaluable in preparing the squadron and in identifying problems. Patrols had learned that the British radios had to be kept dry and were more delicate than the Australian models they had trained with in Australia. Medical problems were now starting to occur and Garland had to remind his men to sterilise their water, to treat scratches and bites straight away, and to take their paludrine tablets to suppress malaria. Garland also strongly reminded his men that there was no excuse for not shaving in the bush.

Meanwhile a number of patrols continued the hearts and minds operations. For example Sergeant Mal Waters, who had served with the British Commandos in Borneo in 1964, was still operating around Kabu and Sakikilo in the HARD STAB area. Between 27 and 30 April Kabu threw open its doors to all kampongs on the Malaysian side of the border to attend a funeral celebration. 'Boy, what a beauty', wrote Waters, 'everyone was drunk from the time it started till the day it was finished, singing songs, cockfights, big feasts, the works. Had to shift camp so as we could get a good night's sleep. We all ventured down to sample the "tapoi", very nice drop if I don't mind saying'. The long period in the bush placed additional demands on the patrol commander. Waters noted that he had to have a little talk about slackening off every now and again. 'I must be doing the right thing, no complaint from the patrol yet. Morale high, we have a few septic sores from leeches, that's all.'

Fifteen kilometres to the east around Saliliran an old Scot, Sergeant Jock Thorburn, and his patrol conducted hearts and minds operations from mid April to the end of May. On 11 May he reported that there had been no enemy incursions, but in the village of Sipital, near the town of Lumbis across the border, about twenty Indonesians had attended a tapoi party. They had placed all their weapons in one room leaving one man to guard them, and the remainder had joined in the festivities dancing and drinking. Thorburn thought that it was a pity there was a border or his patrol could have 'joined in the fun at some future tapoi party and made things a bit interesting'.

Thorburn reported that since the Border Scouts were not available for a six-day patrol, he had hired a local tracker named Bokum. 'He was worth his weight in gold...He worked for the [senior British officer in the area] up until three months ago. Things got too hot for him across the border. It seems that the Indos are after his head. He understands just how we patrol and work and takes his job very seriously...Also a good hunter and can provide a nice meal of fresh venison.'

This was the area where Warrant Officer Thompson had operated early in April, but Thorburn found that the situation in Saliliran was not as clear cut as Thompson had thought. Thorburn discovered that the headman, Likinan, had distributed the civic aid stores only to his own people and not to the 40 or so refugees in the village who had received nothing for twelve months. Other kampongs had also been without stores for twelve months. 'I have talked to the [senior British officer] on this subject', wrote Thorburn, 'but he is the original Town Idiot from somewhere in England and I can't get any sense out of him. I do intend to see that everyone here gets a fair share, refugees etc. Likinan might not like it but it will at least keep some of the other people off our backs.'

In their periodic reports most patrol commanders tended to play down the medical problems of their own men, but Thorburn's patrol medic, Lance-Corporal John Braniff, added a detailed attachment to Thorburn's report of 11 May. He indicated that for the past week all members of the patrol had had minor ailments such as bad headaches, infections and diarrhoea. For the past three days Thorburn had had a

Resupply of C Independent Company of the British 2nd Parachute Regiment by Whirlwind helicopter near Long Banga in north east Sarawak in February 1965. The Australian SAS, particularly Sergeant Arch Foxley's patrol, relieved the paratroops in March 1965. (PHOTO, A. GARLAND)

tender reddish streak running from his groin down the inside of his left thigh, he had had a headache for four days and diarrhoea for two days. Corporal George Baines had had an infected big toe and Private Dan Burgess had also had swelling in the groin. Braniff was treating these complaints with penicillin, codeine and opium tablets. He also had been unwell. 'Felt rotten for about 36 hours—headache, violent vomiting and "shitting through the eye of a needle", but started to recover after that time.'

Braniff, however, did not forget that his task was to help the local villagers, even if he thought some were malingerers. 'If one person gets a bandage on a cut, someone else will come up pointing to a healed scar years old and want it patched up. The same thing when pills, injections or anything else is given out. Just like kids, every bastard wants one.' He added that there were women in each kampong from Saliliran to Ahang who were due to give birth in the next few months

96

and he was grateful that he had been sent a precis on childbirth.

The longest patrol in the hearts and minds programme, and indeed the longest patrol in the history of Australian SAS, was Sergeant Arch Foxley's patrol in the KEEN EDGE area. Aged 27, Foxley had joined the Army in 1956 and had served with 1 RAR and 3 RAR in Malaya before joining the SAS as a corporal in 1962. His task was to render medical and other aid to the Punan groups of people in the area previously occupied by a patrol of C (Independent) Company of the 2nd Parachute Regiment near the River Silat. About 20 kilometres from the border, the area had been largely neglected by the Malaysians and it was important to maintain a presence. The patrol deployed on 29 March and returned on 26 June—a period of 89 days.

Foxley did not complain about the long patrol—after all since he was isolated he had little idea what the other patrols were doing, but in conversation with some Gurkhas at a forward base he learned that other SAS patrols had undertaken cross-border operations and naturally he was disappointed to find that he had been overlooked. When he completed the patrol Foxley learned that he had been kept in the area because he had not been given a security clearance, due to doubts about the political affiliations of a member of his family, and he could not be told about the cross-border operations. As it turned out he later conducted two cross-border patrols and eventually was given a security clearance. In later years he reached commissioned rank and in 1989 was still serving as the quartermaster of the regiment.

The hearts and minds programme fulfilled an important function in establishing a rapport with the local people and in familiarising the Australians with their area. But now, with this added experience, the men of 1 SAS Squadron were to be given the chance to test fully the skills they had developed in years of training in Australia.

6
Across the border
1 Squadron, May–June 1965

Despite the extensive patrolling by the Security Forces and the successful hearts and minds campaign by the SAS and other units among
the native villages, from October 1964 the British strategy in Borneo
was based primarily on the effectiveness of tightly controlled operations across the border into Indonesia. Initiated by General Walker,
these Claret operations were continued by his successor, Major-
General George Lea, who took over as Director of Borneo Operations
on 12 March 1965. As a former commanding officer of 22 SAS Regiment, Lea had a fine understanding of the role of SAS and he was
sympathetic to its needs. An immensely impressive man, huge, energetic but courteous and human, Lea had a clear appreciation of the
political nature of the war. While he realised that the British strategy
was built around keeping the Indonesians on the defensive by the use
of Claret operations to form a *cordon sanitaire* on the Indonesian side
of the border, he understood the delicate political issues at stake and
was careful to ensure that Claret operations did not go beyond the
requirement to keep the Indonesians on the defensive. Although he
was cautious, Lea did not hesitate to act when the situation demanded
it. Thus after a determined Indonesian attack on an isolated and
undermanned company base of the 2nd Battalion, the Parachute Regiment near Kuching on 27 April Lea ordered a step-up in offensive
Claret operations in that area.[1] By May these operations were underway with the Third Battalion, The Royal Australian Regiment (3
RAR), then based near Bau in the First Division of Sarawak, playing
an important role.[2]

On 2 April in the Long Banga area of the Central Brigade front
(opposite the Fourth Division) B Company of the 2/6th Gurkhas

successfully conducted an ambush with Claymore mines, and five days later C Company of the 2/7th Gurkhas attacked Labang (opposite the Interior Residency) with small arms and mortar fire. Although the Indonesians in this area were generally quieter than in the West Brigade area, during April Major Garland began detailed planning for cross-border operations.

Soon after arriving in Borneo he had received a directive from General Hunt setting out the aims of cross-border operations. The first task was surveillance, to detect enemy dispositions and movements, the second task was to ambush enemy track and river lines of communications, and the third task was offensive patrol operations, in order to inflict casualties on the enemy, with the intention of retaining the initiative by forcing him onto the defensive and generally lowering his morale. The latter two tasks were normally carried out by infantry patrols, but from time to time the SAS were involved. Cross-border patrols were limited to a depth of 10 000 yards (9104 metres) from the frontier. Reconnaissance patrols were to be decided by the brigade commander, who would notify General Hunt, but other cross-border operations were to be determined by the Director of Borneo Operations on the basis of recommendations from Hunt and his brigade commanders, on SAS advice, and on intelligence available.

As part of the squadron's preparations, Captain Gollings signalled the Australian Army Staff in Singapore that soldiers engaged on cross-border operations were required to carry $1000 each for use as a bribe if such an action became necessary to escape from enemy captivity. Up to sixteen soldiers could be engaged at any one time. The Australian Army Staff, however, had intended to forward a sum of only $1600 Malaysian and Gollings had to seek the support of General Hunt to obtain the necessary $16 000.

While the SAS had operated across the border from the time when permission had been granted in 1964, both Walker and after him, Lea, tended to use the infantry units for offensive Claret operations for short distances across the border. However, if operations were to be conducted at greater depth beyond the border, the senior officers of the SAS believed that their patrols were the most suitable for the task. There was no doubt that the SAS was still the most effective reconnaissance force in Borneo.

With these concerns in mind, on 14 May Major John Slim explained to General Lea the problems concerning the deployment of the SAS resources available to him. A Squadron 22 SAS was to relieve D Squadron at Kuching on 27 May, but due to other commitments initially it would be able to deploy just 5 patrols, and would only be able to deploy eleven patrols after 1 July. The detachment of the New Zealand Ranger Squadron was due to leave Borneo on 1 June and would not return until 31 July. 1 SAS Squadron formed the largest SAS force in Borneo and could deploy seventeen patrols. After discussions with Garland, Slim recommended that in the West Brigade area A Squadron be reinforced by five Australian patrols until at least 1 July. Furthermore, he estimated that there would be a requirement for two or three patrols in the East Brigade area and they also could be provided by the Australians. These arrangements would leave nine Australian patrols in the Central Brigade area. Slim told Lea that at that time eleven patrols were being used in a hearts and minds capacity and this was excessive, for professional SAS, in an area that could be well covered by infantry or patrols of Special Patrol Companies. On the other hand, Central Brigade and the Australian SAS had planned some excellent future SAS Claret operations, particularly in the Labang and towards Labuk area. Slim hoped that Lea would approve these operations, for which he considered the remaining nine Australian patrols would be sufficient.

This plan would allow the reconnaissance needs to be met until 31 July when it was expected that 1 Squadron would complete its tour. Although the New Zealanders were due to return at that time, the total number of patrols that could be deployed would then fall from 28 to 18. Aware of this difficulty, on 13 April Garland had written to Brigadier Tuzo suggesting that his tour be extended to six months not counting time for acclimatisation, training and movement. He argued that since the members of 22 SAS had engaged in operations in Borneo since early 1963, they had a detailed knowledge of the country and its people, and needed little time on their return to Borneo to get to know their areas of operations. Furthermore, with a requirement to conduct operations in other parts of the world, 22 SAS squadrons arrived back in Borneo 'operationally tired to some degree'. Hence a four-month tour was considered the maximum if effectiveness and efficiency were to be maintained. The Australians, however, were 'not physically, mentally or operationally tired' and could accept a longer tour.[3]

A proposal either to extend the tour of 1 Squadron or to relieve it at the end of its tour had already been forwarded to Australia by Lieutenant-General Sir Alan Jolly, the Commander, Far East Land Forces. The reply from the new Chief of the General Staff, Lieutenant-General Thomas Daly, was to the point:

My agreement to provide the squadron was on [the] understanding that you would relieve 1 Sqn SAS under your own arrangements. Provision of subsequent relief from Australia was to be discussed later on [the] assumption it would not be required for from four to six months after return of 1 Sqn SAS. This latter period is [the] minimum necessary to ensure planned training and expansion of SAS to four squadrons.

Daly explained that while he appreciated the problems faced by Jolly, he was hard pressed to provide the officers and NCOs to meet the introduction of National Service as well as commitments to South Vietnam and Borneo. The existing second squadron of the SAS Regiment was being used to train and provide the nucleus for the formation of the third and fourth squadrons, and to send the second squadron to Borneo in July 1965 would prejudice the overall expansion programme.[4]

Anticipating that he would probably lose 1 Squadron at the beginning of August, Major Slim told General Lea that the limited number of SAS patrols available emphasised the need for him to retain control through SAS headquarters in Labuan. Brigade commanders had 'become possessive' of their attached SAS squadrons and they tended to resent the moving of SAS patrols from one part of the front to another. 'They will not accept that, properly employed, SAS should not become involved in the day to day tactical battle and should always be commanded at the highest level.' Privately, Slim thought that Brigadier Tuzo was more difficult to deal with over SAS operations than Brigadier Cheyne in West Brigade, whom he thought had a better understanding of the SAS.

Slim recommended to Lea that he try to obtain permission to extend the Claret boundary for SAS alone to a limit of 15 000 or 20 000 yards (13 710 or 18 280 metres) to enable the SAS to operate beyond the normal infantry boundaries. In fact, eventually the SAS was to be permitted to operate to a depth of 20 000 yards.[5] Slim also reminded Lea that the SAS was suitable for 'offensive' tasks. 'Small groups correctly inserted can do more killing, cause more havoc and disruption than large bodies of infantry "clattering" through the jungle. Furthermore, it is not difficult to switch SAS effort from one sector to another after saturating a given area. We fully realise you want us, at the moment, primarily in the reconnaissance role, but we hope you will let us "off the hook" occasionally. It is important to "blood" the Australian and New Zealand SAS.'

Lea listened to this advice cautiously and on 14 May issued an instruction detailing the command arrangements applying to special patrol units in Borneo and emphasising that he retained command of the units. He reminded commanders that SAS operations were 'subject to the Golden Rules in as much as they will not become offensive operations without obtaining the sanction of the Director of Borneo Operations through the brigade commander and Commander Land

Forces Borneo'. Proposals could include the engagement of soft opportunity targets during the last 48 hours of a patrol.

After every Claret operation a full report, with sketches, had to be sent to brigade headquarters as well as to SAS headquarters at Labuan. Normally, any contacts during these operations were reported to brigade headquarters only by Top Secret means. Nothing was to appear in routine situation reports that might give any indication that an action had taken place. Indeed unit war diaries usually made no mention of contacts if they took place across the border. If the Security Forces suffered any casualties then it would be necessary for the incident to appear in the situation reports, but in those cases a cover plan would have to be provided.

The fact that cross-border operations by the SAS in Borneo were controlled by standing orders, that had to be read to all patrol members before each patrol, emphasised the degree of caution with which the British approached these Claret operations. As far as the Australians were concerned, many of the orders had been standard practice not only in the Australian SAS, but also in the Australian infantry for many years. Other orders, however, applied specifically to the SAS, for with patrols of only four men greater emphasis was given to security.

In normal circumstances the patrols would be deployed by air to the forward infantry company base nearest to their patrol area, allowing the patrol commander to liaise with the infantry commander. Then the patrol would be lifted by helicopter to an LZ near the border where, if Security Force troops had secured it or cleared it of enemy parties, mines or booby traps, the helicopter would land. However, if it had not been secured the patrol might be required to winch or abseil into the LZ. Once assembled on the ground the patrol would move to its task on foot.

Patrols would last for about fourteen days and all supplies had to be carried. Although the orders stated that the maximum any man could carry in his bergen was 50 pounds (22.68 kilograms), when fourteen days rations were carried, patrol loads sometimes exceeded 50 pounds and loads of 70 to 80 pounds (31.82 to 36.37 kilograms) were known. In these cases food was cached to be collected on the return journey. With the heavy loads, steep terrain and trying climate, six hours of patrolling per day was considered sufficient. At night the patrols would crawl into a thick patch of jungle and erect their hammocks and shelters low to the ground. No lights were shown except those needed to work the radio. Before dawn they would be up, fully dressed with their kit packed. They would then wait quietly for half an hour before moving off to find a place to stop for breakfast. All rubbish would either be buried at least one hour's march away from the camp site or would be carried out of the operational area. Care had to be taken to bury the rubbish sufficiently deeply to prevent pigs scavenging it, and it was wrapped in a polythene bag to reduce smell. Water points could be visited in pairs and only once in the same place.

Security on the move was paramount. Cutting jungle vines and branches was not permitted, tracks were not to be used, and if a track was crossed every effort was made to smooth away footprints or any other evidence. All communication within the patrol was by hand signal or whispered conversation. In his report on his squadron's tour Major Garland wrote that the normal principles of patrolling and cross country movement as taught throughout the Australian Army were observed by the SAS in Borneo. 'Security of the patrol is of utmost importance and it is based, in the jungle, on a small patrol being able to conceal itself, move without leaving obvious tracks and remain completely silent. To be predictable and regular in habits is tantamount to asking the enemy to "shoot you up". In this theatre regular habits and the use of tracks produce only dead soldiers.'[6]

Particular care was taken with dress and equipment. Only issue olive green trousers and shirts, with sleeves rolled down, and jungle hats were worn. An appropriately coloured identification band was sewn all the way around the inside of the hat which could then be turned inside out when patrols were approaching a Security Force location. Except when actually dressing or changing during hours of daylight, boots were always done up and trousers and shirts worn so that troops were ready for instant action. Magazines, water bottles and an emergency pack were carried on the belt which was always worn in daylight hours. It hardly need be added that rifles were carried at all times. The compass was carried in a pocket on the belt but was attached to the clothing, not to the belt, with a cord. The emergency pack contained rations, a shell dressing, paludrine, a clasp knife, matches, tetracycline, salt and an air marker panel. A minimum of 60 rounds per SLR and 100 rounds per Armalite was carried.

Radio communications were crucial to the success of the operations, and each patrol reported with a coded message once every 24 hours at a scheduled time. If the radio failed to operate or if schedules could not be met, a patrol would activate its emergency Sarbe (search and rescue beacon) radio at set times or when an aircraft was heard in its area. Each patrol had a patrol alert codeword which if broadcast meant 'I am being chased and am heading for the Border (or Exfiltration) RV'. In that case a helicopter would be tasked to extract the patrol and it would home in on the Sarbe signal.

After Geoff Skardon's experience in 1964 the contact drills had been revised. On contact maximum fire was to be brought to bear on the

1 SAS Squadron, Borneo 1965. *Left to right:* Lance-Corporal Leo Walsh, Private Charlie Adamson, Sergeant Chris Pope, Lance-Corporal Ron McHugh, Captain Ian Gollings. (PHOTO, L. WALSH)

enemy by all members of the patrol. The patrol would then immediately fall back on the rear man using fire and movement if necessary. Once the front man was level with the rear man, the patrol would withdraw a certain distance, go into a quick defensive position and the patrol commander gave his orders for the next move. If a man was found to be missing and if the enemy situation allowed, the patrol would move back to the scene of the contact in an attempt to recover him.

With only four men, casualties would always pose a problem. As a general rule sick and slightly wounded men had to be helped back to the border. Badly wounded men had to be moved to a safe position from which they could be winched out by helicopter, but only after the Director of Operations had authorised the helicopter to cross the border. If the radio was not working the wounded man had to be left in a safe position with the patrol medic, who would have the Sarbe, while the other two members moved back to the Border RV.

If the patrol became split up the members would withdraw to previously designated RVs: the Emergency RV, open for four hours after an incident; the Troop RV, open for twelve hours and usually the previous night's or midday's stopping places; the Border RV, usually the entry LZ; and the War RV, the nearest Security Force position.

With these orders in mind, at the beginning of May 1 SAS Squadron began its Claret operations, and the first patrol was to the Labang area where the Sembatung River came within about 3000 metres of the border south east of Pensiangan. Labang had been successfully attacked by C Company of the 2/7 Gurkhas in April and the task was to observe any enemy activity.[7] Major Alf Garland led this four-man patrol and in his own words, 'It was not what one would call an outstanding success'. Soon after midday on 2 May they were lifted by helicopter from Pensiangan to LZ Black near the border and they patrolled down steep ridges and through primary forest towards the east bank of the Sembatung. Once there they had to force their way through dense secondary growth from 0.5 to 3 metres high to observe the river. No enemy traffic was seen but on 5 and 7 May the patrol was located by local natives with dogs and each time they had to change location. Garland had no way of knowing if the natives were friendly or whether they would report them to the Indonesians. On 7 May the patrol moved into a position to observe Labang and about twelve persons were sighted in two longhouses. Meanwhile, on 6 May the patrol second-in-command, Lance-Corporal Fred Gosewich, had become sick with a fever and Garland decided to cut short the patrol. They arrived back at LZ Black after midday on 11 May and were back in Brunei Town by 9.30 am the following day.

A few hours after Garland's patrol was inserted at LZ Black, a patrol commanded by Lieutenant Tom Marshall was inserted at the same LZ with the task also of observing the enemy position at Labang. Although they were carrying thirteen days rations, Marshall aimed to keep the load down to no more the 30 pounds (13.61 kilograms) for each man, and their load carrying equipment consisted of one '44' pattern pack and three wicker-type baskets of native origin. Breakfast consisted of half of a bar of chocolate or half of a packet of raisins plus a brew of tea. For lunch they consumed the remainder of the chocolates or raisins and another brew of tea, and for the evening meal there was a tin of meat and vegetables plus tea. Vitamin tablets were taken daily and a wild fruit tree was located which provided a welcome change of diet. When the patrol returned on 14 May they each had lost an average of 6 to 10 pounds (2.72 to 4.54 kilograms). But they had successfully observed fifteen enemy in two main huts at Labang.

As forecast by John Slim, 1 Squadron was required to assist East Brigade, and from 8 to 20 May, as part of Operation UNION, Sergeant Joe Flannery led a five-man patrol inland from Kalabakan towards Salang. They sighted no enemy. Meanwhile, on 13 May a four-man patrol under Lieutenant Peter McDougall returned to the Labang area and began observation of the village. They saw eight enemy at any one time in the Labang camp. On 19 May, while still observing the village, an enemy scout approached the patrol's LUP, but he withdrew quickly and it was thought that he might have seen their wireless aerial. The patrol therefore withdrew from the area with no shots being fired. They moved along the Salilir River and continued their surveillance for a further four days. Brigadier Tuzo was delighted with McDougall's report and he wrote to Garland: 'I doubt if this could be bettered and I would like you to thank them, on my behalf, for a job really well done'.

On 19 May Garland submitted a proposal to Tuzo to extend his reconnaissance patrols in both the Sabah area, where he planned to deploy five patrols, and also opposite the Long Banga District in the Fourth Division of Sarawak, where he would deploy two patrols. This proposal was approved and on 24 May Sergeant Weir's patrol moved out to determine whether the enemy were using any routes from Labuk on the Sembatung River north towards Sabah. His area of operations was to be along the Selimulan River. Further to the east Sergeant Pope returned to the wild 'Gap' area across the border from the area where he had operated in April, while Warrant Officer Thompson's patrol conducted a reconnaissance towards Sinapur, ten kilometres west of Lumbis. In the Long Banga area Lieutenant Roderick and Sergeant Jarvis had the tasks of checking border crossing points and reporting on enemy activity. On 29 May Jarvis reported that just across the border he had sighted two enemy moving west dressed in camouflage suits, wearing caps, but not carrying packs. They were not moving along any made track. Jarvis followed them for a while but eventually lost contact. In this area the border was most indistinct. Indeed on the map provided to Major Garland the original border had been whited-out and had been redrawn in some areas 5000 metres to the east and in other areas up to 5000 metres to the west.

Early in June another series of reconnaissance patrols began in the Lumbis and Labang areas opposite Sabah and in the Long Bangan area across the border from Long Banga in Sarawak. Of these three patrols, that conducted by Corporal John Robinson to locate the enemy at Lumbis and to select a route to lead in a party of Gurkhas for a subsequent attack, was outstanding. Quietly spoken, mild mannered, almost diffident, the 27 year old Robinson was one of the NCOs who had been given command of a patrol at the end of the hearts and minds campaign. This was to be his first operation since receiving his command.

On 4 June the patrol was dropped by Wessex helicopter near the border south west of Saliliran and for the next three days they moved quietly and carefully through primary jungle along a ridgeline into

The airstrip at Long Banga, Sarawak, 1965, which was the liaison post for the 2/6th Queen's Gurkhas and the SAS for patrols across the border. The area around the airstrip had been cleared of jungle, but in the background can be seen the tall trees and thick vegetation that dominated that area of Sarawak and stretched across the border into Kalimantan, Indonesian Borneo. (PHOTO, J. SEXTON)

Indonesia. In late 1964 the 1/2 Goorkhas had destroyed the enemy camp at Nantakor, 3000 metres across the border, and the local natives had evacuated the area.[8] The approach was uneventful except that on one morning the Indonesians in Lumbis fired mortars and 12.7 machine-guns up the river valley to the east. On the afternoon of 7 June they reached the end of the ridgeline which broke into small spurs leading down to an area of open ground and then to the Salilir River, about 60 metres wide, deep and flowing swiftly.

Robinson left two of his men in an LUP to observe the Lumbis area, which consisted of a number of villages, mostly on the other side of the river, and taking his scout, Private Bob Stafford, he moved forward for a closer look. Crawling carefully through the low undergrowth they searched for any sign of the enemy across the river. In the village of Kamong, about 250 metres south of the river, they could see nine enemy soldiers and Robinson spent some time sketching the village. Meanwhile, several canoes carrying natives and soldiers moved past on the river. With painstaking care, the Australians changed their

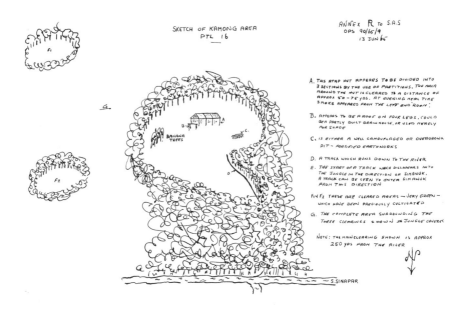

SKETCH OF KAMONG AREA
PTL 16

ANNEX R to S.A.S
OPS 90/65/9
13 JUN 65

A. This map hut appears to be divided into 3 sections by the use of partitions. The area around the hut is cleared to a distance of approx 50-75 yds. At evening meal time smoke appeared from the left end room.

B. Appears to be a roof on four legs, could be a partly built grain house, or used merely for shade

C. Is either a well camouflaged or overgrown pit ~ fortified earthworks

D. A track which runs down to the river

E. The start of a track which disappears into the jungle in the direction of Simanuk. A track can be seen to enter Simanuk from this direction

F1 F2 These are cleared areas ~ very green ~ which have been previously cultivated

G. The complete area surrounding the three clearings shown is jungle covered

NOTE: The main clearing shown is approx 250 yds from the river

~ S. SINAPAR

This sketch, drawn by Corporal John Robinson during his reconnaissance of the Lumbis area on 7 and 8 June 1965, shows the sort of intelligence gathered by the SAS in Borneo. He spent about 24 hours observing the area, not daring to cook a meal or erect a shelter against the rain during the night.

location to get a different compass fix on the village, and since it was now last light they wrapped themselves in ponchos to observe the enemy's evening and morning routine. Not daring even to eat an evening meal, they noted the times that the various enemy lights were extinguished. They estimated that there was a force of about twenty Indonesians but they could not detect any sentries. The next day Robinson and Stafford continued their observation before rejoining the remainder of the patrol and moving swiftly out of the area. They were extracted from an LZ near Saliliran on 12 June.

On 8 June, the same day that Robinson was observing Lumbis, 20 kilometres to the east a patrol led by Second Lieutenant Schuman was in an observation position on the Salilir River. They sighted eight enemy in two boats heading towards Labang. From the beginning of May to the second week of June, 1 Squadron had conducted eleven reconnaissance patrols across the border and Brigadier Tuzo decided that it was an appropriate time to review the squadron's patrol tasks and to initiate more offensive operations. However, the experience of Sergeant Weir's reconnaissance patrol in late May and early June indicated that even without enemy contact, operating across the border in Borneo was an extremely hazardous enterprise.

7
The rogue
1 Squadron: May–June 1965

More than any other unit in the Australian Army, the success of the
SAS depends on the skill, training and determination of the individual
soldier. Much time, effort and care is spent on his selection and
development. Each is a specialist—a signaller, a medic, an explosives
expert, a linguist. Yet the ideal SAS soldier is not a loner. He must
operate effectively within a small close-knit team—the patrol. Indeed
the relationships within the patrol have many of the characteristics of
a marriage. The soldiers live constantly together, rarely speaking,
except in whispers. They know each other's moods, and they rely on
each other for survival. SAS soldiers are therefore outstanding individual operators who nevertheless have a highly developed sense of teamwork and group identity.

Sergeant Roy Weir's patrol was an example of one of these teams.
Aged 39, Weir was one of the most experienced soldiers in the SAS.
He had joined the Army at sixteen and served with the 7th Division in
the Markham and Ramu Valleys and at Shaggy Ridge in New Guinea
in 1943 and 1944 before landing at Balikpapan in 1945. He had
rejoined the Army to fight in Korea and had been an original member
of the SAS. He had been in the first contingent of the Training Team
to South Vietnam in 1962 and had rejoined the SAS on his return in
1963.[1] He was therefore taking part in his fifth operational campaign.

Weir's second-in-command was Corporal Bryan Littler, who had
completed a Malay language course in Singapore during March. Ten
years younger than Weir, he had first met his patrol commander when
he had been called up for National Service in 1955. It was a tribute to
Weir's powerful personality that of the fourteen National Servicemen
in that Western Australian intake to join the Regular Army, nine came

110

from the platoon in which Weir was a corporal. After service with various infantry units Littler joined the SAS in 1961 and met up again with Weir. In Littler's view, Weir 'was probably the best soldier I have ever known, a very strong, silent man with supreme confidence and an easy manner. We got on very well but there was no doubt at all as to who was boss. He was a very private man where his family was concerned and never mixed with us socially'.

The patrol medic was Lance-Corporal Stephen (Blossom) Bloomfield —tough, good looking and always with a grin on his face and a cheeky word. The signaller was Lance-Corporal Paul Denehey. He was a big man, rather clumsy and slow in his movements, but a happy soul who never complained even though he had as a general rule the heaviest load—the radio set.

Although they had trained together in Australia, the hearts and minds operation in the Long Banga area during April welded them into a close-knit team. During hours of patrolling between villages they perfected their hand signals to such a degree that there was no need to ever speak. Two nights before leaving Long Banga they received a further air drop and on this occasion some nice fellow had carefully included a demijon of rum (well padded) and it survived the fall. 'Party tonight!' Roy announced, and that night, well clear of the area and in pitch blackness, their party began. Yarns were swapped in whispers with suppressed laughter. As the night wore on they all drifted off to bed until only Paul Denehey was left. He was completely unaware that his audience had gone to bed and quite happily continued telling jokes until he could not get an answer. It was just one aspect of patrol comradeship.

The patrol had completed its hearts and minds programme in mid May and, as mentioned in the previous chapter, Weir was given the task of determining whether the enemy were using any routes from Labuk on the Sembatung River in Indonesia north towards Sabah. The area was on the edge of the wild 'Gap' area and navigation was bound to be difficult. On 24 May they moved out from their LZ, on the border about 35 kilometres south east of Pensiangan.[2]

For about six days the patrol moved south west along the border which was defined not by one ridge but by numerous ridgelines and spurs running in different directions. The headwaters of the creeks were covered with dense undergrowth, rocks and large boulders and cross-grain movement was extremely difficult. On the seventh day the patrol crossed into Indonesia towards the Selimulan River and on reaching it they patrolled upstream beside the river for two more days. They saw an abundance of wildlife in the area, including pigs, deer and various species of monkeys, both large and small. Signs indicated that native hunting parties had been at work but there was no evidence that Indonesian troops had been in the area. Throughout the patrol it rained daily, generally beginning after last light and ceasing just before first light. As Weir wrote later, 'The combination of rain and pig wallows turned quite a lot of the higher going into a quagmire, making movement both tiresome and difficult'. Even if no enemy were sighted, patrolling in enemy territory was a tense, exhausting business. After ten days in the jungle, eating light rations, struggling through difficult country, constantly alert for enemy or local natives, the patrol was naturally becoming tired.

By the morning of 2 June the patrol had completed its work along the Selimulan and began moving back towards the border, following a ridge line above the river feeling very pleased with themselves on judging the time and distance correctly. The ridge was no more than 2.5 metres wide and visibility was 200 to 300 metres, they were making good time and no trace of enemy activity had been found.

Later that afternoon Littler was forward scout and around 2.30 pm he came across very large tracks. From the sheer size and shape they could only be that of an elephant (gadjah in Malay). Littler continued on for a few more metres until he found a good clear impression and called a halt. They could see that the big toe was facing away from where they were going. 'Good he's going in the other direction, he won't bother us' was Roy's summing up, 'keep going and put a bit more distance between us', and off they went, thankful it was long gone. About 800 metres further on the tracks cut out completely. Some twenty minutes later Littler struck the elephant tracks again and they were fresh! Water was still dribbling into the pad marks, only this time they could clearly see that it was heading in the same direction as they were. Roy Weir decided to take a break to allow it get well ahead of them. Using his camera and Roy's rifle butt as a size comparison, Littler photographed the pad as evidence of elephants in the area; until then they had believed that there were no wild elephants in Borneo. The tracks lasted for several hundred metres then just stopped with no obvious sign as to where it had gone. As far as they were concerned, the ridge was clear and away they went again.

It was now about 4.30 pm and the patrol moved up the side of a steep ridgeline looking for a place to spend the night. As they came onto the ridgeline visibility improved. By now Roy Weir was leading the patrol with Littler behind him in single file followed by Blossom Bloomfield and Paul Denehey. Suddenly Littler saw a huge buff coloured elephant, more than 3 metres tall, coming down a small rise about 200 metres away to their left. He immediately signalled 'off the track!' and the patrol went to the left and right off the track.

This was too good an opportunity to miss, and Littler moved along the ridge about 30 metres getting out his camera as he went. The elephant was still shuffling down the track; abruptly it stopped, looked up, saw Littler, threw its trunk in the air and charged straight at him. 'Run, he's charging', Littler yelled as he turned and ran. He covered about 30 metres when he was confronted by a huge fallen tree about chest height. With his pack on, camera in one hand, rifle in the other he somehow managed to jump halfway over the log. He was lying horizontal on the log when 6 tons of elephant hit. Turning his head in that split second all he could see was a huge eye, centimetres from his, blazing red, with such 'mad hate and ferocity' it chilled his blood. The next thing he knew he was flying through the air. Landing on all fours about seven metres away, his rifle speared into the ground beside him; he rolled and rolled expecting any second to be trampled. Looking back all he could see was the log. He jumped to his feet, grabbed his rifle and a thin stick, and tried to get the mud out of the bore so he could fire.

Looking over the log he could see the elephant moving down its length. At that second Roy Weir fired. The 'crack' of the round was centimetres from Littler's head and the elephant went down on his front knees. Littler's immediate thought was that Roy had killed it, but no, up it got and threw its head back. Roy, from his vantage point on the ridge above, had seen Paul Dennehey run down the side of the log. The elephant had spotted Paul and had taken off after him. Roy had opened up as the elephant closed in on Paul. Blossom Bloomfield now began firing and the bullets could be heard smacking into the elephant; it turned, charged back up the ridge and moved off along it to the left. As it passed some ten metres from Littler, still trying to clear his bore, Blossom fired point blank hitting it right behind the ear, opening it up. 'I thought for sure that would stop him', wrote Littler, 'but no, on he ran straight up screaming and trumpeting towards the ridge. Roy planted a shot straight up his arse as he went past him. I can only describe the sound he made as one of pure rage.' In all a total of nine shots were seen to strike the elephant.

From the moment the elephant charged to its disappearance over the ridge could not have been more than 30 to 40 seconds. Then as the elephant's screams receded Littler heard a groan and the words 'help, I'm gored'. It was Paul! Blossom, the medic, jumped the small end of the log and rushed to help Paul. When he arrived seconds later Paul's pack was half off and it appeared to be squashed on one side where the elephant had obviously knelt when it went down after Roy had fired. As it stood up it had gored Paul under the ribcage on his left side, leaving a hole Blossom could have put his fist into. One tusk on the elephant had been broken off about a third of the way down and it must have used that one to leave such a hole. By now Roy had joined them and they could hear the elephant in the distance, still screaming. Paul was conscious and Blossom covered the wound with a field

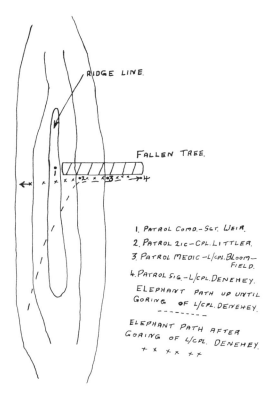

1. PATROL COMD.-SGT. WEIR.
2. PATROL 2ic-CPL. LITTLER.
3. PATROL MEDIC -L/CPL. BLOOM-
   FIELD.
4. PATROL SIG.-L/CPL. DENEHEY.

ELEPHANT PATH UP UNTIL
GORING OF L/CPL. DENEHEY.
- - - - - - - -

ELEPHANT PATH AFTER
GORING OF L/CPL. DENEHEY
+ x + x + x

This sketch, drawn by Sergeant Roy Weir for his patrol report soon after the mission in which Denehey was killed, shows the path taken by the elephant. Littler and Bloomfield managed to move out of the path of the elephant, but Denehey was struck in the chest by the elephant's tusk.

dressing, administering four tetracycline tablets to stop blood poisoning and a shot of morphine for the pain. 'We have to get away from here before that bastard comes back,' Roy announced. They all agreed. It was obvious their little toy guns did not bother elephants; on all their minds was the question, how do you stop it?

Quickly they made a stretcher using two poles and a poncho, lifted Paul onto it, put their packs on and gave Roy their rifles. With Blossom at the front and Littler at the rear they headed for the river far below them where the going would be easier. They had not gone 10 metres when they struck trouble. Paul was a big heavy man and as Blossom stepped down from over a log or rock, Paul scraped the ground causing him intense pain. Roy took the other side of the stretcher from Blossom and between them they lifted him a bit higher as they went down over various obstacles and rocky lips. After about 100 metres they were becoming exhausted, the terrain was becoming steeper and rougher and even though they zig-zagged, choosing the easiest route, it was obvious they could not do it this way. Piggybacking did not work as they needed both hands to climb down. Littler then tried tying Paul's hands over his shoulders onto his belt which

114

freed his hands and distributed the weight. This worked for a while until Paul could not stand the pain. Later they were to learn that two of his ribs had been torn away from his backbone and in this position it was agony for him. They could still clearly hear the elephant trumpeting above them on the ridge, and it appeared to be running up and down the track. Roy decided they had to find a place from which they could defend Paul if the elephant came back. One of the steep ridges with only one way up would do, so with Roy ranging ahead while Blossom and Littler kept Paul going with his arms over their shoulders, they made their tortuous way down. Paul was getting worse and was begging them to stop, but somehow they kept him going until finally they climbed a steep ridge with a good drop at one end and an impossibly steep descent at the other—if it came now at least they had a chance!

With Paul on his hammock and Blossom doing what he could, Roy went back up the way they had come to give them early warning and Littler took a look at the radio. The elephant had completely wrecked the receiver section but the sender part looked serviceable. He set about putting up the best dipole aerial he had ever erected. Towards dark Roy returned and they could still periodically hear the elephant up on the ridge. Paul was apparently sleeping and they sent their message in clear: 'patrol attacked by rogue elephant—one member seriously injured send help LZ!' They sent this message several times. In Perth that night an alert signaller at SASR Swanbourne picked up the message from across the sea, but the patrol had no way of knowing if their message had been received.[3]

In fact the message was picked up by headquarters 1 Squadron in Brunei Town. The operations officer, Lieutenant McDougall, recalled that it 'took some time, several hours in fact, to work out what appeared to be a coherent message—and then we had to rationalise its contents. Goring by an elephant! Was this some type of joke? Why was the speed so slow and the transmision amateurish? Could the patrol have been captured and the transmission be a trick to lure rescue parties into an enemy trap?...Our first reaction was to seek authentication of the identity of the sender and clarification of the contents of the message. We tried to do this during the night but could not make clear enough contact...It was not until the following day that we were sure we knew enough about the situation to make some decisions to provide assistance to the patrol.' Nevertheless, the 1 Squadron patrol log shows that at about 6 pm on 2 June the Gurkhas in the area closest to Patrol 12 were warned that it was in trouble.

Back in the jungle, as darkness fell the screams of the elephant subsided and finally, silence. The silence was the worst as they did not know where the elephant was. Huddled close together in the pitch blackness of the jungle, they combined their knowledge of elephants. They knew that they had poor eye sight and hearing, excellent sense of smell and could move like shadows making no noise. Littler knew they could run fast. He had held the Army record in Western Australia for the 100 and 200 yard (91 and 182 metre) dash for five years and it seemed to him that that huge hunk of meat had covered some 150 metres to his 30. There was no use trying to outrun it. They had already put nine bullets into it with no apparent affect. So in whispers they made their simple plan; if it came they would each put 20 rounds into it and then it was every man for himself.

At about 1.00 am it started. They were sitting back-to-back and Paul was moaning softly in the background. Occasionally they could hear a sound like heavy breathing, then the soft fall of something moving, then a gurgling sound. An hour later it was still there but was moving around making small noises like snapping twigs, and always the gurgling noise. The long patrol and the events of the past few hours were beginning to take their toll, for as they sat there back-to-back Littler could feel the shivers of his two companions, and that bloody noise was beginning to get to them. Was the elephant silently stalking them? Was it waiting for daylight to attack? The gurgling sound could very well mean that it had been hit in the lung. Finally Roy whispered his plan. They were not going to sit here and wait, they would strap their little torches to the barrels of their rifles and at a given signal they would turn them on and blast hell out of whatever. 'Now!' yelled Roy. On went their lights, and for a split second there was silence, no gleaming eyes or white tusks, just a very big very startled pig who left the area, poste haste!

At first light the next morning, 3 June, the patrol moved again and at 9 am they stopped to try to establish communications. Eventually the message was passed to headquarters in Brunei that Denehey had been wounded, but again they could not determine if the message had been received and understood. In fact the message had been understood and Garland had immediately sought approval to send an aircraft across the border. After taking about eighteen hours to secure approval from DOBOPS Garland flew to Sapulot to co-ordinate the rescue.

Meanwhile, Second Lieutenant Schuman, who was already at Sapulot awaiting a helicopter to lift his patrol in to begin another operation, had been contacted by squadron headquarters and told to switch to Weir's frequency and relay the radio signals because communications were weak. Schuman recalled that the first words they could decipher were 'rogue elephant'. He immediately went to the codebook to see what it meant. There was no entry. It was not until the signaller started sending words twice that the full impact of the tragedy started to unfold. Just at that time a senior British officer arrived in a light

After receiving news that Denehey has been gored by an elephant, Major Garland flew to Sapulot near the Sabah/Kalimantan border to coordinate the rescue. Here he is seen on the helicopter pad at Sapulot with a British helicopter pilot.

helicopter. Schuman explained the situation and 'without a blink of an eye' the officer told Schuman to take the helicopter. He flew to the last known grid reference but because of the terrain and weather could not make contact.

However, even if they had made contact the tall, thick trees in the valley would have prevented the helicopter from landing or even winching into the area, and Weir was now faced with the classic situation that had been envisaged in the Standing Orders. He acted strictly according to those orders. Denehey was getting worse and he realised that three men could not carry him through the thick and steep country. He decided to leave Bloomfield with the casualty, and with Littler he would make his way back towards the border in the hope that with his Sarbe he could make contact with a searching helicopter. Bloomfield was instructed to listen for aircraft and if any appeared in the area he was to attract its attention by making smoke with the only white phosphorous grenade carried by the patrol. As Weir and Littler headed back up towards the ridge Littler felt angry—angry at that big bastard of an elephant for what it had done to Paul, and guilt perhaps; he felt that but for the grace of God it should have been him.

They reached the ridge and moved to either side of the animal pad so that they could both shoot with no risk of hitting each other. All seemed clear, after 500 metres the ridge broadened out and they increased their pace. By about 1 pm they were approaching the border and they could hear a helicopter in the vicinity. Weir was about to turn on his Sarbe when from their right front came what Littler described as a 'lunatic, blood curdling scream'. 'My God, that's the mad man from Borneo', thought Littler. They both froze, then they saw the elephant move about 200 metres away to their right front. There was no indecision between them as they both turned left and bolted over the side of the ridge and ran full pelt down the mountain. They ran and stumbled and fell and ran until they could go no further. Looking back all was clear and they kept going, when suddenly from behind them there was a crashing noise. Both spun around and out of the trees appeared five very large red orang-utans. They turned and ran again, straight down the mountain and over the side of a steep bank, landing in the river. Exhausted, they lay there. However, they knew that all they had to do was follow the river and they would cut the ridge where they had come down from the LZ.

Off they went, following the river, trotting where they could. When it was completely dark they slung their hammocks and got what sleep they could. At dawn they were on the move again and it was obvious they were further away than they first thought. An hour later Littler recognised a broken tree and dug up a Mars bar wrapper he had buried at the base of it; the LZ was only two hours away.

At about 9 am they arrived at the LZ to find—nothing—nobody, their message could not have been understood. Hoping against hope, they decided to wait. An hour later they heard a helicopter and attracted its attention with the Sarbe. Nevertheless, it was not until after 3 pm that another helicopter arrived and lifted them out to an LZ closer to where they had left Paul and Blossom. There they found ten Gurkhas from the 2/7 Gurkha Rifles and Sergeant Pope's patrol which had been diverted from its task to help with the search.

Next morning, at 6.30 am Roy left with the Gurkhas to recover the injured man, and Littler was to follow as soon as the helicopter returned with more troops. At 8.00 am it arrived bringing two Gurkhas and two Border Scouts. With these troops and one other SAS soldier Littler set out after Roy. At the first river he lost Roy's tracks and decided to go straight over the mountain to pick up the headwaters of the river and follow it down—rough going but the shortest route. The other SAS soldier kept Littler's spirits up during the agonising hours of the return journey, as the continual trot, walk, trot, walk of the previous day was showing its effects. A little after noon they were making their way down the other side of the mountain through terrible terrain, climbing huge slippery rocks with dense jungle either side. By 4 am the little trickle of water had become a river and they were making good time. Then the forward scout called a halt and pointed out the tracks of a single man. He was wearing jungle boots but the tread was of a type Littler had never seen. He had come down off the ridge, hit the river, had gone downstream for ten metres and back upstream for 30 metres, turned and returned up towards the ridge. To Littler's mind it indicated a scout making sure the river was clear. He put the patrol in a defensive position then did a thorough reconnaissance of the area. It took an hour but he could find no trace of an enemy patrol. That night they camped beside the river. No cutting or moving. The last thing Littler wanted was a fight with an enemy patrol. During the day the smaller Gurkhas had kept lagging behind and Littler had finally told them in Malay that the next one who held him up he would personally shoot. It was an indication of the desperation he was feeling in his effort to find his friends.

By 1 pm the next day, 6 June, Littler realised that the ridges above them were too low, the river too wide, and he knew that he had missed the spot where three logs crossed over each other, marking where he should go up the mountain to their safe, sharp ridge. He would now have to back track. Gathering the patrol together, he explained that up on that ridge was a Gadjah Besar (huge elephant) who had gored his friend. If they were attacked, all were to shoot it as many times as they could. They looked at him as if he was mad and could hardly keep the grins off their faces. Not one of them believed him; they knew there were no elephants in Borneo. Then on the ridge there it was, the great pads of Gadjah Besar. There were some sheepish smiles at Littler. He judged they were about a kilometre from where they were attacked. Twenty minutes later he was standing on the spot where he had first sighted the elephant. Fifty metres further on was where it had begun its charge. Some of the trees it had bowled over were fifteen centimetres in diameter, and not just one or two, but dozens of them. On reaching the log Littler could see that he must have been just too high for the elephant's tusks, so it had butted it, the force of which had broken it through, moved it over a metre and thrown him seven metres away. He found his camera in a small depression on top of the log. The film had been ruined.

Now Littler began tracing their tracks downhill, recovering some of the gear they had ditched. An hour later he lost the tracks completely so he sent the two trackers off to try to pick up their trail again. Half an hour later they returned having found nothing. The rest had jogged Littler's memory so he ordered the patrol to keep well to his rear and set out to where he thought the ridge lay. Bloomfield at this stage would have to be very tired and nervous and Littler could not take the risk that he might fire at the Gurkhas. A quarter of an hour later he spotted Paul's hammock almost hanging over the drop at the end of the ridge. He halted the patrol and from 30 metres away began calling out to Blossom; no answer. At the base of the ridge he called again; still no answer. Roy Weir must have arrived there first so Littler ran up onto the ridge.

'The sight that met my eyes shall remain crystal clear in my mind until the day I die', wrote Littler. 'Paul had left his hammock and begun to crawl; he had moved about twenty feet (six metres), half dragging his trousers off as he went before his strength gave out and his arms were not strong enough to move his body, but he had continued trying, slowly wearing his fingers away as he dug at the earth, trying to get away from the terrible pain deep inside him. Paul's Owen gun was scattered everywhere, every piece of gear that he could reach was torn or broken, and the last scream of agony was on his face. The scene—a nightmare come true—and where was Blossom?'

Littler fired two shots. If Roy was in the area he would reply, so would Blossom. Nothing. Littler sent the patrol further down the mountain and began wrapping Paul up in his poncho. This done he took the patrol down to the river. He could hear the helicopters searching for them in the distance so they lit a large fire trying to get enough smoke above the tops of the trees; no hope, they were too high. Where was Blossom? Then he remembered the lone footprints by the river; he had only just missed him.

Ten minutes later the forward scout pointed out a distant figure and sure enough out of the jungle came Roy Weir and his patrol. Together they went back with ten Gurkhas to get Paul's body. It took two hours to reach the river and clearly no injured man could have taken the pain. Next morning they constructed a temporary LZ and the helicopters lifted them all out to Sapulot.[4]

And so the patrol ended. After the departure of Weir and Littler, Bloomfield had continued to administer medical aid to the injured man, and although Denehey was obviously in great pain he was breathing regularly and there was no apparent infection from the outside. But by about midday on 5 June Bloomfield was running out of medication. Paul, his best mate, was begging him to shoot him because of the pain. Considering that excessive time had elapsed since Weir's departure, Bloomfield decided to set out for help. He administered more first aid, made Denehey as comfortable as possible, and set out carrying the radio. After moving for some distance along the ridge where the incident had taken place he found that he was being delayed by the weight of the radio, and together with the patrol code books, he cached it in a hollow log.

Eventually, on 6 June, looking like he had been caught by every thorn tree in Borneo and also bitten by every insect, he reached an LZ on the border where he met up with a second ground party from the 2/7th Gurkhas. He immediately volunteered to lead the party back to the site, but before they arrived they received news that the other party had found Denehey.

The patrol was deeply affected by the loss of Denehey. Bloomfield spent some days in hospital and Weir, who tended to blame himself, was hospitalised for a few days with exhaustion. Major Garland decided to disband the patrol and reform it with different personnel. After three days' leave Littler was back on patrol with Sergeant Ron Jarvis. Although Garland had made every effort to ensure that his men were cross-trained before they left Australia, he believed that the other members of the patrol had lacked the specialist radio training necessary to ensure good communications. However, the fact that the other members of the patrol could send morse at all enabled them to alert Brunei to their predicament. Whether the radio was actually capable of receiving messages could not be determined because it was still cached deep in Indonesian territory. While Garland thought that Bloomfield's efforts were 'magnificent', he believed that Bloomfield had not been given sufficient medical training to deal effectively with such a serious wound over a long period of time. According to Garland, one reason that it had taken so long to reach Denehey was that the incident had occurred up to 10 kilometres further inside Indonesian territory than the patrol had thought. Such problems with navigation were always likely to occur along the poorly mapped border, and it emphasised the need for all members to carry a Sarbe to guard against the possibility of the patrol being split.

Although the patrol had been well trained to deal with problems likely to be posed by the enemy, they had been shaken by the unexpected encounter with the elephant. An elephant's tusk is really just a big tooth and when he gets old they go bad and ache, this eventually sends them mad. The old bull becomes an outcast—a rogue.

A month later, on 2 July, a patrol led by Sergeant Foxley, courageously accompanied by Lance-Corporal Bloomfield as a guide, returned to the area to recover the radio. All the radio items including the codebooks were recovered and there was no evidence that they had been tampered with. By then the squadron was conducting offensive operations against the Indonesians, but at the time of Denehey's death these operations had not begun. He was the Australian SAS's first fatality on active service, and as the other patrols moved out on their first offensive patrols his death was a salutary reminder of the ever present danger of operating in small groups deep in enemy territory.

8
Contact
1 Squadron: June–July 1965

The key to the success of British cross-border operations in Borneo was not so much that reconnaissance units were able to give early warning of the build-up of Indonesian forces, but that the offensive nature of these operations kept the Indonesians on the defensive and in most instances prevented them from mounting their own operations into Malaysia. On the other hand the British did not want to undertake substantial operations which might result in the Indonesians reinforcing a particular area of the frontier. Small, precise but deadly pinpricks were required, and the SAS was ideally equipped for such tasks. After almost ten weeks of preparation 1 SAS Squadron was itching to begin offensive Claret operations.

On 11 June Tuzo issued Major Garland with a priority of tasks to be undertaken by his squadron by the end of June. Firstly, following the reconnaissances of Lumbis and Talisoi, then being completed by Corporal Robinson and Second Lieutenant Schuman, the SAS task was to guide C Company 2/7 Gurkhas to attack either place, if a worthwhile target was discovered. The second priority was surveillance of the approaches to Long Banga, in order to give warning of any enemy incursions, and the destruction of any small enemy parties which might be encountered. The third priority was the surveillance of the approaches from Agisan, south of the 'Gap', and the destruction of small enemy parties. Fourth was the reconnaissance of the Long Bawan camp and airstrip with a view to carrying out an offensive SAS task there. Fifth was to conduct reconnaissance for any offensive tasks which might be decided on. Sixth was a continuation of the hearts and minds operation in the KEEN EDGE area—the upper Baram area where Sergeant Foxley was still operating. Seventh, operations were to be conducted in the Loop area, on the border about midway between the Bawan and the Long Banga areas; and eighth, hearts and minds operations were to continue in the area south of Pensiangan.

Tuzo also forecast the priorities in the period after 1 July. The first priority was the surveillance of the approaches to Long Banga. The second priority was the surveillance of likely enemy approaches south of Pensiangan and from the Bawan River valley into the Fifth Division of Sarawak. The third priority was to begin offensive operations opposite the Interior Residency and the Fifth Division, with the aim of keeping the enemy in a defensive frame of mind while the 1/2 Goorkhas relieved the 2/7 Gurkhas, and also lowering the enemy's morale at a time when he was changing over his units. The fourth priority was to conduct offensive operations in the Agisan area south of the 'Gap'. And the fifth, sixth and seventh priorities were the same as the last three priorities for the period before 1 July.

These operations got under way with Warrant Officer Thompson's patrol being given the task of locating the airfield at Long Bawan and discovering what aircraft used the airfield and what facilities and defences existed there. The patrol was inserted at the border on the morning of 12 June and by the morning of 15 June had travelled almost 10 000 metres south east into Indonesia towards Long Bawan. That morning Thompson ordered one of his soldiers to return to the border as he had developed a habit of shouting out in his sleep at night and Thompson considered that he could no longer take the risk of this happening as it would jeopardise the patrol's security.

Thompson failed to discover the airfield but on 17 June sighted several enemy in a village. The next morning, at about 8 am, the patrol was establishing radio communications when a group of enemy approached their position. Apparently they had been seen by natives during the previous day and now the enemy were searching for them. Thompson ordered the patrol to pack up and move, but before this could be completed five enemy came into view through the undergrowth 20 to 30 metres away. The leading member of the enemy group shouted in a loud voice and the enemy moved into an extended line in the jungle. The patrol had no alternative but to abandon their bergens and evacuate the area as quickly as they could. However, while clambering up a steep slope the radio slipped from the grasp of the signaller and rolled down the slope towards the position just evacuated. Thompson thought that it was too dangerous to try to recover it and the patrol continued to move away from the area. Two hours later, when they were 1000 metres away, they heard shots and explosions coming from the area of the original contact.

The airfield had not been located and further patrols would therefore have to be sent into the area. In the meantime, Second Lieutenant

Roderick had the task of determining the size and number of enemy in the camps in the Long Medang area, between Long Bawan and the border. The patrol was inserted on the morning of 23 June and they advanced cautiously on the enemy position avoiding booby traps of various types which fortunately they detected. Late that day they sighted their first enemy, and over the next two days they conducted a detailed observation of the half a dozen camps in the Long Medang complex. A total of 24 enemy were observed and Roderick drew a sketch map of the area.

Roderick's patrol had been given a formidable task. The undergowth was relatively light and a carpet of dry twigs and leaves covered the ground. With numerous tracks crossing frequent patches of paddy and waterlogged grassland, it was impossible to carry out a close reconnaissance of the area during daylight, and Roderick considered that at least seven days would be required to complete the inspection of the camps, not counting time to move from camp to camp. Furthermore, the patrol was working against time, for when he had arrived in the 2/7 Gurkha area to begin his patrol, Roderick had been informed that because of an unexpected Gurkha operation he had to delay his insertion date until 23 June and also had to be clear of the area by midday on 29 June.

With this limitation, perhaps Roderick tried to hurry the patrol and on 25 June his leading scout was sighted by an Indonesian sentry. No shots were fired and a subsequent enemy patrol failed to locate the Australians who were able to observe the Indonesian defensive procedures for 'standing to'. But having been detected, the Australians moved out of the area and were extracted from their LZ on 27 June.

During the latter part of June further reconnaissance patrols were conducted in the Long Bangan and Long Medan areas and also in the area south of the Gap, where Sergeant Weir saw thirteen enemy, apparently hunting for game. Farther to the east, Sergeant Ron Jarvis operated under the control of East Brigade in the mangrove swamps where the River Simengaris flowed into the Celebes Sea. On the morning of 27 June Jarvis and his second-in-command, Corporal Littler, left the remainder of the patrol, Lance-Corporal Eddie Chenoweth and Private Monty White, on the edge of the mangroves while they moved forward to establish the patrol's exact position and to site an observation post. Half an hour later two enemy were seen. The enemy disappeared, but soon afterwards a shot was heard. Jarvis and Littler returned to their original position but found that Chenoweth and White had vanished. As it happened, Chenoweth and White had thought that a grenade had exploded and they had returned to the patrol RV. Jarvis and Littler also went to the RV but missed the other two. Both groups were eventually extricated separately, although not before Littler had run across the fresh tracks of an elephant—a frightening experience since this was his first patrol following the death of Lance-Corporal Denehey. Determined to prevent further incidents of this nature, Garland wrote on the patrol report: 'On no account should a patrol break up into two groups in order to establish an OP as was done in this case. If a patrol is split for any recon [reconnaissance] then each group should remain within voice communication distance of each other and preferably within visual distance.'

Thus despite their intensive training, the experiences of 1 Squadron in May and June 1965 indicated that there were still some valuable lessons that could only be learned on actual operations. However, before the end of the month the Australians were to undertake an operation that made classic use of SAS skills. Following Corporal Robinson's successful reconnaissance of Lumbis, Brigadier Tuzo ordered an attack on the village, and Robinson was given the task of guiding B Company 2/7 Gurkhas into position.

On the morning of 21 June the Gurkhas and Robinson's patrol were airlifted into Saliliran, and soon afterwards they moved into the jungle to advance south towards the border, about 5000 metres away. The Gurkhas were carrying two mortars plus 80 rounds of ammunition and eight General Purpose Machine-Guns (GPMGs), but Robinson was impressed by their quiet approach. After patrolling into the area with only four men a fortnight earlier, Robinson now experienced what he described as 'a magnificent feeling of security'. At night the Gurkhas would move off their track by about 100 metres and would leave a small party behind to ambush the track. The only European in the company was the commander, Captain Ashman, and he advised Robinson to stay within company headquarters and not to move around at night.

By the afternoon of 24 June the company had almost reached Lumbis and Ashman halted the advance to enable Robinson to carry out a final reconnaissance of the enemy position. He found that the enemy were still in Kamong but that the ground had been further cleared and two more huts were occupied. Meanwhile, the mortar section began to saw through a number of trees to allow them to fire their mortars. It was raining heavily and they felled trees during periods of thunder so as not to alert the Indonesians.

The company was up at 4 am the next morning and quietly they moved into position. A section of Gurkhas were deployed to each flank to provide protection and the GPMGs were placed in a line on the ridge overlooking the open ground leading down to the river. Robinson was ordered to take a section of Gurkhas about 100 metres forward of the GPMGs to bring fire onto the enemy from a different angle. All were in position by 6 am, but a heavy fog prevented observation until 8 am. Ashman still did not give the order to fire, but at 9 am a gong sounded meal time for the Indonesians. A group of about ten men gathered and started to eat and only then was the order given. Four to six enemy were killed in the first machine-gun burst and the Gurkha mortars quickly adjusted their fire onto the village. The second salvo went through the roof of the eating hut.

The Indonesians were slow to react but soon were retaliating with machine gun and 60 mm mortar fire. After about twenty minutes the mortar was silenced by the Gurkhas' machine-guns and the enemy mortar pit was hit by Gurkha mortar fire. The enemy machine-gun was never silenced, but clearly the Gurkhas were dominating the fight. At 9.50 am the order was given to withdraw and a British 105 mm howitzer at Kabu, across the border, about 10 000 metres to the north west, began to shell the village. One shot landed a little over a metre from the enemy radio shack whose roof lifted and then settled again. The company's withdrawal was conducted at a forced march pace, and before last light they had crossed the border. The Gurkhas ambushed the track from Lumbis but there was no follow up from the Indonesians.

The attack on Lumbis was an outstanding example of co-operation between the SAS and the infantry. The action also kept the Indonesians on the defensive. A week later a Border Scout visited the area and reported that the Indonesians in Lumbis were 'very much afraid'. They were digging more trenches, they did not want to go out on patrol, and they were only waiting to be relieved. Yet it was Garland's style that he did not go out of his way to congratulate Robinson. One of the hardest squadron commanders in the Australian SAS, Garland gave the appearance that he did not like anyone. His saving grace, however, was that he played no favourites. He was extremely fair and was always consistent. No awards were to be issued to any member of 1 SAS Squadron for their performance in Borneo. On the other hand Garland has argued that all his soldiers were well above the average and that they all performed in an outstanding manner.

While the bulk of 1 Squadron had been operating in the Central Brigade area during June, for most of the month four patrols had been detached to operate with A Squadron 22 SAS in the West Brigade area. On 4 June Lieutenant Peter McDougall and Sergeants Mal Waters, John Sheehan and Jock Thorburn arrived with their patrols at Kuching to be briefed by the officer commanding A Squadron, Major Peter de la Billière.

In January two patrols from 22 SAS, one commanded by Sergeant 'Tanky' Smith and the other by Captain Angus Graham-Wigan, had discovered a track from Bemban to Sawah, about 10 000 metres across the border in the First Division area west of Kuching. It was thought that an enemy company was located at Sawah and that troops were periodically sent from there along the track to relieve a patrol at Bemban. It seemed an ideal opportunity to carry out an ambush with a complete SAS troop. Some months earlier Smith had been shown the technique of setting ambushes by linking an array of Claymore mines by an electrical cord. An American invention, the Claymore was a flat rectangular mine which stood vertically just above the ground on two folding scissor-type legs. When initiated, the 1.5 pounds (0.68 kilo-grams) of C4 explosive propelled 700 deadly steel balls forward in an arc sweeping an area 60 degrees wide, two metres high and 50 metres deep. It was designed to cut down an approaching enemy. Until that time the technique of firing a bank of these mines by one electrical cord had not been used successfully in an ambush, but Smith per-suaded da la Billière that such an ambush would be ideal for the track.[1]

During the next seven days Smith taught the Australians his tech-nique of deploying Claymore mines, and they rehearsed and prepared for the operation. McDougall found that the maps and air photographs were not of good quality, and the location of the track was somewhat vague. While the enemy had mortars McDougall would be outside the range of friendly mortars.

On 11 June over a two hour period the sixteen members of the patrol were landed by Wessex helicopter near the border and the troop moved into Indonesia. After several days in primary jungle they passed a small landing stage on a river, which was thought to be a site for a possible new enemy camp. The patrol was not seen by the local men working in the area, and they pushed on reaching their objective on 17 June. The track appeared as if it was in daily use and was cleared to a width of some six metres.

The next morning McDougall began to set up his ambush, arranging seven Claymore mines to cover the killing ground that stretched for about 70 metres. Sentry groups, each with a Claymore, were placed about 25 to 30 metres to either flank, and a protection group including the radio operator was sited to the rear of the position. Commercially purchased National walkie-talkie radios were used for communica-tions between the sentries and the commander, situated in the centre of the killing group.

McDougall and his men now settled down to wait—but not for long. At 9.50 am three enemy moved past his position travelling fast from south to north, but McDougall was after a more rewarding target. Then at 11.20 am eleven local villagers walked past talking loudly. The Australians continued to lie quietly in the jungle, only moving back to a rear area to eat a cold meal after dark. At first light the next day they were back in position. During that day three separate groups of locals moved along the track, and the next day they saw two locals, one carrying a tapoi jar. By the morning of 21 June the Australians had been in the area for four nights. At 11.02 am that morning four locals moved past from north to south and then over the National radio McDougall received a warning that the natives were being followed by fourteen Indonesian soldiers. They looked fresh and well turned out and carried a mixture of weapons, an Armalite, sub-machine-guns and Garrands.

With heart thumping McDougall waited until the enemy were in the centre of the killing ground and gave the order to initiate the ambush. Sergeant Mal Waters pushed down on the initiator lever, and then, in the words of McDougall, 'There was a deafening silence'. Quickly Waters added a spare battery and tried again. Still silence. By this time the enemy had passed through the position and it was too late for McDougall to initiate the ambush with his silenced Sten; in any case he had sited his men too far away from the track for them effectively to cover the area as he had been warned to be careful of the back-blast from the Claymores.

They remained in position until nightfall and then withdrew quickly from the area. By 25 June they were back at Kuching and soon were out on the firing range to find out what went wrong. They discovered that while there was sufficient voltage, there was not enough electrical current for such a long length of wire. They therefore designed and tested a new system. McDougall said later that they had learned an important lesson in not accepting at face value the advice from another army unit, even if it was from the British SAS. McDougall also received another nasty surprise, because all the silenced Stens failed to operate when test fired. Having been kept cocked throughout the patrol the return spring had weakened and could not push the bolt forward hard enough. It was fortuitous that he had not tried to initiate the ambush with his Sten. Meanwhile, two days after the Australians returned, a group of Indonesian regulars and volunteers attacked a Malaysian police post at Seria, 30 kilometres south east of Kuching. Eight civilians and a policeman were killed.

On 7 July McDougall's troop redeployed for a second attempt to ambush the Bemban track. Again they saw small groups of enemy and locals. Indeed, over the next few days they saw a total of nine enemy and 65 civilians. Becoming frustrated, McDougall decided that he would ambush the next party of enemy to pass, even if it was only a small group, but from that time onwards they saw no more enemy. The ambush was not sprung, and the patrol concluded on 20 July. As it turned out, the British SAS was never able satisfactorily to ambush the track. In September 1965 a British SAS patrol with the task of snatching a prisoner was spotted by the enemy and had to beat a hasty retreat.[2]

Despite some uncharitable comments by a few soldiers of 22 SAS, McDougall's lack of success was not due to any amateurish approach. The Australians had never used Claymores before and the British advice had proved incorrect. On two occasions a patrol of over fifteen men had operated for two weeks at up to 10 000 metres across the border in close proximity to the enemy without being detected. Indeed the officer commanding A Squadron, Major de la Billière, thought that in view of the distance from the border and the time taken to get into position it should be kept as an SAS task. Furthermore, the Australians learned many lessons that were to prove valuable not only in Borneo but in Vietnam in 1966.

While McDougall and his troop were operating across the border in the West Brigade area, in Central Brigade other Australian patrols were continuing their operations. Garland was still determined to mount an attack on the enemy airfield at Long Bawan, and on 1 July he sent Blue Thompson back to complete the reconnaissance he had begun a fortnight earlier. By midday on 5 July the patrol had established an LUP about 3000 metres short of the airfield, with the intention of beginning a reconnaissance after dark. The plan never eventuated because at 1.15 pm an enemy patrol approached the Australian position and it became obvious that the Indonesians were tracking the Australians. Thompson and his men waited until the leading Indonesian was about four metres away before opening fire. The first Indonesian was struck by eight SLR and six Owen gun rounds. The second soldier following was hit by ten SLR and six Owen gun rounds. Both enemy were killed instantly. Fire was returned from the remainder of the Indonesian patrol and another group of Indonesians off to one flank also began firing. The Australians briefly returned the fire and then moved quietly away. Some 30 minutes after the contact both groups of enemy could still be heard shooting. Unfortunately the Australians were unable to recover their bergens which were cached in the area, and again Thompson had failed to locate the enemy airfield. It was clear that the enemy were actively patrolling the approaches to the airfield and Garland decided that the next time he would send a patrol of up to a half troop or even a troop who would be more capable of looking after themselves if they ran into trouble.

While 1 Squadron was now conducting a number of ambush patrols, most patrols continued the reconnaissance tasks. Of particular interest was Second Lieutenant Roderick's close reconnaissance of Long Api early in July. The patrol walked for over 300 metres in a stream to hide their approach to an LUP, cached their packs, and then crawled forward to an observation post (OP) to observe the Indonesian position through binoculars. They sketched various groups of huts and counted at least 20 Indonesian soldiers. At about 3.30 pm on 6 July they spotted a group of four soldiers plus ten others dressed as civilians enter the Long Api area carrying loads on long poles. To Roderick it looked as though the Indonesians had shot some wild game. Private Jack Gebhart thought that they were carrying packs but Roderick was adamant that the objects were animals. As the Indonesians came closer it became obvious that they were carrying the four bergen packs that the Australians had cached two days earlier. A short while later mortar rounds started landing in positions which the Indonesians estimated might be used as OPs by the SAS. Clearly the Indonesians were using the local people to locate the SAS, and Roderick decided to withdraw from the area.

On 3 July Sergeant John Pettit led his four-man patrol south from the border to the Salilir River just east of Baluladan. His task was to establish an OP and report enemy boat movement, and he was permitted to ambush any soft opportunity target within his capability in the last 48 hours of his patrol. At 3.30 pm on 5 July he spotted a boat moving upstream carrying six men. All but one were wearing only shorts with the remaining man wearing an olive green shirt. In the centre of the boat was what appeared to be an ammunition box. That evening Pettit was manning the OP when at about 5 pm another boat came into view travelling downstream. The boat was carrying nine men; eight were paddling and the other man, wearing an olive green shirt and trousers and a green cloth cap, appeared to be in command. Suddenly it seemed to swing towards the OP and Pettit deduced that it was going to land near to his position. He alerted the remainder of his patrol and they took up positions. Pettit considered that if the boat came within ten metres of their position they were likely to be seen, and when the boat reached this distance he opened fire on the centre man in the boat. In less than a minute his patrol poured 81 rounds of SLR and 26 rounds of Owen gun into the boat. Not one enemy was able to return fire with the sub-machine-guns they were carrying. Most were either knocked overboard or jumped into the river, and Petit estimated that his patrol had killed seven and seriously wounded two. This estimate was largely confirmed on 17 July when a report was received from the Border Scouts that local civilians had said that five Indonesian soldiers, including a sergeant major, had been killed instantly and that a further three had died of wounds.

Commenting on Pettit's patrol Garland wrote: 'A series of offensive patrols in this area could be rewarded with a number of kills. To this end a further patrol, 12 men strong, has been despatched to the area about Talisoi to carry out the task of ambush against opportunity

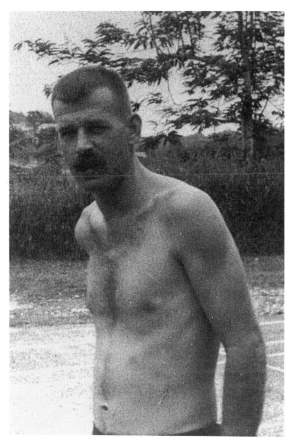

Sergeant John Pettit, who commanded the successful ambush on 5 July 1965 in which at least seven Indonesians were killed. He himself was killed in action in April 1970, during his third tour with the Training Team in Vietnam, and was posthumously mentioned in despatches.

targets using the river.' The patrol was undertaken from 11 to 21 July under the command of Second Lieutenant Schuman, and on two mornings the patrol watched in stunned silence while two Indonesian soldiers bathed and swam in the river opposite the ambush site. Schuman's orders were to interdict Indonesian resupply craft on the river and therefore he did not spring the ambush, but since he had seen no river traffic he decided that from the third morning he would ambush the bathers. The next morning the bathers failed to appear and from that time no target appeared on the river. However, a comment in Schuman's report throws an interesting light on the problems of these patrols. 'It was also most noticeable that a terrific aroma was present if 12 tins of "For You" sardines are opened at one time for meals. The best practice is to stagger meal periods, splitting a patrol of this nature into 6 groups to eat'.

While Schuman was moving into position, a fourteen-man patrol under Second Lieutenant Roderick left an LZ on the border with the

Private Stan Plater and Sergeant Roy Weir at Sapulot near the Sabah/Kalimantan border before insertion for a patrol in July 1965. Patrols flew by fixed winged aircraft to forward bases such as Sapulot and Pengsiangan, and then by helicopter to landing pads near the border.

task of ambushing the Salilir River about 4000 metres to the east. After about two days, as they moved down the steep, jungle-covered ridges towards the river, they received a message to detach four men to observe enemy activity at Labang, and Sergeant Weir was given the task. Roderick was therefore left with ten men, including Sergeant Foxley and Corporal Robinson. Curiously, when Trevor Roderick had been a private soldier he had served with John Robinson who had then been a lance-corporal.

After several days the patrol was in position on the Salilir, and soon after midday on 19 July a boat with five men was seen moving upstream. They thought that four of the men were local natives but suspected that the fifth man was an Indonesian soldier. The boat was carrying what appeared to be about twenty bags of rice, but because of the civilian occupants, no action was taken. At 3.10 pm the same boat and occupants returned, this time moving downstream towards Labang. As they approached the position the occupants, who had been talking loudly, suddenly went quiet and Roderick was concerned that his patrol might have been detected. The next day he ordered his patrol to withdraw, and they occupied another position a few hundred metres closer to Labang. Meanwhile, Sergeant Weir at Labang had reported that he had seen seven enemy organising the despatch of resupply boats up the Salilir.

In the new position the ambush party was concealed in thick jungle on an embankment overlooking about 3 metres of low bracken leading down to the water's edge. Robinson described it as a 'marvellous view' of the river which was about 30 to 40 metres wide. Soon after midday on the following day, 21 July, three boats passed travelling upstream. Each had a crew of five and at least one man in each was thought to be an Indonesian soldier. They were allowed to proceed unmolested.

Then at 12.25 pm Privates Jack Gebhart and Allan Smith, the sentries on the upstream flank, signalled that a boat containing six men dressed in white T-shirts and blue shorts was travelling downstream. It was a prau powered by an outboard motor, but again the Australians were unsure whether the occupants were Indonesians. The boat had reached the centre of the ambush position when a Bren-gunner, Lance-Corporal Chris Jennison, beside Roderick, spotted rifles, webbing and kitbags in the bottom of the boat. He opened fire with Roderick acting as the number two on the Bren. Three rounds from the Bren struck the first Indonesian who was thrown over the bow of the boat. The next Indonesians were also cut down but the last two jumped from the boat and started to swim towards the far bank. Both were killed as they scrambled up the bank. In all, a total of 60 rounds were fired from two Brens, with 52 rounds being fired from the remaining eight SLRs.

The patrol immediately withdrew towards the border and about five minutes later a group of enemy, who had apparently been located just to the north of the ambush site, opened fire into the ambush position vacated the previous day. Ninety minutes later they heard further firing in the area where the ambush had been sprung and it sounded as though an attack was being mounted on the position. They also thought that the patrol was being tracked because a small number of shots were heard each time they stopped to rest and check their position. This sporadic firing continued until about 5 pm. Soon afterwards the patrol halted for the night, but at about 8 pm mortar bombs began exploding to the north. The mortars appeared to be located at Talisoi and the firing continued for about one hour. By 24 July the patrol had reached its exfiltration LZ and the next day they were back at Brunei. They were certain that they had killed all six enemy.

The end of 1 SAS Squadron's tour was now approaching and Garland was determined to finish with a number of large offensive operations. He was still attracted by the idea of attacking the enemy airfield at Long Bawan, but he lacked the detailed information necessary to complete his plans. He had, however, developed a good picture of Indonesian activities at Labang, and on 17 July he submitted to Brigadier Tuzo a plan in which four patrols were to ambush the Salilir River to intercept enemy reinforcements while a further nine patrols attacked the enemy village itself. The final assault group would consist of six patrols led by Garland who would also command the entire operation. In all, he planned to use 52 SAS soldiers. Tuzo approved the plan provided that the Labang garrison contained no more than 25 enemy and, as mentioned earlier, Garland immediately ordered Roderick to divert a four-man patrol from his ambush party to complete a reconnaissance of Labang. On 24 July information was received that General Lea had withdrawn permission for the attack.

Not to be deterred, Garland obtained permission to mount another large ambush, this time on the Sembatung River downstream from Labang. He gave the responsibility to his second-in-command, Captain Terry Holland, who had succeeded Gollings midway during the squadron's tour. The patrol consisted of 35 men, divided into two troops, commanded by Lieutenant Marshall and Warrant Officer Thompson. Unfortunately the enemy chose not to use the river during the period of the patrol, which concluded on 1 August. At midnight the following day the squadron officially ceased operations.

During 1965 the British, Australian and New Zealand SAS killed 107 Indonesian soldiers for the loss of four of their own men. Of that total, the Australians killed seventeen and had one fatal casualty of their own. Although the Australians had been in Borneo for five months, they had spent almost one month training, another month on hearts and minds operations, and three months on cross-border operations. Only about six weeks had involved offensive operations. By comparison, during the year the British had had four squadrons in Borneo and according to Dickens, had been deployed on cross-border operations for almost the entire period. The New Zealanders had been deployed in half squadron strength on two separate tours. During their tour the Australians mounted 23 reconnaissance, 7 reconnaissance/ambush, 2 ambush, 4 surveillance, 1 special and 13 hearts and minds patrols. Patrol durations ranged from 2 to 89 days and were conducted not only in the Central Brigade, but also in the West and East Brigade areas.

The tour in Borneo provided unique experience for the men of 1 Squadron. All hand-picked by Garland during 1964, they were an outstanding group of men, and most were to become senior NCOs during the next few years when three SAS squadrons were deployed to Vietnam. Despite the trying conditions, they had maintained their health and morale. Only two members were evacuated to hospital in Singapore, one with leptospirosis and the other with tinea. There were two cases of malaria, three of dengue fever, three of diarrhoea and four other men were hopitalised as a result of injury.

SAS soldiers who later served in Vietnam were unanimous in their view that operations in Borneo were far more demanding physically than in Vietnam. This was partly due to the difficult terrain, but also to the length of the patrols and the necessity often to live off short rations. The 22 SAS report for the first half of 1965 noted: 'The new methods of operation are even more physically and mentally arduous than those employed in 1963 and 1964'.

The squadron had learned valuable lessons about the role and employment of the SAS. The hearts and minds programme had shown the intelligence value of living among the local natives. The Borneo campaign also demonstrated the importance of maintaining control over the SAS at the highest level. The SAS had provided reliable and timely intelligence by patrolling deep into enemy territory. And when the occasion demanded, it had shown that it could deliver deadly blows to an enemy who thought that he was secure in his rear areas. These achievements had been built on thorough training over many years.

The operational techniques refined in Borneo were to stand the Australians in good stead in Vietnam. In particular it was shown that SAS patrols could be held centrally in the base area and deployed rapidly if helicopters were available. Whereas in training the SAS had emphasised insertion by parachute, the reality of the battlefield was that the helicopter provided infinitely more flexibility.

But there was one matter on which opinion was divided. General Lea was convinced that the largest unit which could perform its task, remain undetected, and be viable in protracted jungle surveillance operations, was the four-man patrol. Any larger group could not remain undetected for long. So long as each patrol had personnel trained to a high standard as signallers, medics and assault pioneers, Garland agreed with the wisdom of four-man patrols in Borneo, but did not recommend that the basic organisation of each troop be altered. Lieutenant-Colonel Brathwaite, commanding officer of the SAS Regiment, injected even more of a note of caution, commenting on Garland's report that the 'benefit of a six-man patrol, if casualties are possible, should not be overlooked'. However, he acknowledged that a four-man patrol enabled more patrols to be deployed.[3] During Vietnam, the size of the patrols was often to vary according to the inclination and experience of the patrol commander. Significantly, many of those patrol commanders who had served in Borneo tended to stick with four men.

Borneo was the first test of the Australian SAS. The men had shown that they could match the skill and flair of the more experienced

British SAS, and they had been willing to develop their own style of operation. And considering the nature of the operations in Borneo, they had learned that they were unlikely to receive any public credit for their work. For men who were justifiably proud of their achievements, their silence over twenty years about their role in Borneo demonstrated their true professionalism.

9
Preparing for action
2 Squadron: September
1964–February 1966

2 SAS Squadron had its origins in mid 1964 when 1 Squadron was warned of its likely deployment to Borneo and another squadron had to be prepared to relieve it once its deployment began. Although this second squadron, 2 SAS Squadron, was raised officially in September 1964, it was obvious that it could not be prepared overnight. 1 Squadron had taken most of the experienced men, and new SAS officers and soldiers would have to be selected, tested and trained. While 2 Squadron, under its new commander, Major James Hughes, selected and trained its own men, the developing infrastructure of the SAS Regiment during 1965 provided an environment in which it could get on with training without devoting unnecessary time to wider matters such as manning and accommodation.

The year of 1965 was of crucial importance in the development of the SAS regiment. Not only was 1 Squadron involved in operations for the first time, but the regiment was consolidated as a unit and preparations were made for another two squadrons to be deployed to separate operational areas in 1966. The change from a company of the Royal Australian Regiment to a regiment in its own right was particularly noticeable with respect to dress.

During the year the head-dress was changed from the red beret to the sand-coloured beret with the metal badge depicting the Flaming Sword Excalibur and the motto 'Who Dares Wins'. Initially the SAS metal badge was worn without a blue shield background on the beret, but eventually the blue background was added to the badge when worn on the beret. The blue lanyard was adopted and worn on the left shoulder, while black Sam Browne belts for officers and warrant officers continued to be worn when the occasion required it. All ranks

Major Bill Brathwaite with the Governor-General, Lord De L'Isle, early in 1965. Brathwaite was Commanding Officer from 11 August 1964 to 28 March 1967 and was promoted to Lieutenant-Colonel in April 1965.

who were qualified wore the SAS paratroop wings, which had extended horizontal wings as distinct from the drooping wings of other parachute units. To complete the ceremonial aspect of the regiment, in May 1965 His Excellency Major-General Sir Douglas Kendrew, Governor of Western Australia, was appointed Honorary Colonel.

While these ceremonial matters are important to the esprit de corps of any military unit, during 1965 they were just a few of the many concerns facing the Commanding Officer, Major G. R. S. (Bill) Brathwaite. Born in January 1920, Brathwaite was aged 44 when appointed to the regiment in August 1964. At the beginning of the Second World War he had enlisted in the AIF and had served with the 2/3rd Field Regiment in Britain, Greece and Crete, where he was wounded at the defence of the Maleme airfield. Returning to Australia, he joined the 2/3rd Independent Company, serving under the legendary George Warfe in the tough Salamaua campaign in New Guinea in 1943. He was commissioned in 1944 and was transferred to Z Special Unit for service behind enemy lines in Borneo in 1945. By the end of the war he had been wounded three times.

138

Brathwaite's last operation is interesting for its similarities with later SAS operations. Known as Operation PLATYPUS VII, it involved the insertion of four parties of between three and five men into the Mahakam River lakes area about 120 kilometres north north east of Balikpapan in Dutch Borneo in conjunction with the landing of the 7th Australian Division at Balikpapan on 1 July 1945. The objects of the operation were: to cover, by visual observation and native contacts the enemy lines of withdrawal from or reinforcements to the Samarinda–Balikpapan area, along the axis of the Mahakam River; to set up escape routes for downed airmen in the Samarinda–Mahakam River lakes area; to organise native resistance; and to cut off food supplies to the enemy.

The first party was inserted by Catalina flying boat on 3 July, but it was not until 23 July that Brathwaite's five-man party, under the command of a British officer, Captain W. J. Martin, parachuted into the Lake Melitung area. During July and August the four parties, all under the command of Captain V. D. Prentice, had several successful engagements with the enemy. They directed airstrikes onto Japanese positions, cut off enemy food supplies, obtained valuable intelligence about Japanese escape routes and, on behalf of the Dutch government, established useful contacts with the native chiefs.[1]

Brathwaite's appointment to command the SAS Regiment therefore provided a direct connection between Australian special operations in the Second World War and the new SAS Regiment. On the face of it, he was well qualified for the task and certainly he believed that he was the most suitable man. After the Second World War he had served in the Citizen Forces in Perth before rejoining the Regular Army in 1951. He had attended the Staff College at Queenscliff in 1955 and had had various operational staff appointments before serving with the SEATO military secretariat in Bangkok from 1961 to 1963. He then became DAQMG on Headquarters Western Command until his appointment to the SAS.

Although Brathwaite was appointed to the regiment in August 1964, he was not promoted to lieutenant-colonel until April 1965 and the appointment was the peak of his career. Despite 25 years of soldiering he still considered himself to be a citizen soldier and claimed that he was not chasing a military career. Furthermore, he had served mainly in Western Australia, was on good terms with the Western Command staff and was well placed to argue for facilities, personnel and re-sources for the SAS. While there is no doubt that Brathwaite had a well-developed understanding of special operations and was able to pass on some of his considerable experience, many members of the unit felt that his age, coupled with his wartime injuries, meant that he was less involved in the unit than they would have preferred.

However, Brathwaite was given great support by his adjutant, Captain Terry Holland, and after mid 1965, Captain Ian Gollings, and by his Regimental Sergeant Major (RSM), WO1 H. J. A. Haley. Born in 1927, Tony ('Trained Soldier') Haley enlisted in the Welsh Guards in 1945 and was discharged in 1948 after service in Germany and Palestine. Joining the Australian Army in 1951, he served with 2 RAR in Korea. An original member of the SAS Company, his previous appointment was as an instructor at the Officer Cadet School, Portsea. He had recently undertaken seven months training in the United States, including a Ranger course. An archetypical British RSM in the barracks area, Haley was an accomplished SAS soldier in the field. He had a good sense of humour but was a strict disciplinarian.

The expansion of the SAS in 1965 matched a general expansion throughout the Australian Army. In addition to the commitment of 1 SAS Squadron to Borneo, in March, 3 RAR was also deployed to Borneo while in June, 1 RAR and attached troops arrived in Vietnam. To allow for these operational commitments three more battalions of The Royal Australian Regiment, the 5th, 6th and 7th, were raised during the year, and the first intakes of National Servicemen began to stream into the new and expanded training units. Then, on 30 April 1965 approval was given to raise a third SAS squadron.

The difficulties faced by the SAS Regiment during this period of expansion were outlined on 10 January 1965 in a letter from the Commander, Western Command, Brigadier Hunt, to Army Headquarters. He stated that the SAS Regiment was committed to maintaining 1 Squadron at operational readiness (they were deployed to Borneo the following month), 2 Squadron had to be built up and trained to operational readiness as soon as possible, 3 Squadron had to be raised by December, 4 Squadron had to be raised by mid 1966, camp duties still had to be performed, the regiment had to provide its own enemy for exercises, individual and specialist training had to continue, selection tours had to be conducted and cadre courses had to be run. The Director of Military Operations and Plans at Army Headquarters replied on 9 March that he appreciated 'the problems and difficulties', but he added that 'with any worsening of the present international situation it will be necessary for 2 SAS Sqn to be prepared for' possible operations in New Guinea. Thus although 2 Squadron might not be fully trained, they nonetheless had to be placed on 21 days notice for movement.[2]

At this time the SAS establishment allowed for a regimental headquarters of five officers and six other ranks, two SAS (or sabre) squadrons each of seven officers and 106 other ranks, and Base Squadron of two officers and 39 other ranks. In all the regiment was supposed to total 21 officers and 257 other ranks. The regiment was, however, deficient by five officers and 65 soldiers, and since 1 Squadron in Borneo was short by only one officer and thirteen soldiers, it was clear that the main burden was to be carried by the remainder of the regiment at Swanbourne.

On 7 April Brathwaite wrote a ten-page letter to Western Command setting out the problems in raising and training his unit. D Troop, which had been left behind when 1 Squadron had deployed to Borneo, was under the command of 2 Squadron and was continuing its training. Meanwhile 2 Squadron had completed a cadre course for 57 men, of whom 50 had qualified. But men still had to be trained as signallers, medical assistants, assault pioneers and Malay or Pidgin speakers. Some also had to qualify as parachutists. He expected that the squadron would be ready for a major exercise in New Guinea or North Queensland in October or November.

Brathwaite planned to start cadre courses in May to provide men for 3 Squadron, and he hoped to begin forming the new squadron after 15 July. But he anticipated that the squadron would not be ready to start troop training until after February 1966, and by then he would need a complete complement of officers for the squadron.[3]

Brigadier Hunt supported Brathwaite and wrote to Army Headquarters to urge that an exercise in New Guinea be approved towards the end of the year. He added that the posting of young officers fresh from Duntroon or Portsea was most unsound. They were less trained in leadership and specialist skills than the soldiers, and leadership was essential in SAS. He thought that troop commanders should be experienced platoon commanders before arriving in the regiment. He also requested that National Servicemen not be sent to SAS. With this background, it is understandable that the Chief of the General Staff denied General Jolly's request during April to relieve 1 Squadron with 2 Squadron as soon as their tour ended.[4]

Later that month Brathwaite outlined another problem. The soldiers of D Troop were continuing their training in case they were required to join their squadron in Borneo, but it was becoming obvious that only a small number of them would be needed as reinforcements. Meanwhile, they were missing opportunities for promotion in 2 and 3 Squadrons where more junior soldiers were already being promoted. Since their morale was declining, Brathwaite decided to transfer selected NCOs and potential NCOs out of D Troop to assist with the formation of the other squadrons.[5]

Despite the regiment's considerable manpower problems, on 19 October 1965 Army Headquarters authorised the raising of 4 SAS Squadron, and the establishment was amended to provide for further administrative staff. To assist with this expansion another major was added to the establishment to become the regiment's second-in-command.

The training of the regiment continued, but as Brigadier Hunt explained in October to the Senior Commanders' Conference, 'in many respects it was running on a shoe string with serious shortages of officers and equipment both from training and operational readiness points of view'. Brathwaite believed that the problem of an insufficient supply of suitable officers was caused by the Director of Infantry's lack of sympathy for the special needs of the SAS—he tended to regard it as just another infantry posting. Indeed the problem in manning the SAS reflected a wide-spread lack of understanding of the role and capabilities of the SAS. By early 1966 the regiment was deficient in thirteen officers, including eleven troop commanders. To meet this shortfall the regiment was sent five officers—three experienced platoon commanders, one a newly commissioned Duntroon graduate, and one an ex-Canteens Service officer, 27 years of age with no prior experience as an infantry officer.[6]

Training during 1965 was conducted with essentially the same accommodation, equipment and facilities as were available to the original company. Equipment was reshuffled between squadrons to enable training to be carried out. The standard of accommodation at Swanbourne at this stage was only barely adequate as there were very few permanent buildings and a major proportion of the unit was housed in second-rate obsolete huts of Second World War vintage.

The major training activity was the preparation of 2 SAS Squadron for service in Borneo. While the deployment of the squadron was expected, it was not actually approved by the government until September 1965, following a request from the British Commander-in-Chief, Far East.[7] The squadron commander, Major Jim Hughes, was born in 1929, had graduated from Duntroon in 1950, and had served with 3 RAR in Korea where he had been awarded the Military Cross for personally leading his platoon in a grenade fight during a Chinese attack. After service in Japan and Australia he was operations officer and then a company commander in 3 RAR on operations during the Emergency in Malaya from 1957 to 1959. From 1960 to 1962 he was an instructor at the Royal Military Academy, Sandhurst. After courses in America he filled a staff posting in Canberra before beginning an eighteen-month course at the Australian Staff College in mid 1963. Although Hughes had attended a basic parachute course and a course at the School of Land/Air Warfare, he had had no previous experience with commandos or special operations. But while attending Staff College he realised that except for the hope of commanding a battalion some years in the future, he had reached the end of his regimental service. The only possibility lay with either the PIR or the SAS, and he duly volunteered for both. He had actually received a posting to PIR when it was decided to expand the SAS, and the posting was altered.

Hughes arrived at Swanbourne at the beginning of December 1964. Small, bird-like, but tough and wiry, he was an outgoing and personable officer who was demanding of both himself and his men. In Malaya he had dealt with both the British and New Zealand SAS and had worked with John Slim. At Sandhurst he had served again with Major Slim, now second-in-command of the British SAS. If Garland could be accused of being influenced towards American techniques, Hughes was more attuned to the British ethos for the operation of SAS patrols. He undertook and passed the SAS selection course, being the second oldest man on the course after his Squadron Sergeant-Major, Warrant Officer Danny Neville, who was returning to the unit after some years' absence.

The general lack of appreciation of the problems of preparing an SAS squadron for operations was shown by Hughes' personal experiences in mid 1965. In May 1965 the DAAG at Headquarters Western Command was posted to Canberra, and Brigadier Hunt was adamant that he needed a DAAG who had attended Staff College. The only suitably qualified officer in Western Command was Hughes, and from 17 May to 1 August he filled the appointment of DAAG as well as trying to command his squadron. Eventually a visiting officer in Perth saw what was happening, phoned Canberra, and Hughes returned to his squadron.

The return of 1 Squadron to Swanbourne in August 1965 enabled 2 Squadron to draw on recent experience concerning operations in Borneo, and further information was forwarded by Captain Mike Jeffery, the operations officer at Headquarters SAS, Far East at Labuan. A Duntroon graduate from 1958, Jeffery had served with the SAS Company from 1959 to 1962 and had recently been ADC to the Chief of the General Staff. He arrived at Labuan on 28 July, just as 1 Squadron was about to leave Borneo. John Slim soon sent him on patrol to gain an appreciation of SAS operations in Borneo. On 19 August he joined a New Zealand patrol with the task of locating an enemy camp in the area of Kampong Segoeman, across the border from the Second Division of Sarawak. They found that the enemy had evacuated the camp.

Jeffery's appointment was part of an effort to give SAS headquarters a Commonwealth flavour, and much of the SAS base was manned by New Zealanders. Jeffery sent monthly reports to Swanbourne, keeping

the SAS Regiment abreast with new developments such as the introduction of lightweight hammocks, lighter climbing irons, and lightweight packs instead of bergens.

When Jeffery noted that a four-man patrol was the most suitable
size, Lieutenant-Colonel Brathwaite was quick to repeat the comments
he had made earlier on Garland's report. He observed that a six-man
patrol had not been tested properly in Borneo and he claimed that the
advantages of a four-man patrol over a six-man patrol were 'fairly
academic'. He concluded that 'regarding the size of patrols', there
could 'be no rule...the OC must be given a free hand to deploy patrols
and or tactical groups of size and composition to carry out his assigned
task'.

Hughes could be in no doubt about what to expect in Borneo, and he
trained his men accordingly. The old ideas of the previous few years of
conducting raids were now taken over by an emphasis on silent and
stealthy reconnaissance. Nevertheless, the weapons skills were not lost.
Hughes gave top priority to navigation, fieldcraft and shooting, with
the policy of 'one shot, one kill'. The men were tough and aggressive,
but it was the disciplined aggressiveness needed by men who would
have to operate behind enemy lines in a political environment in
which orders to refrain from contact might change from day to day.

From 17 October to 30 November 1965, 2 Squadron undertook a
major training exercise in the Lae area in New Guinea. Called LUKIM
GEN, the exercise consisted of three weeks' acclimatisation training in
and around the Lae area, followed by three weeks' patrolling exercise
in the mountains south and east of Wau and Bulolo. The patrols
covered almost 200 kilometres at altitudes of from 1300 to 3300
metres and crossed all types of terrain and vegetation. Brathwaite
believed that if the soldiers could handle the terrain and climate in
New Guinea they would be prepared for Borneo. Hughes would have
preferred the training to have also included a tactical aspect.

During the exercise the squadron experimented with the size of
patrols. Five men were thought to be better than four because in an
LUP four men could face outwards while the patrol commander in the
centre prepared his coded messages. If one man was injured or
wounded with five men there would be sufficient men to carry him.
Furthermore, in setting an ambush the radio operator could be placed
in depth, one man could be deployed on each flank, and there would
be two men in the killing group. Since one man might be sick or on
leave, a basic six-man patrol would ensure that at least five men would
be available for an operation without interfering with other patrols.

By the end of 1965 Hughes had prepared his squadron for operations and on 20 December Army Headquarters issued an instruction
that the squadron was to deploy to Borneo in January 1966. Like

Garland the previous year, Hughes had to decide which of his four troops to leave behind. F Troop lost out. Hughes had put the four troops in competition with each other and he decided that he would not reshuffle personnel between the troops once the time came to leave one behind. Thus F Troop included some excellent men who subsequently became the nucleus of 3 SAS Squadron. Captain Deane Hill, who for short periods had been acting commander of the squadron, was still second-in-command, and Lieutenant Graham Hoffman, a 26 year old Portsea graduate who had served with 2 RAR in Malaya, was operations officer. The Squadron Sergeant-Major, Danny Neville, had served in the Second World War and Korea, as well as with the Training Team in Vietnam in 1964. According to the Team's historian, he was 'well liked, experienced with soldiers and with a humour which belied his determination, he was skilled in weapons, minor tactics and small patrols'.[8] E, G and H Troops were commanded by Lieutenant Ken (Rock) Hudson, Second Lieutenant David Savage and Lieutenant Tony Danilenko respectively. Both Hudson, aged 30 and Savage 27, were Portsea graduates and former soldiers. Danilenko was younger and less experienced. He had graduated from Duntroon in December 1963 at the age of twenty and had joined the SAS in January 1965. The troop sergeants were Doug Tear, John Coleman and Peter White.

On 12 January Major Hughes, Lieutenant Hoffman and an advance party of seven left Perth by air and after a stop in Singapore they were met at Labuan on 14 January by Captain Mike Jeffrey and the Commanding Officer of 22 SAS, Lieutenant-Colonel Mike Wingate Gray. There were several days of briefings in Labuan and on 18 January the party flew to Kuching to be met by Major Terry Hardy, the commander of B Squadron 22 SAS, which 2 SAS Squadron was to relieve in the West Brigade area.

The tactical situation had changed substantially since Alf Garland had left five months earlier. Following the unsuccessful Communist coup in Indonesia on 30 September, President Sukarno's authority had declined, as had Indonesian enthusiasm for Confrontation. So as not to appear as though they were taking advantage of the situation, for a while the Security Forces in Borneo ceased deliberate assaults on enemy positions but continued reconnaissances and ambushes.

Although the activities of the regular Indonesian Army declined, the threat from the various para-military organisations set up by the Indonesians to undertake internal subversion against Malaysia increased, and the main SAS effort was directed towards establishing the location and strength of their bases. The para-military organisations were trained and led by members of the RPKAD, the Indonesian parachute commandos, a force somewhat akin to the SAS.

While B Squadron was continuing operations across the border, often in troop, and at times in squadron strength, it was clear that Confrontation was cooling down. The Security Forces, however, had to remain vigilant against any Indonesian incursions aimed at making contact with the Clandestine Communist Organisation. Hughes was immediately impressed with the efficiency of West Brigade, commanded by that grand infantryman, Brigadier Bill Cheyne. Tough, positive and with a finely developed understanding of SAS operations, Cheyne was delighted to learn that it was intended that 2 Squadron would be available to him for a period of six months.

But it did not take Hughes long to realise that the conditions likely to be endured by his squadron were less than satisfactory. The headquarters of B Squadron were located at 36 Foo Chow Road in Kuching, known as Pea Green House, and the remainder of the squadron was in another house plus several hotels in Kuching. In Hughes' view the houses were 'brothels of the highest order', they were filthy, and unless improvements were made he would not allow them to be occupied by his men.[9] Furthermore, 2 Squadron was larger than B Squadron and there would be insufficient accommodation anyway. Hughes arranged for four more houses to be rented, but in the interim some soldiers would have to live in tents at the training area at Matang, 18 kilometres north west of Kuching.

It would not be until the first week of April that 2 Squadron would be based in one location. Due to operational commitments the burden of cleaning and setting up the new complex would be left to G Troop whose good work enabled the squadron to occupy and operate from an establishment known as 'Earl's Court'. As an Australian correspondent wrote in the 22 SAS journal *Mars and Minerva*, it was 'named such by some Pommy bastard (Note—this is a fair dinkum Aussie word which, when translated, means Noble English Gentleman)'.[10]

But these developments were two months away. The main body of the squadron arrived at Kuching on 28 January and training began the next day. Apart from acclimatisation, the Australians had to become familiar with the British 128 radio sets, Claymores, the SAS explosive and medical kits, the Armalite, the use of the Mark II Sarbe equipment and the latest SAS/RAF abseiling techniques. Soon SAS patrols were operating in the surrounding area where a slight threat from the Clandestine Communist Organisation added a touch of realism to these training patrols and enabled the men to become used to the ways of the local population. Hughes was keen to keep his men on patrol for another reason—it took the strain off his accommodation problems. The Australians also had to take over responsibility for administering the Sarawak Border Scouts, and detachments of Australians assisted 22 SAS who provided officers and NCOs for the Border Scouts.

Pea Green House, the headquarters of 2 SAS Squadron in Kuching, the capital of Sarawak, during the first half of 1966.

The terrain in this area of Sarawak was different to that experienced by 1 Squadron in northern Borneo the previous year. Around Kuching there were paddy fields interspersed with patches of jungle and frequent villages, while towards the border, to the west and south, the country became higher, but did not reach the altitudes found in the north of the island. The border was generally along a jungle-covered ridgeline crossed by a few tracks traditionally used by native tribesmen who moved back and forwards between the two countries. These paths provided possible axes for Indonesian incursions. Across the border the jungle ridges led down to broad rivers and, in the north nearer to the South China Sea, large areas of swamp.

Kuching in 1966 still had the atmosphere of an old river city of the East—something out of the pages of Joseph Conrad or Somerset Maugham. Although there were 60 000 residents, mainly Chinese and Malay with a sprinkling of urbanised tribesmen, there were no nightclubs and people were in bed by 9 pm each night. The outstanding memories of the city were of the deep storm drains, the undoing of many gallant attempts to arrive home safely from a trip to town, sometimes with the British red caps in surveillance.

Soon after arriving in Kuching Hughes was informed that B Squadron had been ordered to carry out a major operation across the border. Brigadier Cheyne had learned that an Indonesian force was assembling at a place called Sentas, on the Sekayan River across the border from Tebedu, and he had tried to disrupt this force with an artillery bombardment and ambushes by the 1/7 Gurkhas. These efforts had proved ineffective due to the uncertainty of the exact location of the enemy camp and poor visibility which prevented accurate aerial spotting. Cheyne had therefore ordered Major Hardy to cross the border in strength to locate the camp. By this stage of their tour B Squadron was considerably depleted in numbers and five New Zealanders were added to his squadron for the operation. Despite the fact that his men had only just arrived in Kuching, Hughes seized the opportunity and gave Sergeant John Coleman and Corporals Frank Sykes and Jeff Ayles the chance of gaining operational experience with the British.

Coleman was one of the more experienced soldiers in the SAS. Aged 31, he had joined the British Army in 1953 at the age of seventeen and had served in the Royal Fusiliers in Egypt and Sudan. After discharge in 1956 he had served on the Reserve with the 2nd Parachute Regiment before being recalled for the Suez crisis. As well as the Suez operation his unit operated against the EOKA (Greek Cypriot) terrorists in Cyprus. Joining the Australian Army in 1958 and the SAS in 1959, he served with the Training Team in Vietnam in 1964 and 1965, and then reverted from warrant officer to sergeant to rejoin 2 Squadron for its Borneo tour.

Coleman recalled that with Sykes and Ayles he travelled from the training camp at Matang to Kuching to be briefed by Major Hardy. The Australians found it 'a bit of an eye-opener in as much as [the British] wore civilian clothes and had by our standards civilian haircuts. But there it finished, as they were thoroughly professional in a casual but efficient way'. Coleman was attached to the troop commanded by Sergeant Dick Cooper who went 'to his own private briefing place, which was much more convivial and told a "bloody" sight more of what to expect'.

On 30 January Major Hardy set off with a force totalling 53 men, plus a platoon of Argylls to guard the crossing of the Sekayan River, the suspected enemy camp being on the far bank. At last light on 3 February the force reached the river crossing point and Coleman was surprised to find that the British did not appear to send out security patrols to protect the crossing. Then, as Coleman wrote later, 'the bloke who was to swim the rope across the river produced a pair of swim fins (flippers) from his back-pack, which I reckoned was a smart move, but when he tied the rope around his waist and started to swim over-arm his feet (fins) breaking the surface, I wanted to depart for

GEN-LAYOUT OF AREA (NOT TO SCALE)

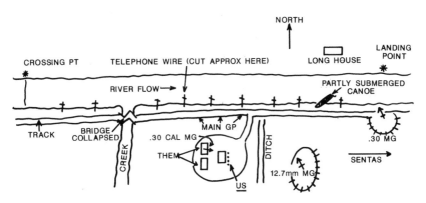

Sketch drawn by Sergeant John Coleman showing the general area of the attack by B Squadron, 22 SAS, commanded by Major Hardy, on the Indonesian camp near Sentas across the Sarawak border in February 1966.

distant parts. The noise was loud! But nobody seemed to mind very much. Also I noticed that the knot that was used on the safety rope wasn't one that could be tightened as successive blokes used the rope. Consequently the blokes who used it towards the end of our mob did more swimming than roping.'

Once across the river the party turned east along a track following the river bank. Although it rained for a while visibility was not bad. The intention was to attack the enemy camp at dawn, but soon after 10.30 pm they unexpectedly stumbled into the enemy position. Sergeant Coleman was with the leading troop which turned right away from the river and silently clambered up a steep track to a clearing with a number of huts. Moving into the clearing Coleman found himself with several other soldiers behind a hut that was open from about waist height. Suddenly the enemy opened fire and the SAS retaliated. The SAS fire seemed to strike a number of the enemy, but the Indonesians reacted fast and from 15 to 20 metres across the clearing began firing with a .30 calibre machine-gun. The SAS had to withdraw.

To cover the withdrawal one of the members of the patrol decided to throw a phosphorous grenade through the open hut towards the main enemy position, but unfortunately it struck the upright on the other side of the open hut and rebounded back into the hut. Coleman immediately realised that the men to his left could not have seen what had happened; he stood up to see where the grenade had landed and to yell a warning. In doing so he exposed his upper body and was struck

149

GENERAL PLAN OF GRENADE INCIDENT:

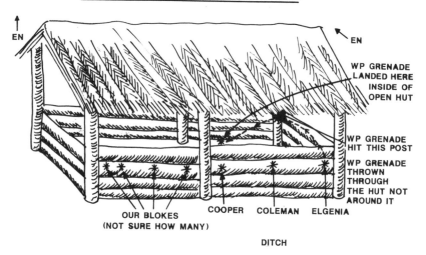

EN

EN

WP GRENADE LANDED HERE INSIDE OF OPEN HUT

WP GRENADE HIT THIS POST

WP GRENADE THROWN THROUGH THE HUT NOT AROUND IT

OUR BLOKES (NOT SURE HOW MANY)

COOPER   COLEMAN   ELGENIA

DITCH

Coleman's sketch showing where the phosphorous grenade was thrown during the attack on the Indonesian camp near Sentas in February 1966. When Coleman stood up to see where the grenade had landed and to warn the men to his left, he was struck on the upper body by the exploding phosphorous.

---

on the head, back and hands by the exploding phosphorous. 'So now', as he wrote later with masterly understatement, 'we were not in a good tactical position, with our hut (our only cover) burning and a few of the blokes on fire, me included'.

The SAS continued to exchange fire with the enemy until the fire slackened, but by now Coleman was 'well alight and hurting'. He told Cooper that he 'was going to put himself out' and ran to a ditch to their rear. Two other soldiers, Sergeant Lou Lumby and Trooper 'Ginge' Ferguson, followed him and tried to put out the blaze with water, but without success. Then, as Coleman recalled, one of the soldiers 'hooked his head up over the ditch and said, "They've gone!"'. Meaning our blokes. This wasn't the best news I'd ever received'. The SAS had indeed withdrawn, and while Hardy ordered artillery to be brought down on the camp the squadron quickly re-organised and discovered that four members were missing—three of them were reported to have been seen dashing into the river ablaze with phosphorous. But the withdrawal had to continue.

Meanwhile, Coleman, Lumby and Ferguson were trying to move along the ditch towards the river, but the undergrowth was too thick and they decided that they 'would do a frontal charge as we thought that they would not expect that. So up we went', wrote Coleman, 'over

150

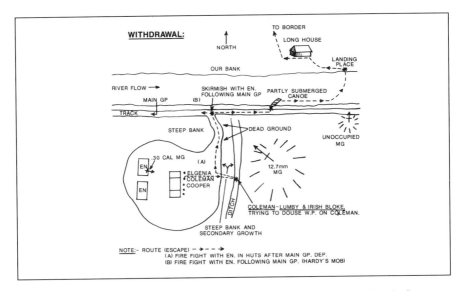

Coleman's sketch showing his withdrawal route after he and two British SAS soldiers, Lumby and Ferguson, became separated from the main body of the British SAS squadron.

the top and with me still on fire and tried to charge past our previous open hut that we'd used for cover, when from a knoll behind us a 12.7 calibre opened up'. They silenced some fire from the original enemy huts and then headed back down the track towards the river. They struck a party of enemy following the withdrawing SAS squadron and there was a 'bit of a skirmish'. With their withdrawal route blocked the three men turned right along the river bank.

The river bank dropped steeply about 4 metres to the water and they noticed a partly submerged two or three-man canoe. It was in a poor state of repair and the best they could do was to put their belts and weapons in and to swim and push it ahead of them. The current was quite strong and they drifted downstream away from the area where the 12.7 machine-gun was still firing. Then, before they reached the home bank, artillery rounds started to land on the enemy camp and also in the river. As they scrambled ashore one of the British soldiers said that he was 'shattered' and was going to sleep right there. Coleman explained to him in blunt language that he would be on his own. The other British soldier agreed with Coleman and they rapidly left that spot.

It was a nightmare journey for Coleman. They came across a longhouse which could have been a base for local scouts and then turned sharply towards the border. For a couple of hours they crawled on hands and knees along wild pig tracks and then rested before first light. Coleman's recall of these events was now 'a little bit hazy' as he 'was feeling pretty crook'. He can remember having his compass in his left hand and insisting on sticking to the bearing. But he was also aware that the other SAS men seemed to know the area well. During the morning they arrived at a Gurkha patrol camp and Coleman recalled that the first thing that the medic did was to offer him a cigarette. 'Before this, I had never smoked, but with the burns and such I sucked the bloody thing inside-out and from that day to this, I smoke.' As Dickens wrote, 'it was a good escape and evasion exercise, which proved once again that all those qualities and skills the SAS are at such pains to acquire are essential if their sort of adventurous operations are to be attempted'. Coleman agreed, adding that he 'was glad to have had SAS trained blokes with me in that situation'.[11]

In fact Coleman, Lumby and Ferguson had reached the border before the squadron, which had passed through the village of Sentas at first light and reached the border RV without further casualty.[12] The operation had been far from satisfactory, but had at least confirmed the enemy's presence in strength around Sentas. And the enemy had certainly suffered casualties, both from small arms and the accurate artillery fire.

While 2 Squadron's training continued at Matang, the headquarters staff were familiarising themselves with operations in the West Brigade area. On 8 February Graham Hoffman and his operations section took over the control of B Squadron operations at Pea Green House and with Hughes he visited the commanding officers of the battalions in the area. They soon had an opportunity to see West Brigade in action, for information was received that the Indonesian RPKAD supported by Chinese irregulars were planning to cross the border in the direction of Tebedu. The 2/7th Gurkhas had their first contact on 11 February. Then on 15 February the Indonesian party was spotted by a Border Scout and Brigadier Cheyne deployed nine infantry companies (from the Argylls, the 1/10th and 2/7th Gurkhas) to track them down. The previous day B Squadron had sent a troop across the border to assess the effect of the raid on Sentas. Sergeant Peter White had accompanied the patrol, which was forced to return when it was unable to cross the flooded river.

In an effort to trap the insurgents, on 19 February Cheyne ordered B Squadron to act as infantry and to take up positions across the enemy withdrawal routes. Despite the monsoon rains, Cheyne brilliantly deployed and redeployed his forces, countering each move of the raiders. The 2/7th Gurkhas had the most contacts, and of the original party of 27 raiders to cross the border only six returned to Sentas, the last three crossing the border on 17 July. The remainder were killed, captured, or died of wounds or hunger in the jungle.[13]

While the work of the Gurkhas was impressive, Hughes was surprised to find that B Squadron appeared to have little idea of how to operate as a normal infantry company as required during this final operation. He thought that B Squadron was tired, and he was a little disappointed by their performance. Hughes had attached five senior NCOs to B Squadron but he observed that their 'tasks should have been done by infantry so our chaps did not learn as much as expected'. By 24 February all appointments on headquarters SAS Kuching were held by Australians, and for a week, while Major Hardy was sick, Hughes commanded both squadrons. On 4 March he wrote to Lieutenant-Colonel Brathwaite: 'Other than RA Sigs I have no suggestions re training as we did not learn anything from B Sqn that wasn't known before. Fact—Not cockiness!!'[14] Clearly, by now 2 SAS Squadron felt that they were ready for their own operations.

10
The Sarawak
operations
2 Squadron: February–July 1966

At 8 am on 24 February 1966 Sergeant Barry Young moved out with his six-man patrol, plus an artillery Forward Observation party, to begin 2 SAS Squadron's operational patrols in Sarawak. Age 26, Young was one of a group of new sergeants who had no operational experience. He had joined the Army in 1958 and two years later, close to his 20th birthday, had transferred to the SAS. He had spent many long hours with patrol commanders from 1 Squadron questioning them about aspects of their tour in Borneo. Young's tasks were to establish one observation post and relieve another along the Sekayan River between Serankang and Segawang, 12 kilometres south west of Tebedu. At Serankang they sighted squads of up to fifteen Indonesian Border Terrorists being trained by members of the Indonesian Army, and in all, a total of 35 enemy were counted. Young's report included detailed sketches such as those of Serankang shown on pages 155 and 156.

This patrol was one of the first of a series along the Sekayan River which followed the border for some 50 kilometres opposite Tebedu. The same day as Young began his patrol Sergeant Cedric (Shorty) Turner led a six-man patrol 4000 metres to the area east of Serankang. Establishing an observation post on 'Enemy Hill' on 28 February they saw a patrol of six enemy, but had no further sightings until they withdrew on 10 March. Meanwhile, on 27 February Sergeant Ron Gilchrist led a six-man patrol to locate a suspected enemy camp on the Sarawak side of the Sekayan River near Kapala Pasang, 12 000 metres west of Serankang. They located the camp, consisting of five recently constructed buildings, but although some civilians were present there were no enemy. However, nearby they spotted a group of twenty persons, six wearing olive green uniforms. They concluded that the

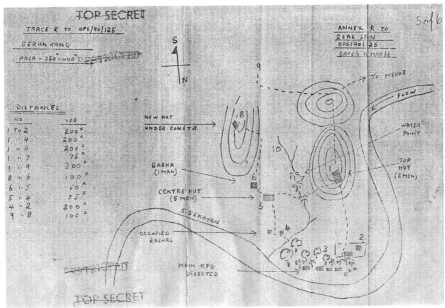

This sketch of Serankang, a few kilometres across the Indonesian border, was drawn by Sergeant Barry Young from an observation post located north of the village, across the Sekayan River. The sketch was one of four drawn to describe the area under observation. 'KPG' stands for Kampong, the Malayan word for village. A basha was a small 'lean to' or tent.

camp was not regularly occupied by the enemy but was visited at regular intervals from the south.

The reconnaissance patrols continued during March and one patrol was to have later repercussions. On 3 March Lieutenant Ken Hudson led a four-man reconnaissance patrol to the area of Kampong Entabang, 6000 metres west of Serankang. They came across fresh enemy footprints which they followed until they struck a flooded river. Hudson decided that it was too dangerous to cross, but the other three patrol members were annoyed, believing that they should have crossed. Hudson was a former soldier who had served in Malaya. His nickname of 'Rock' came not just from the movie star of the same name but from his personality—firm as a rock—inflexible. He was probably well justified in not crossing the river and Hughes advised him that as patrol commander he made the final decisions and should not be unduly influenced by anyone else.

On 17 March Sergeant Peter White led a patrol to the Sekayan and his second-in-command, Corporal Frank Sykes, vividly recalled the operation.

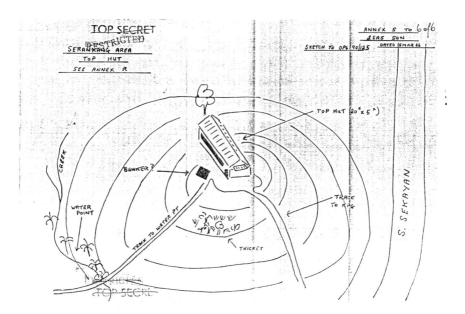

This sketch provides further detail to that given in the previous sketch and shows the 'top hut' located in the right centre of previous sketch.

Our task was to take a Major McCausland (2/7th Gurkhas) into Sentas for a recon so that he could plan a company sized operation on the same place. We had a platoon of his soldiers securing the crossing point on the Sungei Sekayan which was reassuring...

Our selected crossing point was just upstream of Sentas and the plan was to cross these shallows just prior to first light [on 20 March] and walk inland using a small feeder creek to hide our entry as the banks were fairly steep and difficult to negotiate without leaving sign when wet.

The patrol duly crossed the river and as they moved up the creek they heard Indonesian voices. They left the creek, but in the early afternoon struck another creek.

The approach into the creek was fairly steep and we virtually slid into it and the wide-eyed stares at each other told their own story...The Indons had cut out the growth in the creekline and we were virtually in a tunnel and it was obvious that they were using this as a trail. Multi-coloured communication lines were strung through the overgrowth and a branch line disappeared up the bank towards where we had previously heard noise. Intelligence had pointed to a mortar position in that general location and dixie noises confirmed the presence of something up there.

156

While we were digesting this revelation I had a sinking sensation that tactically this was not a good place to be and sure enough the sound of someone approaching from upstream was heard. The steep banks prevented quick, quiet movement so I passed the enemy signal and pressed into the limited growth on the bank.

A heavily laden figure appeared carrying a full basket with the aid of a headband and luckily this prevented him looking up and he kept going until he saw my feet. This caused him some concern which got worse when he raised his head and looked into the end of my AR15. After he stood up and regained a little composure the Gurkha major questioned him in his dialect and it appeared that he was being used as an unarmed forward scout for an Indonesian patrol following him. Various ideas were quickly bandied about and it was decided to let our captive go (hearts and minds) and head back to the crossing point. The unmistakeable slide marks on the bank would no doubt be noticed by the approaching Indons and no-one really felt like sticking a knife into the hapless local.

The move back to the crossing point was conducted with some pace...The move across open water seemed to take for ever and was fairly tiring in the process; no rest was forthcoming on the far side as we linked up with the Gurkha platoon and headed for the border. This was a smart move as it turned out, as a few minutes after leaving the crossing point the mortar rounds started impacting in it.

It was a frustrating time for the men of 2 Squadron, for while they were tasked to observe the enemy but to avoid contacts, in other parts of West Brigade the Security Forces were continuing offensive operations. For example, on 12 March a patrol from 42 Commando Royal Marines located four enemy positions at Sadjingan, about 3000 metres across the border from Kampong Biawak in the Lundu sector. Two Commando companies were flown in and in the early hours of 5 March they assaulted the position. The Indonesians suffered at least twenty casualties with the Commandos losing two killed and two wounded.

Meanwhile in the Bau sector the 1/10 Gurkhas were planning a battalion level operation across the border towards Kindau. The operation was pre-empted when D Company initiated an ambush against a substantial enemy force on 22 March. They inflicted heavy punishment on the Indonesians but lost four dead. As a result the battalion operation was curtailed and on 23 March the remainder of the battalion harassed Kindau and other nearby enemy camps with machine-guns and mortars as well as accurate artillery fire. The operation was, however, shortlived, for that evening all troops were ordered to return to base to allow for political discussions on ending Confrontation.[1] Eleven days earlier General Suharto had replaced Sukarno as the

2 SAS Squadron patrols were often flown by British Belvedere helicopters from Kuching to forward operational bases to begin their patrols across the border.

---

effective ruler of Indonesia. Although surveillance of enemy lines of communication was to be maintained, it was the end of offensive Claret operations during Confrontation.

But reconnaissance patrols were still vital, and on 17 March Hudson returned to the area that his patrol had investigated a week earlier. During 19 March they began to observe Entabang from across the Sekayan River and spotted two enemy moving towards the village. At about midday the next day a boat arrived at the village with three enemy, and a further eight men moved out to help with the unloading. Hudson believed that he had located an enemy base, and despite the fact that it was raining lightly, he decided that during the night they would cross the river, which was from 30 to 50 metres wide, for a closer reconnaissance. They therefore vacated the observation post and with rain starting to fall heavily, moved back several hundred metres to a point nominated as the emergency RV.

It was still raining heavily at 3 am the next morning when they left the RV in pitch darkness, travelling in single file with Hudson leading. Behind him, each man holding the belt of the man in front, were Privates Bob Moncrieff, Frank Ayling and Bruce Gabriel. In a little over half an hour they had reached a creek running into the main river and they began to move downstream. The original plan had been that when they reached the river they would stop and link themselves together by lengths of cord. But no halt was called. Even though it was too dark to see, Ayling knew when they reached the river by the increase in current strength. It was moving extremely fast. He asked Hudson whether he thought that they should reconsider his plans, or at least use the rope, but Hudson was determined to go ahead and told him not to use the rope. Ayling then commented to Moncrieff that this would be a 'hairy one'. Across the river they could see the fires burning in the enemy camp.

With the water tugging at about chest height they tried to wade across, but Hudson turned back towards the bank and moved downstream. He tried to cross again with the same result. By now the current was pulling strongly. Suddenly there was a sharp decline in the river bed. Ayling felt Moncrieff go under and he let go of him so that he could remain afloat. At the same time Gabriel felt Ayling kick as he started to swim, and he too let go.

Ayling was swept into the river and losing all sense of direction, called out, 'Which way?'. A voice like Hudson's replied, 'Over here'. Ayling, who had swum for the Army in inter-service competitions, moved in that direction but could locate no one. He then headed for the bank. He was about to release his equipment and weapon as he was near to drowning, when he hit a log. While recovering his breath he heard someone gasping and he called out, 'Over here Rock!'. He then reached over and managed to grab the person. It was Private Gabriel.

Bruce Gabriel's story was slightly different. After releasing Ayling he found that he was having difficulty keeping his head above water. He tried to jettison his belt and in the attempt he went under. Believing that he would not be able to surface without using both arms, he dropped his rifle. When he reached the surface he could hear someone calling, 'Over here Skipper! Look, here is my watch'. Gabriel could now see the luminous dial of a watch and made towards it. The next thing he knew was that Frank Ayling had grabbed him by the shoulder and pulled him onto a log. They clung to the log for several minutes gasping in air and listening for the other two men. After a period of time they scrambled up the embankment and continued to listen for the others. They were now about 500 metres downstream from where they had first attempted the crossing. In the darkness, above the sound of the falling rain, they could hear the water swirling downstream. Huddled together, shivering from the cold, they smoked a packet of cigarettes and waited about two and a half hours for daylight.

Morning revealed that the river was flowing even more swiftly, with trees and bushes being carried downstream. There was no sign of the missing men. Ayling and Gabriel therefore moved back to the observation post to observe Entabang. Everything appeared normal. By 7.15 am they were back at the emergency RV and for the next two and a half hours they tried unsuccessfully to establish communications. They then headed back to the border RV, arriving there at 5.30 pm on 22 March. Again they unsuccessfully attempted to establish communications. The next morning they had better luck with their radio and at 12.45 pm they were lifted by helicopter back to Kuching.

The missing men presented Major Hughes with a dilemma. Both men were strong swimmers, there was a possibility that they had been carried to the far shore and they might still be alive. But as it was still raining the river was impassable. It would be some days before patrols could reach the area. Nevertheless, on the afternoon of 23 March a patrol commanded by Corporal Jeff Ayles, and accompanied by Private Gabriel, was inserted to search for the missing men. The following evening they heard one shot followed 30 seconds later by automatic fire to the south of their location. Over the next few days they heard a few more shots but could draw no conclusions. The packs and rations left at the original RV were still intact. The river was still in flood, flowing at approximately ten knots and about 45 metres wide. It rained each day of the patrol until they withdrew on 3 April.

Meanwhile, on 24 March, Brigadier F. R. Evans, the Commander Australian Army Forces FARELF, had arrived for discussions. He departed for Singapore the next day and on 28 March a statement was released to the press that two members of the Security Forces were missing in the Bungo Range area on the Sarawak side of the border south of Bau. The plan was that Army Headquarters would inform the next of kin that the men were presumed dead by drowning while crossing a river. Details would not be released to the press, except to confirm that the men were Australians and to give their names. Hughes objected strongly to this aproach and signalled Evans that it was 'not charitable but cruel'. The unit was on active service, both men were volunteers and they understood the dangers. There was no proof that they had died of drowning and initially they should be declared missing on active service. In any case if the next of kin were informed that the men had drowned, eventually the press would find out. Hughes was concerned that unless it was acknowledged that the men had died on active service there might be repercussions on subsequent widows' pensions. However, Hughes was unsuccessful with his representations.

In view of the political implications of the incident the Minister for the Army waived a court of enquiry and death certificates were issued for each member on 1 April 1966. But the military authorities in

Borneo continued their investigations. Border Scouts in the area reported that the locals in the village of Entabang had heard no gossip about the deaths or capture of any white men, and they felt certain that the Indonesians would have boasted to the locals if they had killed any members of the Security Forces. It was never clear why Hudson had decided to cross without a safety line. Perhaps the memory of the earlier patrol when he had turned back at the flooded river pushed him on this time. But his training should have told him differently.

River crossings had been well practised in 2 Squadron. In August 1951, when he had been in Korea, Hughes had lost three soldiers while crossing the Imjin River through faulty launching of boats by the engineers, and as he said later, he 'had a bee in his bonnet about river crossings'. He had therefore arranged several crossings under flood conditions while training in Western Australia and had conducted three more crossing exercises in New Guinea. The squadron had conducted two crossings in February while training near Kuching, and there had been additional crossings ordered by Lieutenant Hudson to settle a troop argument about the relative merits of rope and floatation crossings. In his assessment of the accident Brigadier Cheyne wrote that Privates Ayling and Gabriel had 'behaved coolly and bravely. Gabriel almost certainly [owed] his life to Ayling'.[2]

During April and May 2 Squadron conducted reconnaissance patrols across the whole First Division area, with the average length patrols being 12.6 days. The four-man patrol commanded by Sergeant Barry Young to the Bemban track area from 9 to 18 April was an example. This was the area ambushed by Peter McDougall in June the previous year. Here the flat countryside was cut by wide meandering rivers and the occasional rocky limestone outcrop or Gunong. By 11 April the patrol was in the area of the Bemban track when they found a tree with spikes in it and Private Graham Estella was detailed to climb up to verify the position of a kampong from which childrens' laughter could be heard. From the tree he saw three men dressed in olive greens with weapons approaching down a dry creek bed about 40 metres away. They were accompanied by three or four small children from the kampong. In his haste to climb down Estella fell and fractured his wrist. The patrol moved out of the area and that evening they were extracted by helicopter.

The next day the patrol was re-inserted with a replacement for the injured man, and during the next few days they sighted groups of civilians but no enemy. Then at 6.45 pm on 14 April a shot was fired at them from their front by someone moving quickly through the jungle. The patrol returned fire with a burst of about five rounds, but there was no further action. Probably a local native had panicked on sighting the patrol. The patrol concluded that the Indonesian Army had not re-occupied the Bemban area. Young wrote later that it 'was gratifying...to employ close recce techniques around the kampong areas without being observed'. The 'only blemish' was firing the burst without seeing a target.[3]

Towards the end of April the 4th Battalion RAR replaced the 1/10th Gurkhas and assumed responsibility for the Bau sector and 2 Squadron had the task of maintaining surveillance of the front to prevent interference during the handover. On 10 April a patrol, under the command of Sergeant White, and including another sergeant and six other ranks, was inserted into the area across the border. An LUP was established and two three-man reconnaissance patrols were deployed. One approached Siding and the other sighted a prau moving down the Pawan river towards Siding. This patrol was detected on 15 April by one of the locals who warned two other groups of locals that the Australians were in the area.

Despite the fact that his presence had been compromised, White decided to remain in the area and he issued orders for two patrol tasks beginning on 16 April. However, that morning an enemy patrol estimated at between fifteen and twenty was heard approaching the LUP. Obviously the locals had warned the Indonesians. The Australians withdrew quietly in a north east direction, stopping to set up a booby trap constructed from a 36M grenade. After the booby trap had been set, the forward elements of the patrol moved off before the rear sentries had been withdrawn. They had travelled about 100 metres before noticing that four men were not with the patrol. The men could not be located and White decided to move on to meet them at the border RV. Soon after midday they heard the grenade explode and two hours later there was a single rifle shot 600 metres to the rear of the patrol.

The next morning there was more firing from the rear of the patrol and eventually spasmodic mortar fire began. The firing did not affect the patrol which reached the border LZ during the afternoon of 17 April and was airlifted to A Company 1/10 Gurkhas base at Gumbang. The remainder of the patrol also evaded the enemy, reached the border RV and was lifted to Kuching. The results were summarised in the patrol report by Major Hughes who wrote that 'Sgt White obeyed the rules by avoiding contact. It is still uncertain whether the enemy were visiting Siding or whether the base has now been re-established'. More patrols would be required. Corporal Bill Roods took a patrol back to the area between 27 April and 8 May but no enemy were seen.

Reconnaissance patrols were now deployed to the area across the border to the south of the Lundu sector. Between 9 and 22 May Sergeant Doug Tear saw civilian traffic moving along a track in the

The Australian Prime Minister, Harold Holt, visiting 2 SAS Squadron at Pea Green House, Kuching, on 30 April 1966. On the left is the Director of Borneo Operations, Major-General George Lea. Major Jim Hughes is with Holt, while the line-up consists of Captain Deane Hill (on the left), Lieutenant David Savage, Warrant Officer Danny Neville and Staff Sergeant Ivan Baldwin.

Bergjonkong–Bruang area and heard a mortar firing, but sighted no uniformed enemy. However, Sergeant Barry Young, whose four-man patrol infiltrated across the border on 16 May had more success. Struggling through swamp from 20 centimetres to chest high along the Poeteh River, they patrolled south for three days trying to locate the base-plate of the mortar they could hear firing. This was the area where Captain Robin Letts of D Squadron 22 SAS, later to join the Australian SAS, had been awarded a Military Cross for ambushing an Indonesian boat almost exactly a year before.[4]

Wading for days was a trying experience for Young and his men. Their mission was to establish if the enemy were using the river, but the problem was to find the river in the swamp. The normal night routine was to find a group of trees from which they could sling their hammocks. This involved climbing a tree, attaching one end, repeating the procedure at the other end, and then placing their pack and rifle in the hammock. Then in the dark they climbed the tree again, fell into their hammock and went through the nightly ritual of removing the

163

leeches. Twenty or thirty was not unusual. Before first light they were back in the water with their packs on; breakfast and weapon cleaning taking place standing in the swamp. On one occasion Young left his pack and rifle with the patrol and swam for 150 metres until he found slightly clearer moving water and thus located the river. They were now in the area where in September 1965 the 2/2 Goorkhas and A Squadron 22 SAS had ambushed the Poeteh. The A Squadron ambush had been swift and decisive but the Goorkhas had found themselves in a ferocious battle from which they were forced to withdraw after a number of heroic exploits.

On 21 May, on instructions from base, Young abandoned this area and patrolled north through more swamps parallel to but some distance from the Poeteh River towards Berjongkong. On the morning of 23 May Young's patrol skirted a patch of ladang and approached the Poeteh through thick secondary jungle. Establishing an LUP in the secondary jungle, at about 9.30 am he moved forward with Private Alan Easthope to observe the river. Young instructed Easthope, who was a squadron cook on his first patrol, to watch the approaches to the river while he rested his rifle against a tree and began to take photographs of the scene. Young was on the outer side of a bend in the river and was looking south down the muddy river when suddenly, around the bend from the north on the other side of the river came a canoe carrying five men. The man in the bow of the canoe yelled twice, and the remainder of the men reached for their weapons.

Young spun around and hesitated for some seconds, undecided. Realising that he could not get away, he picked up his Armalite and fired eight single shots from a range of about 25 metres. The first man fell out of the canoe on the left side, the second man was hit in the throat and stomach, and the third man was also hit. Meanwhile, Easthope had moved up to Young and he fired eight single shots from his SLR at the fourth and fifth men in the canoe. When the firing stopped three men were seen floating in the river face down, a fourth man was thrashing about, approximately two metres from the far bank, and the fifth man was not seen again.

The two Australians withdrew, joined up with the rest of the patrol and Young decided to move north rather than head directly east towards the border. Fifteen to twenty minutes later a mortar was heard firing to their rear. They continued trekking north for about 1500 metres and then turned due east heading towards the border. At about 3 pm they heard what sounded like enemy setting up a weapon on a spur running across their front and they had to change course to avoid the feature. Young wrote later that he considered that they were 'very vulnerable, it was impossible to move quietly due to the vegetation, my biggest worry was that we could be tracked or heard by the enemy who had free access to the ridge tracks; we of course were forced to stick to the gullies which were very thick'. At 6.30 pm they halted and established an LUP and half an hour later they heard a shot from about 800 metres to their east which was answered by another a similar distance to the west. Obviously the Indonesians were searching for them.

The LUP that night was particularly uncomfortable; wet and cold, they sat at the ready at the base of a large buttress tree waiting until 2 am to establish communications with their base. Before first light they moved off. Young decided that the enemy could have back-tracked his entry route so he avoided the border RV and continued to walk east, crossing the border at about 1 pm. That evening they established an LUP near Kampong Munti and the next day at about noon on 25 May they were lifted by helicopter to Lundu. Once again Sergeant Young had conducted a most thorough reconnaissance. As Major Hughes commented: 'The contact was not invited. Once the situation occurred the enemy were quickly despatched and the patrol safely withdrawn'.[5]

Young's patrol was one of the last to operate across the border. On 25 May eight Indonesian army officers flew to Kuala Lumpur to begin negotiations and on 28 May a message was received from SAS head-quarters in Labuan that Claret operations across the border were to cease at once. All patrols were ordered to return to base as soon as possible and all future Claret operations being planned or prepared were cancelled. Already 2 Squadron had reduced the number of patrols across the border and attention had been given to hearts and mind operations away from the main populated areas. These operations continued into June. On 15 June Lieutenant David Levenspiel arrived to replace Lieutenant Hudson as commander of E Troop.

Then, at the beginning of June, Brigadier Cheyne received information from captured documents that Indonesian Pasanda (undercover) forces would shortly be moving into Sarawak to gather intelligence and to make contact with subversive organisations in the Bau area. There were two Indonesian parties, Manjap 1 and Manjap 2, and 4 RAR, with its headquarters at Bau, and two companies of the neighbouring 2/7th Gurkha Rifles to the north, were deployed to cut off their return routes.[6] As he had during the Tebedu incursion in February, Brigadier Cheyne ordered the SAS to assist his infantry battalions in rounding up the insurgents. An example of these operations involved the surveillance of the southern portion of the Gunong Gading in the south of the Lundu sector between 12 and 25 June. Four patrols, commanded by Second Lieutenant David Savage and Sergeants Lawrie Fraser, Ray Scutts and Jim Stewart were each given areas of operations. The patrols moved into their respective areas by various routes and methods of entry—helicopter, road and foot—and

Clearing a hut in Sarawak in May 1966. Villagers had warned the patrol that two 'terrorists' had occupied the hut the previous night. The hut was unoccupied but the troops found food and supplies. In the foreground is Private Wayne DeMamiel.

they patrolled these areas looking for hides and any signs of enemy movement. Particular care was taken with the jungle fringe and the neighbouring kampongs in the hope that the enemy would give away their presence in an attempt to secure food. No signs of recent enemy activity were seen in the area, but although they did not realise it at the time, Sergeant Fraser's patrol flushed out six Indonesian Border Terrorists.

On 15 June C Company 4 RAR had a number of contacts with the Indonesian Manjap 2 party along the Raya Ridge on the Sarawak side of the border. Four enemy were killed and two wounded, while two Australians were wounded, one of whom later died. One of the platoon commanders involved, Lieutenant Rod Curtis, was awarded the Military Cross as a result of this incident; fourteen years later he was to command the SAS Regiment.[7] Meanwhile, Cheyne began to tighten the net around the enemy party. In the morning of 18 June three patrols under Sergeants John Jewell, Tom Hoolihan and Barry Young were ordered to cease their hearts and minds operations and to move to Bau where they were to come under the command of the Commanding Officer of 4 RAR, Lieutenant-Colonel David Thomson.[8] There they were joined by Sergeant Doug Tear's patrol. Tear was to command the troop consisting of the four patrols.

At 4 pm the troop was ordered to block a track along the top of Gunong Api at the eastern end of the Raya Ridge, 10 kilometres north west of Bau. They roped down from a Wessex helicopter and were in position by 6 pm. No sooner were they in position than Colonel Thomson ordered the troop to ambush a track 1000 metres west of Kampong Opar, south of the Api Ridge. If need be they might be ordered to come under the command of A Company. Getting off the ridge in the dark was a 'bit of a nightmare', recalled Young. Their route down was a dry creek bed, it was pitch black and they had to negotiate drops of 10 to 15 metres. Cursing and stumbling they reached flat terrain, and short of time they decided to walk along a well-defined track which, although it was dark, showed up quite well due to the white sandy soil.

At 1.30 am Private Wayne (Snow) DeMamiel was leading, followed by Sergeant Young, when they found what appeared to be Indonesian tracks. Moving forward for a reconnaissance they spotted a small hutchie and Young crept forward to examine it. Young can still remember the hair standing up on the back of his neck as they retraced their steps. While Young was explaining to Tear what he had found there were noises to their left and immediately the patrol opened fire. During a pause in the firing Young heard a language that was not Indonesian and called out, 'Who are you?' '2/7th Gurkhas,' was the reply. Two Gurkhas had been wounded, but no Australians. The Gurkhas had sited a machine-gun to fire down the track, but it had not fired. When Young questioned the Gurkha machine-gunner as to why he had not fired, he indicated that he was situated between two SAS

A public relations photograph of a 2 Squadron patrol in Sarawak in May 1966. In the foreground is Lance–Corporal Bob Mutch. Private Frank Ayling is waist-deep in the creek and Lance–Corporal Barry Zotti is coming down the opposite bank. Ayling was mentioned in despatches for rescuing Private Bruce Gabriel from drowning in March 1966.

M16s that had fired about a metre away from his immediate front. Rounds had passed either side of his face.

Later that morning Lieutenant-Colonel Thomson radioed Hughes at Kuching: 'SAS group ordered to move to a stopgap position at night. Gurkhas previous occupant should not have been within 3–4 thousand yards. Communications difficulty. Your chaps did everything right, no blame attached to them at all. Moved quickly and well. Unfortunate but nobody seriously wounded'. Hughes immediately went up in a helicopter with the commanding officer of the 2/7th Gurkhas who conceded that his men had indeed been in the wrong area.[9]

On 24 June a company of 2/7th Gurkhas under the command of 4 RAR contacted the remaining elements of the Indonesian Manjap 2 party which lost three killed and three wounded. Manjap 1 entered Sarawak after Manjap 2, but withdrew in the face of the Security Force activity.

Following the Bau incursion active patrolling was resumed along the Sarawak border but there was no further contact or indeed sign of the enemy. By now Confrontation was declining and a number of 2 Squadron patrols returned to hearts and minds operations. On 21 July the squadron was relieved by D Squadron 22 SAS and on 27 July they flew out of Kuching for Butterworth. They left Butterworth for Perth in RAAF Hercules aircraft on 1, 8 and 15 August. On 11 August the peace agreement officially ending Confrontation was signed by Indonesian and Malaysian representatives.

The operations in Sarawak provided valuable experience for the men of 2 Squadron. They had conducted effective reconnaissance patrols along and across the border in a difficult period while negotiations were underway to end the conflict. Patrolling in Sarawak was physically and mentally demanding and the old lessons of navigation and bushcraft were confirmed. While the Indonesians could react quickly and aggressively they seemed unwilling to take the fight to the Security Forces. The patrols across the border therefore enabled the Australians to gain operational experience in a hostile but somewhat less intense environment to that in Vietnam where many would shortly be deployed. More than that, the work of 2 Squadron confirmed the value of highly trained and hard-hitting long range reconnaissance troops in counter-insurgency operations. The first Director of Borneo Operations, General Sir Walter Walker, wrote in 1975 that Britain could not 'afford to forget the art of hitting an enemy hard by methods which neither escalate the war nor invite United Nations anti-colonialist intervention'.[10] The SAS in Sarawak showed that it could contribute substantially to such a capability.

11
Eyes and ears of the Task Force
3 Squadron: June–July 1966

The Borneo experience was of crucial importance to the development of the Australian SAS. Not only did the soldiers learn the techniques evolved by the British SAS over years of operations, but they also gained valuable operational experience of their own, honing their skills in jungle patrolling and surveillance. These skills were to be well tested against a more determined enemy in Vietnam.

The Borneo experience also indoctrinated the Australian SAS with the belief that their prime role was surveillance—that they were the eyes and ears of the main force. Few Australian officers, however, knew of the SAS's achievements in Borneo and many senior officers were sceptical of the 'super-soldier' tag, often applied in a derisory fashion to the SAS. The SAS had never exercised with Australian troops, other than as enemy or as an independent parachute company, and knowledge of their capabilities was not wide-spread. Thus when 3 SAS Squadron was deployed to Vietnam in June 1966 they found themselves cast in the role of pioneers. They were in a new theatre of war, against a new enemy, and working with troops and commanders who had little understanding of their capabilities.

Despite the fact that members of the SAS had been in the first group of the Training Team sent to Vietnam in 1962, the possibility of sending a complete SAS unit to Vietnam did not arise until the latter months of 1965. Soon after 1 RAR had arrived in Vietnam in mid 1965 the Chief of the General Staff, Lieutenant-General Sir John Wilton, had begun planning to build up the force to task force size, but it was to be some time before further battalions could be raised and trained. In the meantime, following the introduction of US Special Forces into Vietnam, the acting American commander in Vietnam

170

raised the possibility of an Australian SAS squadron being deployed to Vietnam. He was informed that such 'a request would not be welcomed as it could not be met in the light of [Australia's] other requirements'. At that time the decision to send 2 SAS Squadron to Borneo had not been made, but it was accepted that it would have priority over Vietnam.[1] Although Lieutenant-Colonel Brathwaite had been directed to have 2 Squadron ready for operations at the beginning of 1965, in fact it did not complete its training until the end of 1965, in time for its deployment to Borneo in January 1966. Similarly, although 3 Squadron was to be raised by December 1965, it would not be ready for operations until the end of June 1966.

Despite this tight schedule, Brathwaite could be given little information about the likelihood of the deployment of 3 Squadron, for while general planning to increase the Australian commitment began in Army Headquarters under strict security provisions in mid December 1965, detailed planning was not permitted until late February 1966. The final order of battle for the Task Force was only decided at a meeting on 8 March 1966, the day that the Prime Minister, Harold Holt, announced that Australia's commitment to Vietnam was to be increased from 1400 men to a self-contained task force of about 4500. In June of that year 1 RAR was to be replaced by a task force consisting of 5 RAR, 6 RAR, 3 SAS Squadron and various support units. No 9 Squadron RAAF was to provide helicopter support for the Task Force.

Two days after Holt's announcement, at a conference at Army Headquarters at Russell Offices in Canberra, Brathwaite learned that the stores and vehicles for 3 Squadron would be moved on HMAS *Sydney* departing from Sydney on 22 April and 25 May. The main body of the SAS would move in one plane directly from Perth as late as possible in June to allow the squadron to complete its training. The key question was whether 3 Squadron could be ready in time. Brathwaite replied that with minor adjustments the squadron could complete its training cycle by 8 June.

Nevertheless, Brathwaite was not completely confident that his men could be ready in time. While he was sure that the squadron would be prepared to carry out the sort of tasks, such as reconnaissance, surveillance and harassment, for which the SAS had been training since the formation of the regiment in September 1964, he was by no means certain that the squadron would be employed in this way. Indeed, on 4 April 1966 he wrote to the commander of Western Command, Brigadier Hunt, that he had 'gained the impression that it is possible that 3 SAS Sqn could be employed on prolonged co-ordinated infantry company/platoon type of operations on a task force front'.[2] If this proved to be the case the squadron would require further training once it reached Vietnam.

Confusion over the nature of operations to be expected in Vietnam extended well beyond the SAS. Until then 1 RAR, the only Australian combat unit in Vietnam, had operated as part of a United States formation based at Bien Hoa. But when the Task Force arrived in Vietnam it would take over responsibility for operations in Phuoc Tuy Province. Although the Task Force would be under the command of the new Commander, Australian Force Vietnam, Major-General Kenneth Mackay, operational control would be exercised by the American commanding general of II Field Force, and he would determine the general operational mission in Phuoc Tuy. The new Task Force commander, Brigadier O. D. Jackson, had been in Vietnam commanding the Training Team and the Australian Army Force, Vietnam for the past year, and had played little role in selecting the units of his force. Although he had asked Wilton for an SAS element he was not consulted about the order of battle.

It was not surprising that there was confusion about the role of SAS because the Australian forces as a whole had not been given a clear aim in Phuoc Tuy Province. It appears that it was decided to send an SAS squadron to Vietnam for at least three reasons. Firstly, the SAS had been highly successful in Borneo and it was assumed that it could achieve similar success in Vietnam. Secondly, it was impossible to make available any more than two battalions for Vietnam; the SAS squadron would therefore provide the Task Force commander with more flexibility, especially in protecting his operational base. Thirdly, since the two battalions would contain a large proportion of National Servicemen, the balance of regular and National Service combat soldiers would be redressed by the presence of the completely regular SAS squadron.

Without the experience of 1 Squadron in Borneo, 3 Squadron could never have been ready in time for service in Vietnam, and when 1 Squadron returned from Borneo in August 1965 about half of the squadron was posted immediately to 3 Squadron. The three troop commanders from Borneo, Tom Marshall, Peter Schuman and Trevor Roderick, joined 3 Squadron with Marshall becoming Operations Officer and the others commanding J and K Troops respectively. While returning from Borneo, Roderick had been attached to 1 RAR in Vietnam for six weeks to teach patrolling techniques.[3]

Slowly the other senior positions were filled. Captain Geoff Chipman, a 27 year old Portsea graduate with wide infantry experience, was the first officer posted to 3 Squadron, and he became second-in-command. Second Lieutenant Peter Ingram (aged 25), another

3 SAS Squadron officers, Vietnam 1966. *Rear, left to right:* Second Lieutenants Peter Schuman, Peter Ingram and Trevor Roderick. *Front:* Captain Geoff Chipman, Major John Murphy and Lieutenant Tom Marshall.

Portsea graduate who had been in the SAS since December 1964, took command of I Troop. The troop sergeants of I, J and K Troops were Tom Hogg, Jock Thorburn and Des Kennedy respectively. Thorburn and Kennedy had both served in Borneo, while Kennedy had also served in Korea and Malaya, where he had earned a Military Medal, and with the Training Team in Vietnam. Hogg had served in Malaya. It was planned that L Troop would be commanded by a Duntroon officer after graduation in December 1965, but as it turned out, L Troop was designated to remain at Swanbourne once the squadron deployed to Vietnam.

Early in 1966 Major John Murphy arrived to take command of the squadron. Aged 34, he had an ideal background for the task of establishing the role of the SAS in Vietnam. After graduating from Duntroon in 1953, he had served as an infantry officer in Korea following the armistice, and later in Japan. During the Malayan Emergency he was a staff officer on Headquarters Far East Land Forces where he gained an appreciation of SAS operations, and for over three years he served, mainly as a company commander, in 3 and 2 RAR at Enoggera. His commanding officer in 3 RAR had been

Colonel Jackson (commander of the 1st Task Force in 1966), and when Murphy joined 2 RAR he conducted the exercises to prepare 3 RAR for its deployment to Malaya in 1963. In August 1963 he had joined the Training Team in Vietnam, and from February to October 1964 had led the first Australian team to operate with the US Special Forces at Nha Trang. In 1965 he attended the Pakistan Staff College at Quetta and had to return home early when war broke out in August between Pakistan and India. He completed the year at the Australian Staff College at Queenscliff.

With only six months to prepare his squadron for war Murphy had no time to undertake the SAS cadre course, but fortunately some years earlier he had completed a parachute course. His service with the US Special Forces had given him a close knowledge of special operations as well as recent experience of conditions in Vietnam. He quickly stamped his personality on the squadron, not hesitating to replace a number of Borneo veterans who were not flexible enough to accept that they were preparing for a different sort of war. One of his sergeants, still serving as a major in 1988, described him as 'one of the most able commanders I have ever seen in my life'. Geoff Chipman believed the 3 Squadron was 'lucky to have had an officer of the calibre of John Murphy as the initial commander in South Vietnam'.

Murphy was ably supported by his Squadron Sergeant-Major, Allan (Mick) Wright, whom he described as a 'terrific soldier'. Wright had joined the army just before his fifteenth birthday and had served in New Guinea with the 2/3rd Independent Company and the 2/11th Battalion. Quiet but authoritative, in Vietnam Wright was to celebrate the anniversary of his 25th year in the Army. Murphy also described Sergeant Des Kennedy as a 'terrific soldier', adding that he was 'thorough to the extreme, devious in what he proposed to do to the enemy and an extraordinarily patient adviser and helper to all young soldiers (including the officers) and reinforcements. He, together with Mick Wright, were the father figures of the Squadron. No one really knew his age and he was quite coy about it'.

3 Squadron completed its final training for Vietnam on Exercise TRAIIM NAU in New Guinea from 27 April to 26 May 1966. The aim of the exercise was to deploy patrols through forward airfields by helicopter and light aircraft, to patrol and navigate through tropical jungle and mountain terrain, to practise communications and resupply, and to liaise with the indigenous people. The shoe-string nature of the exercise was shown by the fact that there was no RAAF support. The Ansett–MAL company provided one Caribou which proved to be unreliable due to servicability problems, but the DC 3s from Trans Australian Airlines (TAA—now Australian Airlines) and G 13 helicopters from Crowley Airlines were reliable, even though the latter had

Warrant Officer Mick Wright, MBE, Squadron Sergeant-Major of 3 Squadron in Vietnam in 1966 and Regimental Sergeant-Major of the SAS Regiment from 13 November 1967 to 8 September 1970. He had served with the 2/3rd Independent Company in the Second World War.

little lift capability and no winch. The squadron was quartered at Lae in the local showgrounds. Two months after the exercise, with the squadron already on operations in Vietnam, Major Murphy was able to reflect on the value of the exercise, noting that it 'had the added value of binding the Squadron into a cohesive force by developing closer ties and better understanding between all ranks and between sub-units. It would be safe to say that morale of the Squadron, on return from New Guinea, was as high as one could expect from a unit, yet to achieve victory in the field'.[4]

While New Guinea provided valuable training in a tropical environment, the conditions were considerably different from those 3 Squadron were to find in Vietnam. For more than five years the Australian SAS was to operate in Phuoc Tuy Province, some 70 kilometres south east of Saigon, and the men were to come to know its jungle and paddy fields as well as any training area in Australia. Forming a rough rectangle over 30 kilometres from north to south and over 60 kilometres from east to west, the province was bounded by the open beaches of the South China Sea to the south, while to the south west was the mouth of the broad Saigon River lined by countless creeks, tiny islands and stinking mangrove swamps, known as the Rung Sat.

Between the mangrove swamps to the west and the beaches to the east, both areas where the SAS was to operate on various occasions, stretched the peninsula terminating with Cap St Jaques and the port and resort city of Vung Tau. Vung Tau was not part of Phuoc Tuy Province, but became well known to the Australians. Here was located the 1st Australian Logistic Support Group (1 ALSG), and here also the Australians went for Rest and Convalescence (R and C). Reputedly the VC also used Vung Tau as a rest centre, and the city certainly suffered few VC attacks.

The capital of the province, Baria, sat at the junction of the roads leading south to Vung Tau, north west along Route 15 to Saigon, north along Route 2 past the refugee village of Binh Gia and the Courtenay rubber plantation to Long Khanh Province, and east along Route 23, through the rice-growing centres of Long Dien, Dat Do and Xuyen Moc to Binh Tuy Province. Except for the road from Xuyen Moc to the east, the roads were generally through flat, open paddy fields. The villages provided the VC with reinforcements and supplies, and they found refuge in large areas of jungle thoughout the province. In the west of the province were the Nui Thi Vai and Nui Dinh hills. From these mountains the jungle stretched north to the junction of Bien Hoa and Long Khanh Provinces—the Hat Dich (pronounced Hut Zic) area. In the north east of the province, overlapping both Long Khanh and Binh Tuy Provinces, were the May Tao mountains, and in the south, between Long Dien and the sea were the Long Hai hills. These three areas of mountains and jungle had been VC base areas for many years.

The eastern half of the province was bisected by the Song Rai (River) which flowed south from Long Khanh Province to empty into the South China Sea to the east of the Long Hais. Up to 15 metres wide in the centre of the province and over 100 metres wide near its mouth, the Song Rai passed through an extensive area of jungle stretching from the coast to the northern border. VC couriers travelling from the Long Hai Hills to the May Taos therefore had to traverse only a small area of paddy field around Dat Do before entering the cover of the jungle to move unseen to the May Taos to the north east. Similarly, couriers could walk from the May Taos through thick jungle along the northern border of the province, until they reached the Courtenay rubber plantation. Here they crossed Route 2 and were soon in the shelter of the Hat Dich.

The province therefore provided an ideal environment for guerilla operations, but it was also well suited to the sort of operations conducted by the SAS. Like the Viet Cong, the SAS welcomed the cover of the jungle, and except for the three areas of hills, the going was far easier than in Borneo. Furthermore, the distances were shorter. It would no longer be necessary to march for a week to reach the enemy's territory. Centrally based helicopters could cover most areas of the province in fifteen minutes. The jungle still contained its share of thorny bamboo, dense undergrowth, poisonous snakes, malarial mosquitoes, monkeys, deer and even the odd tiger, but it held few terrors for men who had served in Borneo and had trained in New Guinea and the Philippines. Paludrine, taken every day without fail, would generally keep malaria at bay, even if fevers of unknown origin struck irregularly.

The seasons of the monsoon provided different challenges. The dry season from October to May was slightly cooler, but nevertheless still warm, and patrols operating in more open terrain found that their biggest problem was a shortage of water. They would be required to carry eight or even ten water bottles. Dry leaves in the patches of bamboo and along the edge of the jungle made it difficult to move quietly. The paddy fields were dry, yellow and baked hard as concrete, and often a smoke haze hung in the air as the farmers burnt off their fields in preparation for planting in the wet season. Once the 'wet' or monsoon season arrived the SAS patrols had to worry less about water and noise from dry leaves, but the climate became more oppressive. High temperatures and humidity quickly sapped all energy and contributed to heat rashes and maddening tinea. The men of the SAS were particularly susceptible as they did not remove their boots during patrols and were forced to lie motionless, often in water, for long periods. Drenched by the torrential rains, which often fell late in the afternoon, they spent their nights cold and shivering without the benefit of the shelters usually erected by the infantry. Not surprisingly, many men came down with chest and bronchial infections. But these conditions provided the cover for their operations.

Despite the dangers from the terrain, the vegetation and the climate, the most formidable threat came from the enemy. In December 1964 the 9th VC Division had destroyed two battalions of the Army of the Republic of Vietnam (ARVN) near Binh Gia, and by early 1965 the VC had consolidated their strength in Phuoc Tuy. A year later the 5th VC Division was established in the May Tao mountains and seven VC battalions were operating in the province. The VC main force units had developed a series of bases in the jungle and their political cadres had control of most of the villages. As Frank Frost wrote: 'By the time the Task Force arrived in May 1966, the NLF [National Liberation Front] had effectively won the war in Phuoc Tuy'.[5]

The Australian plan was to establish a base at Nui Dat, a small, steep-sided hill 5 kilometres north east of Baria on Route 2, and gradually to assert control by patrolling and dominating the surounding area. In February 1966 the 173rd US Airborne Brigade, with 1 RAR, entered the province and in a series of operations cleared

Nui Dat hill, Phuoc Tuy Province, South Vietnam, looking south south west. On the right is provincial route 2, heading north (right) towards Long Khanh province. From 1967 onwards the SAS squadrons were located on Nui Dat hill, while the remainder of the Task Force was on the flat ground, generally in rubber plantations, to the left and bottom of the photograph.

the area around Nui Dat, including clearing and resettling the VC villages of Long Tan and Long Phuoc. On 24 May 5 RAR began Operation HARDIHOOD when it joined 173rd Brigade to again clear the Nui Dat area. After a number of contacts the area was secured and on 2 June the battalion began to move into its defensive position at Nui Dat.

On 5 June the headquarters of the 1st Australian Task Force (1 ATF) arrived, and three days later the American brigade departed. By mid June 5 RAR had been joined by 6 RAR, a squadron of armoured personnel carriers (APCs), an artillery regiment, an engineer squadron and arriving last, 3 SAS Squadron. No 9 Squadron RAAF, with their UH 1B Iroquois helicopters, was located at Vung Tau but had the task of supporting 1 ATF at Nui Dat. As Lex McAulay wrote: 'The physical conditions in Phuoc Tuy in mid-1966 were appalling. It was the wet season, and rainwater actually poured over the surface of the ground, unable to soak in. An entire military camp had to be built in the centre of an enemy-dominated region. Any construction activity became a mud-wallow'.[6]

It was fortunate that 3 Squadron's morale was high because they might well have been disillusioned by the circumstances of their deployment to Vietnam. The first to arrive was the advance party consisting of Second Lieutenant Peter Schuman, the Squadron Quartermaster Sergeant, Jake Mooney, and three storeman/drivers. Schuman recalled that their problems began when they arrived at Vung Tau. 'It seemed like the whole ship's cargo had been thrown onto the beach. There were stores scattered everywhere, bogged vehicles and little organisation...after a few days work we headed north. On arrival at Nui Dat I was met by the TF [Task Force] engineer who was responsible for the defensive layout and given a piece of the perimeter—about 200 yards [183 metres] long and 400 yards [366 metres] in depth. No one had ever taught me how to defend an area like this with five people.'

The main body of the squadron flew from Pearce to Saigon on a chartered Qantas Boeing 707 on 15 June. Meanwhile Major Murphy and a small group who had taken pre-embarkation leave in the Eastern States left from the RAAF base at Richmond. Murphy recalled the final words from Lieutenant-Colonel Colin East, GSO1 in the Directorate of Military Operations and Plans at Army Headquarters, who saw him off: 'We do not know what you are going to do but we do know that you are not going to be the Palace guard'. It was not certain whether the squadron was to be located with 1 ALSG at Vung Tau or with the Task Force at Nui Dat.

Murphy was clear in his own mind that his major role would be surveillance and he was determined that he would not allow the squadron to be broken up and allocated to the infantry battalions. However, his lack of certainty about what to expect was shown by the 'just in case' items he ordered to be taken. As one SAS officer described it: 'Officers and senior NCOs had been advised to pack their mess dress, some of which had been specially tailored for the occasion and as we were to live under ponchos in a humid environment it was interesting, later on, to view the effects that scarlet trouser stripes had on white linen jackets. The results of their enforced hibernation together were nothing short of startling'.[7] Patrol commanders were also issued with a whistle, an ancient map case and a bicycle lamp, none of which were ever used.

Between 1 pm and 2.45 pm on 16 June the squadron arrived at Vung Tau, and while the soldiers prepared for a rainy night in the sand dunes at 1 ALSG, Murphy went forward to the Task Force base at Nui Dat to meet Brigadier Jackson. At 12.45 pm the next day the remainder of the squadron arrived and in the middle of a tropical downpour were allocated positions on the Nui Dat perimeter. The country was experiencing its worst monsoon for several years and at times between 5 and 8 cm of water covered the ground. The soldiers had only their individual shelters and their weapons pits were soon filled with water, remaining that way until the end of the 'wet'. Indeed if a pit was in a depression, both the pit and the depressions would be filled with water and the pit could not be located except by falling into it.

The squadron had only a few day to acclimatise, and on 20, 21 and 22 June each troop conducted a 24 hour patrol for several kilometres out from Nui Dat. No enemy were sighted. However, on the night of 25 June the squadron received machine-gun and small arms fire on its left flank from the direction of the engineers. The squadron standing patrol reported four or five VC in the paddy field, but they disappeared before effective fire could be directed at them. Next day Murphy declared that the squadron was ready to conduct SAS operations.

During the previous days Murphy had discussed with Brigadier Jackson his policy for the employment of the SAS. Jackson wrote later that 'in the Vietnam setting good intelligence was very scarce indeed. I was more than a little interested to know the whereabouts, movements and habits of the two mainforce VC regiments and one of the local battalions in the Province and nearby areas. I thought the SAS quite invaluable in gaining this sort of intelligence'. He added that it was very tempting to use the SAS in an 'offensive and harassing role but their recce work was far too valuable'. It was therefore with considerable relief that Murphy reported to Brathwaite on 22 June that the squadron was to be 'the eyes and ears of the Task Force', and that this appeared 'to be a major breakthrough'.[8]

Although a signal had been received from Army Headquarters that the SAS was not to patrol in strengths of less than ten men, Murphy argued that 'the Americans are trying it in patrol strengths of 5 and though they are by no means as well trained as ourselves, they are having a fair bit of success'. Jackson relented and permitted 3 Squadron to operate as it had been trained on a trial basis,[9] but Murphy commented later that he thought that the Deputy Director of Military Operations 'was personally opposed to the Special Forces ethos, as were many other senior officers at that time. Our subsequent success caused a change of heart, and he was quite effusive during his visit to the Squadron' later in the year. Murphy also resisted the efforts of the commanding officer of 5 RAR, Lieutenant-Colonel Warr, to have SAS patrols placed under his command.

Murphy continued his report to Brathwaite:

The equipment side of things has stuffed up no end. Our current list of deficiencies includes all accommodation stores, AN PRC 64, URC 10 [radios], cable for generators, exhausts for generators and of all things no OTLP [One Time Letter Pad codes]...I have refused to go on operations until I get it and the Brigadier has agreed. This applies to the beacons also which should not be too hard to get. All USAF [US Air Force] pilots here have one on personal issue. The M16 [Armalite carbine] and M79 [grenade launcher] are not available and I am not sure yet when we will get them either. 5 and 6 RAR are sharing what 1 RAR left behind. These are fairly clapped, US forces are short, so it looks like they will have to come from the States.

At present we are holding a part of the TF perimeter. Are playing infantry soldiers and are digging in madly, wiring and filling sand bags. The main trouble at present is blisters. They are not used to digging. It is also amazing the number that do not know how to fill a sand bag. I have had the SSM give some special instruction on this subject so it's on the improve...I have already spoken to the Commander about moving back to Vung Tau but he will not wear it as yet. He has a lot of other problems at present and it is not politic to push it. When he has properly secured his area and we have completed our tasks close in, it will be worth raising again. The eventual aim is to get further out, but we are limited by TAOR [Tactical Area of Responsibility—out to mortar range] just now. There are lots of things around here which are well worth looking at and the areas are reserved strictly for us. As they are close in we can extricate without too much trouble and it will help the boys to run into form. The standard of training sticks out like the proverbial in just routine stuff around here, and the Commander is indeed pleased to have a bunch of pure pros.

The Commander's plan for this place is to establish a secure base and to work out from there. This area has been owned by the VC for 12 years and they were only pushed out by 173rd Brigade on Operation Hardihood. The Commander intends being slow but sure. We had a provost [military policeman] shot, in the village south of us, supposedly friendly, at six last night. 6 RAR on their first major operation are clearing [Long Phuoc] to our south, about 2000 metres, what is described as the most sophisticated tunnel system yet found in SVN...Lost another APC on the MSR [Main Supply Route] yesterday though no casualties. That is the second in a week. Of the six KIA [killed in action] so far only two can be put down to VC. The people are having a little teething trouble but I expect they will settle down...We were treated to five accidental discharges the day we arrived. Three by the engineers to our rear and two by the Defence and Employment Platoon which was securing the area as we moved in. So far we have, as is to be expected, none. The announced policy on this is 28 days pay first offence.

We have, as can be believed, been very busy with the settling down process and the preparation of this position. We are pretty crowded as all the real estate went to the first here. This was not surprising. The Commander also expressed great surprise that units were told in Australia that they would have a month to settle down. His policy is right in. This has applied to 5 and 6 RAR and to an extent ourselves. Thank the lord for TPNG. Other units who were supposed to collect equipment in theatre have had the same trouble as ourselves. This was predictable. AHQ might well be a million miles away'.[10]

Faced with a shortage of equipment, early in July Murphy flew north to Nha Trang to visit the headquarters of the 5th Special Forces Group where he had served in 1964. While there he arranged to acquire batteries for the AN PRC 64 radio sets, dehydrated ration packs, US pattern jungle boots, and from the US Air Force, three AN PRC 47 high frequency radio transceivers. The first Armalite rifles arrived in the squadron on 4 July.

Meanwhile the squadron had begun its SAS patrols. Brigadier Jackson's plan was that initially his battalions would clear a TAOR out to Line Alpha around Nui Dat. Line Alpha was at approximate mortar range stretching 5000 metres to the west and 8000 metres to the east of the Task Force base. With just two battalions, Jackson could deploy only one battalion on operations while the other defended the base. Thus at the beginning of July, 6 RAR began clearing the tunnels of Long Phuoc and 5 RAR was restricted to TAOR patrols. The task for the SAS was to patrol beyond Line Alpha, and the first target was the grey-green hills of the Nui Dinhs 7000 metres to the south west of the Task Force base. These jungle-covered hills not only provided a secure base for the local VC, but from the peaks of Nui Dinh and Nui Thi they could observe Route 15 running north west from Baria towards Saigon as well as Route 2 and the activities of the Task Force. The problem was how to get the SAS into the area. At that time No 9 Squadron RAAF was still operating under essentially peacetime regulations and the Air Force was loath to move SAS patrols into insecure areas. The result was that the early SAS patrols were deployed by foot, and to reach the Nui Dinh hills they had to traverse 7 kilometres of open paddy, scrub and scattered jungle.

On 30 June patrols led by Sergeants Alan (Curly) Kirwan, Jack Wigg, Tony (Pancho) Tonna, Max Aitken and Peter Healy found their way through the Task Force perimeter wire and set out from Nui Dat. The first sighting was by Kirwan's patrol at 6.30 pm, just on last light, when they saw one VC disappearing into the undergrowth. Establishing an LUP the Australians moved off again soon after 3 am. By 5.25 am they were 2000 metres from the Nui Dinh foothills due west of Nui Dat, and had located an attap hut with two occupants who they observed for the next few hours.

Meanwhile, at about 8.30 am that morning three VC fired on Wigg's patrol about 1000 metres north east of Kirwan. Again, about two hours later they were contacted by four or five VC and they therefore arranged to be extracted by helicopter. There were no friendly or enemy casualties and the enemy had not pressed their attack, apparently content to drive the Australians away from the Nui Dinhs. This contact had been observed by Tonna's patrol through binoculars across a paddy field.

Kirwan's patrol had continued to observe the hut, but at about 10.45 am they began to move out of their position and were seen by two suspected VC, who started to run off. They were ordered to halt several times in Vietnamese, and when they did not stop both were killed. A little later Tonna's patrol contacted four VC and Corporal Norm Bain killed one. The patrol walked back to Nui Dat, arriving at 4.15 pm.

Aitken's patrol to the steep rocky outcrop of Nui Nghe, 6000 metres north west of Nui Dat, saw no enemy. But Healy, who had taken his patrol between Nui Nghe and Nui Dinh, found various enemy camps, and at 4.15 pm on 3 July saw six VC in dark khaki uniforms. A single shot was fired at the patrol at 4.35 pm and a little later they were extracted. Healy reported that the VC appeared to feel safe in this area and he thought that there was a large enemy camp nearby.

These successful early patrols, especially Pancho Tonna's contact, raised the morale of the squadron even further. Later squadrons, however, would have been surprised at the primitive equipment of the early patrols. The main weapons were the SLR and the F1 carbine, which was soon replaced by the Armalite. Heavy barrelled SLRs, capable of firing on automatic, were also carried. But there were no M79 grenade launchers and patrols did not use Claymore mines, even in ambushes. The soldiers wore greens rather than camouflaged uniforms. Communications were a major problem, with early patrols using the A510 radio that came in two separate parts and relied on a different crystal for each frequency. Indeed the SAS patrols in Borneo had been better equipped than 3 Squadron in Vietnam.

The reconnaissance patrols continued during the month, both to the west and east of the Task Force base. One outstanding patrol was that led by Sergeant Tony Nolan to the area of the Long Tan village 5000 metres to the west. Civilians were moving around the village, making it difficult for him to carry out surveillance, but he left his six man patrol in an LUP, moved forward himself and climbed a tree. After observing various civilians for two days at 6 pm on 9 July he observed one armed VC dressed in black. The next morning, lying five metres from a track, he saw four armed VC, and a little later two more armed men and three women carrying baskets and parcels. At about 10.15 a man and a woman were spotted and just after midday Murphy ordered the patrol to return.

Nolan's conclusions were quite significant. He thought that the enemy looked like Main Force VC. They showed no sign of having walked a long distance and he thought that their base camp was close by. It looked as though they were caching their weapons and uniforms on the outskirts of the village and then mingling with the locals in coolie dress. Nolan's comments on the conduct of his patrol were also important. 'For patrols of this nature, six men proved quite adequate to do the task, and quite often fewer men would have been easier to move, control and conceal. In the soft muddy ground and tall grass where it is unavoidable to leave some sign of the patrol's route, the fewer men there are, the less signs are left.'

Murphy had been concerned that because he had to defend part of the Task Force perimeter he was restricted in the number of patrols he could deploy. But on 10 July Brigadier Jackson gave permission to move the squadron to a large area near to Task Force headquarters where they would have a minimum of local defence commitment. Working parties started clearing the area the next morning. Also during that morning the Commander, Australian Force Vietnam, Major-General Kenneth Mackay, visited the squadron to discuss future SAS operations. Murphy had prepared a written brief for Mackay in which he explained the squadron's capabilities. The brief added that the average age of the squadron was 24 years, the average length of service was five years and three months, and 65 per cent of the members of the squadron had had previous operational experience.[11]

During the discussion Mackay showed Murphy a personal letter he had received from General Westmoreland, the Commander of the US Military Assistance Command, Vietnam (MACV), in which he stated that the Americans wished to benefit from SAS expertise and had created a vacancy on their staff for an Australian major to advise them on Australian techniques. To get the task under way Mackay asked Jackson to release Murphy for a couple of weeks to go to Saigon but the Task Force commander refused, and Murphy exceeded his authority by suggesting that either Captains Mike Jeffery or Ian Gollings be sent from the SAS Regiment at Swanbourne to fill the position. As it turned out, neither officer was sent, but Captain Peter McDougall, who had been operations officer of 1 SAS Squadron in Borneo, arrived in September as the SAS liaison officer on HQ MACV. Murphy was also asked to visit Nha Trang to brief Westmoreland.[12]

On 13 July the deputy commander of Project Delta, 5th Special Forces Group, visited the squadron. Under Project Delta, US Special Forces trained South Vietnamese soldiers for clandestine operations into North Vietnam and Cambodia. It was hoped that the SAS would be able to provide instructors for the training camp at Nha Trang, but these could not be spared. However, Murphy proposed an exchange programme by which four to six Australian senior NCOs and soldiers would join the Project Delta team for a period of four to six weeks while US personnel joined the Australian SAS for a similar period. This programme was approved with the first exchange due to begin on 15 August.

Meanwhile, two patrols, commanded by Second Lieutenant Schuman and Sergeant Healy, were given the task of surveillance to the east of Nui Dat. Five thousand metres to the east of Nui Dat was a steep jungle-clad hill, usually described as Nui Dat Two. At 126 metres high, it was slightly higher than Nui Dat and was located 2000 metres north of the abandoned village of Long Tan. On the afternoon of 14 July the two SAS patrols moved out with C Company 6 RAR, which set up an ambush that night near the Suoi de Bang creek, 1000 metres short of Nui Dat Two. Leaving C Company at this point, the two SAS patrols followed the Suoi de Bang north looking for a crossing point, which they found north of Nui Dat Two. Next morning they were spotted by an unarmed man and Schuman believed that they were being followed. Then, just before 10 am, three artillery rounds fell near the patrol and Schuman radioed squadron headquarters to request a check-fire. He was told that no friendly artillery was firing, but as he said later, 'Coordination of fire support was a bit steam driven in the early days'. It was not until 4.15 that afternoon that Schuman believed that they had shaken off their VC shadows.

On the morning of 16 July the two patrols separated and Schuman and his four-man patrol headed south to his patrol area. At 5 pm, 2000 metres north east of Long Tan, they discovered an old enemy camp with the recent boot print of one man heading south west. Next morning they went through their normal drill of 'stand to' before first light, moved some 1000 metres, went into all round defence by sitting back-to-back, and brewed up. After breakfast they headed south. It was a stinking hot, humid day and the patrol SOP was ten minutes movement, stop and listen for three to five minutes, ten minutes movement and finally a five to ten minute break in all-round defence every hour. The vegetation was a mixture of bamboo, head-high scrub and tall timber.

The first indication that something foreign was in the area was a high influx of flies. Often this was from a dead animal, but it could also be from a VC camp latrine. Fully alert, they moved quietly into a large thicket of bamboo. Suddenly, at about 10.40 am they heard a female voice. It was 30 metres to the south. Private Sam Wilson, the medic/scout, and Schuman began to crawl forward leaving Sergeant

Jock Thorburn and Corporal Lyn (Bubbles) Murton, the radio operator, as cover. About 15 metres from the voice they could make out some huts and some movement. Wilson was looking towards his front when Schuman glanced to his right and saw a VC sentry taking careful aim at Wilson. Before the sentry could fire Schuman dropped him with one round from his SLR. The shoot-and-scoot drill then came into play as Wilson and Schuman retreated to their cover. Fire was coming from the camp, there was a lot of confused yelling, but the enemy did not appear to be following up; unable to locate the Australians their fire was high and covering a large arc.

Schuman could never really explain exactly why he then acted as he did. He knew that he had the upper hand, his patrol was organised, he had not taken any casualties and the enemy were disorganised. Quickly he gave his orders: 'Stand up, extended line on me, advance!' The attack went in at a walking pace, firing from the shoulder and waist. Fleeting targets, some naked, were running all over the place. Schuman wrote later that 'Sam Wilson was on my right during the assault. I was yelling orders like "keep up on the right flank" possibly to boost my and the patrol's morale. After the fight Sam said that when he heard me, he looked around and thought to himself "shit, I am the right flank".'

They reached the hut area and were still receiving fire from the far side of the camp so a few grenades were sent over to give the previous landlord his final eviction notice. At this stage the enemy broke contact and the Australians secured the camp. They ratted through the huts, stuffing their packs with a captured National transistor radio and as many documents as they could carry, before setting fire to the camp. The documents later showed that there had been ten enemy in the camp.

Low on ammunition and burdened by equipment and documents Schuman decided to call for an extraction helicopter. They therefore moved about 600 metres towards a large paddy field and found a hide where Schuman could prepare his coded message. The first part of the message was soon sent to squadron headquarters. Schuman was facing Murton checking the encode while the radio operator was sending his morse when the patrol commander glanced up and saw Jock Thorburn levelling his rifle towards his head. For an instant he wondered why Thorburn was going to shoot him. Thorburn fired and the blast was so close that it knocked Schuman's bush hat off. He spun around and saw the first of two enemy go down. The second enemy threw his white hat to the left and dived to the right behind a bush where he was hit by SLR fire. He got up and ran back down the track. Twice more he was hit and knocked off his feet, and then finally he staggered into the undergrowth. By this time the extracting helicopter was clattering over head. They quickly stripped the dead VC of his weapon, a Soviet 7.62 carbine, his belt and a fresh bread roll, and boarded the chopper. As Schuman put it: 'Thus ended a day's work'.[13]

Up to now the SAS patrols had been limited by the fact that they had to move by foot from Nui Dat to their operational areas, and Murphy continued to seek alternative methods for infiltration. In mid July, however, intelligence estimated that an enemy battalion and possibly part of a North Vietnamese regiment were located in the valley between the two hill masses, Nui Dinh and Nui Thi Vai, and 6 RAR was deployed in APCs along Route 15, around the southern end of the Nui Dinhs to sweep the area. Murphy took the opportunity of sending four patrols with the idea that the APCs would move slowly through the scrub with their ramps down, and when they were hidden by the undergrowth the SAS patrols would jump off the back of the moving APCs. 6 RAR moved out on the morning of 16 July and that afternoon Sergeant Healy and his six-man patrol was inserted north east of the battalion base at LZ Horse. On 17 July they located enemy footprints and saw one VC but the unscheduled movement of 6 RAR patrols in front of the SAS patrols and lack of communications with the infantry caused three of the SAS patrols to be withdrawn.

The next morning Murphy arrived to rebrief his patrols, informing them that their task was to destroy isolated observation posts or radio stations in the area. The SAS patrols were also now issued with AN PRC 25 radio sets instead of their A510s. That afternoon Healy's patrol was re-inserted and they moved through thick undergrowth and over fallen trees to reach a spur leading up into the Nui Dinh hills. There were no prominent tracks but the area was crossed by old overgrown timber-cutters' tracks. At 11.45 am on 19 July they located an enemy camp of about platoon strength; the lack of security around the camp and the laughter and loud voices indicating that the VC had enjoyed an unmolested life for some time. It was a classic case for calling in an airstrike and after a careful reconnaissance Healy withdrew 300 metres to radio squadron headquarters. The arrangements were made and Healy was ordered to move a safety distance of 1000 metres from the enemy position.

By 6 pm the airstrike had ended and when Healy moved in to assess the damage he discovered that all the bombs had landed 200 metres west of the target. At 6.30 pm the patrol re-entered the camp and as they crept forward they saw two VC 30 metres away in a tent. One was cooking and the other was lying in a hammock. The patrol opened fire killing both. A nearby tent was also engaged by two members of the patrol and screams of pain immediately came from this tent. The patrol then withdrew, but could still hear the screams up to 400 metres away. However, the patrol moved away quickly as they suspected that

Major John Murphy, Officer Commanding 3 SAS Squadron, briefing a patrol in
Phuoc Tuy in 1966.

they were being pursued by a party of VC. At 8.30 am the next
morning the patrol was extracted by helicopter. A further airstrike was
then made on the enemy camp, but the episode showed that better
means had to be found to enable SAS patrols to direct airstrikes onto
enemy positions. The arrival of ten AN PRC 64 radio sets in the
squadron was expected to solve many of the communication problems
that had arisen when the patrols had operated with 6 RAR.

On 24 July Major Murphy visited Nha Trang to brief senior US
officers on the organisation and role of the SAS. He wrote later to
Brathwaite that the squadron was 'the centre of all attention here at
present. The whole concept, and what is more important, the execution, has captured Gen. Westmoreland's imagination. I spoke for
about thirty minutes at Nha Trang. C in C was there plus 5 x 3 star, 7
x 2 star and numerous 1 star generals. Never seen so much brass. Went
off fairly well and I got plenty of questions both during and after the
affair'.[14] Some days later General Johnson, Chief of Staff of the US
Army, visited the Task Force and was briefed by Murphy.

When he had rebriefed the patrols in the 6 RAR position on 18 July
Murphy had ordered them to try to locate an enemy radio station,
which intelligence sources had located in the eastern part of Phuoc
Tuy Province. As days passed further intelligence was received indicating that the enemy radio operator, code-named 'Fred', had moved

to a new location in the Nui Dinh hills. VC posts in the hills were observing the movement of Australian and Allied forces in the Nui Dat and Baria areas and Fred was believed to be reporting this information to the scattered VC units within the province. The SAS was given the task of eliminating Fred. Sergeant Tony Nolan's patrol was to ambush a track to the north of the suspected enemy location and Sergeant Urquhart was to locate and destroy Fred.[15]

Joe Urquhart had entered the army in 1959 and after service with 3 RAR at Enoggera had joined the SAS at the end of 1960. As a corporal in Borneo he had been second-in-command of Schuman's patrol and had led the patrol when Schuman had been sick. But despite the fact that the patrol had seen the enemy on a number of occasions, it had not actually exchanged fire. He had been promoted to sergeant when 1 Squadron had returned from Borneo. When he set out to locate Fred he was still some months short of his 25th birthday.

By now the RAAF helicopters had accepted responsibility for inserting the SAS patrols, and techniques had been practised. Since No 9 Squadron was not equipped with helicopter gunships the protection of the insertions and extractions was undertaken by US gunships, usually operating as a team of two aircraft known as a light fire team. It was hoped that an RAAF helicopter might be able to insert an SAS patrol by rope into the Nui Dinh hills, but the RAAF could only provide a five cm rope and at last light on 23 July Urquhart and his four-man patrol were landed in the flat, more open scrub south east of the hills.[16] They would therefore have a long approach through difficult terrain and it was likely that Fred would move again in the interim.

Nevertheless, they made good time through the steep country and by the night of 25 July they were high on the northern ridges of the hills. The vegetation was now mainly primary jungle which allowed easy movement. During the afternoon of 26 July they heard voices about 150 metres to the north west and Urquhart decided to carry out a reconnaissance at first light. By about 8 am the next morning they had located a vacated enemy camp that was obviously being rebuilt and following a track with fresh footprints they moved a further 200 metres west where they found a second enemy position. They were looping back towards the main track when Private Keith Burley heard voices further down the track. Then the forward scout, Private John (Dixie) Graydon, who was almost on the track, gave the thumbs down signal. Urquhart signalled an immediate ambush and the four men went to ground.

Seven VC were moving quickly down the track. They were dressed in black pyjamas, two wore felt hats, the first six, including two women, wore sandals and were carrying bundles of building material and food, and the last man wore rubber soled boots and carried a pack. Two weapons were seen, being carried by the first and last men. The party had not come far as although heavily burdened they showed no sign of fatigue or perspiration. They were moving very close together, talking, except for the last man who walked about three metres behind looking from left to right. The Australians lay motionless as the first six VC passed. Then the last man spotted Graydon and started to bring up his weapon. At a range of from one to three metres all four Australians opened fire, with Burley firing his heavy barrelled rifle straight down the track. The last six VC were struck instantly, but the first man was already disappearing around a bend in the track. He too was hit and his feet could be seen sticking out of the undergrowth.

As the Australians moved forward to search the bodies enemy signal shots were heard to the east and west. Urquhart recalled later that 'it all started around us. We were not far off their main camp. The whole world went to shit on us'. The patrol withdrew a short distance from the track, listened to the VC, and then when it became obvious that the enemy were heading down the hill to cut off their escape route the Australians decided to move uphill. Soon the VC signal shots indicated that they had found the Australians' tracks and were following, but the helicopters had been alerted. The Australians continued climbing and after 1000 metres reached a steep razorback with low bushes. Out of the distance came two RAAF helicopters and two armed US Army CH47 Chinooks. The SAS patrol had not used its URC 10 beacon correctly but the helicopters saw its marker panel on top of the ridge which was less than half a metre wide. The first helicopter, piloted by Wing Commander R. A. Scott, hovered over the ridge and a crewman, Corporal Williams, was winched down. He was able to give each SAS soldier a 'leg up' and they scrambled aboard. Williams was then winched back up. As the helicopter moved away from the area it received ground fire but no damage was done.

Urquhart had failed to locate Fred, but it was later confirmed that the women had indeed been part of the VC communication station. Furthermore, the patrol was the first to penetrate the VC rear area. It was a warning to the VC that they could never again assume that their base areas were secure and it had shown that the SAS could move stealthily into occupied enemy positions. But while it might have been admirable to take the fight to the VC in their base areas, at this stage the VC had to be cleared from the area around the Task Force base and this mission was to assume crucial importance during the next month.

12
Long Tan and after
3 Squadron: August 1966–March 1967

The battle of Long Tan, fought for three hours in the pouring rain, and amid the mud and shattered trees of the Long Tan rubber plantation, was the most intense fought by Australians in the Vietnam war. During the battle D Company of 6 RAR held off concerted attacks from up to a regiment of VC before being relieved by APCs and another company of the battalion. Seventeen Australians were killed but 245 dead VC were found on the battlefield.

Beyond the heroism of the battle, however, one of the most persistent stories about Long Tan has been that the SAS discovered that the VC were about to attack the Task Force base only to have this warning ignored by Brigadier Jackson. This story was given further credence in 1985 when an article in the Australian Army newspaper *Army* stated that 'Patrols from the Squadron alerted the Task Force to a heavy enemy build up to the east of Nui Dat just before the Battle of Long Tan'. Terry Burstall, in his book *The Soldiers' Story* published in August 1986, chose to believe this account which he adds was confirmed to him by an SAS soldier who 'had been drinking heavily'. On the other hand Lex McAulay, in his book *The Battle of Long Tan* published four months before Burstall's, states that Major Murphy had refuted the claim that the SAS had located the VC regiment. But there are still members of the SAS, including some who served with 3 Squadron in 1966, who believe that the VC were located. None of these accounts, however, are based on the SAS patrol reports, and the work of the SAS in Phuoc Tuy in August 1966 needs to be examined carefully in the light of these reports.

Brigadier Jackson wrote later that during the period from mid July to mid August, the 'SAS were now being used to search for VC main forces and to give early warning of any major enemy movement through areas where the infantry battalions were not operating'.[1] The crucial patrols were the eleven patrols conducted by the SAS between 31 July and 14 August. The first to depart, at midday on 31 July, was a patrol of seventeen men commanded by Second Lieutenant Schuman. Leaving Nui Dat on foot with C Company 6 RAR they headed towards Nui Dat Two and then moved north, parallelling Route 2, about 3000 metres away to the west. The patrol lasted until 8 August but saw little sign of the enemy; the intelligence assessment was that the enemy had not been operating in the area due to recent operations by 6 RAR. The patrol also provided security for a cordon and search operation of Binh Ba village which 5 RAR and two companies of 6 RAR began on the night of 8 August.

Late in the afternoon of 31 July three reconnaissance patrols were inserted by helicopter to the west of Binh Ba. Sergeant Bob Mossman's six-man patrol was inserted 16 kilometres to the west, between Nui Thi Vai and the Bien Hoa border. The next morning the patrol was seen by one VC; an immediate extraction was arranged during which the helicopter was engaged by two VC. Sergeant Aitken's five-man patrol was inserted about 11 kilometres west of Binh Ba. Shortly before midday on 2 August they sighted eight VC heading north, and in the evening of 4 August the patrol was withdrawn. Second Lieutenant Ingram's four-man patrol was inserted 7 kilometres north west of Binh Ba. Just after midday on 1 August one member of the patrol fired at a party of five VC but the contact did not develop. The patrol was extracted later that day. Mossman's patrol was re-inserted on 2 August, this time to the south of Nui Thi Vai; they saw some sign of the enemy and on 3 August sighted three VC before they were extracted that afternoon.

The next series of patrols, beginning on 7 August, concentrated on the Nui Dinh mountains, with the tasks of locating and destroying enemy observation posts and radio stations. Three patrols moved into the mountains while another searched the area to the north east. On 10 August a five-man patrol commanded by Corporal Michael Folkard contacted two VC high on the jungle-covered southern slopes of Nui Dinh. Both enemy were killed, and a subsequent translation of a captured diary indicated that one VC was a member of an assassination team that had completed a reconnaissance for an assassination task in a nearby village.

Meanwhile, Second Lieutenant Roderick and his six-man patrol had the task of covering the more flat and open ground to the north east of the hills. They saw and heard small parties of men and women, and during the morning of 10 August, beside the Song Sa, discovered a recently used hut with a weapons pit and the beginning of a tunnel.

192

Roderick's problem was that the hut was on the west bank of the river while the most suitable patrol base was immediately opposite on the east bank. Although the river was narrow and deep it could be crossed by a short swim, but enemy activity and the need to maintain security caused it to be a major obstacle. Good cover was available along the banks of the river but elsewhere the terrain was open paddy fields. Roderick decided to send over a small reconnaissance group which, considering the obstacle of the river, caused him some anxious moments. A careful reconnaissance of the enemy position was made and at 4 pm voices were heard. Then through the heavy rain and thick undergrowth a party of nine enemy were seen moving into the hut. On the other side of the river Roderick ordered his patrol to move into an extended line and they all opened fire. Five enemy were seen to take rounds to the head and body and were knocked down by the fire, but the remainder dropped to the ground inside the hut. Concentrated fire was poured into the hut for another minute. Only two shots were returned by the VC and there was no further sound. Hampered by the river, Roderick decided not to take advantage of the surprise they had achieved and ordered a withdrawal. That evening they were extracted back to Nui Dat.

At first light next morning K and J Troops under the command of Murphy, with his squadron headquarters, carried out an air assault on the contact area. J Troop provided blocking groups while K Troop searched the area. The VC had removed the bodies and most of their equipment during the night, but personal clothing, ammunition, grenades and webbing were recovered. There was sufficient evidence in the hut to confirm that most of the enemy party had been killed. During the morning squadron headquarters withdrew to Nui Dat, but Murphy saw the opportunity to deploy two patrols under the cover of a final clearing patrol to be conducted by Sergeant Thorburn's patrol. During the clearing patrol Thorburn spotted two VC lying down pointing their rifles at him. He went to ground just in time to avoid their fire. The two VC escaped the sweep by the remainder of Thorburn's patrol. Two patrols, commanded by Sergeants Healy and Kirwan, deployed to the area and that night four or five VC were seen. The next morning the Australians detained nine men and women.

While these activities were taking place on the plain, Second Lieutenant Ingram's patrol had penetrated the northern area of the mountains and on 12 August they sighted two VC women and located an occupied VC camp. An airstrike on the camp was arranged for the morning of 13 August but Ingram was unable to assess the damage as he had to reach an extraction point some distance to the north. That afternoon he returned in a FAC (Forward Air Controller) aircraft and estimated that 70 per cent of the bombs had been on target. Meanwhile, Sergeant John Robinson's patrol in the eastern area of the mountains had found an unoccupied enemy observation post and saw one VC on the plains to the east of the mountains shortly before they were extracted on 14 August. Their extraction helicopter was fired upon from this area.

By the afternoon of 14 August all the patrols had returned to Nui Dat. During the preceding two weeks they had encountered scattered groups of VC around the Nui Dinh hills but there was no evidence that the VC were preparing an offensive. Meanwhile, the squadron could turn its attention to improving its accommodation in its new position near Task Force headquarters. As Murphy wrote at the time: 'the new set up is far better than the old. We have virtually no defence commitment apart from manning 3 x M60' machineguns. The squadron had to dig new trenches and fill more sandbags, but work was progressing slowly. On the afternoon warning orders were issued to four patrols, to be deployed east of Route 2 between Binh Ba and the Courtenay rubber plantation to support Operation TOLEDO to be conducted by the Americans and 5 RAR later in the month. The patrols were to be inserted at first light on 17 August.[2]

At 2.43 in the morning of 17 August Murphy awoke to the sound of mortar bombs and recoilless rifle shells crashing into the Task Force area. Immediately he took a compass bearing on the primary explosions and sent the directions to Task Force headquarters. Two minutes later the mortar fire began landing in the squadron area. The attack lasted for about ten minutes and about ten rounds exploded in the squadron area. Fortunately, no occupied accommodation tents were struck, but seven SAS soldiers became casualties. Corporal Barrie Higgins, Lance-Corporal Bob Davies and Signaller Ronald McMullen had to be evacuated to hospital at Vung Tau, while Captain Chipman, Second Lieutenant Ingram, Corporal Ken Boag and Private Stan Plater remained at Nui Dat. As it happened, Chipman's wound in the leg was more serious than he thought; he later suffered complications and had to be hospitalised. There was also moderate damage to tents and vehicles. As one touch of luxury, Murphy had arranged for china crockery to be brought from Australia for the officers' and sergeants' mess, and to the delight of the soldiers it was all smashed by a direct hit from the mortars.[3]

With the sound of the Task Force artillery firing at the suspected VC position, Murphy reported to Brigadier Jackson and received orders that the patrols designed to support Operation TOLEDO were to go ahead the next morning—before dawn four patrols were preparing to move out on operations. As it happened, only one patrol, Sergeant Urquhart's, was deployed beyond the area where the enemy mortars were thought to be located. His area was a patch of jungle four kilometres by four kilometres astride the Song Rai, sixteen kilometres north east of Nui Dat. Meanwhile the remaining members of the squadron showed a renewed interest in digging, despite the heavy red mud and the lack of sandbags.

It had been intended that Urquhart would be inserted in the evening of 16 August, but as the US light fire team did not arrive the patrol was inserted at 6.15 am on 17 August. Urquhart soon found himself in an area of heavy enemy activity, and almost certainly the VC had detected his insertion. At about 7.40 am he found signs of movement six hours old heading along a track towards the west, and he thought that he could hear about three enemy in the vicinity. It seemed that small enemy parties were searching for them and the patrol withdrew into a thicket. Urquhart estimated that at least a company of enemy with up to twenty parties was involved in the search; once an area was searched mortars were fired into it. Urquhart's efforts to report his observations to base were thwarted by heavy radio interference and he was eventually extracted on 19 August after he had been located by his Sarbe beacon. His radio was found to be unserviceable.

Meanwhile the other patrols had been inserted north of Binh Gia. At 2.45 pm Sergeant Aitken's patrol saw four VC heading south on a track three kilometres north of Binh Gia. They were dressed in uniform, were spaced tactically at five metre intervals and appeared to be in a hurry. But like Urquhart, Aitken could not establish radio communications. At 3.08 pm Sergeant Kirwan's patrol saw ten VC moving west in a wide-spread tactical formation seven kilometres north of Binh Gia. Two hours later they saw another group of seven VC moving east in the same area. Half an hour later they heard a group of enemy of unknown size moving east.

The 3 Squadron war diary indicates that the next day, 18 August, communications were established with all patrols except Urquhart's, and the diary notes: 'Patrols report frequent enemy movement on tracks'. A close study of the patrol reports show that twice during the morning Kirwan reported voices. Further north, near the Long Kanh provincial border, Sergeant Mossman reported that he had found tracks showing sign of very recent two-way traffic, while closer to Binh Gia Aitken saw two enemy moving north west.

Thus by the morning of 18 August, when D Company 6 RAR was about to move out of Nui Dat on patrol, the SAS had provided little solid evidence that the enemy was massing for an attack on the Task Force. The only intelligence from the SAS was that there was some enemy activity north of Binh Gia—Urquhart had not been able to report what he had found further to the west. By the time D Company initiated its contact with the VC in the Long Tan rubber plantation at

3.40 pm the SAS had discovered little more than they had earlier in the day. Once the battle began the SAS patrols in the field could only continue to watch and observe. They heard plenty of firing by various calibres of weapons, but it was not until midday on 19 August that Aitken saw six VC heading north two kilometres north of Binh Gia. On 20 August both Aitken and Kirwan in the area to his north heard or saw small groups of people. On 21 August Aitken saw five enemy heading south and at 1.15 pm seven others heading north. The latter group was moving fast, almost running; they were tired to the extent that the last man stumbled, groaned and carried on. Later in the day Kirwan saw two more enemy. Both Kirwan and Aitken continued to see small scattered groups of enemy during the following day, and all three patrols north of Binh Gia were withdrawn in the evening of 21 August.

Until the moment of their extraction Mossman's patrol had seen less enemy than the other two patrols, but when they arrived at their LZ a number of single shots were heard in the area. The pilot of the RAAF helicopter, Flight Lieutenant B. MacIntosh, decided that he would have to land despite the enemy fire, and as he descended two platoons of VC to the north and another to the south east of the LZ opened fire with small arms and automatic weapons. The door-gunners returned fire vigorously. The No 9 Squadron diarist wrote that the 'SAS patrol ran to the aircraft firing [their own weapons] as they ran and ducking at times to permit the door gunners to fire over their heads'. As the aircraft became airborne the VC assaulted the LZ from the north and spasmodic fire continued until the helicopter reached 1000 feet. It was an example of the excellent co-operation that was developing between the SAS and the men of No 9 Squadron RAAF. As Murphy wrote to the RAAF squadron commander: '...my blokes think your blokes are doing a tremendous job for them. The joy of seeing the aircraft coming to pick them up really has to be experienced to be fully appreciated'.[4]

Between 23 and 28 August four SAS patrols commanded by Schuman, Hogg, Healy and Thorburn were deployed in the area between Niu Dat Two and Binh Gia to report on enemy movement. All patrols saw or heard various groups of VC. It was thought that Hogg had located the headquarters of 275 NVA Regiment and Schuman's patrol had a contact in which one VC was killed and two others were possibly wounded. 5 RAR was deployed on Operation DARLINGHURST to sweep the area but no enemy were encountered.

While these attempts were being made to locate the enemy involved in the Long Tan battle, Sergeant Pancho Tonna was ordered to return to the Nui Dinh hills to again seek to destroy enemy observation posts and radio stations. As he moved up into the eastern area of the hills he found considerable VC movement. One VC was killed and the patrol withdrew. It was clear that the VC were maintaining a strong screen of local guerillas along the lower slopes of the mountains, and they were not pushed back until 6 RAR mounted a battalion operation to clear the Nui Dinh hills between 8 and 24 September.

With 6 RAR concentrating to the west, throughout September the squadron conducted surveillance over the eastern approaches to the Task Force. There were scattered sightings of VC but no major incidents. The most successful patrol was led by Sergeant Tonna when on 16 September his patrol contacted two VC. Both VC were seen to fall, but only blood trails and drag marks were found. Next day they found a VC camp and during the helicopter extraction four VC fired on the patrol. Gunships engaged the VC and Tonna directed artillery onto the camp. Then on 23 September he captured a suspected VC. On 22 September a squadron patrol of 28 men commanded by Murphy was deployed from Nui Dat to search and destroy an enemy installation located by Roderick. They found 13 recent VC graves, possibly the result of the operations following Long Tan. As Murphy wrote later: 'The twice daily reporting, by six patrols, of a lack of enemy activity, allowed the TF Commander to concentrate on his primary mission without being continuously worried by his open flank'.[5]

At the beginning of October 3 Squadron turned its attention back to the area west of the Task Force in preparation for an operation by 5 RAR to clear the Nui Thi Vai hills beginning on 6 October. Trevor Roderick had the task of conducting a reconnaissance of Nui Thi Vai, while Sergeant Norm Ferguson, who had recently joined the squadron, was to look at Nui Toc Tien. Both patrols approached from the south and saw evidence that the enemy were using the areas. In particular, on 4 October Roderick located an enemy camp suitable for about twenty people in rugged, rocky terrain on the western slopes of Nui Thi Vai. He could hear what sounded like someone operating a radio. While withdrawing from the camp they were seen by a VC sentry and Roderick shot him at a range of five metres. The patrol withdrew south, and the next morning contacted six VC, one of whom was killed just before the patrol was extracted. 5 RAR was about to advance from the south and Lance-Corporal Bill Harris, Roderick's second-in-command, briefed the commander of B Company as to the location of the camp. By 8 October the company was in position to attack the camp, but permission was denied and, without being seen, the company withdrew to allow the camp to be attacked by artillery and airstrikes.[6]

In conjunction with 5 RAR's advance from the south, on 5 October four patrols were inserted into the area stretching eight kilometres across the northern approaches to Nui Thi Vai with the tasks of locating and destroying the enemy in their areas, or as the No 9

Squadron (RAAF) diarist wrote: 'We act as a killer shadow force'. The Task Force commander's diary described them as 'hunter/killer patrols'. On 8 October a six-man patrol commanded by Corporal Dan Burgess north of Nui Toc Tien sprang an ambush on three VC, killing all three. At almost the same time, in light rain, north west of Nui Thi Vai, Peter Ingram's patrol killed three VC, while north east of Nui Toc Tien Corporal Andrew Lennox's patrol located a VC camp. Realising that this was a productive area, on 10 October Murphy sent two more fighting patrols to the north of Nui Thi Vai, and Tonna's patrol contacted four VC, killing one and wounding another.

It will be recalled that it was hoped that the exchange programme with the US Special Forces would begin in mid August, but as it happened the first group of Americans did not arrive until 12 September, and the previous day Sergeants Hogg and Healy and Corporals English and Jewell had left for Nha Trang. It did not take long for repercussions to develop from this exchange, and on 13 October Murphy was ordered to visit Saigon for an urgent conference with General Mackay. Murphy learned that Sergeant Hogg had been slightly wounded in Tay Ninh Province near the Cambodian border, and Mackay emphasised to Murphy that Australian SAS troops were not to be used outside the 111 Corps Tactical Zone.

On 16 and 17 October six patrols were inserted into an area astride the Song Rai, sixteen kilometres north west of Nui Dat. There were a number of contacts and on 17 October Jock Thorburn's patrol killed one and wounded another VC.

The squadron was now required to undertake a number of quite different tasks. The first of these tasks, beginning on 29 October, was to assist 6 RAR with the cordon and search of Hoa Long village. Murphy probably exaggerated when he later described the performance of the squadron in the cordon as 'hopeless', but it underlined the fact that the squadron was not trained to operate as a rifle company.

The next task was to assist 5 RAR with its clearance of Long Son Island, situated in the south west of the province and separated from the mainland by up to five kilometres of mangrove swamps and narrow channels. To deceive the enemy as to the intention to sweep the island, the commanding officer of 5 RAR planned to conduct a cordon and search of the village of Phuoc Hoa, on the mainland coast opposite the island, on the night of 6 and the morning of 7 November. A section of SAS commanded by Second Lieutenant Schuman patrolled the waterways adjacent to the village in assault craft.

On the afternoon of 7 November 5 RAR redeployed by helicopter to Long Son Island and the remainder of 3 Squadron took up its mission of patrolling the waterways between the island and the mainland.

Assault craft training before Operation HAYMAN, around Long Son Island, on the southwest coast of Phuoc Tuy Province, November 1966. It was a welcome change from the jungle, but many soldiers suffered from sunburn and dehydration. The SAS apprehended about 250 persons on the waterways around the island.
(PHOTO, L. WALSH)

Murphy, who controlled the operation, was located on an armed Landing Craft Medium (LCM) from the Regional Force Boat Company. Most of the SAS soldiers were in assault boats but some helped man Popular Force (PF) craft of different types. By the time the operation finished on 10 November the SAS had apprehended about 250 people on the waterways and had effectively sealed the island from the mainland. On 10 November one LCM came under fire from a VC ambush ten kilometres north of the island, but the only casualty was one of the PF crew. The operation was a welcome change after the jungle, but the soldiers suffered from sunburn and dehydration.

On 18 November six patrols were inserted into an area east of Nui Dat to support 6 RAR on Operation INGHAM. Most patrols sighted groups of VC and Mossman's patrol contacted six to nine VC with no clear result. After seeing and hearing various groups of VC for the previous eight days, on 27 November Sergeant Ferguson's patrol killed one VC about a kilometre south west of the abandoned village of Thua Tich.

Following 5 RAR's operations in the Nui Thi Vai area, where possible, all caves and tunnels had been blown and seeded with CS (tear gas) crystals. Before Australian troops returned to the area HQ 1 ATF wished to determine the persistency and effectiveness of the CS crystals and the task was given to the SAS. Murphy considered that it would not be economical to deploy four SAS patrols into the area and decided to do the task himself with the assistance of his SSM, Warrant Officer Wright. Thus on 1 and 2 December he conducted Operation DANGLESNIFF. Murphy and Wright were winched by helicopter individually into eight different areas of the mountains. They walked around, rubbed earth into exposed skin, and concluded that there was no evidence of the CS crystals. Their helicopter was fired at only once during the operations.

Another unusual operation involved the water tower at the abandoned village of Thua Tich, situated in an open grass-covered area 20 kilometres north east of Nui Dat. Observers had spotted what appeared to be a radio aerial on the top of the steel tower and US aircraft had attempted to destroy the tower, without success as evidenced by the bomb craters around it. Indeed there was a standing arrangement that US aircraft returning from operations could expend unused ordnance on the tower with the crew promised seven days in Tokyo if they were successful. Eventually Brigadier Jackson gave the task to the SAS, and on 14 December 34 members of the squadron carried out an air assault on the area. The charges were set by the SQMS, Staff Sergeant Jake Mooney, and his technical storeman, Sergeant Leslie (Dave) Whitaker. The tower was successfully blown, but a close inspection revealed that the suspected aerial was in fact a lightning conductor.[7]

On the afternoon of 14 December, five patrols were inserted into the area to the north of Nui Thi Vai and Nui Dinh to report enemy movement in support of Operation DUCK in which 6 RAR was to clear Highway 15. Further patrols were inserted on 23 December. There were frequent enemy sightings and a number of unnerving experiences. For example, on 20 December a four-man patrol commanded by Lance-Corporal Reg Causton counted nine well-armed and equipped VC. On 24 December Sergeant Kennedy's patrol saw thirteen armed VC, and within the hour saw seventeen more VC; during his three day patrol he saw almost 50 VC north east of Nui Dinh. On 18 December Sergeant Thorburn's patrol discovered an enemy base camp in a cave beneath a jumble of rocks near Nui Toc Tien. They were spotted and opened fire killing four VC. At least four small children in the immediate area were left unharmed and the Australians withdrew as they heard more VC approaching. Four days later, just north of the Nui Dinh hills, a four-man patrol commanded by Lance-Corporal

Ronald Gammie was in an observation post when they saw a VC patrol advancing in extended line towards them. A firefight broke out in waist-high grass and two VC were killed and another wounded. The Australians withdrew under fire from another party of VC.

The new year began with the SAS's first operational parachute jump when, using parachutes borrowed from US Special Forces, the squadron jumped into a dry paddy field. Murphy described the operation as 'largely a sporting gimmick'. For the most part of January the SAS conducted a series of surveillance patrols in an arc stretching from the north west to the north east of Nui Dat out to a distance of about 18 kilometres. As the Task Force commander's diary recorded, during January the SAS operated in the north west, east and south east extremities of the TAOR to try to 'establish suitable targets for further operations'. Meanwhile 5 RAR was involved in a cordon and search of the village of Binh Ba followed by security operations around Hoa Long. There were significant sightings of VC. Pancho Tonna, operating in a four-man patrol north of Binh Gia, saw groups of VC leading several ox carts on 11 January, and the next day contacted a small enemy party, killing one. Over the next five days they saw further groups of enemy including 12 VC on 17 January.

Sergeant Jock Thorburn and his four-man patrol searched the area south east of Binh Gia near the Song Rai. They located an enemy camp and between 14 and 16 January saw over 70 VC and heard other groups. An airstrike was carried out once the patrol withdrew. Sergeant John Robinson took his four-man patrol 9 kilometres north west of Binh Ba; they found evidence of small groups of VC, and on 14 January shot and killed one VC carrying an M1 carbine.

From the point of view of the history of the SAS, however, the most significant operation conducted during this period was the four-man patrol commanded by Sergeant Norm Ferguson, dropped by helicopter into an area about 9 kilometres north west of Binh Ba late in the afternoon of 17 January. The vegetation was variable, ranging from secondary jungle to large open areas covered by vines or thorny bamboo, and the dry vegetation made movement difficult. Before last light they discovered a small building under construction and heard voices. The next morning they saw an armed VC dressed in black pyjamas, carrying a bird in one hand and an M1 carbine in the other, and before midday they located a small VC camp. Four VC men, dressed in black pyjamas, and two women, one wearing a white blouse and the other a blue blouse, were sitting around a cooking pot preparing a meal. The Australians searched the surrounding area, listening to the VC activity which included the firing of an automatic weapon and the sound of a vehicle.

By nightfall the Australians were 1000 metres east of the VC camp. Next morning they could hear laughter 50 metres to their east and they proceeded south east through low scrub. Then at about 10.15 am the forward scout, Private John Gibson, came across enemy tracks and sent back the appropriate signal. Second in line of march was Sergeant Ferguson, behind him came Private John Matten, the radio operator, while bringing up the rear was the 'tail end charlie', Private Russel Copeman, the medic. Suddenly enemy fire opened up from the left (north), and the patrol went into an ambush drill, returning fire with small arms and grenades. It was Matten's turn to give covering fire to Copeman so that he could withdraw, when he realised that Copeman was not there. Screaming out to the rest of the patrol that Copeman had been hit Matten moved forward into the contact area to locate him. He found him in the same place that he had seen him last and furiously began shedding Copeman's equipment while at the same time trying to fire at the enemy. Copeman told Matten to leave him and get out himself, but Matten continued to engage the enemy. He then threw a white phosphorous grenade in the direction of the enemy and, under the cover of the explosion and the drifting white smoke, lifted Copeman onto his shoulder and started to run back through the jungle and bamboo. The rest of the patrol provided covering fire and several enemy were heard to scream as if hit.

As far as Matten knew, the only injury suffered by Copeman was to his right arm. But once they could safely stop to attend Copeman's wounds it was clear that there was a second wound in his middle section. After doing what he could for Copeman, Matten set up his radio and sent the appropriate code words asking for help and extraction. As he was packing up the radio the enemy were heard looking for them and Matten picked up Copeman and continued moving towards the winch point.

It was now over an hour since the contact, but the helicopter arrived quickly and Copeman was winched up first, followed by Matten. Then two VC burst from the scrub into the clearing about 20 metres to the north. The helicopter pilot, Flight Lieutenant Hayes, swung his craft around to give the port door-gunner, Leading Aircraftsman Bloxsom, a clear field of fire while Matten also returned the enemy fire. The first enemy group disappeared from view but two or three more ran into the clearing and Ferguson, dangling on the end of the rope being winched into the helicopter, engaged the enemy, who either fell to the ground hit or were diving for cover.

The enemy had been very aggressive once they had located the SAS patrol and they had not been deterred from attempting to prolong the contact by the presence of the helicopter gunships. That the SAS patrol could be extracted quickly and without further casualties was a tribute

Private Russel Copeman on patrol in the Long Tan area east of Nui Dat, August 1966. He was badly wounded on 16 January 1967 and died in Australia in April 1967, the first and only Australian SAS soldier to die as a result of enemy action. (PHOTO, R. LOGUE)

to the training and determination of the SAS and the close co-operation of the RAAF. Matten's prompt and brave action had been crucial; the helicopter flew directly to the hospital at Vung Tau and Copeman's life was saved by only minutes as he had suffered massive internal bleeding. There was, however, a sad postscript to the story. Copeman returned to Australia and appeared to be recovering from his wounds, but there were further complications and he died on 10 April 1967. He was the first and only Australian SAS soldier to die from enemy action. A capable and intelligent young soldier, according to Murphy, he should really have been at the Officer Cadet School at Portsea.

Three days before this action, on 15 January, Lieutenant Bob Ivey had arrived to take command of J Troop with the intention that he would continue as operations officer of 1 Squadron when it relieved 3 Squadron. Lieutenant Marshall had returned to Australia on promotion to take up a training appointment in the regiment and Trevor Roderick had become operations officer of 3 Squadron.[8] Aged 23, Ivey, a Portsea graduate, had already served as a platoon commander with 3 RAR in Borneo in 1965 where he had conducted Claret patrols. It was planned that initially he would join a patrol as a rifleman, but before he could gain this experience he had an unusual introduction to SAS operations.

It was believed that the VC would attack the town of Dat Do during the Vietnamese Lunar New Year (Tet) early in February, and it was planned to infiltrate three SAS patrols (Robinson, Ingram and Urquhart) into the area south east of Dat Do by landing them from a landing craft on the night of 27 January. The operation would be commanded by Lieutenant Ivey and a US Navy Swift class launch would escort the LCVP. The boats set out from the US Naval base at Cat Lo at 4.30 pm and at 7.30 pm were passing the Long Hai hills when the LCVP received machine-gun fire. The LCVP returned the fire. Ten minutes later those on board the LCVP were informed that the Swift launch was returning to Cat Lo to evacuate a US sailor who had been hit by enemy fire. Meanwhile various enemy signal lights were seen along the coast, and at 10 pm the Swift returned with the news that it had been fired upon again while returning from Cat Lo. The LCVP moved toward shore about ten kilometres east of the Song Rai mouth but when about 25 metres from shore it became stuck on a reef. Two men were seen on shore and further signal lights were seen. The LCVP was extracted from the reef and moved further east towards the next landing point. They had gone about 2000 metres when the Swift came under machine-gun fire. More signal lights were seen on shore, a large search light was directed at the beach from a point far out to sea, and at this stage the operation was called off. Although Swift launches were frequently used in the area LCVPs were not, and this possibly alerted the VC warning system. It is also possible that the operation interrupted a VC resupply being taken in the landing area and the VC could have possibly mistaken the LCVP for one of their own craft.

When 3 Squadron had been deployed to Vietnam it had been intended, possibly because of the experience of Borneo, that its tour would last for only six months. But as the tour had progressed it became obvious that unfavourable comparisons might be drawn between the all-regular SAS and the infantry battalions with their large percentage of National Servicemen. While 3 Squadron had to endure the difficult, wet days of July and August when there were few facilities at Nui Dat, they were still capable of operating beyond the period of six months which had been reached in mid December. On the other hand 1 Squadron had planned on arriving in Vietnam early in 1967, and it was therefore decided to proceed with the relief. One consideration was that February was an ideal month for a new squadron to arrive in Vietnam. The hot, dry weather in Perth was very similar to conditions in Vietnam at that time, thus reducing the period required for acclimatisation. With these facts in mind Murphy suggested to the new Task Force commander, Brigadier Stuart Graham, that there

The SAS lines on Nui Dat hill. These lines were first occupied by 1 SAS squadron when it relieved 3 SAS Squadron in February 1967.

could be a long hand-over of operations to 1 Squadron. It was intended that 1 Squadron would not occupy the old 3 Squadron lines but would develop new accommodation on Nui Dat hill, and the hand-over period would allow this development to proceed smoothly.

Meanwhile, 3 Squadron continued with a series of patrols to different parts of the province. These patrols often included members of 1 Squadron. One of 3 Squadron's most successful patrol commanders was Sergeant Tony Nolan; he had registered numerous enemy sightings but had always managed to avoid contact. His luck ran out (or in) late in February when his patrol was operating near the north west border of the province. During the afternoon of 24 February they located a barbed wire fence running through the jungle; one enemy was encountered and he was killed instantly by Private Fred Roberts.

It is arguable whether this was the squadron's final contact in its tour, because at about the same time Corporal John Jewell led a five-man patrol, including Lieutenant Ivey and Sergeant Thorburn, to the area of the Song Rai south east of Binh Gia. They discovered two small enemy camps and a vacated enemy company position and heard various voices, including the sound of about nine enemy, six males and three females, washing in the river. At 10.30 am on 27 February the Australians spotted three VC, two men and one woman, walking along a track carrying small logs. The patrol had almost concluded and Jewell decided to apprehend the VC. They found that the logs were old and rotten and of no apparent use and thought that they were being carried for 'show'. On the way to the LZ they heard sounds of male voices, chopping and a dog barking and a little later voices about 100 to 150 metres to the east. Suddenly the two male prisoners tried to run away and both were shot dead. On their return to Nui Dat they handed the woman over to the Task Force intelligence unit.

The advance party of 1 Squadron had arrived at Nui Dat on 15 February, on 2 March the main body of 1 Squadron arrived, on 15 March the main group of 3 Squadron departed, and the last group of 3 Squadron left Nui Dat on 24 March. Phuoc Tuy Province in 1966 had posed many new challenges for the men of 3 Squadron. During their tour of nine months the squadron had mounted 134 patrols. There had been 88 enemy sightings with a total of 198 enemy spotted. In 27 contacts the squadron had inflicted 46 enemy KIA, four possibly killed, 13 wounded and taken one prisoner of war.

One of the most important achievements of the tour was the excellent working relations established with No 9 Squadron RAAF. As the No 9 Squadron diarist recorded: '3 SAS and No 9 Squadron had developed an extremely good working relationship together, both units being very proud of the job they had accomplished'. Once the peacetime restrictions were removed the insertion of patrols by helicopters was instituted, although some of the insertions were by single helicopters with no support. The helicopters were always available for extraction, even when the SAS was being hard pressed by the enemy, and in those cases the US light fire teams provided welcome protection.

The achievements of 3 Squadron reflected the outstanding work of John Murphy, who was well liked and respected by everyone in the squadron and whom Brigadier Jackson described as 'an excellent squadron commander and very clear headed about the best use of the SAS in the prevailing conditions'. Under his command the squadron had tried and tested many of the techniques that were to be used by later squadrons, and 3 Squadron developed and gained not only the respect of the other Australian units and their American allies in Vietnam, but according to intelligence reports, also of the enemy. The SAS had proved that they could operate in groups as small as four men in a hostile environment. Indeed their most common sized patrol was four men and the main emphasis of their operations was reconnaissance and surveillance. They had proved that they could provide the Task Force commander with reliable information about the build up of VC base areas. The main trouble was that the battalions reacted slowly to this information. The SAS's ability to act offensively had not been tested fully, but 3 Squadron had created a formidable precedent for later SAS squadrons to follow in the Australian Army's operations against the VC in Phuoc Tuy Province.[9]

13
Fighting for information
1 Squadron: March 1967–February 1968

1 SAS Squadron arrived at Nui Dat in March 1967 with a vastly better preparation than 3 Squadron had managed nine months earlier. It had benefited from its predecessor's experiences in Phuoc Tuy Province and had a clearer idea of its likely roles. Furthermore, the techniques of helicopter insertion and extraction had been developed, the SAS had won the confidence of the Task Force Commander and the other Australians at Nui Dat, and their equipment had been improved and refined. And although about half of the original members of 1 Squadron had transferred to 3 Squadron when they had returned from Borneo in August 1965, it still possessed a wealth of experience, a number of personnel from 2 Squadron having joined 1 Squadron at the end of their tour in Sarawak in August 1966.

When 1 Squadron had returned to Swanbourne in August 1965, it was known that Major Garland would shortly be posted out of the unit, and in October he was replaced by Major Jack Fletcher, GM, a 31 year old Duntroon graduate who had served with 2 Commando Company in Melbourne, with 1 and 2 RAR in Malaya, and had trained in the United States and Britain in airborne, special forces, ranger and commando techniques. Unfortunately Fletcher injured his back and was transferred to the new position of second-in-command of the regiment. In the meantime, Major Dale Burnett had been posted to the SAS Regiment to command 4 Squadron, the raising of which had been authorised in October 1965. He arrived at Swanbourne in January 1966 and soon after was appointed to command 1 Squadron. Captain Tony Danilenko continued in command of 4 Squadron.

Aged 30, Dale Burnett had graduated from Duntroon in 1956 and had served as a platoon commander with 3 RAR in the Malayan

Emergency. He had then undergone six months ranger training at the United States Infantry School before being posted as adjutant and quartermaster of the Sydney Commando Company. His appointment to the SAS Regiment followed a posting as an instructor at the Officer Cadet School at Portsea. Reserved, precise, bespectacled, Burnett had no preconceived ideas on SAS operations. But he had a solid, professional infantry background and he appreciated the value of the training then underway at the direction of Lieutenant Colonel Brathwaite, who was emphasising the basic infantry skills of navigation, shooting, marching and communications.

While there were many experienced soldiers within the squadron, there was a marked shortage of experienced officers. The second-in-command, Captain Terry Holland, had served in Borneo but was soon to be posted from the regiment. There was no operations officer, and of the three troop commanders only one, Lieutenant Bob Ivey, proved to be suitable for SAS operations; he was later appointed operations officer and joined 3 Squadron for the last two months of their Vietnam tour before rejoining 1 Squadron on their arrival at Nui Dat.

In June 1966 two Portsea graduates arrived in the regiment and after their cadre courses joined 1 Squadron. Second Lieutenant Jim Knox (23) became the commander of A Troop, and Second Lieutenant Rick Gloede (23) was given C Troop. Knox had served in the Sydney Commando Company before attending Portsea, and Gloede had previously been a soldier in the SAS Regiment. The next month Second Lieutenant Bill Hindson arrived in the regiment. Aged 23, Hindson already had considerable military experience. After leaving school at 14 he had worked in various jobs for Coles stores in Melbourne for four years until he joined the Army in 1961. He had served in 1 RAR before attending Portsea, graduating in December 1964. Returning to 1 RAR he commanded a platoon in Vietnam in 1965 and 1966, and towards the end of his tour volunteers were requested for SAS. Only two officers, Hindson and Kevin Lunny, volunteered. They were hurried through cadre and parachute courses and in September Hindson joined 1 Squadron to take command of B Troop.

Two of the troop sergeants, George (Chicka) Baines and John (Slag) O'Keefe, had served with 1 Squadron in Borneo, while the third troop sergeant, Ron Gilchrist, had just returned from 2 Squadron's tour in Sarawak. The SSM, Warrant Officer Horace (Wally) Hammond, was a reliable, experienced soldier who had served with the Training Team in 1964 and 1965. In August Captain David Levenspiel was appointed second-in-command of the squadron; he had only joined the SAS in March 1966, but had served from June to August with 2 Squadron in Sarawak.

1 Squadron had the advantage of knowing what was required in Vietnam, but the regiment was still operating on a shoe-string. To provide the administrative and accommodation buildings necessary to house the regiment, in January 1966 a new building programme began at Swanbourne, but this project could not keep pace with the increasing expansion of the unit; nearly half the unit was housed in tents or in sub-standard accommodation.

Another problem was the shortage of Royal Australian Corps of Signals operators, and this hampered the training of unit operators. As a result, in April 1966 Headquarters Western Command recommended to Army Headquarters that a signal squadron be raised to support the regiment. On 31 August Army Headquarters approved the regiment's new establishment, which included Base Squadron, three SAS squadrons and an SAS signal squadron (152 Signal Squadron). There was also a badly needed increase to the equipment tables. The total establishment was now 602, although the posted strength at the end of 1966 was only 408. The first commander of 152 Signal Squadron was Captain Ross Bishop.

Although the Recondo training initiated by Major Clark was no longer held, the selection and training of SAS soldiers was still basically the same. The requirement for twelve months' completed regimental service, however, was dropped to allow for the selection of National Servicemen. Selection was still made by a touring SAS selection board and all ranks were still required to undergo a six-week SAS cadre course, followed by parachute training at Williamtown, NSW. On successful completion of the cadre course at least one member of each patrol was required to qualify as a medical orderly, assault pioneer, infantry signaller, weapons instructor, or linguist (a six-week basic course in Malay, Pidgin or Vietnamese). Once this initial training was complete, soldiers were trained in the specialist skills of small craft handling, unarmed combat, roping and cliff climbing, vehicle driving and shallow water diving.

Coupled with this specialist training were troop and squadron exercises designed to train the SAS soldier in the unit's patrol techniques and minor tactics. Experience indicated that eight months were needed for specialist training together with the necessary troop and squadron training required for operations.

As part of their build-up for operations 1 Squadron took part in three major exercises: Exercise LANDSCAPE from 23 August to 3 September 1966 in the Denmark area; Exercise SEASCAPE, from 21 to 30 September in the Garden Island–Fremantle area; and Exercise TROPIC OP, in New Guinea from 10 October to 4 December 1966. As with 3 Squadron's exercise in New Guinea earlier in the year, 1 Squadron was supported by chartered aircraft. Patrols were deployed in the Wau–Morobe area, and the squadron was based at Lae. Whereas 3 Squadron had been fairly flexible with the size of its patrols, Dale Burnett was probably influenced by the experiences of 2 Squadron in Sarawak and fixed the size of his squadron's patrols at five men.

By the end of 1966 1 Squadron was ready for operations and had begun its pre-embarkation leave. It had been an exceptionally busy year for the regiment, and when the Commander of Western Command, Brigadier N. A. M. Nicholls, visited the regiment on 31 January 1967, Lieutenant-Colonel Brathwaite did not hesitate to outline the difficulties under which he had been operating. He explained that the infantry battalions had been loath to make volunteers available to SAS selection boards, and that the Royal Australian Corps of Signals had not given sufficient priority to the provision of operators to the SAS Regiment. The SAS Regiment equipment table was unsuitable and staff branches and directorates at Army Headquarters failed to understand that the regiment was not just another infantry battalion with a different name. The regiment had to undertake specialist training with a limited manpower pool and these problems were aggravated by lack of adequate notice for the despatch of squadrons on operations, and the lack of a fixed and realistic policy on the length of operational tours. He thought that six months was the optimum. However, Brathwaite was confident of the training of 1 Squadron, as shown in a letter sent to a friend at Army Headquarters on 8 February 1967: 'We are as tight as a drum manpower wise, and will remain so until 3 SAS Sqn gets back into training later in the year. 1 SAS went off in good heart, and if they do not do as well as 3 SAS Sqn I shall be very surprised. It is even just possible that they may even improve a little'.[1]

The main body of 1 Squadron arrived at Nui Dat on 2 March 1967 and began training, briefings, testing weapons, revising signals procedures and practising helicopter drills with No 9 Squadron, RAAF. Shakedown operations began on 8 March and during the next five days the three troops conducted ambush or fighting patrols, each without result. On 13 March the squadron assumed responsibility for the old D Company 5 RAR area on Nui Dat hill. The squadron's first mission was to conduct surveillance across the area north west of Nui Dat out to a distance of 11 kilometres. Meanwhile both 5 and 6 RAR were to operate east of Nui Dat beyond Dat Do. On 18 March five patrols were inserted into the search area, and it did not take long for one of the patrols to find the enemy.

Soon after 11 am on 20 March, the patrol led by Second Lieutenant Jim Knox began to cross an old ox-cart track, about nine kilometres

B Troop, 1 SAS Squadron, Vietnam, 1967. *Standing, left to right:* Ian Bullock, Mick Smith, Corporal Allan Smith, Corporal Bruce Absolon, Gordon Herbert, Gerry Israel, Lance Corporal Terry Ross, Bob (Clicker) Clark, John Ward, Clive Carroll and Peter France. *Front:* Corporal Alan Roser, Sergeant Jack Gebhardt, Sergeant Neville Farley, Lieutenant Bill Hindson, Sergeant John Duncan, Lance Corporal Ian Conaghan. (PHOTO, W. HINDSON).

west north west of Binh Ba, when a group of perhaps ten VC approached along the track. Sergeant Baines was almost on the track and knew that he would be seen. He acted first, at a range of three metres shot the first VC in the chest, and then moved forward to recover the enemy's weapon. Suddenly the wounded enemy started to move; Baines fired a second shot, striking the enemy above the right eye. The Australians then withdrew and were extracted by helicopter at 12.45 pm that day.

The other patrols had no contacts, and between them sighted only one armed VC. When five more patrols were inserted into the same area on 24 and 25 March they obtained similar results; only two more enemy were seen. In early April four patrols had the task of reconnaissance in the Nui Thi Vai and Nui Dinh hills, but again they saw few enemy. Finally on 9 April two patrols were ordered to conduct reconnaissance about ten kilometres north west of Binh Gia. Sergeant Neville Farley's patrol was inserted at 6.30 pm, just before last light, and at 7.20 pm, after moving about 250 metres from the LZ, they came across an enemy patrol and killed one. The Australians withdrew through an unoccupied enemy bunker system.

By mid April 1967 1 Squadron had been on operations for one month, but had little to show for its efforts. Its two contacts had been chance encounters. More importantly, the fleeting sightings of the enemy had contributed little to the Task Force intelligence picture. The problem was exacerbated by the dry weather which made it hard to move without being seen or heard. Indeed, Burnett believed that both 3 Squadron and his own squadron had been very lucky not to have taken serious casualties during the previous five months of the dry season.

Meanwhile, experience from the infantry battalions had shown that the VC often carried considerable information in the form of maps, documents and diaries. Burnett discussed these matters firstly with his troop commanders, and then with Brigadier Graham and the Task Force intelligence officer, Major Jim Furner. When Brigadier Stuart Graham had taken command of 1 ATF at the beginning of the year he had been doubtful of the worth of special forces, his views probably being coloured by an unhappy experience with a commando company in the Second World War. But his experience in Vietnam changed his 'views to a very large extent'. He 'regarded them as fundamentally an intelligence unit...Surveillance, reconnaissance and early warning were their primary tasks'. And as he later commented, 'as far as I can remember I didn't deviate from that. That is not to say that they did not conduct some ambushes but these were subsidiary to their primary tasks and undertaken only to enable the fulfilment of those tasks'.

However, as Burnett recalled, after discussion Graham ordered the SAS to undertake ambush patrols in areas that Furner thought would be productive, and indeed Graham asked Burnett, if possible, to try to capture a VC prisoner. Burnett was pleased to co-operate, but was adamant that enemy bodies would only be searched if the patrol was sure that there were no further enemy nearby. It was a cautious approach but Burnett knew that in a five-man patrol, if only one man was wounded, the whole patrol could be in danger. He had excellent working relations with Graham whom he found would listen to advice and was full of ideas. While Second Lieutenant Hindson found Burnett to be exceedingly quiet and remote, he also found him to be a good planner who selected the right patrols for the right tasks—particularly for ambushes. Brigadier Graham thought that Burnett did a 'particularly good job in balancing the two perpetual concerns—concern for the safety of his troops in a dangerous role and

213

concern for meeting his TF commander's requirements without grossly overworking his limited resources'.

Thus to the traditional SAS task of silent surveillance and reconnaissance was added the additional task of fighting for information. The squadron continued to mount surveillance and reconnaissance patrols, but from then onwards these patrols were matched by an equal number of ambush tasks. It was not a policy of achieving kills for the sake of numbers of kills—as Graham commented, 'that would be a prostitution of a specialised unit's capabilities'—but rather a deliberate policy for obtaining more useful intelligence.

While these plans were being considered, news came from Swanbourne that Lieutenant-Colonel Brathwaite was about to relinquish command. His final message to Burnett, published in routine orders on 14 April 1967, was a fitting introduction to the new emphasis on fighting for information: 'Have sheathed my sword and anticipate early retirement. I will be watching from sidelines and with full confidence. Do not rest on reputation of previous Squadron but keep on improving. Tell all ranks from me to be at all times secure and certain and when necessary sudden and silent.'[2]

On 6 April Burnett gave orders for three ambush patrols (Glover, Hindson and Shaw) to be deployed the following day to the area north of the Nui Thi Vai hills, and the first success was gained by Hindson's six-man patrol on 21 April. The patrol was in ambush on a well-used track running north about 3000 metres north west of the Nui Thi Vai hills when they opened fire on a group of four VC killing two and wounding a third. A wallet and note paper were recovered. By 10.20 am they were back at Nui Dat, and a little later Hindson wrote to his fiancée: 'My hands are still shaking from this morning's efforts'. By 2.05 pm he could write that his hands had now stopped shaking and he could continue his letter.

Meanwhile, two further ambush patrols (Knox and Fraser) had been deployed to the same general area. The morning after Hindson's success Sergeant Gavin Shaw with his six-man patrol was moving into an ambush position 1000 metres north of Nui Thi Vai when twelve enemy were seen walking along the track. The Australians opened fire killing three and wounding another, but the presence of other enemy in the area meant that the bodies could not be searched.

By now Jim Knox's patrol had crossed the border into Bien Hoa Province, north of Nui Thi Vai, and moving through thick secondary jungle had come across many well-defined tracks. During the afternoon of 21 April they spotted two men riding bicycles, but as no weapons were seen the ambush was not initiated. Next morning the patrol moved back to its ambush position and Private Arpad (Paddy) Bacskai described what happened: 'Chicka [Baines] moved forward to a log (quite large) approx one foot [30 cm] from the track, when two VCs came past on bikes. One or both propped their bikes against the log or on the ground and one VC proceeded to piss against the log splattering Chicka on the back—he was squeezed on the bottom side of the said log. All of us were within twenty metres of the log'. The Australians held their breath, not daring to move, but had the presence of mind to notice that the carry tray of one of the bicycles held a package wrapped in green plastic. Again the SAS refrained from springing the ambush; as Bacskai commented, the mission was 'generally stiffled due to lack of comms [communications] and a fear of taking a casualty or unable to do anything constructive with captured VCs. After day 2 or 3 only an ambush concept was entertained as we still had not made contact with Nui Dat. We were also sure that VC had known of our infiltration'. After other sightings and locating an enemy camp, the patrol was withdrawn on 28 April.

These initial ambushes had been initiated by rifle fire and had been markedly more successful than the earlier reconnaissance patrols. However, the squadron had obtained a quantity of Claymore mines, and influenced by the experience of the ex-2 Squadron men who had used the mines in Sarawak, it was decided to use the Claymores in SAS ambushes. At first some of the patrols had no more than two Claymores, but the increasing effectiveness of the ambushes was startling.

Between 26 and 30 April five patrols (Roods, Stevenson, Farley, Hindson and Shaw) were deployed on ambush missions in an arc stretching from the north west to the north east of Nui Dat out to a distance of from ten to fifteen kilometres. The first to gain success was Sergeant Bill (Chester) Roods whose five-man patrol was inserted ten kilometres north west of Binh Ba. On 29 April they fired two Claymores at three VC walking along the track, killing two and badly wounding the third. A few seconds later more VC moved into the killing area and the third Claymore, facing down the track, was fired. The patrol quickly withdrew and was credited with five enemy killed and one wounded.

Bill Hindson's five-man patrol was inserted by helicopter at last light on 30 April into an area only two kilometres east of Roods' contact area. They patrolled through bamboo and scattered tall trees, and on the morning of 2 May were setting up an ambush when they saw a woman wearing a purple blouse, white scarf and black trousers walking along a track from east to west. Not sure if his patrol had been seen, Hindson sent two of his soldiers to follow her. Two hundred metres west she disappeared but the soldiers thought that they could smell a camp. They therefore rejoined the patrol and the ambush was laid. At about 10.45 am three VC entered the ambush and two Claymores and two M79 rounds were fired, killing all three. Quickly the patrol moved out to search for the bodies. They recovered one US M1 carbine, one French MAT49 SMG, a diary and a quantity of documents. By just after midday they were back at Nui Dat.

Due to his early successes, Hindson was given further ambush tasks, and at the end of May he led a ten-man patrol into the area three kilometres north east of Nui Toc Tien. On 2 June they were in ambush beside a jungle track with early warning groups deployed 100 metres to each flank. Hindson later described the ambush as 'something of a fiasco'. The early warning group saw four VC approaching but were too frightened to speak into their radio until the VC had completely passed their position. Of course by this time the VC had reached the killing ground. Hindson was in the centre of the killing group when suddenly he heard the words 'they're coming' boom from his URC 10 radio. He looked up and found one VC watching him. The VC dived behind an ant hill just as Hindson initiated the ambush. Fortunately the Claymores were sighted in front of the ant hill, but the premature initiation meant that only two VC were killed. One VC had been carrying a US M1 carbine and a large dead monkey on a stick. The second VC who was carrying a US .45 automatic pistol, turned out to be a company commander of C240 Mobile Force from Bien Hoa. His pack contained a Sony radio, a US compass, a 1:50 000 map of Phuoc Tuy Province and a quantity of documents. While Hindson was searching him, with a dying gasp he grabbed Hindson's wrist and dug his fingernails through the Australian's shirt into his wrist.

It is not possible to recount the story of every successful ambush patrol, and there were an even larger number of successful reconnaissance or surveillance patrols in which the enemy were sighted but there was no contact. However, the experiences of Sergeant Lawrie Fraser are representative of the sort of ambush patrols conducted during the period. Short, slight and highly strung, in May 1967 Fraser was aged 29 and had already amassed considerable experience in jungle patrolling. After joining the Army in October 1956, he had served with 3 RAR in the Malayan Emergency when it was common for infantry patrols to be divided into small groups to search for the enemy. After returning from Malaya he joined the SAS in early 1960 and when 2 Squadron was formed in August 1964 he was promoted to sergeant to train a new patrol. Fraser commanded his patrol in Sarawak in 1966 where he found the operations to be frustrating but excellent training for Vietnam. As he said later, the difference was like 'chalk and cheese'. In Vietnam there 'was more food, more firepower and more enemy than you knew what to do with. It brought it all together for a professional soldier'.

216

After Lieutenant Bill Hindson's successful ambush near Nui Dat on 2 June 1967. The dead VC, with the dark uniform and the pistol holster beside his right leg, was a company commander carrying valuable documents. Near the head of the other VC can be seen the dead monkey he was carrying on a stick. The light patch in the background is the metre-wide path through the dense jungle. (PHOTO, W. HINDSON)

Of course a high level of professionalism was no guarantee of success. On their first ambush patrol, between 20 and 24 April, Fraser's patrol had no contact with the enemy, but during the extraction three VC fired at the second aircraft. Their second ambush patrol was inserted about 5 kilometres north east of Nui Thi Vai at 6.25 pm on 3 May. The wet season had begun and the LZ was a clearing 150 metres long by 50 metres wide. The central section was covered with long grass and the remainder was grass a metre high. Surrounding the clearing was primary jungle. The ground was swampy, and as Fraser leapt from the helicopter he could see fresh footprints on the ground; the disturbed mud had not settled in one pool of water. As the third member of the patrol alighted from the helicopter a shot rang out, and the last soldier to alight was greeted by another shot. The patrol immediately returned fire and recalled the helicopter. The protecting light fire team engaged the area and within five minutes the helicopter was back on the LZ with the soldiers scrambling aboard. A burst of tracer fire came from the edge of the jungle just as the helicopter became airborne; it was an excellent example of the teamwork between the SAS and RAAF. And when Fraser prepared his patrol report, under the heading 'Condition of Patrol', he wrote: 'A little older'.

Fifteen days later, at 5.20 pm, Fraser's patrol alighted from a heli-copter about ten kilometres north west of Binh Ba to begin the next ambush operation. Moving off from the LZ Fraser found that he had to cross a cleared area with grass about a metre high. He also found an unmarked creek, and his Malayan experience warned him that enemy camps were often found near unmarked creeks. By this stage the first two members of the patrol were crossing a track when suddenly one VC appeared on the track in dark coloured clothing. Private Jim Harvey fired an M79 high explosive round from a range of 30 to 40 metres which detonated on the VC's chest, killing him.

Immediately a firefight broke out between the SAS patrol and an enemy party spread over a frontage of about 50 metres or more. An explosion from a grenade launcher or mortar was heard, and Private Harvey was wounded in the arm. Meanwhile, a heavy machine-gun had opened up. The SAS threw a smoke grenade and recalled the helicopters, but the enemy began moving around its flank in an effort to block the withdrawal route to the LZ. It was only the extremely accurate fire from the light fire team that enabled the patrol to with-draw to the LZ where they were extracted under hazardous circum-stances. One of the Australian helicopter pilots said later: 'We would not have got away with it if it had not been for the magnificent support of the United States helicopters. They got us out of a tight spot'.[3]

Fraser's patrol was back in the jungle from 24 to 27 May, but without contact, and on 7 June they were inserted into an area of thick primary jungle seven kilometres north east of Nui Dat Two. The thick undergrowth made movement difficult and noisy and all the marked creeks contained water. On the morning of 9 June they discovered a north–south track and set up their Claymores in an ambush. At 2.45 pm six VC walked from the north into the ambush area, and remark-ably the first two VC sat down on the track in the centre of the killing ground. The remainder were only a few metres behind when two Claymores were fired. It is not known what happened to the first four VC, but the fifth VC dived off the track towards the patrol while the sixth VC began to run down the track just as another Claymore was initiated. When the dust cleared only one body, that of number five, could be seen; there was a large patch of blood where the VC had sat on the track. A quick search of the body revealed one French 9 mm pistol and a pack containing documents and money. Within half an hour they had been winched out by helicopter. An examination of the documents revealed that they had killed a VC tax collector burdened with the produce of his work, and although somewhat discoloured the money was gladly received by the South Vietnamese authorities. The pack had contained over 300 000 piastres and the patrol subsequently received a reward of A$300. From then on C Troop was known as Currency Troop.

Fraser had no contacts in July, but towards the end of August he was ordered to try to capture a VC. Jim Furner had been urging Burnett to capture a VC and various schemes had been discussed. One suggestion was to use CS gas but there was too long a time delay after a grenade was thrown, so the squadron experimented first with CS gas crystals in a fire extinguisher, and then with blasting CS gas crystals from various containers attached to a tree. During one experiment on Nui Dat hill the drifting CS gas succeeded in putting the Task Force's Luscombe airfield out of action. Fraser preferred to conceal himself beside the track and literally leap up and grab a passing VC. On 31 August while on a 'snatch' patrol his patrol struck a party of three VC. They killed two and wounded the third by firing at his legs, but the wounded VC put up a brave fight to prevent his capture and was eventually killed as well.

1 Squadron mounted a number of snatch patrols, all without success. One example, in which Fraser was involved, was a nine-man patrol led by Rick Gloede south of Nui Toc Tien towards the end of July. Claymores killed one VC, but the other VC whom the patrol had hoped to capture fired back and eventually escaped. Hindson said later that they 'went to the most ludicrous schemes to capture a VC, but you never knew whether the man you had selected to capture was the Far East karate champion or the fastest draw in the North Vietnamese Army'.

Despite the SAS's failure to seize a prisoner, the squadron continued to provide valuable information to the Task Force Commander. The nature of the information included the location of VC base areas, VC movement between their base areas and the villages, assessment of VC morale and physical conditions, the identification of new units, and the concentrations of VC and NVA troops. Sometimes the Task Force Commander could directly task his infantry units to follow up this information. Often the information became another piece in the continuing jigsaw puzzle of VC activities being worked on by the Task Force intelligence staff.

When battalions were given their intelligence briefings for operations they were not always told that specific information had come from the SAS. It is therefore extremely difficult to quantify the worth of SAS intelligence gathering, except to believe the testimonies of the various Task Force and battalion commanders. As the war progressed the battalions realised the value of the information held by the SAS

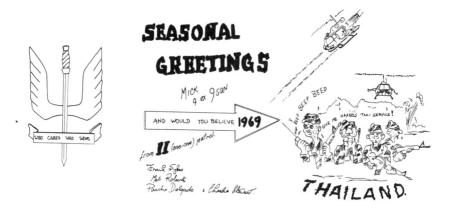

This Christmas card, sent from Frank Sykes's patrol to Flight Lieutenant M. J. Haxell in 1969, illustrates the close bond between the SAS and No 9 Squadron RAAF in Vietnam.

operations office, and often battalion intelligence officers would visit Nui Dat hill to gather further intelligence before their battalion mounted its operation.

Burnett's policy of mounting ambushes to obtain information reached its peak during the middle months of 1967, even though the patrols did not always have time to set their Claymores. For example, an ambush patrol led by Sergeant John (Blue) Parrington shot one VC while crossing a track three kilometres north east of the mouth of the Song Rai river on 27 May, and a ten-man patrol led by Bill Hindson ten kilometres north east of Nui Dat Two was investigating an enemy camp on 20 June when one VC was seen at a distance of 30 metres across a large crater. Sergeant Neville Farley killed him with one shot which hit him in the right eye. In the same area on 25 June Sergeant O'Keefe's ambush patrol shot two and wounded another VC while in a defensive perimeter near to a track.

But it was the Claymore ambush that held out the hope that a complete VC party could be killed and documents recovered, as had been achieved by Hindson and Fraser in May and June. On 3 July Sergeant Ron Gilchrist's patrol was inserted just north of Nui Thi Vai into an area of primary jungle with patches of dense undergrowth. They crossed the flowing Suoi Sao creek and by the morning of 5 July were in position beside an old ox-cart track. Gilchrist set up a Claymore on each flank but had not connected the Claymores in the killing area when five or six VC were seen approaching from the east. He had no option but to initiate the ambush by killing one enemy with rifle fire. The flanking Claymores were fired and another enemy standing in front of a Claymore vanished—his body was never found. Meanwhile the main Claymores had been connected and were immediately fired. Three bodies were found in front of this area, plus the body of the man who had been shot. Three weapons, personal equipment, money, a diary, two wallets and documents were recovered.

Then on 12 August a five-man patrol commanded by Sergeant Ola (Ocker) Stevenson initiated their Claymores on a group of ten VC, 8 kilometres north of Xuyen Moc, killing five. The patrol report noted: 'As patrol was being extracted, the helicopter was fired upon and sustained a few hits'. But the RAAF historian wrote: 'While on the landing pad, with the patrol enplaning, enemy fire erupted and severed the engine oil [line] depriving the captain [Pilot Officer Michael Haxell] of power indication. Because of the position of the enemy and the size of the landing pad, a difficult down wind take off had to be attempted...Haxell was later awarded the Distinguished Flying Cross'.[4]

The close relationship between the RAAF and the SAS is emphasised by comments written by Second Lieutenant Rick Gloede about Mick Haxell:

This totally unflappable gentleman also flew a mission with a patrol I led, lost 'revs' coming in to the Landing Zone, and crashed the chopper. In the process, Ray (Woofer) Neil, who was on the skids as the chopper descended, was hauled back in very quickly, and as the chopper hit the ground, Neil hit the roof, and his rifle was flung out of the aircraft. All members of the patrol were injured (Neil seriously enough to be sent home), but Haxell managed to get the aircraft back into the air. He had no hover capacity, could only turn in an extremely wide arc, and consequently decided to land the chopper on Luscombe Field like a light airplane. Fortunately this was successful, although by the time he had landed the skids were sunk approximately eight inches [20 cm] into the tarmac. I'm not sure whether it wasn't this action for which he received his DFC. All I know is that he was a very fine pilot and an excellent person.

Despite the successes of the ambush patrols, the reconnaissance patrols continued, and often proved more dangerous because the SAS was required to move in close to enemy camps. However, whatever the purpose of the patrol, the SAS was most vulnerable during helicopter insertions, as Lawrie Fraser had found out in April and May. Sergeant Barry (Tex) Glover and his five-man patrol had a similar experience on 21 June when they were inserted at 6.05 pm into an area of defoliated jungle 5 kilometres north east of Thua Tich, near the so-called lakes district. The patrol moved quickly off the LZ and paused on the edge of the jungle to listen. To the west they could hear whistling noises which they took to be VC signalling. They advanced slowly for about 150 metres west of the LZ and from about 30 metres in front came the voices of five VC, men and women, in what appeared to be an enemy camp. The patrol therefore withdrew north and at 6.45 pm in failing light stopped to prepare an LUP. Suddenly two VC appeared five metres in front of one of the SAS soldiers, who killed both with shots to the head and chest, and then saw a third VC. The rest of the patrol joined in firing and as there was no return fire the patrol moved away quickly. They had gone about 100 metres when the last man, Corporal Steve Bloomfield, warned the patrol that the enemy were following them to their left rear. The patrol went to ground and waited. Sensing that the Australians had stopped, the VC also stopped, turned right and advanced in extended line. The enemy force was of unknown strength—perhaps fifteen to twenty VC—and in the fierce clash that followed several VC were killed. By now the patrol was being engaged by about ten weapons at varying ranges so they broke contact, moved away quietly and at 7.30 pm crept into an LUP.

Glover recalled that 'for many hours the area was mortared, rounds coming in so close that dirt and tree branches were landing on us. There was also during the night, a lot of enemy activity and at times it was very close to our position; we were unable to move. During those hours I prayed and no doubt the other patrol members did the same'. The VC were still in the area the following morning and it was extremely difficult to put out the radio aerial to establish communications. The VC were calling out in English for the Australians to surrender and the patrol lay quietly until about midday when an American Tomahawk aircraft flew overhead and Glover activated his URC 10. The extraction procedure was initiated and the patrol was winched out. As Glover described it: 'The helicopters came over, they told me to throw red smoke to indicate my position and the enemy. All hell broke loose—gunships giving us supressing fire and enemy returning heavy fire from the ground. One at a time patrol members were winched out of the jungle (120 feet [36 metre] high trees). As the patrol members got into the slick, one took over the GPMG; others firing SLR, M16 and 79. I was doing the same on the ground. After being winched up into the slick, we waited some distance away and watched two Phantom jets drop 500 lb [226.8 kgs] bombs into the enemy position'. It was thought that there had been at least a company of VC in the area.

The danger surrounding insertions was emphasised again on 23 July when Chicka Baines' patrol shot two VC only minutes after landing 6 kilometres east of Binh Ba. Sergeant Jack Gebhart's patrol had a similar experience on 10 August when they landed 6 kilometres north east of Nui Dat Two. Only 15 metres south of the LZ they struck three VC, killing one and wounding two. It was a tribute to the high level of training and professionalism of the SAS that in all these hazardous insertions they had no serious casualties and inflicted considerable loss on the enemy.

Once the danger of the insertion was overcome the SAS could get down to their business of reconnaissance or ambush. On 27 June Sergeant Frank Sykes's five-man patrol was inserted about five kilometres east of Thua Tich to confirm that a major VC headquarters (possibly 274 Regiment) was located in the area. Sykes later described the patrol:

> The next day was a mixed one to say the least as noises were heard in the distance to the SW but again nothing that could be positively identified. An interesting aspect is that the fairly large grassed hut we located had a tunnel entrance alongside...it was just off the side of the main clearing and had been used recently.
>
> Murphy's law appeared as we hit the oxcart track, as a well-used footpad which paralleled it was our undoing. The scout and myself had moved forward to check things out when voices were heard approaching from the east. There was nowhere to hide and I guess we made some noise moving back to the patrol as there was a lull in the voices, then firing commenced from our patrol. We rejoined the other three members which was only a matter of yards and withdrew.
>
> After a reasonable distance and a dog leg thrown in we stopped for comms and reported the contact. The OC had previously asked us to stay in after contact if possible and as it appeared that no immediate follow-up was taking place we informed HQ that no exfil was required. It wasn't long after closing down that noises were heard to the east from more than one point and our assessment was that a sweep was being organised.
>
> After a fairly quick appreciation of the ground, and we already knew what was to the north from our air recce and infil, I decided that we would have to cross the clearing and hide on the other side. There was sufficient cover from view to cross it if we crouched and this we did using dry fire and movement.
>
> As we climbed out of the clearing up a low embankment a container and fishtrap was noticed in the creek itself and I think everyone then knew that we were in a warm area. Anyhow I decided to stick to the plan and LUP'd to listen and observe as we had a reasonable field of fire over our back trail and the options were limited anyway. I signalled [Lance-Corporal] Harry Devine to move out to the west about 5–10 yards just to ensure that we were not sitting directly next to a track and while he was doing so the world fell in.
>
> As the contact developed firing was heard from the other side of the clearing and I still feel that it was the M79 and XM148 which gave us enough breathing space to withdraw. We hit the centre creekline and moved north fairly quickly with the firing stopping as suddenly as it

started. It was evident that we were not in a good tactical position and that the enemy would reorg and continue a follow-up but the ground itself was not really in our favour.

Communications were established and exfil requested and from a talk that I had with [Lance-Corporal] Sam McDonald later he cut it right to the bone and sent 'Send Help'. Sam was a good signaller and fairly meticulous with his procedure so those words are fairly indicative of the tension at that time. No matter, help soon arrived in the form of a Bird Dog observation aircraft and he co-ordinated the supporting fire from the USAF, gunships and I think artillery.

The cavalry arrived in the form of Sqn Ldr Cox of 9 Sqn and I cannot speak too highly of those guys; he bellied the helo into the secondary growth and literally chopped his way down to us until the skids were chest high. [Private] John Delgado was our shortest member and he was thrown up with some vigour and with all the adrenalin present we were soon inboard and on our way. A yarn with 9 Sqn later brought out the fact that new rotors were required for that helicopter and I think it may have been the incident that Coxie got his well deserved decoration for.

Sykes did not mention in his account that the enemy had been firing a US .30 cal machine-gun and that one VC had been hit in the chest with an M79 round at range of about 25 metres.[5]

One exciting reconnaissance patrol was commanded by Bill Hindson. Burnett described Hindson as a 'bit of a scallywag but a bloody good soldier'. Well built, 1.75 metres tall with dark hair, sharp features and piercing eyes, Hindson led by example and was highly respected. One of his patrol members, Private Noel De Grussa, re-called that he 'personally liked Bill Hindson as a patrol commander. He was not liked by everyone but I found him to have a good sense of humour, which was always good for morale but at the same time he possessed a calm and calculating demeanour'.

The infiltration of the patrol soon after midday on 9 August into an area about four kilometres south east of Thua Tich still stands out clearly in De Grussa's mind: 'The helicopter ride at tree top level was always exciting but shortly before approaching the LZ I remember seeing a number of trees which had been cut down. The trees had been stripped and removed and the stumps had been camouflaged with a few handfuls of sand being placed on the top to cover the bright colour which was easily visible from the air. Shortly after infil, various noises were heard...signal shots, chopping, etc, and we were 100 per cent "on guard". We were always "switched on" when on patrol but this area seemed more tense than usual and we expected a contact at any moment. The jungle was very dense and in parts we were on our hands and knees trying to get through it'. Clearly the enemy were in the area, and as Hindson remarked later: 'they were bloody close'. On 10 August they sighted four VC and again heard shots and voices. It was nerve-racking work. At one stage, while filling their water bottles at a creek, they could hear some VC also filling bottles or washing not 15 metres away around a bend in the creek.

Hindson and his men were still in the area the next day and at 10.15 am, while sitting close together in dense undergrowth, they saw two VC. De Grussa recalled that the first VC 'was about 15 to 20 metres from us, walking in a northerly direction and was moving tactically. That was something that gave us a bit of a jolt because at that stage the most VC we had ever previously encountered were always talking to each other, weapons slung over the shoulders and generally slack. This soldier was dressed in jungle greens, carrying his weapon at the ready and he wore a green hat similar to the Australian issue we were wearing. The front of it was folded up obviously to give him better visibility'. While they were watching the track in drizzling rain, during the next fifteen minutes they counted 63 VC carrying rifles, five machine-guns, and a crew-served weapon, either a mortar or a .50 calibre machine-gun. They were moving very quietly in tactical groups.

De Grussa remembered that 'our eyeballs popped out at what we saw and I think all of us felt a bit insecure at that moment. There were five of us and more than 60 of them and communications were only by morse. We could not communicate with anybody unless we ran out an "end fed" aerial for about twenty metres and then tapped on a morse key for a few minutes. Hopeless trying to do that in the heat of a long running contact.'

The SAS patrol remained nearby chewing a few sweets for lunch until midday when they radioed the information to Nui Dat, and then at 12.55 pm they moved off again on patrol. Their route took them across the track, and when the forward scout, Lance-Corporal Alan Roser, stepped onto the track one VC also emerged about 20 metres away. Roser fired two aimed shots, and as more VC appeared he quickly fired the remainder of his magazine at the light machine-gun that had opened fire on the Australians. He then carried out the correct contact drill, moving past Hindson, towards the rear of the patrol. Hindson was supposed to follow, but he remained in contact with the enemy. Although Hindson fired several magazines at the enemy—he later thought that he had fired nine magazines—the enemy fire was directed at the other members of the patrol.

Eventually the enemy fire stopped, and Roser came forward to assist Hindson and to inform him that De Grussa, the radio operator, was badly wounded in the left thigh. According to De Grussa:

It was a stunning blow, like being hit with a sledgehammer and I did not know initially where I was hit as my left leg from my backside to my toes was completely numbed. I was unable to walk and very concerned about the long range radio in my pack. I shouted that I had been hit and told Gerry Israel to take my pack. I had expected to be left there because of the situation. A wounded member of a five man patrol would render the patrol ineffective if two members were engaged in carrying a wounded [man]. This was something that we had all discussed at one moment or another and in this particular situation I was waiting for Bill to say to me...that he would have to leave me. He did not do this and suddenly all was quiet, the shooting had stopped and Bill decided to take me with him.

Covered by fire from Hindson, the patrol withdrew about 50 metres and then the medic, Private Gerry Israel, applied a torniquet to De Grussa's leg, which was bleeding heavily. They set off again, Roser leading, Israel and Private Pieter Van As carrying De Grussa, and Hindson covering the rear. After a further 150 metres they stopped and Hindson initiated his URC 10 beacon as De Grussa was the only patrol member who knew morse code. Fortunately, after a while De Grussa was able to gather sufficient strength to send a message back to Nui Dat. The shortage of radio operators commented upon by the commanding officer during the previous year was underlined dramatically. De Grussa recalled that 'We waited an agonising hour or so, expecting the enemy company to sweep through at any moment, knowing full well we would have a hell of a time trying to keep them off. Whilst I don't know precisely how the others felt, I just couldn't see how we were going to get out of there.'

Meanwhile, the URC 10 beacon had been picked up by a passing FAC. Within 45 minutes US Skyraiders and F100s were attacking the enemy and RAAF helicopters were overhead. Hindson fired a pencil flare though the canopy; a cable with harnesses was winched down, and De Grussa and Israel were lifted up. Next went Roser and Van As, and finally Hindson was extracted. The enemy were approaching, the gunships were firing and the door-gunner of the helicopter overhead engaged two VC, hitting one. Hindson on the end of the rope joined in the firing and emptied a magazine while holding his rifle by his side—the recoil caused him to spin wildly on the winch cable.

Burnett had allowed his patrols to carry a waterbottle filled with rum in an attempt to counter the bronchitis starting to develop among the soldiers who were spending days in constantly wet conditions. Once the extraction was complete the patrol felt that they had earned a solid drink from the waterbottle, which they also offered to the thirsty-looking pilot. Expecting it to be water, he took a hearty gulp and almost flew the helicopter into the jungle. They landed at Nui Dat at

2.40 pm. The patrol had been lucky to be extracted with only one casualty, and it caused immediate repercussions. The squadron had been contemplating changing to four-man patrols, but after this experience decided to stay with five-man patrols.[6]

The period from May to August had been particularly productive, but early in September the squadron assumed responsibility for patrolling out to Line Alpha within the Task Force TAOR, relatively close to Nui Dat. Many of the patrols during this time included US Long Range Reconnaissance Patrol (LRRP) personnel from 101 Airborne Division. There were few sightings, and Lawrie Fraser reported after one patrol that this type of operation was undesirable for soldiers with SAS training. The long distances covered each day induced bad habits and the soldiers became tired and consequently careless.

Fraser's concern reflected one of the problems of commanding SAS operations. After six months of tense patrolling the soldiers could do with a change of tempo, but less demanding tasks were a waste of SAS training and expertise. Not surprisingly the problem of maintaining morale and effectiveness was exercising the minds of both the higher command in Vietnam and the commanders in Perth.

When 1 Squadron had arrived in Vietnam in March it was planned that their tour would be for a period of twelve to thirteen months, but most expected it to last for no more than nine months, and some claim that they were informed that the tour would end in October. Then in April 1967 the new Commanding Officer, Lieutenant-Colonel Len Eyles, learned that 4 Squadron was to be disbanded. Since this squadron held the reinforcements for 1 Squadron, he decided to send nine of these reinforcements to Vietnam to allow Burnett to employ five patrols per troop. Troop sergeants would then be able to form their own patrols whereas earlier they had acted as second-in-command to their troop commander, or had led mixed patrols.

Eyles also learned that both Burnett and Brigadier Graham favoured a nine-month tour in Vietnam for SAS squadrons. Both Eyles and the Commander of Western Command supported this proposal, pointing out to Army Headquarters that the nature of SAS operations imposed greater strain on individuals, particularly patrol commanders. Experience from Vietnam indicated that patrol commanders had become run down both mentally and physically after about seven months of sustained operations. For many men of 1 Squadron, this was their second operational tour within two years, and along with extensive periods of training they had spent little time with their families. It was possible that after returning from Vietnam some members of the squadron would wish to leave the SAS. If 4 Squadron was reformed, the squadrons could rotate through nine-month tours.

These arguments, however, did not persuade Army Headquarters. The Director of Military Operations and Plans, Brigadier R. A. Hay, accepted that SAS operations were arduous, but observed that the tour for SAS squadrons had to be related to the nature and tempo of operations. 'If Commander 1 ATF flogged them there is no doubt that a six month tour would probably be more than enough for most. This does not appear to be the case at the moment, with most of their patrols being of short duration and relatively close to the Task Force base.' From his discussions with Major Murphy, Hay had gained the impression that 'those showing the greatest strain were former members of 1 SAS Squadron who had already served a tour of duty in Borneo under very trying conditions on long range patrols of quite long duration. The remaining members of the squadron although slightly jaded after quite intensive operations were quite capable of effectively completing a twelve month tour.'

'It is significant', he added, 'that during its tour of duty 3 SAS Squadron sustained only one operational casualty. This would not seem to indicate that members of the squadron were as run down mentally and physically as the W Comd letter would appear to imply, particularly as the squadron killed over 40 of the enemy.' Hay concluded that there was 'no operational reason why the tour of duty should be reduced below 12/13 months', and the SAS Regiment was informed of this decision on 23 August.

Meanwhile, back in Phuoc Tuy Province the squadron reconciled itself to a longer tour; but many soldiers disagreed with the decision on professional grounds. For example, Sergeant Frank Sykes commented later: 'It has been fairly well proven in small unit operation (Special Forces) in a fluid state of war that six months is about the right time for a tour of duty...Personally I believe that as it takes a long time comparatively to train an SAS soldier the aim should be to keep him and develop his potential for further service, as the private soldier of one tour is probably the NCO bracket of subsequent tours, and it was obvious that we were in for a long haul. A twelve month stint just grinds them into the deck.'

At the end of September Headquarters 1 ATF submitted a quarterly summary of its operations and observed: 'Bushmaster style operations have been conducted with a marked degree of success. Harassing and ambush raids deep into enemy base areas and lines of communications have accounted for 27 VC KIA during the quarter and much valuable intelligence has been gained.'[7] But when, towards the end of September, 1 Squadron resumed its ambush and deep reconnaissance patrols they did not quite achieve the same level of success as earlier in the year. The first ambush mission in this new series was conducted by a ten-man patrol commanded by Bill Hindson. The patrol was unsuccessful and the report explained the reason: 'Two members contracted bad coughs and another three members were coughing which ruled out the possibility of setting an effective ambush. It was noticeable that some members of the patrol were showing signs of nervous reaction to this type of patrol, whilst others were becoming careless.'

The style of operations was beginning to change, caused partly by the fact that in October Brigadier Ron Hughes replaced Graham as Commander 1 ATF. Whereas Graham had been content with ambushes to gain information Hughes was keen to use the SAS more offensively. Burnett disagreed, as he explained later: 'Hughes believed patrols should fire on enemy to inflict casualties regardless of the patrol mission, ie patrols briefed for recce/surveillance should achieve kills if opportunity arose. This contravened the basic infantry doctrine of recce patrols v. fighting patrols and could have led to our interest in a particular area being given away (for the sake of body count) which could in turn jeopardise the success of TF operations being planned for that area. Patrols commanders must know whether they are recce or fighting patrols.'

Despite this disagreement, Burnett did not object to his patrols being used offensively as part of a deliberate plan in Operation SANTA FE. Beginning on 26 October, Operation SANTA FE was a Task Force operation conducted in co-operation with US and Vietnamese forces in the north east sector of the province with the aim of destroying elements of the 5th VC Division. Both 5 and 7 RAR patrolled towards the May Tao mountains while the US and Vietnamese forces advanced from the north. The SAS had the task of ambushing the 7 RAR area of operations once the battalion had moved through it, while a few SAS patrols operated in the western part of the province to provide rear security for the Task Force.

Three patrols operating to the west of Binh Ba were particularly successful. Sergeant Shaw located an occupied enemy base camp covering an area of 1000 square metres and on 9 November directed effective airstrikes onto it; about 1,000 metres away, on 17 November, Knox's patrol initiated a Claymore ambush and killed three VC; while closer to Nui Dinh Fraser watched an occupied enemy camp for two days, counting between fifteen and twenty VC.

Meanwhile between 3 and 25 November fifteen SAS patrols were deployed along a line stretching from just north of Binh Gia to the east of Xuyen Moc. On 6 November Sergeant John Duncan's patrol killed three VC in an ambush six kilometres east of Binh Gia. The same day Warrant Officer Wally Hammond's patrol located an occupied enemy camp four kilometres east of Binh Gia and directed an airstrike onto it. Two kilometres south of Thua Tich Sergeant Glover's ten-man patrol, which included Major Burnett and two US Navy SEAL personnel, killed one VC and located an enemy camp on 8 and 9 November. Sergeant O'Keefe located an enemy camp four kilometres east of Binh Gia; Duncan killed one VC three kilometres north of Binh Gia; and Sergeant Farley killed three VC six kilometres south east of Thua Tich. Thus during Operation SANTA FE 1 Squadron accounted for nine VC for no losses. The total casualties incurred by 1 ATF and the Americans numbered five killed and 22 wounded and they inflicted 38 VC killed, 5 wounded and 5 captured.

By now numbers of kills were becoming an important factor in gauging the success of an operation. But the strain was begining to tell on many of the patrol commanders, and this was not helped by a directive from Task Force headquarters that after a successful contact patrols should not be extracted, but remain in the jungle to try for a few more kills. The SAS patrol commanders considered this to be an unsound policy which they tried to avoid.

Also at this time patrols were being inserted at increasingly greater distances from Nui Dat and this created special problems for the RAAF, as explained by the commander of No 9 Squadron in his monthly report for November 1967: 'About 50 percent of the patrols inserted in November operated at ranges up to 25 000 metres from ATF. With the apparently increasing pacification of the province insertions can be expected at even greater ranges as the sphere of influence of ATF increases. When distances of over 35 000 metres are exceeded the squadron expects to have to make special provisions for refuel at secure areas nearer to the insertion point'. The additional fuel was required so that the aircraft could hold in the area for 30 minutes after insertion. As it happened, in December one patrol was inserted 40 000 metres from Nui Dat and an intermediate refuelling point had to be arranged.[8]

During November Second Lieutenant Kevin Lunny arrived at Nui Dat to relieve Lieutenant Ivey as operations officer. Lunny was to continue as operations officer of 2 Squadron when they arrived in February, and his arrival was a welcome reminder that the end of 1 Squadron's tour was approaching.

Throughout December and January reconnaissance and ambush patrols continued across the province. One notable patrol was a ten-man patrol (including two SEAL personnel) commanded by Tex Glover. On 12 December they initiated a Claymore ambush 4 kilometres north east of Binh Gia killing seven and possibly wounding another VC. On 3 January Hindson's patrol killed two VC. On 6 and 7 January Fraser found a cache of VC weapons. On 14 January Duncan killed one and possibly another VC. Hindson had another success, killing four VC six kilometres north west of Thua Tich. And Sergeant Frank Sykes triggered a Claymore ambush north of Binh Gia killing three and possibly another VC.

By the end of 1967 the members of the Task Force were cautiously confident that they had achieved a measure of security in the province. They had swept through some of the VC base areas, had cordoned and searched many villages, and had denied the recently harvested rice to the VC. Much of the success could be attributed to the SAS. A background summary prepared for the media and issued by the Minister for the Army, Malcolm Fraser, did not mention the words SAS, but coyly noted that 'small Task Force patrols continued to probe into enemy-held areas...Viet Cong defectors have revealed that along with major offensive operations, air strikes and artillery, they have great fear of the small, long range patrol which catches them unawares in their base areas and on exclusive supply routes...Travelling on foot, with lightweight equipment and rations, the patrols keep close watch on enemy unit movement and supply or staging areas. Valuable information from captured documents and visual reconnaissance has enabled Task Force planners to mount operations in strength against enemy units.'[9]

To add to these successes, in December 3 RAR arrived at Nui Dat, thus bringing the Task Force to three battalions, and it was announced that the force would soon be reinforced with tanks. These additional troops made it more viable for 1 ATF to be deployed outside Phuoc Tuy Province, and on 23 January 1968 HQ 1 ATF and two battalions, 2 and 7 RAR, moved into Bien Hoa Province for Operation COBURG. With them went Knox's and Fraser's patrols to conduct reconnaissance for the Task Force, with Hindson acting as operations officer; Second Lieutenant Gloede, Warrant Officer Hammond and four signallers were attached to 199 (US) Brigade and 18 ARVN Division respectively as liaison officers. Burnett was 'worried as all hell' about this deployment as the patrols would not be under his command and 'it was always dangerous when you operated' close to larger units. The squadron was almost at the end of its tour and he was concerned at the sort of tasks which might be given to his patrols.

When the VC Tet offensive began at the end of January the Task Force was well positioned to deny their area of Bien Hoa to the VC, but there was little role for the SAS. There were too many VC for the SAS to operate normally. For example, the first attempt to insert Knox's patrol was thwarted by eight VC on the LZ. Two VC were killed and the patrol was extracted 30 minutes later. Fraser's patrol lasted only fifteen minutes longer; they too were extracted after a contact. Finally, in an effort to provide more protection the two patrols joined together and on 29 January patrolled out from 7 RAR battalion headquarters. After half an hour they struck a small VC party. They tried again two hours later and once more were seen by the VC. The next day they tried for a third time and within two hours had come under enemy fire. They returned to Nui Dat on 1 February.

On 6 February the advance party of 2 Squadron arrived and during February some 2 Squadron personnel joined the 1 Squadron patrols around the Task Force base and took part in the cordon and search of Hoa Long village. Most of 1 Squadron was occupied preparing to return to Australia, and tragically, on 13 February, while disposing of grenades, that excellent soldier, Chicka Baines, was killed by a prematurely exploding M26 grenade.[10] Earlier in the tour Private Rick O'Shea had died of encephalitis.

On 20 and 21 February the squadron assisted with the cordon and search of Long Dien, and at 7.30 am on 26 February they departed from Luscombe Field for Saigon. As Sergeant Frank Sykes wrote later: 'It was a fitting end to look back and see the hill a mass of varied coloured smoke as dozens of smoke grenades had been left with time pencils attached'. They arrived in Perth at 7.15 that evening. During the tour the squadron had mounted 246 patrols; had killed 83 VC, and possibly a further 15, and had sighted a total of 405 enemy. Dale Burnett might have been less than completely happy to see the squadron's role changed from fighting for information to harassing the enemy, but he was proud of the outstanding results achieved by his men. It remained to be seen how the next squadron tackled the new situation brought about by the VC Tet offensive.[11]

14
Phantoms of the jungle
Vietnam patrolling: 1966–1968

By the end of 1967 the Australian SAS had built a formidable reputation, which was based largely on two factors. Firstly, it had been
outstandingly successful, both in providing information to the Task
Force and in harassing the enemy. The information was reliable and
the SAS had accumulated an impressive list of kills. Secondly, by
covering their operations with a cloak of secrecy the SAS soldiers had
built up a mystique. The tags of 'super-grunts' or 'super-soldiers' were
often used by ordinary diggers to describe the SAS, and although the
men of the SAS were the first to deny their validity, these tags contributed to their reputation for daring, secret operations behind enemy
lines. Few other Australian officers or soldiers were permitted to enter
the strictly controlled inner sanctum of 'SAS Hill', and the SAS were
tight lipped about their techniques. On leave they stuck together, and
often all that the battalions knew about them was that they were the
people who provided the information on which their operations were
based.

The mystique of the SAS extended to the VC and NVA. One VC
prisoner, interrogated later in the war, stated that he and his unit were
very much afraid of the 'Biet Kich uc da loi'—the Australian Commandos. The prisoner said that they knew that the US and South
Vietnamese had similar units but they were not greatly worried by
them. The Australians, however, were hard to detect and were very
aggressive when contacted. In an effort to locate the SAS the VC had a
standing procedure whereby if helicopters were heard they would send
a three man cell to the clearing in the area to see if the helicopters had
unloaded the SAS.

In fact the VC placed a high priority on stemming the operations of the SAS, causing the Commander, Australian Force Vietnam to comment in a letter to the Chief of the General Staff on 28 June 1968: 'As an example of the present situation in Phuoc Tuy, we are having difficulty in lodging SAS patrols in the "Long Green"; they get "jumped" at insertion on practically every occasion'.[1] One of the Task Force commanders, Brigadier Sandy Pearson, suspected that a Vietnamese radio intercept unit was giving the Viet Cong early warning.

That the VC would go to such trouble to prevent the insertion of SAS patrols was a measure of their concern about their own security, both from the SAS's information gathering and from the swift, violent deaths of small VC parties in their rear areas from an opponent that they almost never saw. It was for good reason that the VC referred to the SAS as Ma Rung—the phantoms or ghosts of the jungle. Indeed even among their own kind the SAS were known as 'can't be seens' or 'walking trees'.

At one time it was rumoured that the VC had put prices on the SAS heads, supposedly up to 6000 piastres (A$60). Another account stated that a specially trained tracker unit was deployed in Phuoc Tuy Province to try to grab an SAS prisoner. The account added that the VC made it known in the villages and hamlets of the south that each Ma Rung was worth $US5000 dead or alive.

Whether these rumours were true or not, and the SAS treated them as a bit of a joke, the Australian higher command also appreciated the worth of the SAS. For example, on 18 December 1967 the Commander, Australian Force Vietnam (AFV), Major-General Tim Vincent, wrote to the Chief of the General Staff that the overall effect of the Task Force as a 'killing machine' was causing him some concern. He pointed out that except when the enemy presented himself for slaughter as he had at Long Tan, the infantry battalions were his least effective killers. He continued:

> Please do not think that we measure overall success by body count kills
> —separation of main forces from the population is important, perhaps
> of paramount importance—erosion of the infrastructure is an essential
> element of success...But what is clear is that infantry on their own in
> search and destroy operations against a dispersed enemy kill at about 3
> to 1. The fleeting enemy and our rifle are too evenly matched. This was
> one of the reasons in asking for medium tanks which can accompany
> the infantry most places with their canister guns. Dispersed or dispersing
> VC can nearly always elude our foot infantry who have insufficient
> immediate contact firepower while on the ground mobility of our
> infantry is no better and usually inferior to that of the VC...On the
> other hand small infantry parties operating where the enemy has

Sometimes, as a result of a successful ambush, the SAS captured underdeveloped film taken by VC soldiers. This photograph taken from captured film shows a typical VC patrol, and gives some idea of the thick vegetation in which both the VC and the SAS operated. At times the SAS were so close to the VC that they could have touched them, and their ability to remain concealed caused the VC to give them the name 'phantoms of the jungle'.

freedom of unobserved movement can reap a comparative harvest. Our SAS now have 81 kills to their credit with one Aust DOW and one Aust WIA. If for every 100 combat infantry slice the Allies could kill 80 enemy a year, as our SAS do, then there would be no worthwhile enemy alive after a year. We would like to do more SAS type ambush patrols but we do not have the Iroquois lift and gunships to do it.[2]

Several weeks later Vincent wrote to the Chairman of the Chiefs of Staff Committee, General Wilton, that he had told the Task Force commander to orient his force more towards SAS type operations: 'if we can get a lien [sic] on more gunships from US Army then we are in business. There is no doubt that the SAS type operation is our most profitable—115 enemy KIA with two friendly wounded and one died of wounds—and we should do more of it. Currently the Task Force is operating by dispersed companies, artillery supported, and this is producing more contacts.'[3]

The idea that ordinary infantry could do the same work as the SAS surfaced at various times. And to some extent it existed among SAS members themselves, perhaps because they realised that in the restricted area of Phuoc Tuy Province they were not operating at long ranges deep behind enemy lines as they had originally been trained. In March the SAS told two visiting British defence attaches that 'any infantry section, given a little extra training and additional helicopter support, could carry out the tasks done by a[n] SAS patrol in Vietnam'. The British officers, however, reported to their superiors that they felt that the SAS was 'being too modest about their successes which are a direct result of the long and arduous training...as well as sound leadership down to the lowest level plus a sense of dedication at least equal to, if not better than, that of their opponents'.[4]

Whatever the merits of Vincent's desire to use infantry in the SAS role, his comments about the effectiveness of the SAS were supported by a report prepared by the Australian Army Operational Research Group for the period from May 1966 to October 1968.[5] The research group reported that the infantry battalions (two until December 1967 and then three) had been involved in 74 per cent of all Australian contacts compared with 24 per cent by the SAS and 2 per cent by armoured units. However, when casualties were examined it was found that of the 410 enemy KIA, the infantry had accounted for 188, the SAS for 173 and the armour for 49. The infantry themselves had suffered 9 KIA and 73 WIA, the SAS had 19 WIA and armour 2 KIA and 22 WIA. The report stated: 'Some SAS contacts were well planned ambushes of large enemy parties, making maximum use of area weapons such as explosives, and claymores and were considerably more casualty producing than patrol clashes'.

The report went on to indicate that the weapons used by the SAS in contacts were as follows: machine-guns and automatic rifles, 35 per cent of contacts; SLR, 52 per cent; Armalite, 66 per cent; M79 (including XM147 and 148), 51 per cent; Claymore, 29 per cent; M26 grenades, 14 per cent; and others, 14 per cent. The SAS had initiated 91 per cent of all its contacts compared with the figure of 84 per cent for all Australian contacts. Of all the SAS contacts 58 per cent were ambushes. The report concluded that 'the higher casualty rate per contact achieved by the SAS unit can thus be attributed to a combination of weapon mix and type of engagement and it confirms that of reported contacts, ambushes produce more casualties'.

During the Vietnam war there were a wide range of views on the most effective way to use the SAS. One possible role was that of training the indigenous people in guerilla warfare. Unfortunately that role was largely pre-empted by the AATTV, although there were usually a large number of SAS soldiers in the Training Team. Opportunities for hearts and minds operations in remote areas, such as they had conducted in Borneo, were restricted by the fact that the Australians were largely confined to Phuoc Tuy Province which was under the control of a Vietnamese province chief. Similarly, political restrictions meant that the SAS could not be inserted in long range patrols along, say, the Ho Chi Minh trail in Laos or Cambodia. It was galling for some SAS officers and NCOs to find that the SAS was not used as a strategic weapon but was restricted essentially to the tactical operations in Phuoc Tuy. Within the province the role of the SAS oscillated back and forth between intelligence gathering and harassment.

While the SAS was generally confined to Phuoc Tuy, by means of their various efforts to train US and South Vietnamese forces in SAS techniques, the men of the SAS had an opportunity of influencing the war throughout South Vietnam. It will be recalled that in September 1966 Captain Peter McDougall had been detached from the SAS Regiment for service with headquarters US MACV and also at the MACV Recondo School that had been formed to teach LRRP techniques. When McDougall's tour had finished in March 1967 MACV had requested a replacement. Headquarters AFV, commented that 'whilst the attachment does not yield a lot for the Aust Army, it appears that it is of real value to the USA' and recommended McDougall's replacement.[6] On 4 May 1967 Major Jack Fletcher, the second-in-command of the SAS Regiment, took up the appointment as the SAS Liaison Officer/Adviser on HQ AFV.

Within the J2 (intelligence) section of HQ MACV Fletcher was held against two vacancies, a Special Forces qualified major and a US Marine Corps major, and he advised Westmoreland's intelligence officer on all matters pertaining to surface reconnaissance and surveillance. At the same time he acted as the SAS Liaison Officer on HQ AFV, advising the Commander AFV on SAS matters and arranging specialist equipment. Although South Vietnamese forces had been training for Recondo operations under American supervision for some time, General Westmoreland was keen for the SAS to become more involved. At the same time General Vincent was anxious for more SAS-type operations in Phuoc Tuy Province and his staff began to consider a scheme whereby the SAS would train South Vietnamese soldiers within Phuoc Tuy Province.

In November 1967 Captain Tony Danilenko replaced Fletcher and was immediately presented with the problem of getting the training scheme underway. In a letter to Lieutenant-Colonel Eyles, the commanding officer of the SAS Regiment, on 7 December 1967 he described the developments:

237

You've perhaps heard that there is a possibility of SAS personnel having to train and use ARVN. I was told the other week to go to Baria and get 30–40 Chieu Hois [returnees] and start training them. This luckily fell through because they were not available in those numbers. On my return I was told to propose a scheme by the next day which I did. That was yesterday. The General's comment was 'Alright I'll think about it' but I believe he took it with him to the conference with Gen Westmoreland. What the decision will be I don't know but I do know that the General [Vincent] wants more SAS patrols whether by integration or whether by my suggested method.

Danilenko's proposal was to establish an SAS LRRP school at the Van Kiep training camp near Baria to be run by experienced SAS personnel who would train three teams of five every six weeks. Eventually the school would set up an ARVN LRRP squadron for operations within Phuoc Tuy under the control of the Australian Task Force. It was intended that the school would operate under the umbrella of the AATTV. This proposal was largely accepted, as Danilenko explained in a letter on 27 December: 'It has increasingly become clear to me from observations made by Gen Vincent that his main aim is [to] increase ATF's SAS capability within a broad framework of training ARVN...However at this stage he has accepted the premise that a large effort on the part of the Sqn in theatre is not desirable. However he almost had second doubts as to whether we could integrate ARVN into patrols just as the chief of staff was poised to sign the covering letter.'[7]

As it happened, it took some months for the LRRP wing to be established, and when it was, it came under the auspices of the AATTV, not the SAS. In the meantime Danilenko arranged a transfer to the AATTV. On 25 April 1968, while commanding a Montagnard Mobile Strike Force company tasked with establishing contact with the 2nd NVA Division as it moved from Laos into Vietnam, he was killed. He was aged 25. Like many other members of the AATTV, he had been involved in activities which exactly fitted the ethos of the SAS. Whether the SAS should have been more closely and formally associated with the AATTV is a matter for some conjecture. If the Australian Army were to again be required to send a training team to a combat zone there would be a case for sending an SAS squadron or troop rather than raising a new force.[8]

The Australian efforts to train the ARVN in SAS techniques will be described later, but the initiation of the scheme was a direct result of the success of the SAS in 1966 and 1967. It would therefore be useful to examine more closely the basis for the SAS success, especially since its techniques changed only slightly during the following years. Despite the hundreds of different patrols mounted by the SAS in Vietnam, they generally followed a standard procedure. Indeed the success of the SAS rested on two apparently contradictory factors. On the one hand the SAS patrol missions were governed by a fixed planning process, and the SOPs were quite specific in their instructions on how patrols were to be conducted. On the other hand the SAS was given a freedom in planning patrols, in selecting equipment and weapons, and in deciding how to tackle tasks that was not given to the soldiers of other units of the Australian Army.

It could work no other way, for SAS patrols were intensely personal. As one journalist wrote at the time: 'They live on their wits, falling back on some of the most intensive training of the Australian Army. They rely on teamwork but depend finally on their own fine-honed reactions'.[9] Although the SAS patrols varied in size according to their mission, generally they consisted of five men; lead scout, patrol commander, signaller, medic and the second-in-command. Patrol commanders were usually lieutenants or sergeants with a corporal, lance-corporal and two privates making up the remainder of the patrol. But to a certain extent the rank structure was immaterial as each man in a patrol was chosen carefully for compatability. 'Discipline in such small tightly-knit groups', wrote one observer, 'is essentially self-discipline and men are not accepted or obeyed for their rank but for their demonstrated ability to perform'.[10]

At the end of 1967 the most common type of patrol was still the reconnaissance patrol which generally operated by stealth and was kept to a strength of four of five men. They avoided contact except for self protection or to take advantage of an unusual opportunity. Typical tasks for this type of patrol were: collecting topographical information on features, tracks and the state of the ground; locating enemy positions; obtaining details of enemy minefields and other field works; observing enemy habits and traffic; locating sites for crossing obstacles; and checking LZs before the infantry was inserted.

Perhaps the next most frequent task was to mount an ambush with the purpose of gathering information in the form of documents from the bodies and to lower the enemy's morale by harassment. Ambush patrols often consisted of ten men, and they were very effective providing Claymores, which covered more killing ground, were used. The patrol usually had two M60 machine-guns, providing greater firepower and security. It was not unusual for a patrol of five or six men to ambush small parties with the aid of Claymores and in these cases control and movement was far better than with larger patrols.

Often patrols were ordered to conduct combined recce–ambush patrols, and in these cases the primary role was reconnaissance while near the end of the mission if a target appeared the patrol commander

This close-up shows the VC's AK47 weapon, but most VC wore uniform, often with webbing.

could mount an ambush. Other patrol tasks included 'snatches' with the aim of seizing a prisoner, demolitions, fighting and surveillance on the flanks of Task Force operations, and following up airstrikes.

The procedure leading up to the insertion of a patrol was standard. About four days before the patrol was due to be inserted the operations office issued a warning order which included the following information. First was the date and duration of the patrol; depending on the task this would normally be between five and fourteen days. Second was the type of patrol—reconnaissance, ambush or a combination. Third was the method of insertion, which was usually by helicopter. Fourth was the number of the zone to be patrolled; usually a zone or area of operations consisted of nine grid squares each of 1000 square metres, surrounded by a no-fire zone of one grid square. Fifth was the time for the squadron commander's brief. Finally they were given the marker panel and smoke codes required for identification during an extraction.

Once the patrol commander had this warning order he informed the relevant administrative elements of the squadron of his requirements for ammunition, rations and maps and he gave his patrol a verbal warning order. The second-in-command of the patrol then started drawing and issuing rations and ammunition.

After further discussions with the squadron commander the patrol commander then gathered as much information as possible about the terrain, possible LZs and the enemy. To find out if a patrol had been in the AO in question he referred to the LZ register which indicated what patrols had previously used it. Past patrol reports and if possible their patrol commanders were consulted. Both the aviation Recce Flight and any infantry battalion that had operated in the area were also able to provide information. Next, accompanied by his second-in-command, the patrol commander conducted an air reconnaissance by Army aircraft of the AO and possible LZs. The patrol commander then informed the RAAF of the LZ, the direction his patrol would be moving, and the time for the insertion.

Despite all these preparations, the patrol commander had to ensure that his men had the day before the patrol completely free for briefings, rehearsals and test firing of weapons. Rehearsals were crucial to the success of SAS patrols. As one former patrol commander, Captain Andrew Freemantle, wrote in 1974, 'it should be remembered that it takes almost as long to prepare correctly for a patrol as it does to carry it out. It is fair comment that a patrol's success bears a direct relationship to the amount of its previous preparation, planning and rehearsal'.[11] Finally, some time within 24 hours before insertion the patrol commander, his second-in-command and the RAAF flight leader would complete another reconnaissance to confirm the primary and secondary LZs. Conducted by RAAF Iroquois, this would normally be a quick overflight so that any VC observers would not be alerted.

The five methods for insertion were helicopter landings, rapelling, tracked vehicles, waterborne and foot, but the usual method was by helicopter landing. On the day of insertion the patrol attended a pilot's briefing about half an hour before take-off time to ensure that everyone was familiar with routes and immediate drills. Held in a tent, and later a hut beside Kangaroo Pad at Nui Dat, the briefing was conducted by the RAAF flight leader, but often the SAS operations officer attended to give a last minute intelligence summary.

To insert a five-man patrol five helicopters were used. The lead helicopter (Albatross) flew at about 2000 feet some distance from the LZ and controlled the remainder of the flight, consisting of a light fire team of two helicopter gunships for protection, and two 'slicks' for insertion, one to carry the patrol to the LZ and the other waiting at a predetermined area in case one of the other aircraft was forced to land.

When the flight was about 6000 metres from the LZ the lead aircraft directed the slick and the two gunships to descend to treetop level and make their approach. The gunships were about 800 metres line astern. Travelling at over 100 knots at treetop height it was extremely difficult for the pilot of the slick to locate and land quickly at the LZ, so the lead aircraft directed him towards the LZ and advised when he should begin decelerating for landing. If possible the slick landed on the LZ without circling. Sometimes the slick did a 180 degree turn to slow down while all eyes searched the jungle for enemy. The slick landed quickly, or perhaps hovered just above the ground, the patrol alighted, the helicopter lifted off and then joined on the end of the gunship flight which would just be passing overhead. Once on the ground the patrol maintained radio contact with the lead aircraft for between 15 and 30 minutes (depending on the flight time from Nui Dat) after insertion in case of enemy contact. SOPs required the aircraft to remain in the general area until 20 minutes had passed. Once the patrol was deployed two slicks (UH 1Bs or UH 1Hs) and a light fire team were available in case an emergency extraction was required; often these aircraft were given other tasks within the Task Force AO, but SOPs allowed for their rapid despatch if required.

About two days before the scheduled extraction the patrol commander informed squadron headquarters of his intended extraction LZ and the timings were confirmed. The patrol arrived at the LZ with sufficient time to secure the immediate area and informed base of their arrival. When the lead aircraft was in the general area he contacted the patrol by radio and requested them to produce a panel or mirror. Once the gunships were ready for their runs the lead aircraft told the patrol to throw smoke for identification and to show the gunships exactly where the patrol was on the ground. The slick then moved in, picked up the patrol and returned to base, along with the lead aircraft and gunships.

Crucial to the safety of the SAS patrol was its ability to call for an immediate or 'hot' extraction. When the patrol radioed to SAS headquarters that it required an immediate extraction, SAS headquarters telephoned Task Force headquarters which informed the Air Transport Operations Centre (ATOC), collocated with the headquarters. The ATOC radioed the designated flight leader for that day who began directing the necessary gunships and slick aircraft to a holding area. The aircraft then approached the SAS's AO and spoke to the patrol by radio, asking whether the slick could land or whether it would be a winch extraction. The enemy situation was discussed and the patrol commander indicated whether he required suppressing fire from the gunships. Once smoke was thrown the LZ was identified by panel or mirror, and the gunships and slick went 'down the mine'.

Hoist extractions were extremely difficult and trying for both the pilot of the slick and for the flight leader. Flying at above 1500 feet the flight leader was responsible for the whole operation, including the direction of the gunships. Meanwhile, the slick pilot had to keep his craft steady while his crewmen, often under fire, winched up the SAS patrol. The door-gunners tried to engage the enemy who sometimes tried to move as close to the winch point as possible to avoid the fire from the gunships during their firing passes around the slick. While the extraction was in progress, the co-pilot of the lead aircraft would be busy organising artillery to fire on the area once the extraction was complete. He would also call up an airborne FAC to direct any fighter ground attack aircraft that might be available. Since the SAS had presumably located the enemy it was a good opportunity to put in an airstrike on the enemy position before they withdrew.

To an overwhelming extent, the success of SAS operations depended on the co-operation of the RAAF; indeed it was the RAAF who provided the 'air' part of the Special Air Service. One pilot with over 30 years' service in the RAAF recalled that he had never seen closer co-operation between two Services than he experienced in Vietnam. Air Vice Marshal John Paule, who commanded No 9 Squadron in 1968 and 1969, agreed with that assessment. Most of the SAS patrol commanders knew the RAAF crews personally, and if an SAS patrol had a few days free from operations or training a friendly Iroquois crew would fly them down to the RAAF base at Vung Tau where spare beds could be found. Often the SAS patrol and the Iroquois crew would enjoy a night out together in Vung Tau city.

The reliance of the SAS on the RAAF is shown by the statistics for the month of July 1967, when No 9 Squadron conducted sixteen SAS insertions, nine routine SAS extractions, one emergency SAS extraction, two emergency hoist SAS extractions, nine troop lifts for 2 and 7 RAR, 30 medevacs (Dustoffs) and four night medevacs. As early as July 1966 the RAAF squadron commander had noted: 'The most difficult and dangerous tasks have been those associated with the insertion and extraction of SAS patrols into enemy held territory'.[12]

Once they had returned, patrols were debriefed and then prepared their equipment for the next mission, in case they were required to assist another patrol or there was some other emergency. The squadron maintained two stand-by patrols, each of five men, at all times ready to go to the assistance of deployed SAS patrols if necessary, or to execute the destruction of downed aircraft anywhere in the Task Force AO.

The SAS allowed each soldier a high degree of individuality in selecting his own equipment. The dress was either the US pattern camouflaged 'cam suits', or the normal Australian olive-drab 'greens'. After a while it became obvious to the SAS that the VC were presented with a considerable identification problem in that they had to be

243

careful not to shoot other groups of VC, while the Australians could be sure that if they stayed in the designated AO anyone they saw would be enemy. Consequently, rather than wearing their bush hats, some Australians tended to wear scarfs around their foreheads in the VC fashion and allowed their hair to grow a little longer than normal. A number of Australians owed their lives to hesitation on the part of the VC as they tried to identify an SAS soldier in the gloom of the jungle. All parts of the equipment worn or carried were checked for rattle or shine. Personal camouflage, including applying camouflage cream, was carried out just before leaving for the pilot's brief. It was maintained throughout the patrol. Boots were either Australian, British jungle or American pattern. Some soldiers wore green mesh gloves with rubberised palms and index fingers. These camouflaged the hands and protected them from thorns.

Each patrol carried a variety of weapons according to the mission and the wishes of the soldiers. Because of its stopping power, most soldiers preferred to carry an 7.62 SLR adjusted to fire on automatic; sometimes with a shortened barrel. Other patrol members were happy with the M16 Armalite, while a popular weapon was the M16 with the M79 grenade launcher underneath, the early version being the XM 147 or 148 and the later version the M203. For specific tasks the GPMG M60 or the heavy barrelled SLR could be carried. But it was possible to obtain other weapons such as a silenced Sterling sub-machine-gun, and even a silenced Thomson sub-machine-gun. Sometimes patrols carried a shotgun, particularly for firing at an enemy's legs on a snatch patrol. At one stage an automatic shotgun with fleshette rounds was tested, but it was too difficult to control.

The most important pieces of equipment besides the patrol weapons were the wireless sets. The most widely used set, especially at long range, was the AN PRC 64, a lightweight high frequency set that could be used for both voice and morse. The next set in order of priority was the ground-to-air ultra high frequency URC 10 which had both a beacon and voice capacity. It was used in case of emergency and also sometimes on extraction. Two URC 10s were carried within the patrol, one on the frequency of the supporting aircraft and the other on the international (Mayday) frequency, capable of being picked up by most aircraft. The third type of set carried by many patrols was the AN PRC 25, a very high frequency voice radio used as a ground station or for ground to air communications. The combination of all sets gave the patrol many avenues of communication in time of trouble or if one was unserviceable. Communications to base were by morse in cypher using an OTLP.

A patrol signaller described the signalling procedures:

We had two routine signal schedules to keep each day, and these were at different times for each patrol out on operations. This allowed our base station at Nui Dat to deal with only one or two patrols at any one time and provided for an orderly flow of traffic. Of course, a listening watch was maintained at Nui Dat at 24 hours a day for non-scheduled traffic.

Radio communication with our base was via morse code using the AN/PRC 64 set. To establish comms a patrol would go into a defensive LUP with the sig in the centre. The patrol commander would write his message into the code book, and encode it using a random letter transposition. The sig and another patrol member then ran out the dipole antenna through the trees, trying to get it as high as possible. This required walking out of the LUP about fifty feet [15 metres] in opposite directions hooking the green plastic-covered wire over the highest reachable vegetation. Then the 64 set proper was broken out of a pack and the antenna and morse key connected, the antenna was tuned and communication attempted. It was not unusual for the first or subsequent attempts to elicit no response, requiring a re-orientation of the antenna (bringing it in from where it was and run it out again on another compass bearing) or even a complete change of location.

When Nui Dat acknowledged, the coded message was transmitted in morse, confirmation of receipt received (hopefully) or perhaps a request to send again or some or all of the message. When transmission was complete and acknowledged, the antenna had to be retrieved and the equipment packed away. The whole process at best took fifteen minutes and sometimes much longer. During this time the patrol was particularly vulnerable. Firstly some movement was required (movement was the most common give-away in those conditions); secondly the patrol was effectively split in two during the antenna rigging and retrieval; and thirdly we had equipment out of packs preventing quick movement. Perhaps most importantly, at least one and sometimes two patrol members were directing their attention to tasks other than ensuring the security of the patrol. We did use three letter code words for some standard situations, including contact, which cut out part of the decoding process.

The actual conduct of the patrols was guided by detailed SOPs, although these could be varied for specific tasks. The SOPs began by setting out the daily routine. Under normal circumstances a patrol was to be up, fully dressed with 'huchies' (tents) down before first light. In fact few SAS patrols constructed any sort of tent, preferring to sleep wrapped in a piece of plastic, or if available, in American sleeping shirts. Often the rain would cease during the night and wet tents would shine in the moonlight. A quiet half an hour was to be observed before moving into or out of an LUP.

To supplement the SOPs the SAS produced a memorandum outlining the methods and sequences of patrolling in Vietnam. The memorandum began by stating that a patrol should see or hear the enemy before it was heard or seen. This demanded a high level of individual and collective skill, and fieldcraft became a paramount factor. The memorandum continued that SAS patrols were normally conducted in single file, avoiding open areas where possible and staying well clear of tracks. Arm and hand signals were used whenever possible and immediate obedience to these signals was important. Obviously silence was essential at all times not just in refraining from speaking but in moving through the jungle. Undergrowth was parted carefully and dry leaves, sticks and rotten wood had to be avoided. To achieve success, the SAS fieldcraft had to be superior to that displayed by the VC. As one officer who served in Borneo and Vietnam wrote, 'If the enemy in Borneo had been anything like as cunning in his use of camouflage and concealment as he was in Vietnam I believe we would have been really pushed to find him'.[13]

Where possible, patrols moved in tactical bounds which varied, but in close country could be less than 20 metres. Often patrols moved only 1000 metres per day moving for half an hour and watching for up to twenty minutes at a time. Around midday the patrols usually observed about two hours of 'park time'—the period when everyone, the VC, the NVA and the SAS, took it easy in the heat of the day. During this time the SAS would deploy near a track where they could continue to observe if it was being used by the VC.

Night movement was not employed in most areas because a patrol could quite easily walk into an enemy camp without realising it or inadvertently stop near a track used by the VC. At night the patrol crawled into an LUP relying for security on concealment in a thick patch of jungle. No sentries were mounted as the effectiveness of the patrol would be diminished due to lack of sleep. An LUP was selected in an area that provided concealment from the ground and overhead. Night routine was observed. Patrol members remained in position until well after last light and were called in when the patrol commander felt that the LUP was secure. A short discussion generally took place and then the patrol would settle down to passively gather information on VC activities, taking bearings and estimating distances to signal shots and mortars. It was normal for someone to be awake most of the night, albeit unintentionally through someone snoring or because of an uncomfortable position.

Both the SOPs and the SAS memorandum were adamant that SAS patrols were not to move on tracks. Furthermore, efforts had to be made to disguise or hide signs of movement. One technique was to wear footwear similar to that used by the enemy, but there were

This is the sort of sight an SAS patrol might see either while lying in ambush or during the reconnaissance of an enemy area. The first VC is carrying an AK47, while the second has an RPG rocket launcher.

drawbacks to this technique as the sign might later be misinterpreted by friendly forces and the size and depth of the imprint might not fool an enemy which included accomplished trackers. But the most important technique was the maintenance of strict track discipline; leaves and twigs were not to be broken and obviously no litter was left. When crossing established tracks all signs of crossing were obliterated by the rear man.

While on patrol the SAS patrol members observed with all their senses, for not only were they required to recognise signs of movement, marks, tracks and broken vegetation, but they needed to pick up strange smells such as tobacco (menthol was used extensively by the VC), cooking and wood smoke. Since it was not possible to search a normal SAS AO in five days the patrol had to select a route where it was likely that they would locate the enemy. Shots fired by an enemy, it they were concentrated in one area, were a reliable indicator of enemy activity.

If a patrol located an occupied enemy camp the first action was to determine the approximate strength of the camp without carrying out a detailed reconnaissance and to inform squadron headquarters. Then,

Indoctrination lectures were a feature of VC camps. It is unlikely that the SAS would have crawled close enought to take this photograph, but the reconnaissance and surveillance of enemy camps was one of the main roles of the SAS in Vietnam, and often patrols spent extended periods actually hiding inside occupied enemy camps.

as directed by Task Force headquarters, the patrol would be instructed to either conduct a close reconnaissance in preparation for a company attack or for information, to mark the camp with delayed action smoke for an airstrike, or to continue the original mission. Patrols waged a constant psychological war against the VC, and if they entered an enemy camp would sometimes leave leaflets telling the VC that they had been visited by the SAS. On some occasions the leaflets included the name of a local VC personality.

If a patrol thought that it was being followed there were certain actions that could be taken. The enemy normally used signal shots to maintain contact between themselves and these sometimes indicated that a patrol was being followed. The patrol commander then decided whether to try to shake off his pursuers by repeatedly changing direction or to mount an ambush. If night was approaching the patrol would move into an LUP. The members would sit back-to-back and they would only fire when it was obvious that the enemy was virtually going to walk over the top of the patrol.

248

Considering the small size of the patrols it was crucial that contact drills be well practised and effective. The basis of all contact drills was the production of maximum fire in the area of the enemy followed by a rapid withdrawal using fire and movement. As one SAS officer wrote, the one shot one kill idea was suitable for some situations, but 'Vietnam proved conclusively the effectiveness of using the maximum possible firepower...immediately contact is initiated'.[14] It saved Australian lives and caused the enemy more casualties. The heavy volume of fire was assisted by the short barrelled SLR which fired at a slower rate than the normal SLR and with a louder sound. Indeed it sounded just like a .50 calibre machine-gun and when he heard it the enemy would think he was facing a larger force. After the contact the patrol would withdraw quickly for 30 to 50 metres and then proceed slowly and quietly for another 100 metres and hide. While withdrawing, the patrol would often release a CS grenade or fire an M79 CS round to discourage close follow-up by the VC. This technique never failed to confuse the enemy. As soon as possible the radio operator informed the base that they were in contact and the RAAF was put on stand-by for a rapid extraction.

The mere description of SAS techniques does not adequately convey the discipline, skill and nerve required for a successful patrol. It does not describe the effect of eating cold meals in pouring rain. Nor the extreme loneliness of patrolling for five days with barely a word to the other patrol members, communicating only by hand signals. A reconnaissance patrol could involve hours, even days, of lying, standing or sitting motionless, observing the enemy. Often the enemy were so close that they could easily be touched. During one ambush an SAS soldier had to lean backwards so that the muzzle of his rifle did not actually touch the VC he was about to shoot. In such circumstances the SAS soldiers were protected by their superb camouflage which involved the frequent application of thick, greasy camouflage cream on all exposed skin, regardless of the tropical heat. But all the time they had to be ready to spring into instant and violent action. One patrol stood for four days up to their waists in water in a swamp close to a VC base camp watching the VC move back and forth. During the wet months patrols would return like drowned rats. Boots would be pulled off to reveal leeches too bloated with blood to move, or sometimes severe cases of trench foot.

In the jungle of Phuoc Tuy Province the SAS trademarks of training, preparation, planning, discipline, patience and swift violent action were brought together in their most effective manner. By the end of 1967 the phantoms of the jungle had exerted a psychological dominance over the VC. The question in the minds of the commander AFV and the Task Force Commander was how best to take advantage of this dominance. The question would be answered in 1968.

15
Dominating the rear areas
2 Squadron: February–October 1968

It was ironic that in 1966 and 1967 the SAS argued strongly that its main role in Vietnam should be reconnaissance and surveillance. After all, its original purpose, as envisaged by David Stirling in 1941, was harassment of the enemy in his rear areas. But whatever the wishes of the SAS might have been, by the end of 1967 the Australian high command was determined to use the SAS more offensively.

As described in the previous chapter, General Vincent was impressed by the high level of kills achieved by the SAS and wanted more SAS-style operations for this reason. Brigadier Ron Hughes, who took command of the 1st Australian Task Force on 20 October 1967, agreed with Vincent's emphasis on the offensive use of the SAS. He wrote later that when he took command the Task Force had been in Phuoc Tuy for sixteen months and 'there was little that was not known about the topography of the province—there was little need for employing the SAS to gain topographical information although undoubtedly Intelligence [staff] updated their information after debriefing every patrol. SAS [patrols] were rarely if ever employed in or near villages. VC jungle camps were also very seldom more than overnight bivouacs so there was little opportunity to employ the SAS on gathering information on the enemy. This was probably different when David Jackson first employed the SAS in Phuoc Tuy.'

Hughes explained that during his tour he saw 'the SAS primarily as a means of harassing the enemy in his base areas or on his routes to and from his source of supply, ie the local population. The kill rate achieved by the SAS was very gratifying...I did not view the SAS as an intelligence gathering organisation, rather as a reaction force to intelligence gathered by other means.'

Although 1 Squadron had begun these offensive operations in November 1967 they did not become the SAS's main role until four months later when 2 Squadron arrived towards the end of February 1968. During January No 9 Squadron was required to assist the infantry battalions deployed out of the province on Operation COBURG, and therefore SAS operations had been curtailed by a lack of support aircraft. Then during February SAS operations were restricted due to the change-over of squadrons.

The commander of 2 Squadron, Brian Wade, had no hesitation in accepting an offensive role, for while he had a considerable SAS and special forces background, he appreciated that the SAS could not afford to be dogmatic. After graduating from Duntroon in 1956 he had a short period with a National Service battalion before becoming one of the first officers to join the SAS Company in July 1957. As an exchange officer with the 17th Gurkha Division in Malaya in 1960 and 1961 he had gained valuable experience of jungle operations, experience consolidated by a posting to the Jungle Training Centre in 1961 and 1962. During 1962 and 1963, as a member of the AATTV, he had been one of a team of four ex-SAS members at the Ranger Training Centre in South Vietnam. He was then posted to 2 RAR before departing for the USA in November 1964. While there he undertook training with the US Rangers, attended pathfinder, special warfare, diving and halo (high altitude, low opening) parachute courses, and was attached to Special Forces, the US Marines and the Air Assault Division. He also visited Britain for training with the British SAS and the Royal Marine Commandos. He took command of 2 Squadron on its return from Sarawak in August 1966, a little over two months after his 31st birthday.

Although many members of 2 Squadron transferred to 1 Squadron, which was about to depart for Vietnam, and others transferred to the Training Team or were discharged, a hard core of NCOs remained and they formed the basis of the squadron as it rebuilt for it first tour of Vietnam. By July 1967 the squadron was almost up to strength and like previous squadrons it conducted its pre-operational training in New Guinea. But for the first time there was a tactical aspect to the exercise with the enemy being provided by the PIR.[1]

By December 1967 the key appointments had been filled and the squadron was ready for operations. The second-in-command was Captain Ian Wischusen, aged 29, who had been a warrant officer in the 2nd Commando Company before joining the Regular Army as a canteens officer in 1963. He had transferred to the SAS in January 1966. The operations officer, Lieutenant Kevin Lunny, 23, was a Portsea graduate who had served as a platoon commander with 1 RAR in Vietnam in 1965 and 1966. Like Wade and Lunny, the SSM, Jim

McFadzean, was also about to begin his second tour of Vietnam, having served with US Special Forces while a member of the AATTV in 1965. He had earlier served in the Second World War and in Korea.

As always, the selection of the troop commanders was a delicate task. Second Lieutenant David Procopis had served as a soldier on operations with 3 RAR in northern Malaya in 1964. Now aged 24, he had graduated from Portsea at the end of 1965. Second Lieutenant Ron Dempsey, also a 24 year old Portsea graduate, had CMF Commando experience. While Wade had no worries about Procopis and Dempsey, he could not decide on the third troop commander. The choice was between Lieutenant Gordon Simpson, 23, who had graduated from Duntroon in December 1966, and a 25 year old Portsea graduate who had been in the SAS six months longer than Simpson. Wade found 'Sam' Simpson to be an unlikely SAS officer and to be 'a little mouse in training', but after careful consideration selected him. The older Portsea graduate was extremely disappointed at being left behind as reinforcement officer, and later in the tour, when one of the officers was a casualty and another officer was sent as replacement, he resigned in despair.

The problem with officer selection for the SAS was that there was no honourable way out if an officer was seen to falter, for to falter was seen as failure and failure in the SAS was not tolerated. But at a time when the Army was expanding to nine battalions it was extrememly wasteful of good officers. The reinforcement officers for 1 Squadron in its 1967 tour and for 3 Squadron in its 1969 tour also resigned when they were not selected for service in Vietnam.

The three troop sergeants, Peter Sheehan, Jim Stewart and Peter White had served in Borneo, as had most other NCOs in the squadron.

When 2 Squadron arrived in Vietnam the province was still feeling the final stages of the VC Tet offensive and there was little time to settle in before patrols were deployed. The first patrol, a reconnaissance mission, was deployed on 5 March, and before the end of the month 29 patrols had been inserted. However, as the squadron war diary noted: 'Operations during the month have been difficult due to the dryness of the area. Movement is very noisy and close recce of VC positions almost impossible in some situations. The situation should improve during April when some rain should occur.'

Undoubtedly the success of the 'tractor job' on 19 March, as described in Chapter One, raised the morale of the squadron, and during March there were two more successful contacts. On 16 March Sergeant John Coleman's patrol contacted three VC near the Song Hoa river 4 kilometres south of Xuyen Moc, killing one and possibly another. And on 29 March Second Lieutenant Procopis encountered three VC while investigating an ambush position 12 kilometres north west of Nui Dat. He and his scout killed two VC.

But it was only in April that the squadron began its ambush patrols in earnest. Also, by this time Wade was ordering what he called recce–ambushes; patrols conducted a reconnaissance of an area for several days and then towards the end of their mission mounted ambushes on likely tracks.

One of the early reconnaissance patrols was commanded by Sergeant Ray Swallow and was inserted into an area 7 kilometres north east of Binh Gia. Three months earlier, when he had been a reinforcement in 1 Squadron, Swallow had operated in the area and during a contact had dropped a 30 round magazine behind a tree. He returned to the contact site, found his magazine, and nearby set up another ambush. In an effort to hide the patrol's presence, Swallow hoped to use a silenced Sterling, which before leaving Nui Dat he had stripped and distributed throughout his five-man patrol in case it was needed. On 11 April two VC approached about 3 metres apart. Swallow killed the second man with the silenced weapon but the first VC saw what had happened and escaped.

Rather than call for an extraction Swallow decided to remain in the area and the patrol went into an LUP about 300 metres to the south. It turned out to be 'a dreadful night'. The five SAS soldiers sat back-to-back in thick scrub while throughout the night six VC searched the area, coming as close as 15 metres. One of the SAS soldiers was suffering from asthma and was wheezing heavily, and when they were extracted the following morning Swallow vowed never again to take him on patrol.

Another successful patrol commander was Sergeant Ray Scutts who killed two VC in a Claymore ambush about a kilometre east of Swallow's contact area on 16 April, and recovered two diaries. On return from the patrol Scutts was promoted to Warrant Officer Class Two and about a week later led a nineteen-man patrol across the Bien Hoa border, 9 kilometres north of Nui Thi Vai, with the mission of ambushing Fire Support Base Dyke, recently vacated by 3 RAR. Intelligence had revealed that co-ordinated enemy attacks could be expected against Saigon and the Bien Hoa–Long Binh complexes, and 2 RAR and 3 RAR had been ordered north to occupy ambush and blocking positions. The aim of the ambush was to engage VC who came in to forage for any food, ammunition and equipment that might have been left behind.

As expected, late on 25 April small groups of VC began approaching Fire Support Base Dyke and the next day three VC were seen. The SAS patrol was not in a good position to engage them and so Scutts sent a five-man patrol to approach from a different direction. After about 120 metres they came across another party of four VC and all four were killed by M16 and M79 fire. Meanwhile the original three VC were engaged by the remainder of the SAS patrol and these three were possibly killed.

Following the success of this large patrol, and with two battalions still deployed in Bien Hoa Province, in early May, Wade deployed an even larger patrol by APC to the area north east of the Nui Dinh hills. Commanded by David Procopis, the patrol numbered twenty SAS personnel and ten members of the LRRP company of the 9th US Division, who had joined the SAS on exchange. Scutts and Sergeant Vern Martin had taken their patrols to Bearcat, in Long Khanh Province, as part of the exchange. Procopis recalled that in effect it was a platoon operation and many of the SAS NCOs and soldiers were not trained in platoon tactics. While troop level operations of this nature had been included in predeployment training, he found that he still had to teach fundamentals such as harbour drills and platoon defensive layouts. In his view, the patrol should never have been attempted. Procopis established a series of firm bases and sent out smaller parties, sighting various groups of VC over four days until there was a contact on 14 May in which one VC was killed.

Meanwhile, on 12 May the main body of the Task Force had redeployed from central Bien Hoa to the northern border of that province where they established Fire Support Base Coral. They were soon in contact with regimental sized groups of NVA regulars. This commitment demanded the support of most of the Australian helicopters and the SAS was forced to continue deployments within Phuoc Tuy in larger patrols by APCs or on foot. Also during May five patrols were extracted and four cancelled due to bad communications. Whereas in April 2 Squadron had deployed 32 patrols and had accounted for sixteen VC killed with possibly five others killed, in May the squadron deployed only twelve patrols for one VC killed and 24 sightings.

While a number of 2 Squadron patrol commanders had considerable success during the first three months of operations, it soon became obvious to Wade that Lieutenant Sam Simpson was 'emerging as the most aggressive and best patrol commander'. Indeed his performance was a complete reversal of his work in training in Australia.

On 6 April his five-man recce–ambush patrol was inserted into thick secondary jungle in Bien Hoa Province, 12 kilometres north of Nui Thi Vai and they soon came across enemy signs. The next day they closed in on an enemy camp, hearing voices, chopping and an ox-cart moving along a track at 11.40 that night. In the morning they crept into the enemy camp and placed two white smoke grenades with time pencils to guide in an air-strike. But the smoke grenades failed to explode and the air-strike was called in by voice. They continued the patrol and on 9 April discovered another recently vacated camp. Finally on the morning of 10 April one VC passed within 15 metres of the patrol and he was killed. The patrol then came under fire; they withdrew and were extracted.

The patrol began another recce–ambush task on 22 April when they were inserted into the area known as the Long Green—a stretch of jungle running west to east, south of Route 23, six kilometres east of Dat Do. At 4.05 on the afternoon of 23 April the patrol had finished its radio schedule and was preparing to move when Simpson saw two VC. One was aiming a rifle at a patrol member and the other was 1 to 2 metres to the left of the first. Simpson engaged both VC with his XM 148, putting a burst of M16 fire into the first enemy's chest and neck and a similar burst into the face of the second. The second was also struck by an M79 cannister round in the abdomen. The patrol then withdrew using fire and movement as further enemy were in the area.

Three days later Simpson's patrol was inserted deep into Bien Hoa Province near the Long Khanh border. HQ 1 ATF had just been deployed to Fire Support Base Hunt, four kilometres to the north, and Simpson's task was to detect enemy movement to the south of the fire support base which was protected by 2 RAR. When they were inserted at 4.25 pm on 26 April Simpson's patrol was about 40 kilometres from Nui Dat. By the time they had moved 120 metres through an open area with deadfall that made movement difficult they had seen five VC but had managed to avoid detection. Then another VC emerged from a clump of bamboo. He was dressed in black, was carrying a KAR 98B and spotted the SAS patrol. He was immediately engaged and was hit in the chest and stomach, being thrown back into the bamboo. Fortunately the helicopters were nearby and the light fire team brought in suppressive fire while the patrol was extracted.

As had been proved in Operation COBURG, if there were too many enemy in the area the small SAS patrols could not operate effectively. Within a space of seventeen days Simpson's patrol had had three successful contacts. Simpson wrote later: 'As we gained experience we became more aggressive and cocky. I personally thought there was much more we could do in eliminating minor VC elements by direct means rather than acting as the eyes and ears of the task force...The folly of youth!!' In the view of a fellow officer, David Procopis, 'Simpson personified the unit motto: "Who Dares Wins". His success was, I believe, a mixture of his belief in the motto, and his desire to be the best'. Corporal Danny Wright thought that he was a 'damn good soldier', while Private Adrian Blacker admired Simpson because 'he was willing to have a go'. It seemed to Blacker that some of the older patrol commanders were a bit wary of having a contact.

Simpson had no patrol tasks during May, but on 31 May he was warned for a recce–ambush mission in the Nui Dinh hills. In fact Brian Wade had to be somewhat circumspect in briefing Simpson, for although intelligence had located an enemy radio station which Brigadier Hughes wanted captured, he could not tell Simpson the source of the information. Wade told Simpson that an enemy camp was suspected in an area about a kilometre square; he was to attack it, seize the enemy's radio, and if possible capture the codebooks. In view of the nature of the mission, and the fact that there were supposed to be several hundred VC in the area, the strength of the patrol was increased to six men and it now consisted of Simpson, his second-in-command, Lance-Corporal Tom Kerkez, his signaller, Private John Harper, his medic, Private Dennis Cullen, and two riflemen, Lance-Corporal Gary Lobb and Private Barry Spollen. Simpson decided that the patrol would carry three M72 light anti-armour weapons in case of bunkers, and Lobb chose to carry an M60 GPMG modified with a canvas magazine and a sling. In all, Lobb ended up carrying 700 rounds for the M60, six M26 grenades, a CS gas grenade, a smoke and a white phosphorous grenade, a Claymore, seven waterbottles and five days' rations plus the standard SAS equipment. Indeed each patrol member would have been carrying close to 100 pounds (45.5 kilograms); each man carried two weapons. The patrol was planned with great care and at least four reconnaissance flights were made over the area, dropping propaganda leaflets for deception.

The insertion also employed deception, involving a team of ten helicopters. While one group of helicopters conducted a dummy insertion elsewhere, at 7.30 am on 8 June the patrol was dropped by helicopter onto the end of a steep spur and set out for the suspected enemy camp. The country was steep and mountainous with large rocky outcrops and low vine entanglements. There was no water. The next day they heard voices to their south west and when they crawled into an LUP at 4.30 pm they could hear the rattling of utensils about 150 metres to the south west. That night they could hear music on a radio from the same area.

Next morning they could still hear the VC. They were now about 200 metres from their objective and they began to clamber up a steep rock-covered slope. At about 8 am Spollen slipped and found himself jammed between two trees. Suddenly two 'miserable-looking VC' passed them 30 metres below. Once they had passed, the other members of the patrol pushed Spollen out from between the trees. Then the weight of his load became too great for Lobb; he too slipped and found himself spread out on the rock like a starfish. The sling swivel on his M60 snapped and he caught the sling as the M60 clanged

Equipment carried by Lance-Corporal Gary Lobb for the attack on the enemy radio station near Nui Dinh, southwest of Nui Dat in June 1968. It can be seen that he carried seven water bottles (two on the webb belt), an M60 machine-gun, two smoke grenades, six M26 high explosive grenades, a CS grenade, a Claymore mine, an M72 light anti-armour rocket launcher, and a considerable quantity of link ammunition for the M60. (PHOTO, T. BOWDEN)

and crashed on the rocks below. He had to sling the weapon like a pendulum across to Spollen, who could hardly suppress his laughter at the ridiculous situation. The patrol reached a small ledge and then it started to rain. They had just gathered their breath when they heard more music and voices. The advance continued up the slope and at 12.35 pm they heard the enemy radio playing 'The East is Red'. By 1 pm the scout, Cullen, and Simpson had sighted the enemy camp and had identified fifteen VC.

The original plan had been to spend a day observing the camp, but Simpson decided that he could not risk being discovered and he decided on an immediate attack. He called Lobb forward and the rocky area was so congested that Lobb had to crawl over Simpson and Cullen to get into position. He had tied the sling of his M60 back onto the weapon, but now that it was shortened it was hanging tightly around his neck. It reminded Lobb of a St Bernard dog with a cask under its neck. Following Lobb was Kerkez, and together they crawled about ten metres to the left onto a higher rocky ledge where, about seven metres away, they could see at least four VC in the enemy camp. They set up the M60 and watched with astonishment while two VC argued heatedly over who owned a hammock space. Meanwhile, Simpson and the other three patrol members moved out into extended line. At 1.15 pm, on a signal from Simpson, Lobb opened fire with a 100 round belt from his M60, and as it finished Kerkez inserted a 250 round belt before joining in with his automatic SLR. Together they swept the camp back and forth, receiving only sporadic return fire.

By now Simpson's group were in the camp and with M16 and M79 fire they advanced through the enemy position. Simpson recalled:

The assault was great fun...We killed about six plus [VC] in the first round and grenaded all bunkers on the western perimeter. From the noises in the bunkers they were definitely occupied. Dennis Cullen pushed forward about ten metres in front and was held by about four VC firing on the assault party with M1 carbines and M3 SMGs. This confused us initially as I thought the support might have mistaken us for VC. We grenaded this trench then neutralised the area behind with M79 rounds. The results were unknown so I decided we had better find the radio before a counter attack was organized. Lobb and Kerkez kept fire up to our line of advance. From the sounds further down the spur line it seemed that a flanking party was setting out.

I stumbled across the radio roughly in the centre of the camp. It was sitting on a rough table under a camouflage of hessian and undergrowth. Fortunately Lobb and Kerkez had missed seeing it otherwise there would have been little to pick up. I grabbed the radio under one arm and stuffed all the documents I could find down my tunic. Fire was now coming from across the valley. Yellow tracer usually signified an RPD [light machine-gun] in the vicinity so I ordered withdrawal. Several of the more agitated VC were now moving towards us but they were dealt with by Cullen...The only bodies we claimed were the ones we had shot again to ensure they did not come to life behind us during the assault.

In fact the patrol report recorded that eight VC were killed with possibly three more. In addition to the radio, one M2 carbine, one 45 pistol, two packs containing documents and one pack of clothing and 50 rounds of ammunition were captured. Lobb recalled later that he had never felt so elated in all his life, but Simpson brought them back to realities. 'Right, now we have to get out', he said. They moved south, set up their radios and requested extraction. At 3.30 pm they were winched out by helicopter while a regimental fire mission shelled the remains of the enemy camp.[2]

Five days later, on 15 June, Simpson's patrol was on another mission, this time back in the 'Long Green'. It was relatively open country

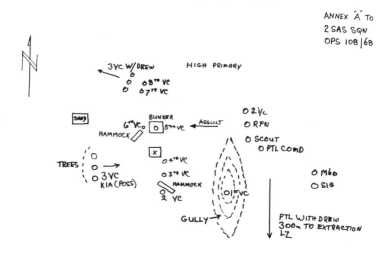

ANNEX 'A' TO
2 SAS SQN
OPS 108/68

Original sketch map drawn by Lieutenant Sam Simpson showing the assault on the enemy camp near Nui Dinh on 10 June 1968 in which the VC radio was captured. Lance-Corporal Lobb occupied the position marked 'M60' on the right of the picture while Lance-Corporal Tom Kerkez was in the position marked 'Sig'. They provided supporting fire across the gully into the VC camp, while the remainder of the patrol, Simpson, Cullen, Spollen and Harper, marked as 'Ptl Comd, scout, rfn and 2i/c', assaulted the camp from right to left.

with low, new growth. Movement was easy over the light, sandy soil and visibility was 100 to 150 metres. That afternoon they decided to establish an LUP in thick bamboo with visibility about 10 metres and at 5.15 pm they saw movement in the scrub 10 to 15 metres away and heard further movement in a nearby creek. Suddenly five VC appeared 8 or 9 metres north of the patrol. The SAS opened fire with M16, M79, M60 and grenades. Fire was received from the north west and the patrol withdrew 200 metres and established another LUP. While communications were being established three more VC were engaged. At 5.40 pm the patrol was extracted by helicopter. They were credited with 6 VC killed plus two more possibles.

June and July were busy months for 2 Squadron. From 6 to 10 June it provided a guard for the Prime Minister, John Gorton, in Saigon, and two patrols (Procopis and Sheehan) plus a liaison officer's party were deployed with 1 RAR in Bien Hoa Province. This additional activity was reflected by the commanding officer of No 9 Squadron in his report for June 1968. 'During recent weeks there has been an increase in VC activity in the TAOR and adjacent areas. This has increased the frequency of LRRP contacts with the enemy. During the month, seven LRRPs were extracted under fire. One of these extractions occurred at night with the aid of a C47 flareship in addition to the LFTs and FAC...During the month a total of 119.05 hours were flown in support of LRRP insertions and extractions.'[3]

By June 1968 the VC had become used to the SAS insertion techniques and were trying to anticipate the arrival on obvious LZs. An example of the VC's approach was shown by the experience of John Coleman's patrol in the Long Green in late May. The area was partly defoliated with undergrowth from 1.2 to 2 metres high and visibility was 200 to 600 metres. A group of up to twenty VC formed a line and then a semi-circle and advanced on Coleman's patrol. The patrol was extracted by helicopter but the gunships received automatic weapon fire as they made their pass.

The squadron ranged far across the province. On 1 June Sergeant John Agnew's patrol located an enemy camp 3 kilometres north west of Binh Ba and observed nine VC digging and chopping wood. They submitted a detailed plan of the enemy camp which was engaged by helicopter gunships and artillery.

At 10 am on 17 June Warrant Officer Ray Scutts' patrol fired on two VC approaching along a track near the Firestone Trail in Bien Hoa Province just across the border from the north west corner of Phuoc Tuy. As the two VC fell dead a third VC opened fire with an RPD light machine-gun from the left flank, hitting Corporal David Scheele in the vicinity of the inside of the right knee cap, while the rest of the patrol came under heavy fire. In full view of the enemy Lance-Corporal Gary Lobb attempted to engage the enemy 15 metres away but his SLR malfunctioned; coolly he cleared the stoppage and then returned fire, hitting the VC in the chest and neck. The patrol then withdrew towards the north for about twenty minutes. Suddenly Scutts saw a VC wearing a tiger suit approaching from the north. The VC seemed confused, mistook Scutts for another VC and spoke to him. Scutts replied by engaging him with his M16. Another VC appeared and was killed by SLR fire from the remainder of the patrol. At 11 am the patrol was extracted having killed five VC.

Sergeant Peter White led a ten-man patrol to the same general vicinity a few days later. Interestingly, the patrol included Warrant Officer Ray Simpson, (soon to be awarded the Victoria Cross), who was an orginal section commander in the SAS Company in 1957. He was on leave from the AATTV and preferred to go on patrol with the SAS rather than remain at the R and C centre at Vung Tau. They located an enemy camp which they assaulted, killing two and wounding two more VC.

Meanwhile, Second Lieutenant Ross Hutchinson of 161 Recce Flight had mentioned to Sam Simpson that he had seen truck tyre tracks on Route 329 north west of Xuyen Moc near the abandoned village of Thua Tich. Simpson confirmed this sighting himself and Brian Wade agreed to allow him to ambush the area. By the evening of 5 July Simpson and his six-man patrol were in position on Route 328. Simpson conducted his reconnaissance for the ambush and, as he described later, 'decided to have a "smoko" before setting up the ambush. I think we each had about six Claymores and two M60s and were generally loaded for the assault on Hanoi. As we were finishing coffee along the road strolled' about 30 VC, dressed in an assortment of greens, khaki, sandals and some wearing sweat bands. Their weapons included two RPDs, ten AK47s, SKSs and M1 Garrands.

'They saw us first;' recalled Simpson, 'in fact they saw Paul Duffy first who was urinating against a tree. The fire fight that followed would have enthralled John Wayne. Fortunately we had brewed up in our ambush position so our flanks were covered. Unfortunately, though, our Claymores were still in the manufacturers' packs. It was like a rabbit shoot complete with hysterical rabbits. The end of the column of VC (estimated at about 80 plus) decided to roll up our flanks and ran straight into Adrian Blacker's M60. I decided there were far too many for any further heroics and withdrew to a pre-planned LZ. By this time it was dark and Dennis Cullen "volunteered" for the job of holding the strobe light to bring in the choppers.' The extraction was particularly difficult and Blacker recalled that 'it was just like a big fireworks display with the gunships rolling in from the north to the south down the road, going west over the old Vietnamese gardens of Thua Tich, and back in again from the north'.[4]

After the extraction the commanding officer of No 9 Squadron wrote to the commander of the US 118th Assault Helicopter Company to thank him for his help. The flare ship requested by the RAAF had not arrived but the light fire team put its ordnance exactly where requested. Miraculously, no Australians were killed, although one M79 was damaged. Thirteen enemy had been killed.

Five days later Wade sent Sergeant Ray Swallow back to almost the same area. The patrol established a Claymore ambush and at 7.30 pm on the evening of 11 July killed five VC. They captured four weapons, clothing and documents. The patrol withdrew in the dark to an LUP and spent a harrowing night with VC searching for them all around their LUP.

Further to the west, Sergeant Jim Stewart and a patrol of eleven men was ordered to again ambush the old Fire Support Base Dyke position in southern Bien Hoa Province. One VC was killed at 4 am on 26 July and two more entered the ambush area at 10.30 am; one of these was killed and the other wounded. It became obvious that the enemy were closing in for an assault, and as the VC came closer the rear protection

Claymores were fired, killing two more VC. Meanwhile the helicopters had been summoned, and during the extraction another VC was killed by a door-gunner.[5]

June and July had been remarkably successful for 2 Squadron. Indeed, they had conducted 34 patrols, had killed 50 VC and possibly another eight, and five Australians had been wounded. Three of the Australians were wounded in an unfortunate incident in mid July. Many of the SAS patrols inserted into the Long Green had been intercepted by the enemy and it was decided to set a trap for them. For five or six days a troop of APCs was to patrol through the western end of the Long Green returning at last light, and on their last patrol D Company 1 RAR, commanded by Major Tony Hammett, was to dismount behind cover and remain in position. Meanwhile an SAS patrol would be inserted and would make its way towards the infantry company. Hopefully the SAS would be followed by the VC who would then be dealt with by the infantry. The operation was to be commanded by headquarters 1 RAR, located at the Horseshoe, north of Dat Do, and another company with APCs was kept on standby at the Horseshoe. It was always dangerous to have the SAS operating close to other troops but to try to ease this problem Wade had sent Sam Simpson's patrol on R and C with D Company and he believed that they were the best patrol for the task.[6]

Wade was aware that after Simpson's heavy contacts during the previous month he should have been given a rest but he decided that Simpson would conduct this patrol and then be given a considerable period of rest. In any case, most patrols were becoming tired and were suffering from various complaints. For example, after one patrol by six men towards the end of July two members had chest complaints, one had a leech in his penis, one had very badly cracked feet and one had a strained back. Indeed the following month Wade organised a programme by which each patrol spent seven days at 17 Construction Squadron at Vung Tau where they undertook a special course to improve their health and physical condition.

At 9 am on 13 July Simpson and his seven-man patrol were inserted into the Long Green and no attempt was made to conceal their presence. They were heavily armed, but everyone was very nervous. If they struck any enemy force they would have to withdraw towards the infantry company. Within half an hour they had sighted one VC. Fifteen minutes later they killed the VC and almost immediately saw four more VC. These were engaged and the patrol withdrew to establish communications with the infantry company about 800 metres to the north. In case the SAS's radio was hit the company had been ordered to advance at the sound of contact, and soon after, at 11.40 am three more VC were engaged. Within a short time the company

Lieutenant Sam Simpson with some members of his troop, shortly before his last patrol in July 1968. *From left:* Privates Adrian Blacker, John Harper, and Tony Bowden, Simpson and Private Dennis Cullen. (PHOTO, T. BOWDEN)

had joined the SAS patrol. In the view of Lance-Corporal Lobb, who was a member of the patrol, the first day of the operation had been a real foul up; he had not seen any VC and he remained convinced that some of the firing had been nervous and quite unnecessary. The SAS was not used to operating with other friendly groups in the vicinity and he suspected that some of the supposed enemy movement was in fact the movement of the company. On the other hand elements of the company fired on the SAS patrol putting a bullet hole through Tom Kerkez's pack. At this stage there was some discussion in the 2 Squadron operations room about whether to extract Simpson's patrol but the commanding officer of 1 RAR was reported to have said: 'Take away the bait and we'll lose the fish'. As Simpson commented later: 'This method of operation had no place in SAS'.

At 6.35 the next morning the patrol moved out of their LUP and began patrolling south with the company following, one tactical bound behind. It was hot, humid weather, and soon after 10.30 am the SAS struck an old paddy field with an abandoned hut near the edge of the jungle. Simpson decided to skirt the paddy field and the patrol began

264

to enter thick secondary jungle along the left hand edge of the field. Dennis Cullen was leading, followed by Simpson and then Lobb. At 10.37 am they came across the junction of an old tank track and a small foot pad. While Simpson paused to decide which direction to take Lobb, struggling with an M60, moved forward to try to persuade him to go back into the paddy field—after all, they were supposed to be trying to attract attention. There was a flash and a bang and Lobb went into an immediate ambush drill. He and the rest of the patrol thought that a Claymore mine had been fired. Within seconds they realised that it was an anti-personnel mine. Simpson called out to the patrol to stay where they were, but ignoring the danger from other possible mines, Lobb hurried forward to find that Simpson had been badly wounded—one foot and lower leg had been blown right off and the other lower leg was shattered. Lobb had just administered morphine when Cullen, the medic, arrived, bleeding from the head. Together, they attended to Simpson who was quite conscious. Indeed, Simpson said to them that he could see that his legs were shattered but asked them to check that his testicles were still there. They were. When Lobb and Cullen could not remember the right military terminology to describe the time that they had administered morphine (so that they could write it on Simpson's forehead) the injured officer calmly helped them out. Both Cullen and Lobb had been wounded but that had not prevented them from rendering immediate aid to Simpson.

By 11.15 am an American Dustoff helicopter had arrived and at 11.45 Simpson was on the operating table at the 1st Australian Field Hospital at Vung Tau. At 1.30 pm both Simpson's legs were amputated below the knee. The loss of Simpson came as a considerable shock to the squadron and Corporal Terry O'Farrell recalled that they heard the news that night while attending the movies at Ocker's Opry House: 'it was a real blow'. While in hospital at Vung Tau Simpson was treated by Major Bill (Digger) James who had lost one leg and part of another in Korea. He had a considerable impact on Sam's morale and his future rehabilitation.[7]

Some appreciation of Simpson's reputation can be gauged by this comment in the No 9 Squadron RAAF war diary: 'An unhappy day for SAS and also for Squadron, as "Sam" is highly regarded both as a Patrol Leader and as a man. A visitor to the Mess on a number of occasions, he is well known to Squadron members'. The commander on No 9 Squadron, Wing Commander John Paule, recalled that Simpson's injury was a very salutory reminder that the SAS soldiers were operating in dangerous situations and that their lives depended to a large extent on the RAAF.[8]

Lieutenant Sam Simpson with members of his troop, on leave from hospital in Vung Tau, July 1968.

While training in the United States in 1966 Wade had learned the techniques of rope rapelling from helicopters and of extracting personnel from the ground by the use of half a dozen hanging ropes. The US method involved the use of the McGuire rig. With Sergeant Peter White and Corporal Scheele, Wade spent some time at Fairbairn airbase near Canberra developing a technique that involved no equipment other than that used for the rapelling insertion itself. The soldiers would simply fix a Swiss seat, a lightweight series of cotton ropes, around the waist and upper legs, attach their Karabiner hooks to a loop at the end of the rope and the helicopter would then lift off with the soldiers dangling 30 metres below.

2 Squadron had been trained in this technique at Swanbourne, but when they arrived in Vietnam the commanding officer of No 9 Squadron held the view that, as the technique was not included in squadron SOPs, it was not to be practised. At Wade's request Major-General A. L. MacDonald, Commander of the Australian Force Vietnam, consulted with his RAAF deputy and directed the SAS and No 9 Squadron jointly to work up a demonstration. By 15 July the SAS and the RAAF believed that they were ready for a demonstration. Both high speed and low speed 'drill' demonstrations were carried out and

The Long Range Reconaissance Patrol (LPRP) Wing at Van Kiep, near Baria, was designed to train South Vietnamese soldiers in SAS techniques and was manned by former SAS NCOs. *Left to right:* Warrant Officers Roy Weir, John Pettit and Kevin Mitchell. (PHOTO, R. WEIR)

MacDonald, accompanied by Wade and the RAF air attaché from Saigon, were lifted off to experience the technique. MacDonald's remark was typical: 'What a wonderful way to inspect the camp!' The No 9 Squadron war diary had an equally appropriate comment: 'The aircraft was reported to have been very stable—as it should have been with that amount of brass suspended beneath it's centre of gravity. Thereafter, although not used frequently, rope extraction was accepted as an operational technique. The first such extraction took place on 29 July when six soldiers were lifted out of the jungle and carried for 20 kilometres at an altitude of up to 2500 feet. The soldiers were exhausted at the end of the flight, but the advantage was that the actual extraction had taken less than two minutes.[9]

Another development during July was the beginning of the patrols from the LRRP Wing at Van Kiep near Baria. A typical patrol consisted of about eight Vietnamese soldiers commanded by one of the Australian instructors. The Chief Instructor was Captain Stan Krasnoff, an infantryman, but his four warrant officers, Shorty Turner, Roy Weir, Sonny Edwards and Merv Cranston all had SAS experience.

Warrant Officer I. L. (Sonny) Edwards at the LRRP Wing, Van Kiep, with some of his students, 1968.

The patrols were included in 2 Squadron's patrol programme and were inserted in the normal SAS manner by Australian helicopters. The Australian insistence on high standards was shown in a letter from Roy Weir to his wife in June: 'We have a Vietnamese general paying us a visit tomorrow to see how our little army is getting along. I figure he's in for a bit of a shock, on account of out of the forty we started off with, we now have twenty six. The way it's going we'll be lucky to have a patrol of five each by the time we've finished training them'. The Vietnamese patrols had few contacts during their training patrols, but on 28 July two LRRP patrols commanded by Turner and Edwards were together when they were attacked by ten VC. At least one VC was killed, but when a further 20 VC were seen the LRRP patrols withdrew and were extracted by helicopter.[10]

After the busy period of July the majority of patrols during August were devoted to track surveillance and there were considerably less contacts. Corporal Danny Wright's surveillance of Route 328 north west of Xuyen Moc between 15 and 21 August was typical. On 16 August his patrol found a north–south track used by a small group of people 300 hundred metres east of Route 328 and that evening they counted a group of 70 VC moving rapidly north along Route 328. All

VC wore mixed black and green uniforms with hats and about twenty carried packs. All carried weapons although only two RPDs were seen.

On 13 August Second Lieutenant Terry Nolan arrived to replace Sam Simpson as commander of E Troop. Nolan, aged 23, had begun work as a bank clerk in Sydney but had found that he was enjoying his weekends as a member of the 1st Commando Company more than his work in the bank. When a fellow member of the Company, Jim Knox, said that he was going to enter the Officer Cadet School at Portsea, Nolan followed suit and both graduated in July 1966. Both Knox and Nolan were posted to the SAS, and after a period of time with 4 Squadron and then as SAS Regiment Liaison Officer he joined 3 Squadron which was working up for its second tour of Vietnam. The order to proceed to Vietnam to replace Simpson came as a considerable surprise as he was not the reinforcement officer for 2 Squadron, but after two years in the regiment he was one of the most experienced troop commanders. Tall, strongly built, thoughtful but with a sense of humour, Nolan was a future commanding officer of the regiment. It was difficult to take over from the popular Simpson, but Nolan achieved the task and won the loyalty and respect of the soldiers.

On 3 September another officer, Second Lieutenant Peter Fitzpatrick, joined the squadron to command the signals troop. With the formation of an SAS Regiment signals squadron it was now possible to form a dedicated troop for each SAS squadron. Fitzpatrick had graduated from Portsea in December 1966 at the age of 23. The success and safety of the SAS patrols relied heavily on the work of the signallers manning the command post signals centre. Furthermore, when the operational patrols were short a member, often the signallers would make up the numbers. Fitzpatrick erected a balloon above the squadron base to increase the height of the radio antennas, and after that it was possible for the closer patrols to speak directly to base on the 25 set. The time and effort saved was greatly appreciated by the patrols who otherwise would have to send their messages by morse on the 64 set.[11]

The emphasis on track surveillance continued during September and in fifteen reconnaissance patrols 138 VC were sighted in a total of five sightings. Five ambush patrols were mounted. Three VC were killed and possibly another four. A ten-man ambush patrol commanded by Lieutenant David Procopis near the village of Lang Phuoc Hoa in the west of the province demonstrated the problems of operating near to the civilian community. On the afternoon of 2 September the patrol was deployed by helicopter to Phu My on Route 15 and there they boarded APCs and travelled down the road towards Lang Phuoc Hoa, dismounting on the move in low scrub 1000 metres east of the village.

They then moved back closer to the village through paddy fields and low scrub and established their ambush. Procopis had been instructed to ambush the road and he had only one man facing to the rear. At midnight six VC dressed in black and carrying a weapon similar in size to an RPG, the VC rocket launcher, moved to the rear of the ambush site, but the rear sentry hesitated to move before the VC had passed and consequently the ambush was not initiated.

Before first light the patrol withdrew east into thicker vegetation where they established an LUP for the day, and that night they returned to their ambush position, this time with the patrol members facing alternately front and rear. At about 9 pm six people were seen to the rear of the ambush site at a range of 20 metres and the patrol engaged the area with small arms and M79 fire. A fierce contact ensued which continued for the next hour and twenty minutes. During the firing Procopis was in radio communications with Major Wade at Nui Dat who quizzed the patrol commander on his location. 'Hang on', replied Procopis, 'I'm reading my map by the flash of machinegun fire'. After a short while the SAS came under mortar fire but when Procopis reported this to Nui Dat the firing ceased and it became obvious that the Australians were probably in contact with a friendly force. In fact the US advisers with a Vietnamese Popular Force (PF) company had reported that they were in contact with a company or more VC, and when they called for artillery support it was seen at Nui Dat that the co-ordinates were inside the patrol zone allocated to the SAS. Procopis had reported that he was in contact with about a platoon of enemy.

At 10.20 pm Procopis established voice contact with a US adviser and insisted that he send someone across to identify himself. The US captain, who was wounded, sent his sergeant; as Procopis recalled, he was 'the bravest man I have ever seen', walking across the paddy field in the dark with his hands held in the air. Seven PF soldiers and one US adviser had been wounded. There were no SAS casualties. An investigation revealed that the PF company had incorrectly been given clearance to operate in an area normally allotted to the PF but allocated to 1 ATF at that time.

During September reconnaissance flights by 161 Recce Flight reported that drag marks at the mouth of the Song Rai river indicated that sampans were using the river at night, and this information was confirmed by a number of SAS patrols which had either spotted sampans on the river at night or footprints nearby. Wade received a verbal tasking from 1 ATF to ascertain what the sampans were carrying, and almost in jest Lieutenant Kevin Lunny, the operations officer, suggested stringing a tennis court net across the river to ensnare one of the craft. The river was far too wide for a tennis net to be of any

Preparing for Operation Overboard, near the mouth of the Song Rai, Southeast Phuoc Tuy, October 1968. Second Lieutenant Terry Nolan, with hand outstretched giving instructions. Facing Nolan in the centre of the picture is Private Kevin Smith and to his right is Sergeant Frank Cashmore. The soldier on the left is carrying flippers in his pack.

practical value and instead a number of large fishing nets were purchased locally. In addition a tiny two-man rubber inflatable survival boat was obtained from the US Air Force for use in the proposed operation. Techniques for erecting the nets across the river were developed and finally rehearsed in the swimming pool at the Peter Badcoe Club at Vung Tau. Security requirements prevented the SAS from giving any explanation of what it was doing and it can only be speculated what the other residents of the Badcoe Club thought of their strange behaviour.

Major Brian Wade decided to lead the operation himself and on the afternoon of 30 September he and a small reconnaissance party were infiltrated by helicopter. Three days later they were joined by the main body for the operation. The plan was that Sergeant Frank Cashmore and a three-man patrol would proceed south towards the mouth of the river as an early warning group while Sergeant Bernie Considine and his four-man patrol travelled north up stream also to provide early warning. Both groups were equipped with Starlight Scopes so that they could provide good early warning and could observe the actions of the sampan crew in the ensuing contact. Wade, with the SSM, Jim McFadzean, Kevin Lunny, Sergeant Peter Sheehan and three other SAS soldiers would form the near bank group about 1.5 kilometres from the river mouth on the east bank. Second Lieutenant Terry Nolan, Sergeants Vern Delgado and Mick Ruffin and four others became the far or western bank group. Nolan's task was to send a couple of swimmers with wetsuits and swimming fins across the river with a line. They would then pull across the remainder of the group in the inflatable boat and finally would stretch the fishing nets across the river supported by two nylon ropes so that the net formed a vertical barrier extending from just below the surface of the water to about 2 metres above it. The sampan crew would thus have to move to one of the banks to get around the net.

Even at this early stage the operation was almost compromised when a sampan travelled downstream with a crew of two close to the western bank. Once they had passed, the Australians could begin to cross the river which was about 100 metres wide with the tide flowing upstream at about seven knots. In the dark of the night it took Mick Ruffin, one of the swimmers, four attempts to struggle across the river against the current, and it was midnight before Nolan and his men were all across the river.

The far bank group was now in a miserable position as the bank, which by day had been above the river level was constantly covered by water. The shorter members of the group were forced to perch in the stinking mangroves like a flock of vultures in a tree, while Nolan, being taller than most, found himself standing chest deep in muddy water holding a silenced Sterling, aware that one grenade could ruin his family prospects for life. Then the tide turned, and down the river came pieces of driftwood, vegetation and debris, all of which caught in the net. By 3.45 am the net could no longer take the weight and the bottom supporting rope snapped with a loud report leaving the net suspended from the remaining rope and trailing on the surface of the river.

Half an hour later they saw a sampan about five metres long approaching upstream from the mouth of the river about ten metres from the west bank. The fact that the VC were using the western bank came as a surprise as all previously observed traffic had been along the eastern bank and the SAS plan had been devised accordingly. Fortunately there was sufficient flexibility to accommodate the unexpected turn of events. The boat had a crew of two, one of whom was noticeably larger than the others, and the larger person was operating a sweep oar. Although the night was dark Nolan was close to the surface

The north warning group for Operation Overboard at the mouth of the Song Rai. *From left:* Corporal Bob Nugent, Lance-Corporal Neale Flemming, Private Bob Burton, Sergeant Bernie Considine and Signaller Kerry (Curley) Clarke.

of the river and the two figures were silhouetted against the sky. As soon as the boat struck the net Private Dennis Cullen ordered them in Vietnamese to stop. Immediately the larger VC dived into the river. Nolan shot the second VC who fell into the river and was not seen again. It was thought that the first VC might have made the river bank some distance to the south west although this was never confirmed either way. Cullen leapt into the river to grab the boat but was swept away by the current and was rescued by Mick Ruffin. Eventually the group dragged the boat to shore and found that it contained fishing gear plus a small quantity of medical supplies.

A discussion then ensued by radio between Wade and Nolan. The squadron commander wanted to remain in position for another night but Nolan objected that it was almost dawn, and as it must be assumed their presence was compromised, his men would be sitting ducks on the open mud flats during the day. There was also the probability that the net might not survive through to the following night. They decided to end the operation and the captured sampan offered a convenient means for returning to the east bank. A safety line was attached to the sampan but as it reached the centre of the river it became difficult to control in the fast-running water. Suddenly, with the current pulling the boat downstream, the safety rope wrenched the stern out of the boat, which promptly settled in the water and sank gracefully stern first. The patrol members in the boat made the bank with the assistance of the swimmers but two radios and six weapons were lost.

Thus 'Operation OVERBOARD' ended on a less than successful note. The SAS could easily have ambushed the sampan but their aim had been to capture a boat and its crew and that is why the VC had been challenged. Wade did not think of the operation as a failure, arguing that a sampan had indeed been captured and searched. In his patrol report he recommended a similar operation during the dry season when the river current would be less, but the operation was never again attempted.

It was also Lunny's last patrol. A few days later he completed his tour and was replaced by Lieutenant Tony Haley who two years earlier had been RSM of the regiment. Haley was the first SAS NCO to be commissioned in the regiment and, he said later, that he accepted the commission in order to show the younger NCOs that they could set their sights on a commission and not necessarily leave the SAS family.

Apart from Operation OVERBOARD, the main mission of the SAS during October was to intercept the VC Rear Services echelons and this programme got underway on 5 October when a seven-man patrol commanded by Corporal Allan Murray ambushed three ox-carts 2 kilometres north of Nui Thi Vai. Eight VC and five oxen were killed. The patrol came under enemy fire and was extracted under the protection of helicopter gunships. Three days later a ten-man patrol under David Procopis ambushed a VC party carrying canteen goods 4 kilometres north of Thua Tich. Six VC were killed and 70 kilograms of dry goods destroyed.

While Procopis' patrol was returning, Sergeant Peter Sheehan's patrol of ten men was inserted 12 kilometres north east of Xuyen Moc, near the Binh Tuy provincial border. They watched various groups of VC pass their ambush position and were mindful of Wade's instruction that if only two or three VC were engaged with a silenced Sterling they should not consider their patrol to be compromised. If such a contact occurred they would not be extracted because of the long distances for the helicopters to travel and the shortage of flying hours available to the squadron. About midday on 11 October three VC sat down on the track at the left hand end of the ambush position and Sergeant Frank Cashmore thought that he could deal with them using his silenced Sterling. Crawling forward through long grass until he was 10 metres from the VC he began to empty his magazine at them. After a short burst the Sterling had a stoppage and the VC ran off. Sergeant

Sheehan yelled 'Get them', and Corporal Terry O'Farrell fired a Claymore. Before the dust of the explosion could settle Cashmore hurried forward following the enemy's tracks and 30 metres away, hiding behind a tree, he found a young woman who miraculously had not been harmed. Two VC had been killed by the Claymore. The patrol was extracted and the woman was turned over to Task Force intelligence. Ten days later Sheehan led the patrol back to the same area and in a Claymore ambush killed seven VC carrying supplies.[12]

Meanwhile, Sergeant Vern Delgado had taken a seven-man patrol into the lakes district in the north east of the province. Soon after 6 pm on 16 October they were eating their evening meal when they spotted five armed VC following their tracks. Corporal Terry O'Farrell distinctly recalled Corporal Ron Harris shouting that one of the VC appeared to be a European. In the ensuing firefight three, possibly four VC were killed. In the failing light the patrol withdrew to an LZ in a lake but could not establish communications to arrange an extraction so they formed an LUP on an island about 75 metres out in the lake. During the night the enemy continued to move around the lake and O'Farrell waded towards the shore and set up a Claymore on a log above the water. Radio communications were not achieved until near dawn and soon the helicopters were on the way. But while they were speaking by radio to a light aircraft the VC came up on the radio with the words: 'Don't worry Aussies, we're going to get you'. The patrol was extracted soon afterwards.

By the beginning of October it was clear that the squadron would soon pass 100 kills. A celebratory barbecue was planned but by the time it was held later in the month the number of kills had reached 125.[13] The day-to-day humour also made life more bearable—incidents such as the night Jim McFadzean decided to instruct all the junior NCOs in the workings of the .50 cal machine-gun mounted in the squadron OP. The weapon, which was of unknown origin having been obtained by 'trade' with an American, had not been fired for quite some time and had only been cleaned superficially. When McFadzean went to cock the weapon, the cocking handle came off in his hand. The problems increased when the cocking handle was repaired; the gun started to fall apart, and finally during a sustained burst the gun mount came off its base and fell over. Another time one of the signallers decided to smoke out his bunker to flush out the wildlife suspected of being in residence. He did not realise the difference between a smoke grenade and a white phosphorous grenade, and the bunker, which was in the middle of the squadron area directly outside Brian Wade's office, provided a spectacular setting for the

Patrol 15, 2 SAS Squadron, Vietnam 1968. *Left to right:* Sergeant Peter Sheehan, Private Barry Spollen, Private Kevin Smith and Lance-Corporal Ian Franklin.

exploding grenade. While Wade was a demanding commander operationally, he understood the need for relaxation in the camp environment and was somewhat forgiving of some of these activities.

During October 2 Squadron accounted for 33 VC and sighted a further 108. Moreover, Warrant Officer John Pettit, with a patrol of Vietnamese from Van Kiep, possibly killed three VC in October, and another three in November. By this stage the staff at Van Kiep included Warrant Officer Kevin Mitchell who had also served in the SAS. By the end of October 2 Squadron had been on operations for eight months and had accounted for 126 VC. But it was doubtful if this pace could be maintained. The time was right to reconsider whether the squadron should continue to try to dominate the enemy's rear area or whether it should again give greater emphasis to reconnaissance.

16

A matter of training
2 Squadron, October 1968– February 1969

2 Squadron's successful ambushes during the wet season of 1968 should not overshadow the fact that the SAS was still the main reconnaissance force for the Task Force. This point was emphasised when Brigadier Sandy Pearson succeeded Brigadier Ron Hughes as Task Force commander on 20 October 1968. Pearson wrote later that he was not very impressed with the capabilities of the SAS:

> The selection of personnel is good. They seek and get the best sort of individual physically and mentally and train them well. I know from Indonesia, from Singapore and my days as [Director of Military Intelligence that] those who served in Borneo performed very well indeed, provided excellent information and harassed the opposition greatly. These were particular circumstances.
>
> I also saw the SAS in training from time to time in circumstances varying from individual training to specialised sub-unit training. My first contact though in any sort of personal way was on one of our big exercises on the Colo–Putty area where the SAS had been given several days to reconnoitre a route for 1 RAR. This they botched. My own intelligence people did in two hours what the SAS claimed could not be done and that after several days.

Pearson had been commanding officer of 1 RAR which in the early 1960s was one of the large Pentropic battle groups. As recounted in an earlier chapter, there were other views about the role of the SAS in this exercise, and Brian Wade, who was on the exercise as part of the enemy force, thought that Pearson's views did the SAS 'an injustice'.

Pearson wrote:

277

In Vietnam, some of their operations appeared to be successful. That is, they provided information and contributed greatly to harassment and attrition. But we were never successful in contacting enemy—based on that information—by normal infantry. What happened too, was that the time spent on the ground between insertion and extraction by the SAS patrols got progressively shorter. In fact, they were requesting almost immediate extraction—in some cases minutes only—which made the whole operation farcical and expensive in helicopter hours...

These sorts of results if not confirmed at least inferred to me that there must be some question mark over the value of special forces on a permanent basis. Specialists with special training are certainly needed from time to time for particular tasks but to tie some of one's best troops up awaiting those tasks is questionable to me. Because you cannot allow the special forces to sit around when the particular task is not required.

In Pearson's view, the SAS was used primarily for 'gathering information, carrying out reconnaissance to find large enemy forces which should be handled by normal units'. Even then, he thought that 'SAS intelligence was variable', and he admitted 'to some scepticism at certain times as to the veracity of some information'. With these views about the worth of the SAS, Pearson took a stronger hand in the direction of the SAS and was loath to tie up as much of his helicopter support as Hughes had. As a result, on occasions Wade tried to keep patrols in the field after they had had a contact. The patrol commanders, however, resented this approach believing they they were in the best position to judge the seriousness of their position.

An example of a reconnaissance patrol at this time was Terry Nolan's patrol to an area 6 kilometres east of Binh Gia. On 5 November the five-man patrol went into a night LUP and heard the faint sound of music to the east plus various weapons being fired. Next day they located an enemy camp and Nolan and his scout, Denis Cullen, conducted a close reconnaissance of the enemy camp, spotting five VC, two thatched huts, enemy packs and clothes on a line. That evening the patrol withdrew into a night LUP nearby and Nolan planned his attack. Next morning the patrol entered the enemy camp along what appeared to be the enemy's escape route—a crawl path through deadfall from aerial bombing. Cullen and Nolan were leading and when they reached the entrance to the camp they came across one VC sleeping in a hammock strung across the path. Quietly they crawled underneath the hammock and proceeded towards the camp. Suddenly a burst of fire cracked over Nolan's head. The sleeping VC had awoken just as the third patrol member had reached him and the burst of fire had been directed at the VC. Then confusion set in. Cullen and Nolan became separated from the patrol; they engaged and shot three VC but were not able to inspect or search the camp. The patrol withdrew after setting a time delay smoke marker for an airstrike which went in that afternoon.[1]

Reconnaissance patrols such as these contributed substantially to Task Force intelligence, and it was partly as a result of SAS patrols in the area east and north east of Binh Gia that the Task Force conducted Operation CAPITAL between 13 and 30 November. The aim was to destroy the elements of 84 Rear Services Group and 274 Regiment in the area and to find and destroy enemy camps, caches and cultivation areas. During the operation 26 enemy were killed, six wounded and one captured. Many caches and camps were found and destroyed.

Meanwhile, the SAS mounted a number of ambush patrols such as the ten-man patrol commanded by Vern Delgado in the area north of Nui Thi Vai. On 15 November the patrol directed a Cessna aircraft onto two VC ox-carts which were engaged with rockets. Next day three VC were killed.

Brigadier Pearson claimed that about this time he was forced to take the squadron out of operations for about one month for retraining. There is no record of this incident in the squadron commander's diary and many members have no recollection of such an event. On the other hand Private Adrian Blacker recalled that the squadron was withdrawn from operations for two weeks. A careful examination of the patrol tasks reveals that no patrols were deployed on operations for four days in the period 22 to 25 November, for six days in the period 3 to 8 December, and for five days in the period 22 to 26 December. From 9 to 22 December twelve patrols were involved in a squadron operation, Operation STELLAR BRIGHT. If Pearson was thinking of STELLAR BRIGHT as a training exercise, then the period from 3 to 26 December approximates to one month of training.

However, Brian Wade recalled that one reason for STELLAR BRIGHT was the shortage of helicopter hours, and the whole squadron was deployed by APCs to the area south east of Binh Gia where Nolan had located the enemy camp. It was thought that various groups of VC were operating in the area and the squadron was directed to saturate the area. If contacted, patrols were to fall back to the squadron firm base. Squadron headquarters was deployed to the field and about fifteen patrols moved through the area. According to the Task Force operation instruction, the objective of the operation included surveillance, testing communications, familiarisation with seismic detector equipment, and testing 'administrative procedures for protracted patrol activities involving major elements of the squadron'. The experiment was only moderately successful with three contacts and five VC killed. Wade commented later that he did 'not consider that Operation STELLAR BRIGHT was a good example of the employment of

SAS. The nature of the operation was such that it could have been perfectly well performed by a well trained infantry company, utilising normal company communications and techniques'.

While the squadron was in the field, on 12 December 4 Troop New Zealand SAS Rangers arrived to join the squadron. Commanded by Lieutenant Terry Culley, the New Zealanders were allocated to Australian patrols to gain experience before beginning their own patrols.

The introduction for one New Zealand soldier was to be particularly trying. From 30 December to 4 January eight patrols were deployed on Operation SILK CORD in AO Dookie, south east of Xuyen Moc, in response to local intelligence reports suggesting that the enemy were using the many deserted beaches to land arms and stores by boat for the VC D445 Battalion. The most interesting patrol was that commanded by Sergeant Mick Ruffin. Aged 26, Ruffin had enlisted in the army in 1960 and had seen operational service with 2 RAR on the Malay–Thai border in 1962. After returning to Australia he had joined the SAS and had served with 2 Squadron as a patrol second-in-command and commander in Sarawak. He was promoted to sergeant in 1966 and by the time he arrived in Vietnam in February 1968 was a confident and experienced patrol commander. Private Dennis Mitchell, a 22 year old reinforcement who had arrived in Vietnam a month earlier, has left a vivid account of Ruffin's patrol:

> The patrol was a reconnaissance and intelligence gathering operation, the object being to get in, get the information and get back out again without the enemy knowing that we had been there. Consequently, we equipped ourselves for a 7 to 10 day patrol without resupply, and with a mainly defensive ordnance selection.
>
> The five man patrol was commanded by Sergeant Mick Ruffin, the 2ic was Lance-Corporal Mick Honinger and Private Brian (Blue) Kennedy was the medic. These three were the experienced patrol members, having been in-country for about 10 months. Sergeant Fred (Bad News) Barclay from the NZ SAS troop was with us for familiarisation before he started to take his own all New Zealand patrol out. Fred got his 'Bad News' nickname following his participation in a series of patrols which all encountered fierce contacts, of which this was the first.
>
> I was the new boy, and would perform the signalling tasks, and carry the 25 set, a VHF voice communications radio. The 64 set, the mainstay of patrol communications, was with Mick Honinger; Blue Kennedy and Mick Ruffin had the URC10's, a handheld UHF emergency beacon and ground to air voice comms radio.
>
> We were inserted by 9 Sqn RAAF slicks with 22 patrol at about 1430 hrs on 30 December, 1968. Both patrols moved east together for about an hour, until it was clear that the infiltration was successful. At about

1600 hrs we separated, 23 patrol moved south into the grid squares assigned to us, and 22 patrol moved on to their area of operations. We patrolled routinely for about another hour, during which we kept the signal 'sched' using the 64 set, the only radio we carried capable of reaching our base at Nui Dat. This was my first use of the morse key in operational conditions, but it was as routine as it had been in training.

We moved further south until we hit a line of craters from an old B52 strike and then doubled back parallel to our path and established an LUP for the night at about 1700 hrs. The night was quiet with no activity either by friendlies or the enemy.

The following day was equally uneventful as we patrolled about 2000 metres of mainly primary jungle that had been defoliated some time ago. Regrowth was well under way and the going was sometimes difficult as we encountered thick underbrush. We saw no sign of any enemy activity all day, and LUP'ed for the night in a thick growth of underbrush. I remember thinking about all the New Year Eve parties that would be underway in other places, but consoled myself with the thought that there would be a lot of heads feeling worse than mine the next morning. On New Year's Day 1969, we went through the usual patrol routine—awake before first light, alert as the day emerged then move a couple of hundred metres before breakfast (a brew and biscuits). Then we headed southeast continuing our search of likely watering spots for enemy signs.

Around 1400 hrs we came upon a running creek not marked on the map, crossed it and commenced to climb up a gradual slope out of the creek bed in our normal patrol order of scout, commander, sig, medic and 2ic. As the slope levelled off about 200 metres above the creek we found our first enemy sign. A ceramic 3 gallon water jar with a black plastic sheet over the top of it stood against a tree. This was my first ever sight of anything even vaguely like enemy activity, and my concentration level went from 99.5 to 100.

Within a few more yards we came to the first bunker, and very cautiously reconnoitred the perimeter of what was a large enemy camp. Satisfied that it was unoccupied, we moved through the camp noting that there were 6 bunkers joined by well formed communication trenches laid out in a perimeter capable of holding a company-sized group. The bunkers had not been used for a while but in the centre of the camp were several shelters made of bush cut from the jungle. These shelters contained shell scrapes and had been occupied within the last week.

As we moved out past the eastern perimeter of the camp Fred Barclay, who was scouting, came to a well-used track running north/south. He called the patrol commander up to inspect the track, leaving myself, Blue and Mick Honinger about 20 metres to their rear covering them.

Being the furthest forward I was the first to see two VC moving north on the track about 30 metres to my right. They were about to stumble onto Fred and Mick Ruffin, who were hunched right down on the track's edge inspecting it. There was no time to attempt a covert warning, and a shout would assist the VC as much as it would help us. There was nothing for it but to engage them, which I did with a full magazine of 3–4 round bursts. Blue joined me as he located the targets. Within 5 seconds the VC were clearly dead on the ground, and Mick Ruffin and Fred Barclay were moving towards them under the cover of our weapons to conduct a body search.

They didn't get there, however, as 4 more VC were coming up the track from the same direction. Mick Ruffin used his 'under and over' to lob a M79 round at them and signalled us to withdraw. The VC were not intimidated by the M79 and opened up on us with small arms fire. Plenty of tracer, particularly green, came flying past us as the VC very aggressively advanced on us at the run.

Ruffin recalled that the 'amazing thing was they didn't seem to hesitate in assaulting us. This didn't do much for our morale as they seemed extremely aggressive and confident...They simply seemed to throw caution to the wind and of course that was their undoing'.

Mitchell takes up the story again:

The patrol swung into a smooth fire and movement routine as we returned fire and withdrew back towards the camp. These VC were very committed and charged onto us allowing Fred Barclay to kill one with three aimed single shots to the body and Mick Honinger to leave another on the ground for dead.

As soon as visual contact was broken we ceased fire and swung south to break the contact, moving relatively quickly to build up a buffer of space against the pursuit we expected would follow. Mick Ruffin kept us moving at a fast but sustainable and reasonably quiet pace for about 10 minutes and then swung west, zigzagging away from the sounds of follow-up that could be heard to the east.

All the time we were moving across multiple well used tracks and then through another extensive enemy camp which, although thankfully unoccupied was obviously receiving considerable and current attention. The camp was bristling with bunkers and trenches, both new and under repair. Large trees had been felled and used to support substantial earthworks, and some of the bunkers had concrete structural components. From the state of the works and the wear on the tracks and the jungle floor it was clear that this was the base for a large enemy force who had been in occupation very recently and who would be reappearing at any time.

Try as we might we couldn't break out of the complex, which we were keen to do as it might have started spewing VC out of those bunkers at any moment. Our way to the east was blocked by the now occasional sounds of follow-up, we knew that the area to the north was more of the same and both southerly and westerly bearings kept us within the VC complex.

It was an anxious time, but finally we returned to primary jungle at around 1600 hrs and after re-crossing the creek we could no longer hear any sounds of pursuit. Our patrol was now well and truly compromised and the time had come to get out while we hopefully still could.

We formed an LUP to establish comms, and I went to work setting up the 64 set to request extraction. I had just come back from running out the antenna when we again heard the enemy moving and talking to our east. Mick Ruffin told me to send our code word for 'contact'— M E L (Mel was Mick's wife's name, Mike Echo Lima; dah dah, dit, dit dah dit dit—I've forgotten every other morse symbol except those three). Nui Dat acknowledged my initiating transmission, and I got away about ten MELs before the insistent signals for me to get ready to move got through to me. I abandoned the antenna (we carried a spare), closed up the pack and took my place in the patrol. The thickest bush was towards the south-west, so that's the way we headed. We hadn't waited for an acknowledgement from Nui Dat of our contact code so I didn't know if it had been received. It had!

At about 1700 hrs the OC flew over in a light aircraft and we got voice communications with him via the 25 set I was carrying. Mick Ruffin gave him a sitrep and requested extraction. Unbeknown to us, our colleagues in other patrols were also in contact. As we had at least broken off our fight the OC decided to direct his available resources to those still in contact and said to Mick 'NO EXTRACTION TONIGHT—LUP'. What Mick heard was 'EXTRACTION TONIGHT—LUP', a simple error which given the state of our stress levels and the noise of the aircraft was understandable.[2] Mick acknowledged the transmission and closed down the 25 set, giving us the thumbs up to indicate the agreement to our extraction that evening.

We consulted the pictomap, identified a suitable LZ within walking distance to the northwest and headed off to what we thought was a helicopter ride back to Nui Dat. No sounds had been heard from the VC for some time and I was starting to feel quite pleased with the way things had gone. My baptism of fire had been faced and I thought I had acquitted myself quite professionally. We were out of a sticky situation and the slicks were probably already in the air to come and get us out.

As Ruffin explained later, the usual procedure was for a light fire team to be despatched to assist a patrol requesting extraction and by selecting an LZ in a reasonably open area the enemy would have to expose themselves to get close enough to take them out. This plan was destined for disaster as the light fire team was not sent.

Mitchell continues his account:

By the latter half of 1968 some patrols in Vietnam were being extracted by hooking a Karabiner (a metal hook attached to a body harness) to a loop on the end of a rope dropped by a helicopter through the canopy to the SAS below. It was a dangerous technique but undoubtedly saved the lives of a number of patrols.

We arrived at the chosen LZ at about 1745 hrs, and hearing 'Hueys' in the distance which we thought would be for us (in fact they were in support of our other patrols) moved onto the LZ to be ready for 'Albatross Leader'.

The LZ was an oval shaped clearing about 500–600 metres across, covered in dry grass about 2–3 feet high. An old road, long since grassed over, ran northwest/southeast through it. We found an old dried-up mud wallow in which the grass was sparse and which formed a small depression in the ground about 3–4 metres across. At its deepest it was about a foot below the level of the surrounding ground. We settled into defensive positions ready with panel and smoke to signal the choppers.

Shortly before 1800 hrs Fred Barclay picked up movement in the treeline, and then we could hear stealthy movements through the grass towards our position. M26 grenades were readied as we started to assess the difficult position we might be in if the choppers didn't appear soon. Of course they were never coming, but that thought hadn't yet entered our heads.

284

An enemy soldier raised himself immediately in front of Fred Barclay apparently to orientate himself, and Fred shot him through the head with a single aimed shot. Two more appeared beside the dead VC, and one attempted to throw a grenade. However they appeared to panic when they saw how close we were (about 15 metres) and they dropped very quickly back to ground under our fire. The grenade exploded harmlessly closer to them than to us. Two M26 grenades were thrown into the spot where they had gone to ground and as they leapt up to escape, the grenades exploded almost simultaneously throwing the two VC in opposite directions, the tail end of which Blue and I saw as we raised ourselves to fire at anything still moving.

Unfortunately, there were still three of the enemy left untouched and they opened fire on the two of us kneeling up from a range of about 15 metres. All of the rounds went high but we did get down onto the ground very quickly. Two more M26 grenades were thrown into the spot they were last seen but they were already up and running back to the treeline, under covering fire from a single enemy soldier within the trees. Mick Ruffin fired several M79s at the now re-united group without verifiable result.

The enemy then commenced a sporadic pattern of fire into our position. We had arranged our packs to provide some cover, but the fire passed harmlessly, if noisily, a foot or two overhead. They then became much more accurate with rounds impacting between us and kicking up dirt in the depression. To achieve this improved accuracy one had climbed a tree and had a clear view of our position. He was again engaged by M79, with which Mick was looking for a tree burst. This encouraged him to abandon his perch but in doing so he was exposed to our riflefire and was hit, falling some 5 metres to the ground, where he was left by his companions.

Mick Ruffin told me to get on the 25 set to raise the choppers, because it was clear that they needed to arrive soon to catch the enemy on the ground. At this stage we were confident of defending our little dry mud wallow, believing that we only had to hang on until the gunships arrived to take out the remaining three VC with rocket and mini-gun fire. Although we could occasionally hear a Huey, I could not get any response to my VHF or UHF calls, even after about five minutes of trying.

Things then took a sudden turn for the worse when about 50 enemy emerged from the trees in an assault line about 130 metres to the south-east. As we moved around in the depression to face this new situation a large detonation occurred about 20 metres to our north, which peppered us with debris. Mick Ruffin told us forcefully to make sure we threw those M26's further out, but the puzzled expressions returned to him and a second explosion confirmed that it was in fact incoming mortar fire.

They kept up the mortar bombardment for about 10 minutes, with occasional bursts of automatic rifle fire. We returned fire with both M79 and short bursts, although our ammunition supply was starting to become a concern. The mortar fire was quite poorly directed, despite the best efforts of one of the enemy who was clearly the formation's commander. Just one round in our little depression would have settled the matter immediately, but between the packs and the slightly lower ground level we were untouched, if showered with dirt.

For our part, we were alternating between kneeling up to get off a few quick rounds at the assault line (and checking that they weren't any closer) and hugging the base of the depression. I was also working the 25 set to get any sort of an answer, and we had one of the URC 10 beacons going—still without response.

What to do?

As much as we would now have liked to 'bug out', it seemed most unlikely that we could all get away. The treeline was 150 metres away over open ground, which had to be covered under fire from 50+ enemy who obviously had the ordnance and the intention to make that option unlikely to succeed. On the other hand, the gunships could be appearing at any moment (we thought) and the fire we were returning seemed to be ensuring that the enemy assault line did not advance. It seemed a question of what would happen first—reinforcements or an accurate mortar bomb, but we thought that defending our position was the best available option. We didn't exactly vote on it—Mick Ruffin put it forward as his intention and we all assented without question.

The situation then deteriorated further as the enemy group brought up a tripod mounted medium machine gun, with which they engaged us with long accurate bursts. We directed our fire at the MMG [medium machine-gun] group with M79 and aimed single shots, as this new threat was making things extremely unpleasant for us. I don't know if this upset their accuracy, but although the air seemed full of the crack of passing rounds, none of us was hit.

Kneeling up to fire at the MMG or assault line brought forth heavy automatic rifle fire and RPG rockets, and the mortar rounds continued to fall haphazardly. This all seemed a bit like 'Russian Roulette', but was essential to keep tabs on what they were doing and to discourage any advance.

We continued like this for another fifteen minutes, although it seemed very much longer. Our ammunition situation was becoming uncomfortable; the M16s with their last or second last magazine fitted, and the SLRs with less than 60 rounds each. All of the M79 rounds had now been expended, and we had two M26 and two white phosphorous grenades remaining.

We kept the radios working all of this time but still without response. The expectation of the arrival of the Albatross flight was turning more to hope (and in my case prayer—Blue Kennedy and I both faithfully promised to attend church the next Sunday if we were still around). The one improving factor was that the light was beginning to fade. If the helos didn't arrive, and if the enemy didn't get lucky with a mortar of RPG round we may have been able to defend the depression until it was dark enough to slip away.

The VC may have had the same idea, because with a sudden intensity of small arms fire and much shouting and yelling they began their long expected assault. The MMG now fired continuously and the mortar and RPG rounds arrived at 5–10 second intervals.

As we all faced the assault to try to break it up with return of fire, we saw two groups of about 10–15 enemy moving quickly down both flanks. Within a minute or so we would have been encircled, and the previously disregarded option of running the gauntlet of the open ground became marginally more attractive than staying put. We simply would not have survived if we had stayed.

Mick Ruffin told the others to leave their packs, but told me to take mine as it held the 25 set. As I lay down backwards onto it to get into the shoulder straps, a mortar round landed close to the lip of the depression and blew all the others over. They were up again immediately, heading off towards the treeline, with myself bringing up the rear.

Mitchell continues the story:

I simply cannot describe the amount of ordnance directed at us. In the dusk, the volume of tracer going past me was simply continuous, all the way from knee height to 10 feet [three metres] overhead. And the noise of the passing rounds and the RPG and mortar impacts seemed like a massive string of immense fire-crackers. It seemed impossible at the time, and inconceivable now, that five men could cross 150 metres of open ground under such fire from so many weapons without one gunshot wound. Of course, one would have been all that was necessary, as we would not have left a man, wounded or dead, and returning to pick him up would without doubt have accounted for the rest of us.

I fell further behind as the heavy pack made movement more difficult, and as the first of my fellows reached a fallen tree on the edge of the clearing and turned to give covering fire, a mortar round impacted immediately behind me. I remember sailing through the air and landing heavily on my chest, winding me a bit. My companions thought the worst (they subsequently told me) but I was able to get up again. As I stood the left shoulder strap of the pack failed and it fell to the ground knocking my rifle from my hand. I had to bend to retrieve the pack, only to see the rear facing section lacerated with rips, and water leaking from a holed container. I think that Mick's insistence that I take the pack saved me from a similar fate. I abandoned the pack, attempting to fire a burst into it to destroy the radio, but only two rounds remained in my weapon and I think only one of those hit the pack.

I made the fallen tree where the rest of the patrol was intermittently firing single shots and shouting at me to 'come on, Mitch'. I'm not sure whether it was encouragement or abuse.

The enemy's barrage was now less fierce, and we quickly moved into the treeline on a bearing directly away from the main group. A large amount of dressed timber was hidden just inside the trees, and we had to negotiate a course around it. Under the canopy the light was fading fast and although firing continued in our general direction none now came near and it gradually diminished.

We continued to move until midnight, quickly at first but then into a normal patrol pace. At the night stop we redistributed the ammunition remaining and tried to assess the damage. Fred Barclay had a head wound from shrapnel, as did both Micks. I had a bit of a peppering around the back of the neck, but none of these were serious or debilitating.

The next morning saw us moving about 200 metres before a light helo out looking for us responded to our URC10 call. In about half an hour we were on an LZ and the slick was on its way in to extract us. We arrived back at Nui Dat just a bit too late for breakfast.

Brian Wade recalled that he had 'held grave fears for the safety of the patrol, and at first light on the day the patrol was eventually extracted I commenced searching the area in a helicopter. After some time I eventually spotted the patrol through the trees and dropped some ammunition to them together with a message indicating I would direct them towards the nearest suitable landing zone and arrange extraction. It was a very "relieved" patrol that was eventually extracted by the ever reliable 9 Squadron RAAF'.

Ruffin's recollection was similar: 'I tried to make contact with [the possum helicopter] on the internal frequency. He did not acknowledge but stayed in the area so I realised he could hear me. I told him that if he could hear me, to do a right hand turn. He did the most beautiful 90 degree turn you have ever seen in your life. I then directed him overhead until he saw us and he dropped a bandolier of 5.56mm and in the bandolier were handwritten instructions to move to a clearing to our west and standby to be extracted'. Before receiving the ammunition resupply the patrol had left just seven rounds of 5.56 ammunition and one white phosphorous grenade.

Mitchell was left with two outstanding thoughts about this contact. Firstly, 'how good the training was that equipped me to handle a situation which was far from the classic contact drill, and how well the patrol functioned together despite the fact that it contained two "strangers". The quality of that training I have no doubt kept us alive'.

The second thought was 'how the brain slowed down the action so that it was almost like slow motion, with every frame in a sort of brilliant clarity. Thus were the right decisions able to be made within the very short span of time available as measured in seconds, but ample time when measured by "brain time".'

It had been a 'near run thing', and Ruffin's patrol emphasised the fact that the commander on the ground was the best judge of whether an extraction should be ordered. Small SAS patrols had no option but to flee if confronted by a superior enemy force. Once they had expended most of their ammunition, and their presence had been compromised, there was no value in their remaining in the jungle. It might have caused a heavy demand on helicopter hours, but this was the penalty to pay for inserting small groups behind enemy lines.

2 Squadron conducted a number of successful patrols late in December and during January, and one of note was Terry Nolan's patrol east of Nui Dinh in which they entered a VC camp and shot three VC. Another successful patrol was conducted by Sergeant Bernie Considine near the Firestone Trail in Bien Hoa. Considine's patrol was the climax of a series of reconnaissance patrols across the north of the province aimed at finding a large VC stores complex. Indeed Corporal Danny Wright thought that Brigadier Pearson had demonstrated 'a brilliant use of the SAS' in locating the complex, in that various SAS patrols had observed a trail apparently running along the northern border of the province. By counting the numbers of VC and observing their dress, the intelligence staff were able to home in on their base. Considine became involved in a series of contacts with a numerically superior force within the base camp complex. While he was manoeuvring his five-man group across a creek nine enemy approached from the rear. Lance-Corporal Kim Pember and another soldier engaged the enemy and immediately killed three of them. Shortly afterwards the rest of the patrol came under heavy automatic fire while crossing the creek. Pember and another patrol member crossed the creek and with accurate and sustained fire covered the remainder of the patrol as they made their crossing. An airstrike was conducted onto the VC base.[3]

As mentioned, Considine's patrol was one of a series of patrols deployed to the southern Bien Hoa–Firestone Trail area in mid January to support Operation GOODWOOD. Task Force intelligence believed that 274 VC Regiment would soon start moving through the Hat Dich area to interdict Route 15 and to attack the Long Binh–Bien Hoa complexes for the 1969 Tet offensive. The operation began on 11 December 1968 and continued until 19 February 1969.

On 10 January Sergeant Alan Stewart and his six-man patrol were inserted into the north west area of Phuoc Tuy, 10 kilometres west of

Ngai Giao, with the mission of locating and reporting enemy movement through the area. On 12 January, during one half hour period, his patrol observed 47 VC and NVA carrying three .50 calibre heavy machine guns, three tripods, four 60 mm mortars with ammunition, ten Rpds, and six canvas-covered weapons, possibly RPGs. The first man was dressed in greens with a green bush hat, was about two metres tall, Caucasian, carrying an M1 or M2 carbine, and was followed by a radio operator. The VC were moving slowly in groups of ten to fifteen. A light fire team was called and engaged the enemy groups. Stewart had located a major enemy route and over a period of three days reported information on the movements of 163 enemy. His patrol was extracted on 14 January to allow the area to be engaged with air and artillery. On 22 January he returned to the area and once again began reporting enemy movement. On 26 January he set an ambush on the trail but decided not to engage the enemy as groups of up to 80 were moving past his position. In all 151 enemy passed in a little over half an hour. Despite the fact that his patrol was located only a few metres from the trail Stewart remained in position and continued to report back knowing that the slightest mistake by any member of his patrol would have resulted in a contact with a numerically superior force.

The patrol was extracted on 27 January and the next day he guided W Company 4 RAR/NZ back into the area. 4 RAR/NZ had just been deployed to the west around old Fire Support Base Dyke. On the night of 29 January W Company was attacked by a company-sized group of VC whom it was thought had mistaken the New Zealand company for an SAS patrol. Spooky (Dakota gunships) and Australian and US artillery were used to support W Company in repulsing the attack. The VC must have suffered heavy casualties although they had removed most of them by dawn. It was an impressive demonstration of the way SAS could be used to locate the enemy and guide in larger forces. And the attack on the company demonstrated the nature of the threat faced daily by SAS patrols.[4]

The incident seems to contradict Brigadier Pearson's assertion that 'we were never successful on contacting enemy—based on that information—by normal infantry'. Indeed Wade wrote later that 'it was certainly a standard part of the patrol briefing that, should a worthwhile target be observed, appropriate notes were to be taken to assist patrol members should a "guide" task eventuate. There were a number of occasions during the squadron tour that I recommended follow-up action after a patrol was extracted, but the problem generally was the non-availability of an infantry company to conduct the task'.

While Stewart was conducting the first reconnaissance patrol it was decided to deploy more SAS patrols to the area. Initially the plan was to conduct reconnaissance patrols, but on 16 January on orders from Task Force headquarters the plan was changed. One of the patrols, that commanded by Warrant Officer Ray Scutts, was given the task of observing the Firestone Trail on the night of 17 January with the intention of establishing an ambush, along with other patrols and an APC troop, on the night of 18 January. It was expected that the enemy would consist of groups of up to 30 or 40, preceded by scouts moving five to ten minutes in front of the main group.

With this change in plan Scutts's patrol was increased from five to six men and Corporal Ronald (Harry) Harris joined the patrol from another troop. A highly experienced SAS soldier, Scutts had served in Borneo and had been in Vietnam for over eleven months during which he had participated in 26 patrols, five resulting in contacts. Twice he had received minor wounds. Corporal Harris was also a well-trained SAS soldier. Since joining the squadron as a reinforcement in June he had been on twelve patrols, four involving contacts. Being Aboriginal, Harris had dark skin, was about 1.68 metres tall and was not unlike a Vietnamese in stature. While all members of the squadron knew him there was always the possibility of him being mistaken for an enemy. Unlike all other members of the squadron, he did not wear camouflage cream on his face.

At 10.00 am on 17 January 1969 the patrol was dropped by APCs on the Firestone Trail ten kilometres north of Nui Thi Vai and by 1.45 pm they had established an OP 50 metres to the north side of the trail with an LUP 10 metres further north. As the afternoon passed Scutts became concerned that the LUP was in a fairly open patch of jungle and he moved its location a further 15 metres north. Soon afterwards Corporal Harris and Signaller Allen Callaghan moved out to relieve Corporal Paul Duffy in the OP. They had been there only a short while when at about 4.30 pm Callaghan saw two VC with what appeared to be camouflaged weapons advancing along the edge of the trail 30 metres away. At the same time Corporal Gary Lobb in the LUP heard movement south on the trail and also movement to his north east. He alerted the rest of the patrol in the LUP.

Meanwhile, in accordance with earlier instructions Harris left the OP to return to the LUP to give details of the sighting to the patrol commander who had been ordered to report such sightings immediately to squadron headquarters. Scutts said later:

At approximately 1630 hours I saw movement to my west which was in the area of our insertion track from the APC drop off point. I first saw old faded Tiger-suit trousers with the movement going west to east. The figure was about ten metres distant and I thought he was wearing a brown pack. The figure kept moving and went to the area of our first LUP where it stopped and suddenly swung around. I saw a black face and hands with no facial camouflage cream and the barrel of a rifle. As a result I instinctively fired as a reflex action.

As soon as I had seen the movement I gave the thumb down enemy sign to those in the LUP. At no stage did I think the movement was that of Corporal Harris. I had fired two single sense of direction shots from my M16 Armalite.[5] There was no returned fire and then Private [Kevin] Tonkin said 'where is Harry?' It was only then that I thought that the movement could have been someone other than a VC. The patrol moved forward to do a body search of the VC but realised that it was Corporal Harris.

Harris died just as Corporal Duffy reached him. Dustoff was requested and the patrol was extracted at 5.45 pm.

The incident was thoroughly investigated and no criticism was levelled at Scutts. Indeed the investigating officer, Major Brian Lindsay of 4 RAR/NZ, reported that he believed 'it was as a result of his experience that WO II Scutts took the action he did. A less experienced man may not have. Scutts knew, however, from experience of the requirement for reacting with fast and accurate fire to an enemy who is posing an immediate threat to the security of his six man patrol. No action by Cpl Harris indicated that he was friendly; on the contrary, his direction of movement and sudden body movements combined with the colour of his skin and no camouflage paint indicated him to be enemy in a situation where reflex action was required'. Scutts deeply felt the impact of the tragedy but there were no recriminations within the squadron. Where men lived on their nerves in constant danger from the enemy it is remarkable, and a testimony to their training, that there were not more accidents.

The squadron only had a few more contacts before completing its tour. On 30 January, while supporting Operation GOODWOOD Sergeant John Agnew's patrol was engaged close to its LZ near the Bien Hoa–Long Khanh provincial border 30 kilometres from Nui Dat and was extracted after 25 minutes having killed one VC. Agnew sustained mild shock after being knocked to the ground by enemy fire striking his pack. In the same area a twelve-man ambush patrol commanded by David Procopis was interrupted by five VC while laying an ambush on 4 February. One VC was killed. Three days earlier Terry Nolan's patrol had killed one VC ten kilometres north west of Binh Ba.

One of the last patrols was a seven-man patrol commanded by Frank Cashmore, and including Brian Wade and Jim McFadzean plus a senior Task Force officer who should not have been present but wanted to see how the SAS operated. The patrol, to the area 5

Corporal Ronald Harris, just before the patrol on which he was killed, January 1969.

kilometres south east of Binh Gia on 3 February, recovered an eight horsepower Fulperland diesel motor that had previously been found during Operation STELLAR BRIGHT. The patrol was marked by its excellent navigation which took it straight to the location.

2 SAS patrols continued until 14 February and on 21 February the squadron departed for Australia. Shortly before the last patrol was withdrawn the ammunition bays directly below the squadron area exploded. The squadron area was showered with white phosphorous and high explosive grenades and 20 pounder tank shells were seen cartwheeling along the road from Nadzab pad. As the incident occurred in the late afternoon many soldiers were caught in the showers or in various states of undress. Most of the patches of grass and bamboo around the area were set afire and an all out effort by the squadron members was required to avert total destruction. It was just as well that the troops were not required to take shelter because some of the squadron bunkers were filled up with masses of stereo equipment and other valuables ready for return to Australia.

SAS: Phantoms of War

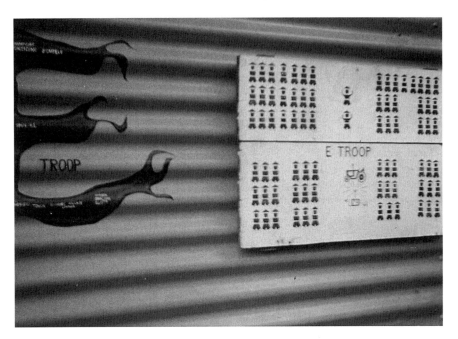

The tally board of E Troop, 2 SAS Squadron at the end of their 1968–1969 tour.
Notice the 'tractor' and 'radio'.

Without doubt 2 Squadron had succeeded in dominating the
enemy's rear area. Ranging to all parts of the province, and to neighbouring provinces, they had accounted for 151 VC with a further 22
possibles. Enemy had been observed on 134 occasions with a total of
1462 being sighted; 812 of these enemy had been sighted in January
1969 alone. One Australian had been killed and of the seven wounded,
two—Simpson, and Sergeant Ray Swallow who received internal injuries during a Claymore ambush—required medevac to Australia. The
results achieved by 2 Squadron reflected the fact that main force
enemy units had entered the province in strength. It was fortunate that
the SAS had not suffered more casualties. However, it was not just
luck, but the result of the high level of training the squadron and
individual soldiers had undergone before commitment to operations.

By the time 2 Squadron departed the nature of the war was changing
once more. During 1968 the Task Force battalions had deployed outside the province on many occasions, not just on the better known
Operation COBURG in January and February, for the Coral–Balmoral
battles in May and June, and for Operation GOODWOOD in December
1968–February 1969, but also in lesser operations across the border

2 SAS Squadron waiting for transport to return to Australia, February 1969. Standing, Warrant Officer J. H. McFadzean and Major B. Wade. *Sitting from left:* T. O'Farrell, K. L. Smith, K. S. Cartmell, T. J. Collins and D. H. Wright.

into Bien Hoa and Long Khanh Provinces. Sometimes the SAS supported these operations with reconnaisance patrols; often the SAS was left to carry the fight to the VC within the province. During 1969 the Task Force would leave the province less frequently and as the nature of the operations changed the SAS would also revise its approach.[6] If 2 Squadron's response to a changing situation was a reflection of the level of training in the SASR, the next squadron could expect similar successes.

17
The regimental base
Swanbourne: 1968–1972

By 1968 the SAS operational squadrons in Vietnam had earned the high regard of the rest of the Australian Task Force as well as the fear and respect of the VC. But this effectiveness was not just the result of actions in Vietnam; it was also the result of thorough training at the regimental base at Swanbourne. Once Army Headquarters decided in 1967 that operational tours of Vietnam would last for twelve months, a fixed programme could be formulated to ensure that one squadron was trained and ready for operations at the end of each year. But this was not the regiment's only responsibility; it had to be prepared to supply reinforcements at any time to the squadron operating in Vietnam and, due to the nature of the SAS, it had to be ready for any other contingency that might arise. Thus to a certain extent the regiment was constantly on a war footing. The responsibility for selecting the personnel, training the reinforcements and squadrons, maintaining the skills necessary for other contingencies, and instilling the ethos of the SAS, rested ultimately with the commanding officer of the regiment.

In Lieutenant-Colonel Len Eyles, who at the age of 41 had returned as commanding officer on 29 March 1967, the regiment had a commander with an appreciation of the type of man suitable for the SAS and a knowledge of personnel procedures to match his earlier infantry experience. In the six years since he had commanded the SAS Company he had attended the Australian Staff College, had been on the headquarters of the 28th Commonwealth Infantry Brigade in Malaysia, and had recently been Assistant Adjutant General at Headquarters Eastern Command. He was ably supported by his RSM, that grand old SAS soldier, Warrant Officer Class One A. K. (Mick) Wright, who had been SSM of 3 Squadron during its 1966 tour of Vietnam. He had replaced Tony Haley when the latter had been commissioned in November 1967.

Despite the hard work of many within the regiment, it had taken a number of years to bring the unit to full effectiveness. For example, it was not until March 1967 that Base Squadron received a major as commander. The first squadron commander was Major E. W. (Wally) Marshall, who at the age of 53 was in the twilight of his career. He had won his Military Medal with the 2/4th Independent Company in Timor in 1942. Base Squadron's responsibilities went beyond the present Base Squadron's orbit of armourers, mechanics, drivers, medics and cooks. It included a signals component and was responsible for running the selection courses. With his cadre staff, Wally ran up to four selection courses each year in order to reinforce the sabre squadrons as they returned from Vietnam. Furthermore, there was a requirement to provide a ten-man pool on short notice to reinforce the squadron in Vietnam.

By the beginning of 1968 the training cycle for each SAS squadron had been refined into five phases. The first phase consisted of post-operations leave and re-assembly and lasted for up to two months, usually March and April. The second phase was a period of from eight to ten months (May to February) and was taken up with re-forming the squadron, individual training of new soldiers, and cross-training of the more experienced soldiers. The third phase began in February of the second year and lasted until August (six to eight months); it was devoted to collective training at patrol, troop and squadron levels. Between September and November there was a squadron exercise in Papua New Guinea, and the final phase of about two or three months began on their return in November; it involved the final preparation and leave before the squadron embarked for Vietnam in February.

To provide a fully manned and trained squadron every twelve months meant that there was a continuing requirement to select and train new soldiers, and thus each year an officer of the regiment, usually the commanding officer, conducted up to six selection tours of Australia during which volunteers from serving soldiers appeared before a selection board consisting of the SAS officer and a psychologist. Understandably, there was considerable resistance from units as they were asked to give up their best soldiers. The Director of Infantry argued that the units would eventually get them back as NCOs, but they never wanted to leave the SAS once they settled in. The acceptance rate at this stage averaged about 35 per cent of applicants. The applicants were then required to attend a regimental selection course designed to teach them the basic knowledge needed to be an effective

SAS soldier and to assess their acceptability for SAS work. About 65 per cent of applicants usually passed the course. By way of example, the first selection course of 1968 began with 30 soldiers; 18 completed. In the second course 11 of the 16 starters completed; in the third course 20 out of 23 completed; in the fourth course 16 out of 28, and in the fifth course 16 out of 19. All officers, other than squadron commanders and those with previous SAS experience, were also required to undergo the selection course.

From the selection course the normal progression was to a basic parachutist course and then either to a squadron, a reinforcement troop or to an appropriate trade. Allocation to squadrons normally took place during the second phase, the individual training period. Once in the squadron the new SAS soldier was taught skills in navigation, contact drills, tracking, unarmed combat, driving, free fall parachuting (recommenced in 1970, having begun originally in 1962), small craft handling, underwater diving, canoeing, cliff climbing and helicopter rope drills. During the height of the Vietnam war those skills relevant to operations in Vietnam were given priority, leaving little time for amphibious and climbing activities. In addition to this training each soldier was required to qualify in as many of the particular skills of signaller, medic, assault pioneer and linguist as possible. Highest priority was the signals course which lasted six weeks and was designed to achieve a proficiency of ten words a minute sending and receiving morse. Next priority was the medic course which involved five weeks at the School of Army Health in Victoria and three to four weeks attachment to a hospital in Western Australia. Lower priority was given to the assault pioneer course (two weeks) and the language course (three weeks).

So as not to undermine the training of the squadron preparing for operations it was found necessary to provide a separate stream of reinforcements for the squadron in Vietnam and to meet this requirement a shortened programme was arranged. As a very minimum soldiers took ten weeks to complete recruit training and ten weeks in corps training before they were eligible to apply for SAS. The selection course took five weeks, the basic parachute course, four weeks and the patrol course, four weeks. Thus in theory, a recruit could be a trained SAS soldier 36 weeks after joining the Army. Inevitably, however, the soldier would not be required as a reinforcement as soon as he finished this training, and thus he was sometimes able to complete a specialist course as a signaller or medic. This shortened programme was particularly suitable for regular soldiers who had joined the Army for only three years or for National Servicemen. Soldiers had to have at least fifteen months' residual service when they joined the SAS, and National Servicemen were therefore usually required to extend their service beyond their normal term of two years so that they could spend one year in Vietnam. Until 1969 there was no formal procedure to allow National Servicemen to join the SAS and only a small number succeeded in doing so. For example, in the period from July 1967 to August 1968 a total of 56 National Servicemen applied, seventeen were accepted for the selection course, six completed and four re-engaged and were posted to SAS.

In support of this increased activity, during 1967 and 1968 training facilities and accommodation were improved as new buildings were constructed. The building project was spread over five stages with the final stage being opened by the Minister for the Army, Philip Lynch, on 27 April 1968. In all, the project cost $4.2 million and provided accommodation for 500 soldiers, NCOs and officers. The building programme included brick accommodation and administrative buildings for each squadron, a new Sergeants' Mess, parade ground, assembly hall, two 25 metre ranges and a swimming and diving pool. On completion, Campbell Barracks was one of the most modern barracks in Australia—concrete testimony to the fact that the regiment had been accepted as a vital part of the Army's order of battle.

According to the rotation scheme, 3 Squadron was nominated to relieve 2 Squadron in Vietnam in February 1969 and during 1968 the squadron trained intensively for their second operational tour. Their new officer commanding was Major Reg Beesley who, at the age of 30, took command in October 1967. Short, aggressive, enthusiastic and personable, Beesley, known as 'the Beast', had graduated from Duntroon in 1959 and had served in 3 RAR before joining the SAS at the beginning of 1961. During almost four years with the company he had served as recce officer, platoon commander, adjutant, second-in-command and squadron commander before being posted to the 28th Commonwealth Brigade in Malaysia in 1964. His second-in-command, Captain Ross Bishop, was a 25 year old Duntroon graduate who had originally joined the regiment as signals officer in mid 1966 and had transferred to infantry in November 1967. The operations officer was the former RSM, Lieutenant Tony Haley, while the SSM was Warrant Officer Des Kennedy, now preparing for his fifth operational campaign. He had been one of the father figures in 3 Squadron's first tour in Vietnam and Beesley later described him as an exellent soldier—'a quiet, father to the lads'.

The three troop commanders were young, recently graduated officers. The most experienced was Terry Nolan who had joined the regiment in July 1966. As recounted in the previous chapter, he went to Vietnam as a reinforcement officer in August 1968 and would rejoin 3 Squadron when it arrived in February 1969. The other troop commanders were Second Lieutenant Nick Howlett who, at the age of 23,

Left to right: Lieutenant-Colonel Len Eyles, Commanding Officer of the SAS Regiment, Lieutenant-Colonel John Slim, Commanding Officer of the British 22 SAS Regiment, and Major Reg Beesley, Officer Commanding 3 SAS Squadron, at Campbell Barracks, Swanbourne W.A. 1968. Slim had a part in the raising of the 1st SAS Company in 1957 and later commanded the Australian SAS Squadrons in Borneo in 1965 and 1966.

had graduated from Portsea in December 1966, and Lieutenant Chris Roberts, who had graduated from Duntroon in December 1967, one month short of his 23rd birthday. Roberts was a future commanding officer of the regiment. Of medium height, stocky, fair-haired and labouring under the nickname of 'Sniffy', he was a thoughtful, tough and determined, perhaps even stubborn, officer who would not curry favour with his men even if such an approach might have lost him some popularity. Beesley gave him command of the most difficult of his troops. Like Nolan, the signals troops commander, Second Lieutenant Peter Fitzpatrick, was already in Vietnam. When Nolan left for Vietnam in August 1968 he was replaced by Second Lieutenant John Ison, a 21 year old Portsea graduate who had served two years with the PIR. Ison was to remain in Australia and join the squadron in Vietnam at the end of Nolan's tour.

The troop sergeants were John Robinson, Tony Tonna and Ray Neil. All three had served with 1 Squadron in Borneo. Robinson and

Tonna had been sergeants with 3 Squadron during its first Vietnam tour during which Tonna had been awarded the Military Medal. Neil had joined 3 Squadron in Vietnam as a reinforcement towards the end of 1966 and had completed his tour with 1 Squadron.

Although 3 Squadron's training was planned to have them ready for operations by the end of December 1968, for a few hours in June 1968 it looked as though they would have an early introduction to action. At about 9 pm in the evening of 5 June the Western Australian Police Commissioner phoned the Commander Western Command, Brigadier Nicholls, and stated that he was speaking with the personal authority of the Premier. He explained that over 500 prisoners confined in the Fremantle gaol had gathered in the exercise yards of the gaol for several hours, were creating a disturbance and were refusing to return to their cells. There had been a few minor injuries to both warders and prisoners but with the help of 60 police it was expected that the warders would be able to control the situation. However, there was a possibility of a mass breakout by a large group of prisoners, including a significant number of dangerous criminals. He requested that the Army make available about 100 men in case of emergency.

The SAS Regiment and the nearby 22 Construction Squadron could muster only about 50 soldiers between them, and these men were placed on stand-by. Reg Beesley was nominated to command any troops provided by the Army and he visited the gaol in civilian clothes to make contact with the Senior Police Inspector, the Commissioner's on-the-spot representative. At about 11.30 pm the Police Commissioner telephoned Brigadier Nicholls that about half the prisoners had returned to their cells and there would be no requirement for Army assistance. Despite the fact that one newspaper stated that SAS 'troops were rushed to the gaol in a convoy of trucks', the troops never actually left the barracks area.

The following week 3 Squadron travelled south to conduct a patrol course which included a night crossing of the Collie River. On the night of 14 June two patrols arrived at the river and Private Trevor Irwin had the task of swimming across with a light line which would then be used to haul the main rope across. About 45 metres wide, the river was flowing slowly but was extremely cold and discoloured due to recent rains. Visibility was poor and there was intermittent rain. Irwin was a strong swimmer, but suddenly in the middle of the river he called out 'Help me' in a loud voice. Beesley, who was on the bank, shouted 'Get him' to the personnel in the safety boat, and immediately the area was lit by vehicle lights. There was no sign of Irwin. By this time Lance-Corporal Tom Bradshaw had dived from the boat into the river but Irwin could not be found. Search efforts continued for several hours during the night until the police, who had been called, recommended that they stop until dawn. Irwin's body was recovered by SCUBA divers from the Collie Civil Aid Team the following morning. The court of enquiry considered that it was 'possible and quite probable that Pte Irwin sustained an attack of cramp in the stomach or legs'. No negligence was attributed to any of the responsible officers and the court concluded: 'Bearing in mind the importance of realism in training when units are about to be committed to operations...there is no need to modify training procedures or to modify existing equipment especially for units such as SAS Squadrons'.

The patrol course on which Private Irwin had drowned had been introduced by Beesley with the dual aims of ensuring that soldiers would become effective patrol members and giving NCOs the opportunity of gaining qualifications for promotion. The patrol course was based on the old Recondo course which had lapsed during the Borneo period.[1]

The training of 3 Squadron continued with an exercise in late July in which patrols were deployed by helicopter from the British Commando ship, HMS *Albion*, to the Lancelin area. Then towards the end of September the squadron began its final exercise in Papua New Guinea. Lieutenant Roberts was highly impressed by the standard demanded by Beesley. Ambush and contact drills were practised with live ammunition and the final exercise in Papua New Guinea was so realistic that Roberts found that his first patrols in Vietnam seemed no different to those in training. Major Mike Jeffery's company of the PIR provided an enemy force which operated like the VC and obtained full co-operation from the local villagers. Thus the SAS found itself exercising in an unfriendly environment—an excellent preparation for Vietnam.

While 3 Squadron was completing its training, decisions were being made in Canberra and Wellington, New Zealand, which would have a direct bearing on the operations of the squadron in Vietnam. As early as February 1967 the New Zealand Army had recommended to cabinet that a rifle company and/or an SAS troop be sent to join the Australians in Vietnam, but later that month, after careful consideration, the New Zealand government had announced that for the time being only an infantry company would be deployed. The New Zealand Chief of the General Staff did not favour the idea of using his SAS personnel as infantry reinforcements and was concerned at the lack of opportunity for their operational employment, with possible consequences for their future in the New Zealand Army. As a result, with the approval of his minister, in April 1968 he approached the Australian Chief of the General Staff, Lieutenant-General Sir Thomas Daly, about the possibility of a troop of 30 New Zealand SAS soldiers joining the Australian squadron in Vietnam. Daly welcomed the proposal and when he visited New Zealand in July he inspected the New Zealand unit. On return to Australia he wrote to his New Zealand counterpart: 'I was tremendously impressed by your SAS Squadron; it has all the earmarks of a very high-class unit and I hope that before very long you will be able to find suitable employment for them'.

In October 1968 the New Zealand cabinet approved the despatch of 26 SAS personnel to Vietnam in December. Designated 4 Troop NZSAS, they soon afterwards departed for Terendak in Malaysia where they practised helicopter drills, familiarised themselves with the weapons used by the Australians such as the M60, and completed their jungle patrolling training. The main body arrived in Vietnam on 12 December 1968. As recounted in the previous chapter, they joined some of the patrols conducted by 2 Squadron, and during the handover period with 3 Squadron in February 1969 maintained the SAS operational patrolling commitments. There was little problem with co-operation with 2 Squadron as both parties knew that they had to work together for only a short period, but co-operation with 3 Squadron was to present a bigger challenge.[2]

The training cycle of 1 Squadron matched that of 3 Squadron although it began a year later. The new officer commanding, Major Ian (Trader) Teague, 33, assumed command of the squadron in June 1968. After graduating from Portsea in 1955 he had a number of postings in Australia before serving with 1 RAR in Malaya during the Emergency. Soon after the raising of the SAS Company in 1957 he had expressed an interest in joining it, and had completed a parachute course. From December 1961 to April 1964 he had been adjutant and training officer of the Sydney Commando Company and had then joined the AATTV in Vietnam. There he was attached to the Combined Studies Division in Quang Ngai Province where he was responsible for attacking the VC infrastructure. He achieved some notable results and extended his tour until December 1965.[3] He was then an instructor at the Officer Cadet School, Portsea until he joined the SAS.

Teague's second-in-command was Captain Craig Leggett. Aged 24 when he took up his appointment in mid 1968, he was a Portsea graduate who had served with 1 RAR in its first tour of Vietnam. The operations officer was Lieutenant Mike Eddy, a 23 year old former National Serviceman who had served with 2 RAR in Vietnam. The SSM was Warrant Officer Edward (Snow) Livock, a long-serving SAS soldier who had served mainly in quartermaster appointments but had been on operations with 2 RAR in Malaya in the 1950s. The squadron signals officer, Second Lieutenant Brian Schwartz, had arrived in Vietnam in August 1969 to replace John Fitzpatrick, and would continue there when 1 Squadron arrived in February 1970. The three troop commanders were Second Lieutenants Robin McBride (19) and Lloyd Behm (21), both of whom had joined the SAS direct from Portsea in July 1968, and Lieutenant Charlie Eiler (22) who joined the regiment in April 1969 after graduating from Duntroon the previous December. For the first time an SAS squadron was to begin operations without a single troop commander who either had previous operational experience or had spent time as a private soldier. As usual, the troop sergeants were men with a wealth of operational experience. Ron Gilchrist, John O'Keefe and Lawrie Fraser had each served in Borneo and Vietnam.

Lieutenant-Colonel Eyles' training directive for 1969 was concerned specifically with preparation for Vietnam, and stated that emphasis was to be placed upon 'aspects relevant to current operations in Vietnam, including a detailed study of the nature of, and methods used, in those operations'. The directive went on to list the subjects to be emphasised. These were:

a. Patrol techniques, and conduct, including the ability of each individual to take over all aspects of conducting a patrol, and each role within the patrol.
b. Shooting. The minimum standard required is First Class in all weapons. Emphasis is to be placed upon rapid and accurate reaction to close quarters, upon ambush fire control, upon fire and movement to break contact, and night firing.
c. Navigation is to be brought to a high individual standard using bearing and distance methods with a minimum of natural and artificial aids.
d. Close recce, against a live enemy.
e. Ambushing, with and without Claymores.
f. Current radio and beacon equipment, and communicating.
g. Rapelling from helicopters and other drills.
h. Tracking.
i. Languages, Vietnamese and contingency [that is Pidgin and Malay].
j. Study of current operations and the enemy in South Vietnam.
k. Foreign weapons.
l. Grenades.
m. Night operations.
n. Mines and booby traps.
o. Air recce.
p. Target grid procedure.
q. Tropical health and hygiene.
r. Physical fitness.
s. Swimming.

While there was no obvious need for parachute training for Vietnam the regiment continued the training because it built up the soldiers'

304

confidence and gave them a sense of identity. In any case, it was important to keep alive the specialist skill that might be required at a later time. The Director of Military Operations and Plans at Army Headquarters, Brigadier L. I. Hopton, set out the requirement in September 1969: 'Whilst there has been no proposal, so far, for the SAS Squadron in Vietnam to conduct parachute operations, this does not necessarily imply that there is no requirement for the SAS Squadron to be parachute trained, with the potential for combat descents...there is an operational requirement for all three SAS squadrons to be available for parachute operations. These operations are not necessarily in support of the Vietnam war'. He accepted that with the problems of manning the regiment there might be some personnel who had not had sufficient time for parachute training, but they all had to be volunteers for parachuting. There was also the matter of esprit de corps. As he put it: 'Parachuting tends to bind unit members together as a team, due to their sharing the same hazards'.

The hazards of parachuting were brought home graphically on 12 August 1969 when 1 Squadron conducted refresher parachute continuation training at a DZ at Bindoon, about 50 kilometres north of RAAF base Pearce. Simultaneous double door exits were being undertaken from a C130 Hercules aircraft, and during their third descent for the day Lieutenant Charlie Eiler and Sergeant John Grafton collided after leaving the aircraft. Eiler passed through Grafton's rigging and Grafton ended up, with a collapsed canopy, suspended below Eiler. Eiler's canopy supported both men for some time but progessively deflated and by about 130 metres above the ground it was useless. Grafton activated his reserve parachute but it became entangled shortly before the two men struck the ground. Eiler died on impact. Grafton died in hospital a few days later.[4]

Lieutenant Zoltan (Zot) Simon (23 in 1969) was transferred from 2 Squadron to 1 Squadron to replace Eiler. Simon and Eiler were close friends, having graduated together from Duntroon in December 1968, and both having a Hungarian background.

In December 1969 Lieutenant-Colonel Eyles' tour as commanding officer came to an end and Lawrie Clark arrived as the new commanding officer. Nine years earlier he had taken over from Eyles as officer commanding the SAS company. In the intervening time he had served in Vietnam with the AATTV, had been on the staff of the Jungle Training Centre at Canungra, had been chief instructor at the Officer Training Unit at Scheyville, and had recently been GSO1 (Land/Air Warfare) in the Directorate of Military Operations and Plans at Army Headquarters. Clark's first impression when he reached the regiment, was that it 'was a competent, organised group who knew what they were doing'. But he realised that he was not a commander in any operational sense; rather his chief responsibility was training the squadrons and reinforcements for Vietnam. Soon after taking command he was reported in a local paper as stating that the SAS's record so far was not based on just good luck. 'It is good professionalism', he said. 'It proves that our training gives the men the necessary stamina'. He added that the men of the SAS were 'not super grunts or rough, tough morons. My men are well trained soldiers with tough minds'.

During Clark's tenure as commanding officer an increasing number of National Servicemen joined the regiment, and it was clear that without these men the regiment could not have continued to maintain a squadron on operations. Clark 'picked over' the National Service applicants 'very carefully' and the successful ones proved to be 'very good'. One concern, however, was the standard of junior officers. Despite the fact that the officers straight from Duntroon had generally proved to be quite capable, he believed that officers should spend some time in another unit before joining the SAS. He managed to persuade the Military Secretary, and from December 1969 officers were no longer posted straight from Duntroon. It did not matter quite so much for Portsea officers as they spent longer as subalterns before becoming due for promotion to captain. Clark wrote later that 'the weakest link in SAS had always been the inexperience of junior officers. A highly trained SAS trooper can just not relate to a raw officer'.

Reg Beesley later expressed surprise at Clark's remarks, noting that he might have been 'out of contact with the young officers due to the growing independence of squadron commanders whose responsibility it was to train young officers as regimental officers and patrol commanders. As operations later proved, they acquitted themselves well whilst some of the old and bold didn't fare that well.'

Robin McBride, a young officer at the time, was impressed by 'the absolute professionalism of the unit's sergeants and other experienced soldiers who recognised the awesome task confronting junior officers (regardless of place of graduation). As a consequence, they made enormous efforts to provide both theoretical and practical advice drawn from their own vast experiences to ensure their new troop commanders were well prepared. This is one of my most vivid recollections of my formative years in the Army...I am convinced that the motive here was not so much that we junior officers would not embarrass our senior NCOs with our conduct, but for the pride of the squadron and the regiment we would simply get it right.'

One of Clark's early challenges was to ward off a threat to the regiment's continuing occupancy at Swanbourne. In the Federal election of October 1969 the Liberal–Country coalition government lost heavily in Western Australia and the new Minister for Defence, Malcolm Fraser, began to discuss the possibility of basing additional

Army units in Western Australia. On 21 November 1969 the Deputy Chief of the General Staff, Major-General S. C. Graham, wrote to other members of the Military Board that consideration was being 'given to the advisability or otherwise of stationing an increased proportion of the Regular Army in Western Australia'. He wished to explore the implications of such a proposal, but it seemed to him that if a task force were to be based in Western Australia the SAS Regiment would have to move east, perhaps to Enoggera in Brisbane.[5]

In fact Army policy was to aim for greater concentration in the Eastern States, and nothing much eventuated for some months. But in July 1970 the Chief of the General Staff, General Daly, directed that a study be initiated 'based on the establishment of three battalion task forces in the Townsville, Holsworthy and Perth areas and the removal of the SAS Regiment from Swanbourne to Enoggera'. The study progressed slowly, and when on 23 October 1970 the Directorate of Staff Duties issued instructions to various branches to examine the implications for SAS training, by coincidence, the instruction was signed by Len Eyles, the previous commanding officer of the regiment. As was to be expected there were a wide range of training factors for and against the transfer of the SAS to Enoggera. But the issue was decided on wider grounds. Daly vigorously opposed the idea, believing that it was politically motivated. The Secretary of the Department of Defence, Sir Arthur Tange, agreed with Daly's arguments, if not with the slowness of the Army bureaucracy, and the move was defeated.[6]

Clark's earlier appointment as GSO1 (Land/Air Warfare) at Army Headquarters gave him a keen appreciation of possible changes or extensions to the role of the SAS. Indeed during 1969 there had been discussions at Army Headquarters as to the possible amalgamation of the SAS and the Commando units into a Special Action Force Group that would be responsible for the SAS Regiment, the Commandos, the attached signals squadrons, the Base Squadron and a Special Warfare Training Centre. At that time the roles of the SAS Regiment and the Commandos included unconventional warfare; that is, guerilla warfare, the organisation of rescue, evasion and escape and sabotage. Since an Australian doctrine for unconventional warfare had not been developed, various US pamphlets were obtained for use by the SAS and Commandos.

One problem in refining the role of the SAS was that it was achieving success in Vietnam in the medium reconnaissance role. But as Colonel Peter Seddon, the Deputy Director of Military Operations, wrote in November 1969: 'It seems to me that giving the responsibility in war for medium recce to the SAS is rather like putting Rain Lover or Vain [famous race horses] behind the plough'.[7]

While the role of the SAS was 'to seek out, often by deep penetration, report on, harass and disrupt the enemy as well as working with irregular indigenous forces', the Army Headquarters instruction covering the role of the SAS added other tasks. These included providing advisory teams for training of indigenous personnel in friendly countries, conducting refresher and continuation parachute training, initiating research, development and user trials, preparing operational and training doctrine, maintaining area studies, and advising on, supervising and conducting special and unconventional warfare training for the Australian services. To meet these requirements, in July 1970 Western Command submitted a detailed proposal to change the establishment of the SAS Regiment, removing the training responsibility from Base Squadron and forming a training squadron with amphibious, parachute, reinforcement, demolitions, unconventional warfare, climbing and operational research wings. Changes were also proposed to the Base and Signal Squadrons.

In addition, it was also proposed to increase the size of the SAS squadrons to formalise the operational arrangements then pertaining in Vietnam. Whereas under the present establishment each troop consisted of three patrols each of six men and a troop headquarters of an officer, NCO and a driver, it was proposed to provide for five patrols each of five men, with the troop commander and the troop second-in-command (troop sergeant) each having his own patrol. It was proposed that the squadron operations officer and the troop commanders be upgraded to captain, that the troop seconds-in-command be upgraded to staff sergeant, and that SAS private soldiers be given the rank of trooper.

The proposal to change the establishment met both support and resistance in Army Headquarters. The Director of Military Training acknowledged that the SAS Regiment might have to conduct courses for non-SAS personnel, but thought that the proposal to raise an SAS training squadron was 'premature'. The Director of Military Operations and Plans supported the proposals but the problem was finding the additional personnel.

In the meantime the SAS Regiment continued a wide range of activities. In April 1970 Lieutenant-Colonel Clark wrote a paper advocating increased attention be given to long range vehicle patrol training in Western Australia. He wrote that long range vehicle patrols achieved the following aims:

    a. Development of expertise in map reading in ill-defined areas, cross country navigation by day and night, using compass, sextant and stars.

b. Practice in long range communications, with the forward stations continually on the move.

c. Experience of vehicle maintenance under difficult conditions.

d. Living off the land, frequently employing survival techniques.

e. Development of independence, determination, initiative and teamwork. For the commander, there is experience both in leadership under difficult conditions and in the practice of sound administration, staff work and maintenance of morale.

f. Formulating [sic] of vehicle mounted patrol tactics, particularly movement and harbouring.

Clark explained that vehicle patrols in Western Australia met conditions similar to those on active service in that the environmental conditions provided a natural enemy, there were man-made hazards such as vehicle breakdowns and loss of fuel supplies, and there was the satisfaction of exploring new areas. In August 1970, 29 men of 3 SAS Squadron led by Captain Ross Bishop parachuted into Meekatharra and set off in vehicles for Wiluna, Carnegie, Charlie's Knob and Marble Bar. During the sixteen-day period in the Gibson Desert four Land Rovers were parachuted in.

Also in August 1970, 2 Squadron achieved the first squadron massed parachute jump. Two C130 aircraft were used to insert the squadron onto barren inland sand dunes to begin an exercise at Pemberton. Lieutenant-Colonel Clark led the squadron out.

At this stage 2 Squadron was preparing for its 1971 tour of Vietnam. The squadron commander was 31 year old Major Geoff Chipman who had been second-in-command of 3 Squadron during its 1966 tour of Vietnam. Since then he had spent twelve months in Britain attached to the Royal Marine Special Boat Squadron and had then been training officer of the regiment at Swanbourne until he was promoted to command 2 Squadron on 16 June 1969. His second-in-command was Captain Robin Letts who had joined the Australian Army from the British SAS earlier that year. Aged 28, he had been awarded a Military Cross for operations in Sarawak in 1965. The squadron operations officer was Lieutenant Joe Flannery, who had served as an NCO in Borneo with the SAS and with the Training Team in Vietnam, and the SSM was Warrant Officer Mick Grovenor, who had served mainly in quartermaster postings in the regiment. The squadron signals officer, Second Lieutenant Chris Lawson-Baker, arrived in Vietnam in August 1970 and would remain there when 2 Squadron arrived in February 1971.

Two of the troop commanders were young and inexperienced. Second Lieutenant Brian Jones joined the regiment direct from Portsea in June 1969 at the age of 20. Second Lieutenant Brian

Patrol extraction in the Sepik District of Papua New Guinea during 2 Squadron's pre-Vietnam exercise in 1970. Training in Papua New Guinea was vital preparation for operations in Vietnam.

Russell, who joined the SAS in January 1970, was the first National Service officer in the regiment. However, Lieutenant Andrew Freemantle (aged 25) already had considerable military experience. He had served as a platoon commander in Borneo and Cyprus with the Royal Hampshire Regiment of the British Army before transferring to the Australian Army specifically to seek action in Vietnam. As usual, the troop sergeants were well experienced; Ray Swallow, Frank

310

Cashmore and Paul Richards were all about to begin their second tour of Vietnam. Richards was a former Commando who had spent a year in Vietnam on CMF full-time duty during which he had managed to join the SAS for one or two patrols.

Exercise SIDEWALK, 2 Squadron's final exercise before deployment to Vietnam, took place from September to November 1970 in the Sepik and border area of New Guinea—the first time the SAS had exercised in that area. Under the direction of Captain Letts and Sergeant Danny Wright, the procedure for operational free fall insertions was perfected during the exercise.

Well before 2 Squadron was deployed in February 1971 it was clear that the commitment to Vietnam was winding down. In April 1970 the Australian Prime Minister, John Gorton, had stated that a battalion would be withdrawn and not replaced at the end of the year. Then in December the New Zealand government announced that it would be withdrawing its artillery battery and SAS troop, and the latter finished its tour in February 1971. In March 1971 the Prime Minister, William McMahon, announced a further withdrawal of 1000 men in the next three months. And finally in August 1971 it was advised that the Task Force would be withdrawn by the end of the year.

With the prospect of operations ceasing there would no longer be a requirement to maintain three SAS squadrons and it was decided that the personnel for a training squadron could be found by disbanding one of the SAS sabre squadrons. Since 1 and 3 Squadrons were at a reasonable strength and were closely involved in training, and 2 Squadron in Vietnam would lose a good number of men—including its National Servicemen—when it returned from Vietnam in October, Lieutenant-Colonel Clark decided to disband 2 Squadron. Thus a training squadron was added to the establishment, although the proposal to raise the rank of troop commanders to captain and troop seconds-in-command to staff sergeant was not accepted. The new establishment came into existence in the last months of 1971, thus setting the framework for SAS activities in the post-Vietnam period. But before moving on to the post-Vietnam era it is necessary to go back and follow the course of events in Vietnam during the last three years of the SAS's commitment.

18
The lure of the
May Taos
3 Squadron: February
1969–August 1969

For six years the VC strongholds in the Nui Thi Vai, Nui Dinh, Long Hai and May Tao mountains dominated the attention of the Australians at Nui Dat. Overlooking the Task Force base and close to both Route 15 and the provincial capital of Baria, the Nui Thi Vai and Nui Dinh mountains, both to the west of Nui Dat, were the first target for the SAS in July 1966. Before the year was out the two Australian battalions of the Task Force had swept through the hills, and other battalions returned in subsequent years. The VC were never completely eliminated, but the threat from that area declined substantially.

The Long Hai hills, in the south of the province, were usually the responsibility of the Vietnamese Regional Forces, and were not entered by the Australian battalions in any concerted fashion until February 1967. At periodic intervals the Australians cleared the hills, but usually suffered heavy casualties from mines which all too often had been lifted from the Australian minefield running from Dat Do to the sea. Eventually the Long Hais were criss-crossed by engineer trails, and they were always vulnerable to artillery and airstrikes. The Minh Dam Secret Zone, as it was known, in the southern Long Hais always remained a VC sanctuary, but the entry and exit routes into the surrounding villages could be ambushed effectively if troops were allocated to the task. While the Long Hais were an important refuge for the VC in Phuoc Tuy they were somewhat isolated from the neighbouring provinces.

But the May Tao hills were different. Straddling the north east corner of the province, they were just outside the 1 ATF TAOR and they were almost 40 kilometres from Nui Dat. Located in the centre of a vast tract of jungle, they could not easily be surrounded and isolated as could the other two hill systems nearer to Nui Dat. The May Taos and the surrounding jungle therefore provided a perfect haven for the VC and NVA main force units operating not only in Phuoc Tuy but also in Bien Hoa and Long Khanh Provinces.

Although the May Taos proper were outside their TAOR, the Australians could interdict the VC routes to and from the mountains, and Brigadier Sandy Pearson had been aware of this possibility when, towards the end of 1968, he had ordered a series of SAS patrols to operate along the northern border of Phuoc Tuy Province. The VC were using that route to move from their May Tao bases to the Hat Dich area of Bien Hoa. Then, as related in Chapter Sixteen, the Task Force was deployed on Operation GOODWOOD in December 1968 to interdict VC movement from the Hat Dich to the Long Binh–Bien Hoa military complexes.

By the time Major Reg Beesley and his advance party had been joined by the main body of 3 Squadron on 21 February, the 1969 Tet offensive was well underway. It was only a shadow of the 1968 Tet offensive, but again the main body of the Task Force left the province for operations in Bien Hoa. By mid February both 4 RAR and 9 RAR were involved in Operation FEDERAL, east of Long Binh in Bien Hoa Province, while 1 RAR had just returned to Nui Dat to be relieved by 5 RAR, then preparing for its first shakedown operation west of the Task Force base.

Fortunately, the presence of the New Zealand troop enabled the SAS to continue its operational patrols during this tense period. For example, on 22 February three New Zealand patrols were deployed to the southern Bien Hoa area. The next morning, at 3.05 am 3 Squadron 'stood to' as news was received that the VC were attacking Binh Ba, Phu My, Baria and Hoa Long. The 1 ATF Ready Reaction Company was deployed to engage the enemy in Baria. During the morning of 25 February two New Zealand patrols spotted groups of enemy, with Sergeant Fred Barclay's patrol reporting 70 VC (50 armed) moving west, ten kilometres north of Phu My.

Before leaving Australia, Beesley had been informed that Brigadier Pearson had definite views on the use of the SAS, and on his arrival he soon found himself at odds with the Task Force commander, who wanted to deploy the whole SAS squadron on 'fan' type patrols. Beesley argued strongly that he should be allowed to conduct his operations in the manner in which he had trained his squadron—namely inserting patrols separately from Nui Dat in a fashion designed to cover a target area over an extended period. Pearson relented and told Beesley that he had six weeks to produce results using his SAS method.

Major Reg Beesley, Officer Commanding 3 SAS Squadron, at Nui Dat chatting to Lieutenant-Colonel John Murphy, a former commander of 3 Squadron and then Military Assistant to the Chief of the General Staff, 1969. To the left is Sergeant John Robinson who had served with Murphy in Vietnam in 1966.

Pearson directed Beesley to conduct extensive reconnaissance in two areas, the Hat Dich in southern Bien Hoa, west and south west of the Courtenay rubber plantation, and along the northern Phuoc Tuy border, cutting the likely routes from the May Taos to the Hat Dich, determining the locations of the supply routes for the 84th Rear Service Group. A specific task was to find the exit and entry points of the VC Ba Long training area near the Mao Taos. While Beesley might have disagreed with his Task Force commander over techniques, he admired his approach. As Beesley wrote later: 'In fairness to Sandy Pearson, I would say he was the first Commander who had clear ideas on what he wanted from SAS operations—the method to achieve this was the difference. Sandy would brief me regularly indicating his future concept for operations and his intentions of the force at large, and leave me to appreciate the intelligence requirements, establish a plan, which after being cleared by him, would be executed'. Pearson's aim was to obtain intelligence upon which he could plan and conduct battalion operations.

Beesley fully agreed with Pearson's emphasis on reconnaissance. Indeed one of his first acts after arriving at Nui Dat was to go around the squadron base kicking down the 'kills boards' erected by the previous occupants. 'I thought it was a bit immature and unprofessional', he observed later. 'We were not there to kill people but to gain information'. As the year progressed only a handful of people in the squadron—himself, the second-in-command, the operations officer and the SSM—knew the kill tally.

One of Beesley's early problems was to establish a sound working relationship with the New Zealanders. He was 'astonished' to find that since their arrival in December, they appeared to have 'developed a little camp of their own'. To Beesley's mind, their situation was similar to that of V Company 6 RAR/NZ, and he brought the matter to a head by placing all the vehicles in the squadron, including the New Zealand vehicle, under the control of squadron headquarters. It was not just 'bloody-mindedness' which caused Beesley to determine that the New Zealanders should not be allowed to task their own patrols, but a belief that all patrol tasks should be co-ordinated and directed by squadron headquarters. Before long Beesley's aim of integration was achieved successfully. He 'believed that SAS, whatever country, could operate together and that such a "family" had little room for paranoic megalomania. Furthermore, it was possible that, operationally, the need could have occurred with little notice. Hence the need to unzip inflexible minds! Overall, I was most impressed with the Kiwis' sense of purpose, their dedication and professionalism. Initially they lacked mature leadership and an understanding of an environment which was very different to that experienced in Borneo.'

The first non-New Zealand patrols from 3 Squadron were deployed on 28 February, and the reconnaissance of the two target areas continued for the next six weeks. During that time eleven reconnaissance patrols were inserted into the Hat Dich. This area was gently undulating with patches of light to medium primary jungle and areas of secondary growth. Large areas of bomb damage were encountered making movement difficult and noisy. Water was flowing in most creeks. Banana plants were found occasionally bearing fruit, and wildlife in the form of pigs and monkeys was abundant in the area. Between 25 February and 27 March there were seven enemy sightings and a total of 81 enemy were observed, including the 70 on 25 February. Throughout the duration of the patrol, shots and bursts of fire were heard and movement was observed, particularly near the Firestone Trail.

The only patrol to contact the enemy was a six-man patrol commanded by Second Lieutenant Terry Nolan, who had stayed in Vietnam when 2 Squadron had returned to Australia. On 20 March his patrol was inserted about nine kilometres slightly east of due north from Phu My and during the next 24 hours they heard numerous shots. Then in the afternoon of 21 March female and male voices were heard approaching the patrol. One armed VC started searching the foliage near the patrol while Nolan and his scout, Private Colin Jackson, lay motionless. When they heard the VC disengage the safety catch on his AK47, Jackson shot him at a range of four metres. The patrol then withdrew and established a night LUP. During the next few days the patrol heard more shots, discovered various abandoned enemy installations and saw enemy footprints. Late in the afternoon of 27 March the patrol was crossing a track when it contacted two VC moving along the track. Both were killed. Suspecting an enemy camp in the near vicinity, and after suppressing the area with small arms and M79 grenades, the patrol withdrew. They completed the mission and were extracted to Nui Dat on 29 March.

The patrols deployed to the Hat Dich during the first six weeks used a system that Beesley described as 'saturation reconnaissance'. The concept was to take a large area, assign a number of patrol AOs (3 x 2 kilometres), and on the extraction of one patrol insert another patrol with an AO covering that part of the previous patrol's AO that had not been searched. This achieved two things: it provided a measure of deception as extractions had a definite signature; and secondly, it enabled coverage of an area for a longer period, an important factor since enemy intelligence changed regularly.

Also, during the first six-week period sixteen patrols were deployed to the areas west, east and south east of the Courtenay rubber plantation, astride the route from the May Taos to the Hat Dich. The first sighting, three VC east of the Courtenay rubber plantation, was on 3 March, and from then until 6 April there were 23 enemy sightings resulting in 104 enemy being observed. The enemy were dressed in a mixture of blue, faded grey and black; weapons varied from AK47s, SKS's and M1 carbines to Chicom pistols. Movement was usually in groups of three, but sometimes up to ten, and they were moving fast, closed up and quiet. An occupied camp was located 300 metres south of the Courtenay rubber plantation 1500 metres east of Route 2. In six contacts a total of seven VC were killed with four possibles.

Significantly, the most successful patrol east of the Courtenay rubber plantation was the one deployed closest to the Phuoc Tuy–Long Khanh provincial border. On the afternoon of 31 March Sergeant Tony Tonna's six-man patrol was inserted 1 kilometre east of the rubber plantation 2500 metres south of the border. The terrain included sparse forest and very thick secondary growth consisting of tangled bamboo, vines and thorn bushes. By last light the patrol had moved south only about 400 metres, and next morning at 8.35 they observed nine VC with mixed uniforms moving north on a foot track. They established an OP and for the next 24 hours heard numerous groups of enemy moving north, east and west.

Late in the afternoon of 2 April they established another OP about 100 metres further south and promptly observed three VC moving south. The first carried an RPD, the second an M1 carbine, and the third was a nun dressed in light royal blue habit. They were moving quietly 3 metres apart. In the next 90 minutes the patrol observed a further 17 VC. During the following three days they heard numerous shots and sounds until in the afternoon of 5 April, having moved about 400 metres north east, they observed eight VC moving west on a track. Suddenly two VC stopped while a third VC advanced on the OP. At a range of 8 metres Private Barry Williams fired four aimed shots with his SLR. The patrol successfully evaded the VC and next day, 200 metres to the west, struck a well-used track. Their presence had now been compromised and Tonna decided to set an ambush. Within half an hour the ambush was sprung on four VC, all of whom were killed. In a little over an hour the patrol was extracted by helicopter. During a period of six days they had sighted 44 enemy.

Beesley's final task was to find the exit/entry point for the VC Ba Long Province, and as he said later, the successful patrol 'was a gem. I suppose I did a lucky appreciation in looking for a needle in a haystack. I tasked Sergeant Ned Kelly's patrol to approach the [swampy] area [just south of the Courtenay rubber plantation] from the south using the [Suoi Tam Bo] river as the axis. His patrol effort was superb—they spent days living in trees and eventually stumbled on an HQ location, medical elements (nurses) and a large cache containing beaucoup rice. At one stage he was some five metres from two VC, male and female, enjoying their sex. He stayed in location observing and reporting, and eventually [on 27 March] married up with elements of V Company 4 RAR who were deployed by Pearson to destroy the site. Kelly was by far my best *reconnaissance* patrol commander who later did an excellent close reconnaissance of supply convoys near the Long Hais...I put him up for a decoration—however, the "rations" for SAS were tight.'

By now Pearson's six-week trial period had passed, but he was obviously satisfied with the SAS performance because he told Beesley that he was free to continue operating in the manner for which he had trained. With the conclusion of the Task Force operations in Bien Hoa Province and the return of the battalions to Phuoc Tuy, Pearson ordered a number of SAS reconnaissance patrols to other areas of the province. For example, on 9 April a five-man reconnaissance patrol commanded by Ned Kelly was inserted into the Long Green, four kilometres east of Dat Do. During the first four days the patrol observed 24 VC and heard other groups. Then on the morning of 13 April they saw fifteen ox-carts carrying a total of 95 VC (including ten women); fifteen to twenty were armed with an assortment of weapons. By the time they were extracted on 15 April they had observed a total of 175 enemy who appeared to be establishing a supply dump for use by a large enemy force. As a result of this patrol 5 and 9 RAR were deployed on Operation SURFSIDE to clear the Long Green and search for the headquarters and two companies of D445 Battalion. Only small groups of enemy were found but one ambush caught the commander of one of the D445 companies.

Meanwhile, 4 RAR was deployed into the Hat Dich, and in conjunction with this operation a number of patrols were inserted into the Nui Thi Vais. Between 15 and 23 April a patrol of New Zealanders commanded by Sergeant Bill Lillicrapp killed two VC and located an enemy base camp under construction. Soon afterwards 5 RAR began operations at the base of Nui Thi Vai. As the battalion historian wrote: 'The Nui Thi Vai feature had not been cleared for years, and virulent rumours attesting to the enemy strength and fortifications to be found in the mountains, were heard'. Although not without their own losses, by the end of the month the battalion had cleared both the Nui Thi Vai and Nui Dinh hills.

At about this time the squadron began to sponsor a number of patrols from the LRRP Wing at Van Kiep. Like the Van Kiep patrols the previous year, these patrols were all led by former members of the SAS, including Warrant Officers Jack Wigg, Barry Young, John Grafton and Clem Kealy.

After about six weeks of operations Beesley re-assessed the effectiveness of his patrols, and concluded that they had been too short in duration and deployed too frequently, thus wearing out the patrol members. After a number of experiments he decided to extend the duration of patrols in the dry season to ten days, with patrols in the wet season extended to twelve days. This would allow several days 'to get the smell' of the jungle, and would lessen the drain on scarce helicopter hours. He also instituted the resupply of patrols by rope from a helicopter, the whole process being completed in 45 seconds. It was surprising to find that this innovation was resisted by some of the 'old sweats' in the squadron who claimed that patrol security would be breached.

Beesley was concerned that operations were following a set pattern, and as US helicopter support was decreasing there would be a greater drain on scarce Australian helicopter resources. Consequently, he assembled the whole squadron in a 'Chinese parliament' and pointed out that if a patrol called for an immediate extraction when it was not absolutely necessary, the result might be that another patrol in a truly desperate situation might not be extracted. In such a situation, Beesley argued, the first patrol might be responsible for the demise of the second. His argument was accepted. The decision to request extraction still rested with each patrol commander, but he asked them to be completely objective and responsible in making their decisions. Beesley later commented that it was a gamble which paid off; however he 'sacked a patrol commander who cried wolf for an immediate extraction. He awoke the morning after the insertion to note that he was surrounded by tracks. On hearing this, I thought that the VC had armour, and then realised that he meant used foot tracks'.

All the SAS squadron commanders during the six years of operations in Vietnam recognised that their main role was not to join their men on patrol but to direct and co-ordinate the operations of the patrols from squadron headquarters. However, unlike most other SAS squadron commanders in Vietnam, Beesley believed that from a leadership point of view it was essential that on occasions he should lead his own patrol on operations. The previous squadron commanders had all accompanied patrols on operations and had commanded some large special patrols, but they argued that a patrol was a small close-knit team that needed to train and work together over a considerable period to be fully effective. Beesley believed that if he could not lead his own patrol he should not have been a squadron commander. Furthermore, his second-in-command, Captain Ross Bishop, also had his own patrol. Beesley saw Bishop as the alternative squadron commander, rather than merely his chief administrative officer. Bishop found 'The Beast' to be 'a dedicated professional soldier, conscious of his responsibility and dedicated to excellence. He pushed hard but was highly respected'. And this view has been confirmed by several other soldiers in the squadron.

Beeesley put his policy into practice on 2 May when his five-man patrol was inserted into an area on the Phuoc Tuy–Bien Hoa border north east of Nui Thi Vai. The patrol was generally uneventful until about midday on 9 May when voices of two people were heard. A firm base was established and Beesley went forward with Private Keith Jackson to observe two enemy who were fishing in a creek. Meanwhile another VC, carrying an AK47 and three magazines, approached the firm base and at a range of 10 metres was shot dead. Following his own policy, Beesley did not call for an immediate extraction, but completed the patrol and was extracted by rope the following day.

The squadron continued its interest in the approaches to the May Taos and on the afternoon of 2 May a five-man patrol commanded by Lieutenant Chris Roberts was inserted two kilometres south east of the Courtenay rubber plantation. For the next day and a half the patrol

319

observed or heard small groups of VC moving from the rubber plantation to the east, but on the morning of 4 May, after they had spotted fresh footprints on a track 'the patrol 2ic passed a message', wrote Roberts later, 'that a party of enemy were following us, and urged us to hurry along. After about 150 to 200 metres I attempted to prop but the 2ic urged us to move on as we were still being followed up. At the time I felt we were withdrawing in indecent haste, as I could not see the enemy nor observe any movement behind us. The 2ic, however, was sure that the enemy were onto us and that we should move back...To this day I believe I was wrong drummed'. Although years later Roberts recalled that he doubted whether they were being followed, the patrol report indicates that the patrol heard groups of enemy and sighted one. However, two of the Australians were suffering from medical problems, and in view of the possibility of a contact, Roberts requested extraction. Back at Nui Dat he indicated that he was prepared to go back into the area, although not to the same LZ. The patrol was re-arranged, the two men were left behind, and Sergeant Fred Barclay, who was recovering from an illness and wanted to test his fitness, joined the patrol. Next day a four-man patrol comprising Roberts, Barclay and Privates Terry Hancock and Mick Malone, was re-inserted into the same area.

Again small groups of enemy were observed, with the largest group numbering nine, including a female with a large pink scarf. At 3.40 pm on 7 May the patrol engaged two NVA main force soldiers, killing one. But a larger NVA force reacted swiftly to the contact. Although they could not locate the Australians it seemed that they were attempting to drive them towards a prepared NVA ambush position. At 4.30 pm the patrol called for an immediate extraction. About an hour later the helicopters were overhead and the Bushrangers (Australian helicopter gunships) began their suppressing fire. Various groups of enemy were seen to be advancing and the patrol joined in, firing their rifles and M79s until they had attached their Karabiners to the dangling ropes. Suddenly, under the cover of the smoke grenades released to guide in the helicopters, four enemy were seen closing in on the LZ. They were engaged and one was shot at a range of 15 metres. The extraction was completed safely, with a helicopter door-gunner claiming one VC killed. The patrol concluded that there was possibly a large NVA camp in the vicinity and that there had been heavy movement from this area to the east.

Four days after the return of Roberts' patrol Sergeant John Robinson was inserted with a four-man reconnaissance patrol north east from Roberts' contact area with the mission of locating enemy camps near the Phuoc Tuy–Long Khanh border. For the first four days they saw little sign of recent enemy activity, but at 10.30 am on 15 May they came across a track which they estimated had been used within two hours. The track was about 650 cm wide and had been worn to a depth of about 150 cm. In the next two hours they heard groups of enemy numbering in all about 50 people and Robinson and his scout, Private Keith Beard, crept forward for a close reconnaissance of an enemy camp located where the track crossed a creek. As they crawled closer they could see a small native cookhouse occupied by a male and two females. There was a large native hut with sleeping areas for eight hammocks and eight ground areas. To Robinson, it looked as though it was a staging area and he carefully sketched the camp.

Determined to obtain a better view Robinson withdrew and decided to swing around the camp and approach from the south. They had gone only a short distance when they struck another track, and as they were preparing to cross it five VC appeared, moving towards the camp. At a range of 15 metres Robinson engaged the first VC with an aimed shot to the chest while Beard shot the second VC at a range of 10 metres. Immediately the other VC returned fire, but in pouring rain the Australians withdrew about 500 metres and established an LUP for the night.

Next morning the patrol headed back towards the enemy location and at 10.10 am struck a Y junction in the track running towards the enemy camp. They had just crossed the track when they heard enemy approaching on both arms of the Y. While listening to about seven to ten people walking along one track, they observed another nineteen VC on the other track, some pulling 7.62 mm light machine-guns on wheels. Then another group of six VC were seen, and as they passed the patrol's location one VC detached himself and approached the patrol. He was 5 metres away when he stopped, dropped his trousers, and started to squat. Suddenly his eyes lit up and he grabbed for his rifle. He was immediately shot by the patrol signaller, Private John Dodd.[1] Within seconds the SAS patrol found itself engaged by two groups of enemy, and the Australians divided into two teams, each of two men. Robinson seized the initiative and with Private Nicol McKelvie, engaged the larger enemy group armed with three light machine-guns mounted on wheeled carriages, killing two and inflicting heavy casualties. Having surprised and confused the enemy he ordered a rapid withdrawal.

The Australians withdrew only about 100 metres and found themselves in an occupied enemy company defensive position. They had time to observe five bunkers, a long open weapon pit, a small control pit with wires and detonating cord running out to home-made claymore mines, a well-used latrine and a track system. Fortunately, as the Australians ran into one end of the camp they could see the occupants running out to another flank, weapons in hand, towards the Y junction and the scene of the recent contact where firing was still continuing.

The patrol kept moving away from the area and after about 300 metres Robinson told Dodd that it was time to obtain communications. The patrol stopped, formed an LUP and Robinson began to run out the aerial. As he strung the wire out he came upon another track. He took off his pack, attached the wire to his pack, and then went back to ensure that Dodd could obtain communications. He then returned to the track to keep watch. No sooner had he arrived back at the track when he spotted nine VC approaching their position. He fired a complete magazine down the track, killing one and possibly another enemy, while the remainder of the patrol moved forward to support him. They maintained their fire for a few minutes and then Robinson gave the order to retire: 'Go now and I'll cover you'. He personally covered the rest of the patrol as they gathered their equipment from the LUP. Robinson tried to recover his own pack by pulling the radio aerial but the enemy saw the movement of the pack and riddled it with bullets. Beard returned to assist Robinson; they fired a couple of M79 grenades, and together they rejoined the patrol, where Robinson organised such effective and controlled fire that he was able to prevent the enemy from following them up closely. He recalled later that they were 'taking a tremendous amount of fire—twigs and pieces of tree were falling all around'.

Withdrawing to the west they came across an open area with a number of bomb craters, and about 70 metres across the open area they sank into a large crater. 'We were just about physically exhausted and were very low on ammunition', recalled Robinson. They threw out the radio aerial but could not obtain communications. In the jungle to the north, north east and west of their bomb crater they could hear the enemy searching for them. But the earlier attempt to obtain radio communications had been successful. By about 1 pm the helicopters were overhead, and while a heavy fire team suppressed the edge of the jungle, the four men were lifted out by rope. As the RAAF historian wrote: 'The hazardous extraction was a success because of the heavy suppressive fire by the gunships in spite of enemy return fire which was kept up until all helicopters had moved 2000 metres from the pick-up-point'. About 8 kilometres away the lift helicopter, piloted by Flying Officer T. Butler, Royal New Zealand Air Force, landed, the men clambered aboard, and they were soon back at Nui Dat.[2]

While Robinson was searching the northern border area, Sergeant 'Windy' McGee and his New Zealand patrol were engaged in a reconnaissance of the southern approaches to the May Taos, about 7 kilometres south east of Thua Tich. Between 9 and 19 May they sighted nine VC and five ox-carts. Three oxen and one VC were killed in an ambush on 18 May.

322

By this time Beesley and Brigadier Pearson had agreed that SAS patrols should be allowed to ambush the enemy if a suitable opportunity was presented. Although there was the danger that SAS ambushes could alert the enemy to the fact that his area had been penetrated, the enemy often knew from helicopter activity that the SAS was in the area and it was good to keep the enemy off balance. Furthermore, the infantry battalions could not be deployed to all parts of the province at once and the SAS could maintain an Australian presence in the areas where the battalions were not operating. In line with this change in policy, and following the sightings by Robinson, Terry Nolan was ordered to conduct an ambush with seventeen men in the area about a kilometre east of the Courtenay rubber plantation. Until this time most patrols in this area had been chased out by enemy contacts or had been detected on insertion, and the patrol was dropped off by APC after a deception operation mounted by APCs, tanks and helicopters firing into a previously located bunker system.

After only a couple of hours the patrol had located a well-defined track and in the evening of 18 May they withdrew into a night LUP. The next morning they established a five-man OP near the track and within half an hour 61 enemy had been observed dressed in grey uniforms and moving quickly and quietly at well spaced intervals. Throughout the day more enemy were observed, including some women, many walking wounded VC and six wounded VC on stretchers. Unknown to Nolan at the time, the 33rd NVA Regiment had recently attacked the Long Khanh provincial capital, Xuyen Loc. A US force had almost been overrun, but the NVA had taken considerable casualties in the assault. The NVA regiment with its wounded was now withdrawing towards the May Taos.[3] Nolan's men observed that most of the large groups of enemy were moving east towards the May Taos, while small groups, apparently guides and escorts, were moving back to the west to pick up new parties from the 33rd Regiment.

Nolan was faced with a considerable problem in mounting an ambush. His patrol was an amalgam of four patrols and its ability to fight was fairly limited; if it found itself in a large firefight it would be much harder for seventeen men to escape and hide than for five men. To set an ambush Nolan would have to deploy warning groups to each flank and his killing group might number less than ten men. If these men were spaced at 10 metre intervals his ambush would cover 100 metres, and in any case he only had sufficient Claymores to cover 100 metres. But if the enemy party numbered fifty, each spaced at 10 metre intervals, it would stretch for half a kilometre. On the other hand it was questionable whether it was worth risking Australian lives merely to ambush one of the VC escort parties of three or four men.

The OP continued to observe the enemy and over the next three days saw women carrying white packs resembling ice packs; they saw enemy dressed in black, khaki, grey, white and blue; some wore basic webbing, others had packs, some had bush hats and there were even two white pith helmets; they saw an enemy soldier with a wooden leg carrying an RPD machine-gun; there was an enemy on crutches; and they observed one enemy wearing black nun's habit carrying an RPD. A little later another VC was seen in nun's habit. On another occasion one VC walked to within five metres of the OP and defecated. Enemy weapons included AK47s, SMGs, RPDs, RPGs, M1 rifles and SKSs. Meanwhile, a reconnaissance party sent out from the LUP had discovered another track 80 metres south of the LUP showing signs of having been used by the VC. Considering the large number of VC Nolan radioed Nui Dat that in his view a company of infantry should be inserted to conduct the ambush. However, the rainy season was well underway and it was quite difficult to get the message through to Nui Dat; eventually, perhaps 24 hours later, the reply was received that Nolan was to mount the ambush himself. Sergeant Tony Tonna, second-in-command of the patrol, recalled whispering to Nolan on receipt of this news that there were 'more of them than there are of us'.

Nolan began to prepare for the ambush but by chance knelt on a root, dislocated a cartilage in his knee and on the evening of 23 May was extracted by a Sioux helicopter from an LZ about 300 metres away to the north. It was now Tonna's responsibility to mount the ambush, but at 7.40 am on the morning of 24 May, the OP reported that they had seen 97 well-armed enemy. Half an hour later they reported a further 118 enemy preceded by an advance party of three men moving 15 metres apart. One enemy from the main group moved towards the OP and dry retched violently.

Tonna decided that the only option was to spring the ambush on one of the smaller groups approaching from the east, and in the afternoon of 24 May a Claymore ambush was initiated on four VC travelling about fifteen metres apart. Three VC were killed, but as the ambush party of nine men moved forward to search the bodies the fourth VC, who had been wounded, opened fire on Private Ray Mickleberg, who was knocked to the ground by the impact of bullets striking his pack. The rest of the patrol suppressed the area but the enemy escaped. Next morning the patrol was extracted by helicopter. In eight and a half days they had observed a total of well over 780 enemy. Almost 600 had been travelling from the west to the east and about 185 had been going in the other direction.

Although it was not quite so clear then as it was later, the contacts during May were an indication that the VC had entered Phuoc Tuy 'in strength to demonstrate their ability to conduct operations freely in

the villages and hamlets which lie along National Route 15 and Provincial Route 2 and to disrupt the normal government administration
and control over the people of the province'. Brigadier Pearson was
not aware of the VC intentions, but acting on SAS intelligence, at the
end of May he deployed the newly arrived 6 RAR on Operation
LAVARAK into the northern part of the province, mainly on the western
side of Route 2. The operation lasted for a month, and before it was
over a company of 5 RAR and two troops of tanks had been reacted to
clear a strong enemy force out of the village of Binh Ba. Elements of
33 NVA Regiment, 274 VC Regiment, 84 Rear Services Group, the
Chau Duc Company and VC local forces were contacted, and after the
operation the enemy withdrew north west into Bien Hoa Province,
and north east towards the May Taos.[4] Over 100 VC were killed
during the operation. As part of the VC offensive, during May and
June the 74th NVA Rocket Artillery Regiment was responsible for a
number of rocket attacks on the Task Force base at Nui Dat. Most of
the rockets landed in the area of the Task Force rubbish dump, but
several landed on 3 Squadron's lines without damage. On 7 June one
exploded above Corporal Bob (Buff) Lorimer's tent. The tent was
riddled with shrapnel and Lorimer was evacuated to hospital with a
small piece of shrapnel just below the heart. He was extremely lucky
not to have been killed.

The battle for Binh Ba and the operations in early June marked the
end of the main force operations in Phuoc Tuy for some time. Indeed,
even before the final main force operations began the Task Force had
changed its concept of operations to concentrate on pacification
—interdiction of the local VC as they entered the villages in the east of
the province. The directive for this change in operational priorities
had been issued to 1 ATF by Lieutenant-General Julian J. Ewell,
commanding general II Field Force, on 16 April 1969. As a result, in a
period of three and a half months from May to August 1969 Brigadier
Pearson attempted to reduce the presence and influence of the VC by
close patrolling and ambushing near the populated areas stretching
from Dat Do to the sea, and by improving the effectiveness of the local
forces in the area. The battalions involved incurred considerable numbers of casualties from mines, causing some concern in Canberra.

3 Squadron was not closely involved with pacification, but in addition to its reconnaissance tasks across the province, was required to
maintain an Australian presence in those areas where the battalions
were not operating. One large recce–ambush patrol was commanded
by the squadron second-in-command, Captain Ross Bishop. His ten-
man patrol was inserted seven kilometres north of Phu My, in Bien
Hoa province, on the afternoon of 8 June, and within an hour they
had sighted their first VC. They did not find a suitable target for an

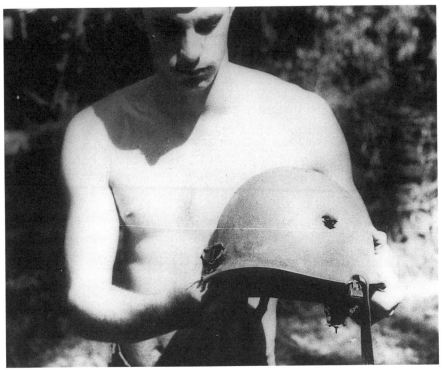

Private Barry Williams holding a helmet damaged during the rocket attack on SAS Hill, June 1969. This was the last major Vietcong attack on Nui Dat during the time the Australian Task Force was in Phuoc Tuy Province.

ambush until 12 June when their OP observed sixteen enemy on a track and an ambush position was occupied. That evening the Claymores were fired on four VC, all of whom were killed. The next day the patrol received a resupply of Claymores by rope and they remained in the area for another five days observing enemy movement.

While Bishop was returning from his patrol in the west, Second Lieutenant Nick Howlett was preparing to take another ten-man patrol to the north east of the province. Early in the afternoon of 24 June his patrol was inserted four kilometres north of Thua Tich, near Provincial Route 330, which in this area was little more than a track. They found no sign of recent enemy movement until 28 June and at 7.45 am on 29 June set an ambush on a well-defined foot-track. At 8.45 am the ambush was initiated on three enemy. Two dead were searched but the third, who had been in front of a Claymore, was never found. On 4 July the patrol was extracted by helicopter and landed 2 kilometres north at an LZ secured by C Company 9 RAR. Howlett and two other patrol members guided elements of C Company to the area where one VC had been sighted the previous day, but there was no further sighting.

Although in different parts of the AO, both Bishop's and Howlett's patrols had been marked by similar characteristics. Both had been ten-man recce–ambushes; both had remained in the area after a contact, and both had continued for long periods—nine and ten days respectively. Lieutenant Chris Roberts' patrol in southern Bien Hoa Province had the same characteristics. It was inserted on 5 July and on 10 July, while moving forward to investigate an unusual 'ant hill', Roberts found himself in an enemy fire-lane looking at five or six VC sheltering under a poncho about 8 metres away. Suddenly a VC appeared from the ant hill which was actually an enemy bunker. Roberts was signalling directions for his patrol to assault the camp when another VC moved out of the camp to urinate and spotted the Australian. Roberts fired an aimed burst from his M16 at the VC, fired an M79 round into another enemy group, and then the patrol withdrew using fire and movement. A light fire team was directed onto the enemy camp, and the SAS patrol remained in the general area for a further six days.

Beesley's determination that patrols should remain 'in' after a contact was emphasised on 31 July when a five-man New Zealand recce–ambush patrol commanded by Sergeant Fred Barclay contacted three VC in the southern part of the Nui Dinh hills. Soon after the contact the patrol discovered an enemy camp and Beesley decided to take a patrol of eleven men, mainly New Zealanders, to join Barclay's patrol and to attack the camp. Communications were poor and the two patrols did not join up until the morning of 2 August. When they reached the enemy camp no enemy were present. By chance, Beesley stopped one soldier just as he was about to step on an M16 mine.

During June the squadron conducted eleven patrols. A total of 157 enemy were sighted in 26 sightings and there were five contacts resulting in eight VC being killed plus possibly two more. The results for July were similar, with 308 enemy being sighted and five killed. After the relatively quiet months of June and July there was more activity in August with a series of patrols across the approaches to the May Taos. On 5 August a six-man New Zealand patrol commanded by Corporal Percy Brown was inserted 8 kilometres east of Thua Tich. The next day Lance-Corporal Sam Peti killed two VC with single, aimed shots to the chest. Almost simultaneously a grenade exploded six metres away, slightly wounding two of the patrol. Six days later they contacted three VC. At least two enemy were killed and the patrol was extracted the next day.

Meanwhile, on 5 August Chris Roberts' patrol of five men had been inserted four kilometres south east of the Courtenay rubber plantation. In a detailed description of the patrol, Roberts wrote:

We were lucky to get out without casualties. Just prior to departing on this patrol John Robinson gave me a St Christopher medal imploring me to wear or at least carry it on the patrol as we were going back to the area...

On the morning of 10 August we had discovered an occupied [company-sized] camp and that night we LUP'd quite close to it. We could hear music from an enemy radio plus singing and yelling until 10 pm that night. Next morning two of us crawled forward for a close recce but found our way barred by a deep gully full of broken bush and branches. However, we could clearly see movement on the other side of the gully. After watching it until lunchtime, I withdrew the patrol and decided to move around the southern perimeter, note the tracks leading into the camp and try and recce it from the west. We crossed several tracks in a short distance and then moved through some fairly open scrub and big trees. Suddenly [Private] George Franklin, the scout, signalled three NVA aproaching from the opposite direction. I thought we were well hidden by some large trees but George called out 'Dong lai' and immediately shot the first soldier...I was pretty pissed off as I thought his actions were premature and remember asking 'Why the hell did you do that George?'

According to the patrol report, however, the first enemy had raised an AK47 to fire at the patrol. Roberts continued:

We were quite close to the camp and there were several well worn tracks to our rear so we withdrew to the south for about 100 metres and then went into LUP for about five minutes. Having heard nothing I decided to move back to the east, move north over the ridge, cross the next creek and then move west again and approach the camp from the north. As we moved out we crossed another well worn track and then moved downhill passing through an open area and crossed a small gully. About five to eight metres past the gully we came to a thicket caused by fallen timber and thick secondary growth. George said he couldn't get through, and as we were looking for a way around it, my new 2ic [Lance-Corporal], Garry Daw, signalled enemy.

Looking around I saw a large party of NVA coming down the slope behind us. There was at least a platoon and probably more, and they were led by two trackers. They were about 40–50 metres away and moving fairly quickly. It was the first time I *really* felt true fear—it went through me like bad acid and I could tell that this was going to be a bad contact. To try and force our way through the thicket would have been suicide. I decided to fight it out there, hoping to surprise them in the first burst of fire, and in the ensuing confusion move north towards the high ground around the thicket.

The 2ic engaged the trackers at about 5–10 metres and all hell broke loose. The NVA put in the best counter ambush drill I have ever seen. They immediately engaged us with MG, RPG and grenades and the noise was quite overwhelming. I thought that the casings ejecting from George Franklin's rifle were hitting me in the cheek and as I turned to talk to him I noticed that in fact, they were pieces of wood flying off the tree between us.

The 2ic and the next man tried to crawl back but the third man in line appeared to be in shock or overwhelmed by the sheer intensity of the fire and the explosions going off around us. He may well have been stunned by a close explosion. I remember [Lance-Corporal Michael] 'Cowboy' Mellington, a national serviceman from 152 Signal Squadron, calmly standing there and engaging the enemy as if he were on the range back home. He was a very steady soldier under fire. Things were looking crook. Garry and the second last fellow couldn't get past the fellow in shock, there were rounds kicking up dirt around them, and an enemy assault force had moved onto the high ground on our right flank some 50–60 metres away. Cowboy Mellington very calmly continued to engage the enemy while I ran a few paces to my right front, grabbed the fellow by the harness, and dragging him back behind me yelled out to the patrol, 'Follow me! Follow me!'

I ran towards the assault as this appeared the best way around the thicket. George Franklin burst through it leaping feet first over a log, which surprised me. At this moment I realised that the patrol might be split up although we had only come about 10 metres from the contact area. So I told George to head east and pushed the trooper after him. I then counted the other three through and thankfully confirmed that we were all still together. By this stage the enemy assault was moving down the hill from the north about 30 metres away and looked like overlapping us. I fired a long burst towards them and then ran after the patrol at a very fast pace. I was pleased to see that no-one had dropped his big pack and it was amazing how fast we could run with them on when we had the incentive to do so. I recall thinking 'I hope your bloody medal works, John!'

We moved fast up a slope to the east and then made our way obliquely to the top of the ridge. About 100 metres from the contact area I was about 15 metres behind the patrol when I thought I heard an Australian voice shouting behind me. I looked around and saw [Private] Mick Malone caught up in some very nasty vines about 10 metres away and having great difficulty in disentangling himself. I ran back towards him and could see some NVA ascending the slope some distance away. Running past Mick, I clipped on a full magazine and fired a long burst at them followed by at least one M79, which I think was white phosphorous. Mick told me he couldn't disentangle himself from the

Helicopter extraction, 3 SAS Squadron, Vietnam 1969. During most extractions the lift helicopter landed briefly while a light fire team (consisting of two helicopter gunships) prepared to engage any enemy that might be threatening the extraction.

vines, so I grabbed his webbing and pulled as hard as I could. Unfortunately I dragged him through the vines and cut his face and hands badly. We then moved on and caught up with the patrol.

We moved reasonably quickly for about 1000 metres before stopping to establish comms. We got a quick message away in clear and then moved again. [Beesley] came out in a chopper and authorised extraction after seeing an enemy group following us up about 400 metres behind us, and another group in the creek to the south and parallel to us. We came out through a very narrow opening in the trees some 1500 metres or so from the contact area. Probably a lukewarm extraction.

Beesley later admitted that he had given Roberts some of the most difficult areas for his patrols, adding that it was 'good leadership and judgement on his part plus some pure luck [that] contributed to his survival'. He thought that Roberts was 'an excellent patrol commander'.

About three kilometres to the south of Roberts' contact area, Warrant Officer Eric Ball from the New Zealand troop led a five-man patrol (including one Australian) on a nine-day patrol which found a

The lure of the May Taos

Warrant Officer Tony Tonna, the intelligence NCO of 3 SAS Squadron, pointing to the approaches to the May Taos on the the north east border of Phuoc Tuy while briefing a patrol, 1969.

---

foot-track being used as a resupply route, and located a VC training area. On 14 August they sprung an ambush on two VC, killing both.

About this time Second Lieutenant John Ison arrived to relieve Terry Nolan, whose tour had ended. Nolan had proved to be an excellent patrol and troop commander. A little later Second Lieutenant Brian Schwartz replaced Peter Fitzpatrick as commander of the squadron's signal troop.

During August the squadron started to mount patrols across the Phuoc Tuy border into Binh Tuy Province to explore the south west approaches to the May Taos. Sergeant John Jewell's five-man patrol was inserted eight kilometres south east of Nui May Tao on 12 August and heard various forms of enemy activity; shots, a chain saw and a truck engine. On 19 August eight shells, suspected to be from an armoured vehicle operating to the north west, landed 100 metres from their night LUP. Before dawn the next day they heard a vehicle

331

approaching along a track and mounted an immediate ambush; they were too late to initiate the ambush. Just before midday they again heard the vehicle approaching, and at noon a 1.5 tonne truck, with two enemy in the cabin and four standing or sitting on the tray, drove into the killing area at about 24 kilometres per hour. The patrol engaged the vehicle with rifles and M79 fire. Three enemy fell from the tray and the vehicle stopped after travelling about three metres. The patrol poured fire into the truck for a full two or three minutes before withdrawing. A little later a light fire team set the vehicle ablaze and reported seeing three VC lying on the ground.

Before Jewell had withdrawn from Binh Tuy Sergeant Fred Barclay had had another contact eight kilometres east of Thua Tich. In a fierce battle two enemy were killed plus possibly another, while Barclay was slightly wounded in the leg and chin.

Meanwhile Beesley had decided to insert a patrol to demolish a log bridge on the track along which the truck destroyed by Jewell had been travelling. The task was given to Lieutenant Chris Roberts with a nine-man patrol plus a six-man security group commanded by Sergeant Fred Roberts. The bridge was located ten kilometres south east of Nui May Tao and crossed the Suoi Tram river, which had steep embankments over 10 metres high. The whole operation took exactly 46 minutes. The troops leapt from their helicopters onto a nearby LZ, the security group took up its position and the demolition party began work. The bridge was blown, and just as an afternoon storm broke, the security group was lifted out by helicopter at 5.11 pm. Chris Roberts' patrol remained in the area in ambush positions for a further seven days, but without success. During his air reconnaissance on 23 August Roberts had seen the burnt-out shell of Jewell's truck. Observing from the helicopter after he was extracted on 6 September he noticed that the vehicle had been removed.

The events of the month had shown that the VC were sensitive to patrolling on the approaches to the May Taos and the time was approaching when not only the SAS but also the infantry battalions were to give closer attention to the VC stronghold. In anticipation of this change in emphasis, towards the end of August Brigadier Pearson halted operations around Dat Do, but he would not have the chance to direct operations in the May Taos; his tour ended on 1 September 1969.

19
Into the May Taos
3 Squadron: September 1969–February 1970

Brigadier Stuart (Black Jack) Weir took command of 1 ATF on 1 September 1969 and immediately indicated that he was keen to get away from the villages and to take the fight to the VC main force units deep in the jungle. He explained later that soon after he arrived he found that there had been an agreement between the Provincial Senior Adviser and the Province Chief that the Task Force would pull out of the populated areas and that these would be taken over by local forces. 'That appealed to me because that was the proper function of the Task Force...to get out after the regular VC and knock them out, eliminate them and separate them from the population.' But it took some time before operations could be mounted against the May Taos. As he described it: 'If you [the VC] were on the tri-province boundary nobody ever came near you because the most difficult thing to do was to co-ordinate operations with people in other provinces, and I had the greatest difficulty in getting co-ordinated operations against the Ba Long Headquarters and all the base units they had up in the May Taos'.[1]

On another occasion Weir wrote that 'by 1 Sep 69, the full task force had been back in the province for over three months and the major task was to destroy the enemy May Tao base and deprive the local VC of whatever support they might have been getting there. It was clear from SAS contacts and sightings early in September that there was enemy movement from the coast to the May Tao base, and still some from the western part of the AO to the base. SAS patrols continued to be deployed into these areas to gain information. I saw this as the main role for the SAS squadron, but I made it clear that I expected them to inflict casualties on the enemy when opportunities arose'.

An example of one of the patrols between the May Taos and the sea was Sergeant Fred Roberts' five-man patrol which was inserted on 3 September across the Binh Tuy border ten kilometres south east of Nui May Tao. In a contact with five VC on 12 September three were killed.

But the other parts of the province were not ignored, and on 16 September Sergeant John (Johno) Johnston led a five-man patrol of New Zealanders five kilometres west of Ngai Giao. On 20 September they initiated a Claymore ambush on three VC after which they found one body. On 24 September they saw over 60 VC and the next day received a resupply of Claymores by rope from a Sioux helicopter. On 27 September they sprang another ambush, this time on fourteen enemy, killing at least five. In a period of over eleven days they had sighted 107 enemy. Sergeant Arch Foxley also operated in the same general area and on 22 September his patrol ambushed two VC.

Despite the success of Johnston's and Foxley's patrols the focus remained primarily on the May Taos, and the task of locating and harassing the VC 84 Rear Services Group in what Beesley called the Ba Long Training Centre, north west of the May Taos, was given to one of the squadron's outstanding patrol commanders, Corporal Michael (Joe) Van Droffelaar. He had already patrolled the area on several occasions and knew it quite well. Of Dutch extraction, but actually from England, Joe Van Droffelaar had only recently joined the SAS. After various jobs such as spray painter, window cleaner and brickyard attendant, he had enlisted in the Army in 1966 and the following year had served in Vietnam with 2 RAR, completing the tour as a section commander. He reverted in rank when he joined the SAS in January 1968 but by the time 3 Squadron began its tour of Vietnam in February 1969, he was second-in-command of a patrol. Of medium height, slim build and with fair hair, Joe quickly showed that he was a calm and resourceful leader and by September, aged 23, he had command of his own patrol. A persistent stutter in his speech did not seem to interfere with his performance in the field.

Before leaving Nui Dat for this patrol Joe had an unusual experience. He was just about to step aboard the helicopter when Brigadier Weir's Land Rover was seen approaching the LZ. Weir asked who was in charge, and when Van Droffelaar identified himself, the brigadier simply stated: 'you kill em, soldier'. It seemed to sum up Weir's approach to the war.

At 10.30 am on 20 September Van Droffelaar's five-man patrol was inserted seven kilometres north west of Nui May Tao in flat, sandy terrain with defoliated primary jungle and B52 bomb craters. They patrolled west, and for the first six days saw little sign of recent enemy activity. It rained constantly each day. Then late in the afternoon of 26

September, near the Suoi Trong, a tributary of the Song Rai, they struck a well-defined foot-track about a metre wide. The surface was worn bare and hard and fresh footprints were just visible. Van Droffelaar established a night LUP with the intention of mounting an ambush the next day.

After a cold, wet night the patrol began to return to the track, and at 8.35 am, as they approached the track they sighted eight well-armed NVA at a range of about ten metres. The Australians froze, hoping that their camouflage would be successful against the background of the jungle, while for a full ten seconds the NVA stopped, looking in their direction. Suddenly one NVA started to lift his AK47 and Van Droffelaar and his forward scout, Private John Cuzens, shot three NVA with well-aimed shots to the chest and throat. Private Les Liddington killed a fourth. A group of NVA then tried to outflank the patrol to their right and they were engaged by the patrol second-in-command, Private David Fisher; one enemy was seen to fall as if hit in the body. Under Van Droffelaar's direction the patrol withdrew using fire and movement. The enemy expected the Australians to withdraw through thick jungle, but Van Droffelaar chose a more open route and they covered about 300 metres before being located by the enemy.

By now the patrol had reached the Suoi Trong and as they scrambled up the bank two rounds from an RPG slammed into the bank. Van Droffelaar was pushing his medic up the bank and received slight shrapnel wounds to the face. But the explosion was deafening; he felt an agonising stab of pain in his ear and blood started to run out of it down the side of his face. Looking around he could see about 30 enemy in a semi-circle sweeping towards the creek. Again he directed a withdrawal using fire and movement, and once they reached the cover of the primary jungle he ordered the patrol to stop firing and to remain motionless, standing back to back in a thick clump of vegetation.

Away to the east they could hear firing from another group of NVA and an enemy officer was blowing a whistle, directing the advance against the SAS. 'Gents, we aren't going to go no where', stated Van Droffelaar quietly to his patrol. Nearby he could see a small opening in the canopy where some artillery had exploded some months earlier, and he realised that they could be extracted from this area. All around they could hear the enemy firing single shots, trying to draw their fire. It was now about 11 am and Van Droffelaar told the radio operator, Les Liddington, that he had to obtain communications with Nui Dat. Liddington looked at his patrol commander as though he was mad; he had no intention of running out the aerial the normal ten metres. He therefore threw the wire only about two metres and began to send his message, in the excitement forgetting to give his callsign. Fortunately, the alert operator at Nui Dat, Sergeant Barry Standen, picked up the weak signal and recognised the message as coming from Van Droffelaar's patrol.[2]

The patrol continued to stand back to back with the enemy moving all around. Quietly they attached their Swiss seats and waited for the helicopter. Suddenly the enemy came closer. 'Joe, we've got to move', whispered one of the patrol members. 'No', said Joe, 'we'll fight it from here'. He pulled out an M26 grenade and put it in his top pocket. Just then he heard the helicopters in the distance, vectoring in on the URC 10 beacon. 'About time', he thought, although it was less than half an hour since they had sent their message. It started to rain again.

He let the Albatross lead come closer, not wanting to speak on his URC 10 in case he was heard by the NVA. 'Bravo Nine Sierra One One, this is Albatross Leader', came the voice over the radio. Joe tried to muffle the sound of the radio. He waited until he could actually see the helicopter and then replied, 'Albatross, this is Bravo Nine Sierra One One'. The Albatross asked the enemy situation and Joe replied, 'There are about twenty Victor Charlie to the...' Joe was inflicted with a bad stutter and could not pronounce the word 'east'. The Albatross asked Joe to 'Say again'. He tried again and still could not pronounce the word 'east'. By this stage some of the patrol members were becoming a little impatient. Finally Joe picked up the radio and informed the Albatross that the VC were 'opposite to west'. 'You mean east', replied the helicopter. 'Roger', said Joe.

Two or three minutes later the gunships started their run. Van Droffelaar threw a smoke grenade, the gunships opened fire, and the lead helicopter, piloted by Flying Officer Michael Tardent, skilfully dropped its rope 20 metres down into the gap in the canopy. There had been no enemy fire for some time, but now the patrol thought that they saw incoming tracer fire, and just before clipping on their Karabiners they delivered a long burst of fire towards the suspected enemy location. They were lifted about ten metres off the ground and Paul Saxton became caught in the fork of a tree. Van Droffelaar spoke on his URC 10 and the helicopter lowered sufficiently for Saxton to extricate himself.

As soon as they were clear of the jungle the gunships moved in and the helicopter gathered speed. The ropes were all at different lengths and Private David Fisher was on the longest rope.[3] Suddenly the other members realised that Fisher was missing; he had fallen from a height of about 30 metres.

The helicopter travelled about two kilometres, landed and allowed the four remaining patrol members to scramble aboard. They then returned to the site where they thought Fisher had fallen but could see no sign. By the time the helicopter reached Nui Dat Major Reg Beesley

Preparing to leave Kangaroo Pad at Nui Dat for a patrol insertion, 1969.

was preparing to lead a patrol to find Fisher. Meanwhile Captain Ross Bishop took off in a Sioux accompanied by a light fire team, and he continued searching from about 1 to 3 pm.

At about 4.30 pm Beesley and a nine-man patrol rapelled into the jungle to begin the search, and next morning they were joined by C Company 9 RAR. The company sighted three VC and killed one before it was relieved on 1 October by B Company 6 RAR. Fisher's body was never found. Aged 23, he was a National Serviceman whose tour had two months to run. A subsequent investigation found that it was likely that in the heat of the moment he had attached his Karabiner not to the correct loop at the end of the rope, but to the false loop created where the free end of the rope had been taped back.[4]

Despite the unnerving aspect of Van Droffelaar's patrol the SAS continued to focus on the May Taos, and John Robinson led a four-man recce–ambush patrol to an area five kilometres south west of Nui May Tao. On 29 September he and his second-in-command, Corporal Jim Phillips, were conducting a reconnaissance when they observed four enemy, two in a hut and two nearby. Robinson and Phillips stalked the group and engaged them, killing three and wounding the other. The information gained confirmed the presence of a large

Private Lawrie (Burma)
Mealin, 3 SAS Squadron
1969. He was killed in a
training accident in 1976.

enemy medical and support installation and helped in planning a large
Task Force operation later in the year.

A few days later, Sergeant 'Windy' McGee took his five-man patrol
of New Zealanders six kilometres north west of Nui May Tao and
between 10 and 15 October sighted a total of 173 VC. On their last
morning they found a group of five VC advancing in extended line
towards their position. In the ensuing contact two VC were killed but
the New Zealanders had to abandon their packs to escape. They were
extracted by rope at 10.30 am but as the helicopter lifted off its engine
failed. The patrol was dragged a distance of at least 60 metres along
the LZ before the helicopter released the ropes. The aircraft crashed
100 metres from the patrol and the patrol rendered first aid to the
helicopter crew before the entire party was extracted by another helicopter half an hour later.

The subsequent investigation showed that the helicopter had allowed the hanging soldiers to start swinging like a pendulum. When
they moved outside the centre of gravity limitation of the aircraft it
became uncontrollable. For this reason the RAAF eventually decided
the cease rope extractions and to revert to hoist extractions. Although
rope extractions shortened the time the SAS spent on the ground, the

aircraft had to lift the soldiers vertically until they were clear of the jungle, and this took some time. Then with the soldiers hanging below, the aircraft had to move relatively slowly to avoid the pendulum effect.

At about this time there were a number of changes in personnel in the squadron, and on 1 October Lieutenant Mike Eddy arrived to relieve Tony Haley as operations officer.

By the end of October 1969 3 Squadron had been on operations for eight months and Major Reg Beesley prepared a five-page training memorandum which reflected the squadron's experience to that time. He began by observing that while there had been little change to the concept of operations, 'the deployment of patrols on recce tasks has been more closely related to the Task Force Comd's aims and priorities for gathering intelligence. Recce ambush patrols are designed to maintain the offensive presence of 1 ATF in areas where elements of 1 ATF have not operated for some time'. Whereas previously patrols had been extracted after a contact the 'current policy is for patrols to remain in the AO after contact and continue their mission'.

His views on the enemy were interesting:

> In most cases the enemy's first reaction is to dive for cover and return fire simultaneously. The initial fire is normally high and inaccurate.
> Main Force units, when travelling in force, react quickly with fire and movement. Follow up, in most cases, does not extend over 200 to 300 metres. Of late, it has been noted that the enemy is prepared to protect his major L of C by small screening groups patrolling on either side of the track in conjunction with squad size clearing patrols.
>
> With all major supply routes, small advance and rear parties are used. These move approximately 10 to 15 metres ahead of the major group, are well spaced and search their areas in detail. Supply groups normally travel heavily laden and well bunched presenting a softer target.

Beesley observed that the use of 'early warning by flank parties using URC10 squad radios or vine communications allowed [the SAS] better and more effective use of fire power. In the main, enemy movement is along tracks... From a study of contacts and subsequent follow up and reaction, it has been noted that such action is relative to the amount of ammo carried, which, in the main does not exceed 40 rounds'.

Beesley's assessment of the enemy's tactics was confirmed by two ambush patrols early in November. On 29 October a twelve-man patrol commanded by Sergeant Bill Lillicrapp was inserted about seven kilometres west of Nui May Tao. The patrol was formed from Lillicrapp's New Zealand patrol and Sergeant Van Droffelaar's Australian patrol. Van Droffelaar was second-in-command. The next day Sergeant Fred Roberts led a twelve-man patrol into an area four kilometres north west of Nui May Tao. Both Lillicrapp's and Roberts' patrols were to ambush the same VC suppy route following the line of Provincial Route 330 which was little more than a track in this area. Beesely had suggested to Roberts that if he had a successful contact he was to withdraw from the track for a day or so and then to return and ambush it again. Fred Roberts was one of the squadron's outstanding young patrol commanders. Promoted to sergeant at the beginning of the tour, he was short, quiet, serious and professional. All three sergeants had previously operated in the area. It was expected that groups from the VC 84 Rear Services Group would be carrying large quantities of supplies and that they would be escorted by small parties of NVA.

Lillicrapp's patrol did not find a well-used track until the morning of 31 October, but within a few hours observed 14 VC moving rapidly from the west. The first five carried weapons while the remainder were unarmed and carrying packs and bags. Van Droffelaar was amazed to observe that their clothes appeared clean and pressed. The next morning the patrol began to set up their ambush. They were carrying eighteen Claymores and had spent considerable time at Nui Dat rehearsing so that they could lay the entire bank of Claymores in less than 40 minutes. At about 9.40 am they heard Claymores being fired three kilometres to the north east in the direction of Fred Roberts' patrol and soon afterwards three armed VC were seen moving rapidly from the west. Lillicrapp held his fire hoping to obtain a more substantial target.

Roberts had arrived in his area a day later than Lillicrapp but his patrol located a well-defined track on the afternoon of 31 October and set an ambush the following morning. At 9.38 am two armed VC passed along the track but Roberts held his fire. Two minutes later a carrying party of four VC, including two women, arrived and the ambush was initiated. The search party found three bodies and one blood trail. Two packs were recovered and Roberts later described his efforts to dispose of the food stuffs they captured:

> As there was far too much to carry we decided to use a gas grenade with a time delay to contaminate the food so it couldn't be used. Consequently, once the food had been stacked over the grenade we patrolled away into the wind and the 2ic, who was at the rear of the patrol, snapped the time delay pencil as he left the site. After we had covered about fifty metres and some minutes after the grenade had ignited, a great deal of noise, similar it seemed at the time to a herd of elephants, was heard from the rear of the patrol and as we turned to see what was happening the 2ic flashed by at a great rate of knots followed in turn by the rest of the patrol peeling off one by one from the rear. Then it hit me; the wind had changed and we had gassed ourselves and

Major Reg Beesley congratulates Sergeant Fred Roberts in December 1969 on the award of the Distinguished Conduct Medal for conducting two successful ambushes during the one patrol. *From left:* Sergeants Chris Jennison (with drink can), Mal Waters, John Jewell (looking away from the camera), Roberts, Ray Neil, and Beesley. (PHOTO, R. NEIL)

for a short period the entire patrol, eleven soldiers and one pretender, covered more ground than would normally be done in an hour of movement.

The 'pretender' referred to one patrol member whom Roberts had been asked to take along, and in his view did not contribute positively to the patrol.

At 7 am the following the morning Roberts' patrol occupied another ambush position about 500 metres south east of their first position. At 7.50 am four VC moved into the killing ground and Roberts inititiated the ambush by firing his M16. All four VC were killed. Two AK47s, an RPG, an M16 and two large packs were recovered. While the patrol was withdrawing they heard the sounds of a heavy load being dragged along Route 330 plus the voices of several persons. The track complex made it inevitable that there would be another contact and Roberts obtained air support from a heavy, a medium and a light fire team before being extracted by rope at 11.20 am.[5]

341

The success of Roberts' patrol had an immediate effect on the conduct of Lillicrapp's patrol, still in position three kilometres to the south west, because apparently the VC decided to send clearing patrols through the jungle parallel to their main supply route. Lillicrapp's ambush was divided into two groups with Van Droffelaar's patrol on the right, or eastern, flank of the killing ground and the New Zealanders on the western flank. At about 8.30 am on 3 November the right flank group observed five enemy moving in extended line towards the ambush position. Just then the surface of one of the Claymores set in front of an ant hill flashed in the sunlight. It was obvious that it had been seen by the enemy approaching from the unexpected direction and the right flank group initiated three Claymores and began firing. Almost immediately Lillicrapp fired the twelve Claymores in the killing ground. Van Droffelaar recalled that they then came under intense enemy fire from what appeared to be a platoon of NVA. He tried to put on his pack but it was shot out of his hands. He picked up his URC 10 and it too was shot from his hands.

By now the NVA were moving to the right around the patrol and were mounting a determined attack. Van Droffelaar can distinctly recall firing up at what seemed to be an angle of 30 degrees to catch the NVA who were jumping over their fallen comrades. By this time Lillicrapp's patrol had moved to support Van Droffelaar's, and using fire and movement the combined patrols withdrew away from the track. It seemed that even more enemy were heading towards the contact site but the SAS evaded the enemy who could be heard firing at clumps of jungle. By chance their radio beacon was picked up by a South Vietnamese Bronco fixed wing ground support aircraft, which had seen the white smoke from the white phosphrous grenades attached to the Claymores in the killing ground. The Bronco was carrying napalm and after seeking permission from the SAS dropped the napalm on the white smoke. No further firing was heard from the NVA and at 11.10 am the patrol was extracted by helicopter.

A few days later, on 2 November, Sergeant John Robinson led a six-man patrol (including three soldiers from 6 RAR) back into the area four kilometres east of Binh Gia. Spotting, on 8 November, a woman in black tending a fresh garden, he labouriously crawled forward through thick, dry undergrowth until he was only a few metres from her. He then stood up and apprehended her. At the same time the remainder of his patrol moved forward, and after searching the area found an M1 carbine under a mat covered by cashew nuts. It turned out that the woman was from North Vietnam and was president of the Quang Giao women's guerilla committee. Quang Giao was a hamlet three kilometres east of Binh Gia. The subsequent intelligence gathered was of great value in determining the location and activities of an active enemy unit.

By this time Brigadier Weir had received permission to enter the May Tao mountains and on 1 December 6 RAR/NZ (ANZAC) began Operation MARSDEN to clear the enemy out of the area. The operation lasted until 28 December and was highly successful. The battalion lost three killed but killed 22 and captured over twenty VC. More importantly, they captured numerous weapons and large quantities of ammunition, medical supplies and food. The battalion located a large VC hospital, a workshop and a weapons factory. As the battalion history, written in 1970 observed, 'Operation MARSDEN dealt a considerable blow to the VC morale and efficiency in Phuoc Tuy Province. The last remaining VC secret zone in the 1 ATF TAOR had finally been entered and cleared, the bulk of the VC medical supplies and equipment had been captured and their replacement could only add to the problems of an already disintegrating logistics system...The VC had been dealt a serious blow to his supply system and he has been unable to develop any large offensive operations in Phuoc Tuy since'.[6]

The SAS did not play a large part during 6 RAR's operations in the May Taos, but soon after the operation Brigadier Weir stated that its success was a direct result of information from SAS patrols. One SAS patrol conducted in conjunction with the operation was a recce–ambush mission deep in Long Khanh Province, eleven kilometres north west of Nui May Tao. The six-man patrol commanded by Corporal Jim Phillips was inserted on 2 December and contacted what appeared to be a lone VC on 5 December. The patrol then came under a heavy volume of fire. They managed to break contact and on 7 December observed a number of VC. On 8 December they set an ambush along a well-used foot-track and killed at least four VC. It was an excellent effort by a young patrol commander.

Meanwhile, Weir had given Beesley the choice of three operational tasks: to take over Long Son Island and destroy the VC infrastructure, thought to number some 30 VC; undertake independent operations in the close TAOR near Nui Dat and destroy all enemy moving in the area; or to conduct a parachute insertion. Beesley selected the third option and during Operation MARSDEN 3 Squadron conducted an operational parachute jump into a grassy area 5 kilometres north west of Xuyen Moc. A ten-man pathfinder group, led by Beesley, parachuted in on the afternoon of 15 December and the next morning about half of the squadron jumped from Caribou aircraft onto the LZ. From there three patrols were deployed into AOs by APCs while two more walked into their areas. The remaining patrols were extracted by

Preparing for 3 SAS Squadron's operational parachute jump in December 1969. On the left are Major Reg Beesley and Warrant Officer Des Kennedy. (PHOTO, J. FRAZER)

Chinook. While Operation STERLING, as it was known, provided the opportunity to practise parachute operations, it also served as a deception for the insertion of the five patrols. The last patrol returned to base on 23 December, the squadron having sighted two enemy with no contacts. At about this time the troop of Kiwis was relieved by another troop commanded by Captain Graye Shatkey.

Following the conclusion of Operation MARSDEN the SAS returned to the May Taos to determine if the enemy had re-occupied their camps and to harass any enemy they found. Sergeant Ian Stiles was inserted five kilometres north west of Nui May Tao and on 8 January killed two VC; Sergeant Chris Jennison's patrol was deployed two kilometres south of Nui May Tao and on 11 January killed one VC, and Sergeant Clem Dwyer was inserted about a kilometre further south, where on 11 January his patrol killed six VC with small arms fire in an ambush. They were engaged by another group of enemy and the scout, Private Wladyslaw Kaczmarek, a recently arrived reinforcement, killed another VC with a burst to his chest. The patrol withdrew using fire and movement. Corporal Peter Bye's patrol operated 7 kilometres west of the May Taos and on 9 January killed two VC.

344

Captain Ross Bishop gives a last-minute briefing to Corporal Jim Phillips at Nui Dat in late 1969. In the background is Private Dennis Mitchell. Note the bulge under Phillips' shirt from the full water bladder carried at the beginning of a patrol in the dry season.

Corporal Jim Phillips was given the task of moving right into the mountains. On 2 January his patrol was inserted 5 kilometres west of Nui May Tao and for the next five days walked steadily east into the mountains. By 8 January they had located three abandoned enemy camps, one of regimental size covering an area 300 by 200 metres. On 9 January they set an ambush beside a track showing signs of recent use, and at 9.05 am three enemy moved into the killing ground. When Phillips initiated the ambush the Claymore firing device failed to function. The enemy heard the noise and immediately turned towards the patrol's location. Meanwhile, two more enemy entered the killing ground and Private Barry Williams observed a further ten or more enemy moving towards the ambush position from the left flank. Despite the enemy searching three metres from his position Phillips calmly replaced the Claymore firing device with another and sprang the ambush, killing five enemy and inflicting heavy casualties. During the rapid reaction by a larger enemy force of platoon size, the patrol received heavy machine-gun and automatic rifle fire from three flanks. Phillips rallied his men, issued precise orders and organised such effective fire and movement that the patrol was able to break contact without sustaining casualties. Disregarding his own safety, while exposed to hostile fire, he covered the withdrawal of a man entangled in heavy undergrowth. The patrol was withdrawn late that afternoon. During the period from 8 to 11 January the SAS had killed seventeen VC and possibly a further four in and around the May Taos, favourably matching the effort of 6 RAR during Operation MARSDEN and continuing to keep the VC off balance.[7]

During the early part of Phillips' patrol Brigadier Weir had been absent in Saigon and Major Beesley had arranged with Lieutenant-Colonel Kahn, commanding officer of 5 RAR, to place a company of his battalion on stand-by to follow up any sighting by Phillips. Weir arrived back about the time of Phillips' contact and was adamant that the patrol would remain in the area. Beesley took a deep breath and said 'no'; Phillips had been in the jungle long enough and he intended to withdraw the patrol. Weir was furious but Beesley remained unmoved, expecting to be relieved of his command. Next day Weir visited squadron headquarters for a cup of tea and made no further mention of the incident.[8]

3 Squadron continued to maintain surveillance over the approaches to the May Taos and on 13 January 1970 Sergeant Graham Campbell led a five-man patrol of New Zealanders into Binh Tuy Province twelve kilometres south east of Nui May Tao. On the morning of 14 January Campbell and his scout moved forward to observe a track they had spotted the previous night. As they advanced Campbell accidentally snapped a stick and the two men halted and listened. Forty-five minutes later they resumed their advance and suddenly were engaged by about five enemy at a range of about ten metres. Campbell was struck in the head by an AK47 round and the scout returned fire, calling for the remainder of the patrol to close up. They organised a withdrawal and under cover from a light fire team were extracted at 10.40 am. Campbell, one of the finest patrol commanders in the squadron, died in transit.

On 21 January Beesley returned to Australia to attend a number of courses in Britain, and Captain Ross Bishop assumed command of the squadron for the last few weeks of its tour. Towards the end of January Bishop was informed by staff of the Task Force intelligence unit that they had a VC defector who was prepared to become a double agent and needed to be inserted back into the May Taos. Sergeant Fred Roberts' patrol was given the task, and they wore greens instead of their normal camouflage suits, while the RAAF changed its techniques for the insertion of the patrol into an area about eight kilometres south west of Nui May Tao. Roberts wrote later: 'During the time the agent was with us we changed much of our daily routine because we were very skeptical, initially, of not only his sincerity but also of the cover story he had been given to explain his long absence from his regular unit. However, we were all moved by his conduct—especially his farewell salute to the patrol which even Tony Haley would acknowledge as being copy book stuff—and felt after he had left he was, indeed, a friendly operator who trusted us far more than we trusted him'.

The squadron had only two more contacts. On 27 January an OP from Sergeant Arch Foxley's patrol, just east of the Courtenay rubber plantation, sighted four enemy following the patrol's tracks, and after the ensuing contact two bodies were recovered. And on 8 February Sergeant Ray Neil's patrol 7 kilometres south of Xuyen Moc killed two VC. By now the advance party of 1 Squadron had arrived, and the last patrol of 3 Squadron was extracted on 14 February. The squadron departed for Australia on 18 February.[9]

During its tour the squadron had mounted 230 operations, and had had 78 contacts in which 144 enemy were killed plus possibly 32 more. A total of 2551 enemy had been observed during 372 sightings. But the statistics do not tell the full story. The SAS had provided extensive and detailed intelligence on VC routes and activities throughout the province. In particular, the SAS had provided the key information on which the successful operation in the May Taos was based. By the end of 1969 the Task Force dominated the province to a greater extent than at any time in the previous four years. From the Nui Thi Vais to the May Taos the VC had been chased from their base camps. It now remained for the Task Force, assisted by the SAS, to maintain Australian dominance in Phuoc Tuy during 1970. The VC would, of course, return to the May Taos, but the distant mountains had now revealed their secrets. And in achieving this success, the SAS had played a major role.

20
Flexibility and frustration
1 Squadron: February 1970–February 1971

When 1 SAS Squadron returned to Vietnam on 18 February 1970 for its second operational tour it was obvious that, with the decrease in VC activity, it was facing the prospect of a frustrating year. SAS operations in Vietnam had always been surrounded by a measure of frustration. An SAS patrol might lie in hiding for days watching a VC camp, only to find that when it directed an airstrike onto the camp the bombs or rockets missed by hundreds of metres and the VC escaped unharmed. Or the patrol reported a VC concentration in an area and there was no infantry force available to follow up the sighting. Even when the infantry was available its presence was often detected by the VC who melted away. Some infantry commanders even doubted whether the SAS had actually seen any enemy.

In a broader context, it was frustrating for the SAS patrols to be employed on reconnaissance tasks relatively close to the Task Force base when they considered that their true role was long range reconnaissance deep in enemy territory. The SAS, however, realised that the political conditions of Australian involvement in Vietnam precluded true long range reconnaissance and they made the most of their opportunities in Phuoc Tuy Province. For the young SAS troopers (for they were now called troopers rather than private soldiers) the tour of Vietnam offered the chance to test and further develop their professional expertise. But after hearing of the experiences of the previous operational tours, the new soldiers of 1 Squadron were disappointed to discover that the tempo of the war had decreased markedly by the time they arrived. Clearly the VC were on the defensive, for during the preceding months they had been hammered by the operations of the Task Force in their base areas, and these operations continued during the early months of 1970. Indeed, soon after the squadron commander, Major Ian Teague, arrived he was advised by Brigadier Weir that 'you will have to be flexible and adapt to a number of different tasks that I will want you and the squadron to do'.

As it turned out, the style of operations conducted by 3 Squadron, particularly the recce–ambushes, were continued by 1 Squadron during the early months of 1970, but the number of contacts decreased markedly and Sergeant Lawrie Fraser, returning to Vietnam for his second operational tour, recalled that he 'found Phuoc Tuy far more subdued—there was a lot less enemy activity'. His patrol went for months without any contact and with barely an enemy sighting. Realising the frustration of his soldiers he studied the intelligence summaries in an effort to find a likely area, but after several months of operations his patrol was still without success.

While 1 Squadron completed its acclimatisation and in-country training, the New Zealand troop maintained the continuity of SAS patrols, and the squadron's first operational patrol was deployed on 1 March. By the end of March all patrols had been inserted at least once. Generally there were few sightings or contacts, but when on 7 March Sergeant Jack Gebhart's patrol was inserted nine kilometres south east of Nui May Tao, in Binh Tuy Province, they contacted three VC within four minutes of arriving on their LZ. Gebhart killed one and the others disappeared. More enemy were thought to be in the area and the patrol was withdrawn. Gebhart's contact and a similar contact by Lieutenant Zot Simon's patrol east of Xuyen Moc on 25 March, in which one VC was killed, was to set a pattern which persisted for most of the year.

Six days after their first contact Gebhart's patrol was inserted a kilometre east of Thua Tich. They established an OP in thick secondary undergrowth to observe an abandoned fire support base, now covered by long, dry grass, and at 1.10 pm on 16 March Trooper Jim Raitt sighted two enemy moving into the area. When they stopped four metres from his position he engaged both with automatic bursts from his M16. The patrol remained in position and the next day was reinforced by A Company 8 RAR in APCs, which carried out a series of sweeps for no result and left the area in the afternoon of 17 March.

Within two hours the SAS patrol sighted another VC who appeared to be searching for them. The enemy soldier withdrew and just on last light the SAS detected an unknown-sized group of enemy moving towards them. In failing light Corporal Ian Bullock and Warrant Officer Snow Livock, the SSM who had joined the patrol as medic, engaged two VC at a range of five metres. The patrol withdrew to the west and found that on a front of 200 metres a grass fire was burning towards their location. They avoided the fire, plus two others which had been lit near their location, and within the hour were reinforced again by 8 RAR in APCs. The patrol joined the infantry and left the area the following morning, again having found no enemy. The incident was a dramatic example of the difficulty faced by the infantry in reacting to SAS sightings, and results such as these were disappointing for all involved.[1]

The squadron maintained its effort on the approaches to the May Taos, particularly the approach from the coast, and at 3.10 pm on 2 April Sergeant Alan Roser's five-man patrol was inserted across the border in Binh Tuy Province, about five kilometres from the coast. It was flat terrain with low defoliated secondary vegetation and visibility varying from 10 to 100 metres. Within two minutes of alighting, Trooper Graham Smith observed five enemy on the edge of the LZ and with two aimed shots appeared to kill one enemy soldier carrying an RPG. Smith then engaged a second enemy who also was hit. The three remaining enemy moved into extended line and a firefight developed between the two opposing patrols. Another enemy was seen to fall and at 4.08 pm the patrol was extracted by helicopter. If possible the SAS usually fired first at any enemy carrying an RPG, as they had found from experience that the RPG tended to be more dangerous than the enemy's AK47.

Meanwhile, an eight-man ambush patrol commanded by Sergeant Ron Gilchrist had been inserted by APC south east of Xuyen Moc, about two kilometres from the sea. On the morning of 3 April their ambush was sprung on one VC but almost immediately the patrol came under fire from two more VC. Trooper Neville Rossiter was wounded in the left thigh and right groin, but the patrol second-in-command, Lance-Corporal Allan Gallegos, initiated the Claymores placed for flank protection, killing both VC. Other patrols went closer to the May Taos, and on 13 April Sergeant Ross Percy's patrol killed three VC ten kilometres south east of Nui May Tao, deep in Binh Tuy Province. Two days later Lieutenant Zot Simon's patrol killed one VC six kilometres west of Nui May Tao while leaving their LZ.

Although the SAS continued to concentrate on the approaches to the May Taos, since early February elements of 8 RAR had been operating in and around the formidable Long Hais. The battalion's involvement with the Long Hais began on 10 February when C Company, with tank, APC and mortar support, deployed to the western side of the Long Hai hills to protect quarry operations being undertaken by 17 Construction Squadron. C Company established Fire Support Base Isa near the quarry site and ambushed likely enemy routes leading from the hills. After several contacts the whole battalion became involved in an operation lasting until 3 March. A Company 8 RAR remained in the Long Hais to protect the quarrying operation, and the company commander soon became suspicious of enemy activity in the fishing village of Ap Lo Voi, barely a kilometre south west of Isa.

Sergeant Alan (Junior) Smith with a six-man patrol was given the task of conducting a surveillance of the seaward approaches to the village. On 14 April the patrol was carried by helicopter to Isa, and was then transported by APCs back along the road, through the town of Baria, and then south along Route 15 towards Vung Tau. Three kilometres south of Baria, where the road crossed a causeway, they transferred to a recce boat, and at night moved through the mangroves and waterways back to a position from which they could observe Ap Lo Voi. In this area the Rach Sua Lap was about 500 metres wide flowing south when the tide was rising and north when falling. The SAS patrol concealed itself in mangroves and thick marshy reeds and over a period of five nights observed sampans moving in and out of the village. With the exception of one craft sighted at 3.30 am on 16 April, which was moving rapidly, without lights and with a very quiet motor, they concluded that all sampans sighted were fishing vessels breaking curfew. A month later the VC ambushed a Popular Force group in the village and were themselves ambushed by the 8 RAR assault pioneer platoon as they were leaving the village.[2]

While these minor operations were taking place near the Long Hais, on 19 April the Task Force, with three battalions and support units, began Operation CONCRETE in the Xuyen Moc district with the aim of destroying D445 VC Main Force Battalion and its base areas. In support of this operation a number of SAS patrols were deployed towards the May Taos and various groups of enemy were sighted. On 30 April Sergeant Gebhart's six-man patrol located a suspected enemy camp about eleven kilometres south of Nui May Tao, but found that they were being followed by an enemy force of unknown size. Despite their efforts to throw off their enemy tail, such as calling for support from a light fire team, late in the day the enemy appeared to be closing in on their location and a contact was initiated. Two VC were killed but others continued to follow up the patrol and at 7.15 pm Gebhart requested extraction.

Within 35 minutes two Bushrangers were over the patrol's position where a firefight was still underway. The RAAF report described the outcome: 'While intense suppressive fire was laid down with mini guns, rockets and twin M60 machine guns, two more aircraft from the [RAAF] squadron arrived, one to provide illumination and the other to extract the patrol'. The aircraft, captained by Flight Lieutenant Bill Robertson, 'went into a hover at 50 ft just above the tree tops. Using a searchlight to determine their position, they began the dangerous task of winching troops out. During the 15 minute operation, with the hovering aircraft a "sitting duck" for enemy fire, three other squadron

SAS: Phantoms of War

A patrol from 1 SAS Squadron returning to Nui Dat. On return, the helicopter carrying the patrol landed at the small Nadzab Pad, near the SAS lines, high on Nui Dat, while patrols beginning a mission usually took off from the larger Kangaroo Pad, some distance away, as the crews of up to five helicopters needed to be briefed. In the distance can be seen Luscombe Field, the main fixed wing airstrip of the Task Force base. (PHOTO, R. MCBRIDE)

helicopters arrived at the scene to standby in case the extract helicopter was shot down. At one stage, while Leading Aircraftsman Jones was operating the aircraft winch, the first soldier to be rescued took over the starboard side machine gun and laid down suppressive fire on the enemy positions while the others were being hoisted through the jungle foliage. The entire operation, from the first call for help, until the troops had been returned to Nui Dat took just under two hours'. It had been the first night-time 'hot' extraction for more than a year and was only the third performed by No 9 Squadron since it had begun operations in Vietnam more than four years earlier. It was notable that the patrol was extracted by winch, as following the death of Private Fisher in September 1969, the hot extraction technique used by 3 Squadron was no longer acceptable to the RAAF.[3]

Early in May 8 RAR was redeployed from the eastern side of the province to conduct operations west of Route 2. As the 8 RAR history recorded: 'By the 25th May a calm had settled over the southern area of Phuoc Tuy, attributable largely to the tight 8 RAR and 7 RAR control of the Vietcong access routes to the population centres. Even D445, which intelligence estimates located to the north east of Dat Do with some elements still in the Long Hais, seemed unusually quiet'. In these circumstances, it was the role of the SAS to give warning of VC developments from unexpected quarters, and on 24 May a five-man patrol commanded by Second Lieutenant Lloyd Behm was inserted just across the Long Khanh border from the north west corner of Phuoc Tuy. For five days the patrol found no sign of recent enemy activity, but on the morning of 29 May came across a freshly cut track

352

and established an ambush. At 3.30 pm the ambush was sprung on three VC, but while the patrol was searching the bodies another enemy group was heard moving towards the position and late that day the patrol was extracted. Two companies of 8 RAR were reacted to the area and killed another VC, identified as belonging to 274 Regiment.[4]

As in previous years, the SAS squadron continued to task patrols from the LRRP Wing at Van Kiep, and as in the past, these patrols were commanded by ex-SAS soldiers. Captain Doug Tear, who had been a sergeant with 2 Squadron in Borneo, was chief instructor of the wing and he led a number of patrols, as did Warrant Officers Ola Stevenson and Tom Hoolihan and Sergeant Bob Broadhurst, all of whom had had operational tours with the SAS, either in Borneo or Vietnam. They were assisted by Signallers Alan Gronow, Malcolm Hollaway and Tony Jacques who were detached from 1 Squadron.

In a Training Information memorandum written on 22 April Major Teague summarised his operational concept: 'SAS patrols are deployed generally on reconnaissance tasks related to the Task Force Commander's aims and priorities for gathering intelligence. Many of these patrols combine an ambush task which is designed to maintain the offensive presence of 1 ATF in areas where infantry battalions are not currently operating. A few patrols have been employed on purely ambush tasks and on straight surveillance'. In writing the memorandum Teague became concerned at the rather stereotyped nature of the SAS patrols and to introduce variety decided to land a number of patrols from small craft along the Phuoc Tuy coast.[5]

On the afternoon of 30 May four SAS patrols assembled at Cat Lo, near Vung Tau, and embarked on a US Navy coastguard patrol boat. By 9.50 pm the patrol boat was south east of Xuyen Moc and the first patrol was despatched to shore in a Gemini assault craft manned by SAS soldiers who then returned to the patrol boat. The patrol boat moved slowly west along the coast launching the other patrols at regular intervals until the last patrol was launched at 12.40 am about a kilometre east of the mouth of the Song Rai. All four patrols were inserted safely.

Sergeant 'Junior' Smith commanded an eight-man patrol, including Major Teague as a rifleman, which was inserted due south from Xuyen Moc. They sighted two enemy on the afternoon of 31 May and another person, possibly riding a bicycle along Route 328, on 3 June. Sergeant Tom Dickson's five-man patrol of New Zealanders was inserted near the mouth of the Song Rai and discovered three recently used VC camps plus caches of ammunition and food. Sergeant Don Fisher's patrol, inserted about five kilometres east of the Song Rai, killed two enemy on 31 May. On 4 May the four patrols rendezvoused on Route

SAS: Phantoms of War

328, six kilometres south of Xuyen Moc, and were extracted by helicopter the following morning. The information from these patrols seemed to indicate that the VC were establishing a 'winter camp' south of Nui Kho, and on 19 June 7 RAR mounted a four-day operation through the area.[6]

While these activities were taking place south of Xuyen Moc, Sergeant John O'Keefe's five-man patrol was inserted just south of the province border, eight kilometres west of Nui May Tao. For several days they heard the sound of chainsaws and shots and located two enemy camps that had not been used for some months. On the afternoon of 6 June they found another unoccupied enemy camp and stopped to establish communications with Nui Dat. At 5.45 pm three VC were seen following the Australians' tracks into the LUP and were engaged by the SAS at a range of about two metres. As the patrol moved forward to search the bodies another group of VC was heard approaching the position. At 6.15 pm the patrol was extracted by helicopter while a light fire team put down suppressing fire.

As part of the pacification programme, throughout the first six months of 1970 the Task Force operations had progressively moved closer to the villages, and the effect of these operations was shown by the VC reaction. For example, during April or May the VC D445 Battalion received an instruction from the Ba Long Provincial headquarters 'to attack the Pacification Teams, Rural Development Cadres, PSDF [Peoples Self Defence Force], and villages and hamlet officials within the Dat Do/Long Dien districts'. The task was to be carried out during the wet season. This information, which the Australians received from Hoi Chanhs (returnees) during April and May, resulted in the Australian battalions concentrating even more heavily on operations around the villages. The SAS was not involved in these pacification operations but was required to maintain a wide surveillance over the areas closer to the provincial border.[7]

This change in operations coincided with the handover, on 31 May, of command of the Task Force from Brigadier Weir to Brigadier Bill Henderson. The new Task Force commander had definite ideas on the role of the SAS, as he made clear some years later:

> On arrival in theatre I observed the manner in which [the SAS] were being deployed and was not impressed. Essentially they were being employed in an harassing role and emphasis was on 'body count'. In my view this role was not productive. Much planning effort and employment of limited resources was involved resulting in the insertion of a patrol using gunships, lift helicopters and chase aircraft—the result possibly two kills, no information and the requirement for the expenditure of additional resources for the subsequent, 'hot extraction'.

Flexibility and frustration

I discussed this situation with the OC, Major Teague, and told him I
wanted reconnaissance and surveillance which would lead to improved
intelligence on which to deploy units of the Task Force. He was not
terribly happy, so at his request I spoke to the whole of the SAS
[squadron] assembled at Nui Dat and explained my requirements. The
squadron got the message and from that time onwards we received
greater benefit from their operations. As a result of change to
reconnaissance and surveillance there was an immediate result whereby
we were able to target [companies] and [battalions] against enemy
concentrations instead of having 'hit and run' actions against small
enemy groups which effectively dispersed him and made it much more
difficult to come to a worthwhile result. In my view the [squadron] in
this role made a much more worthwhile contribution to the [Task Force]
role. In addition, this method of employment led to the discovery of
arms caches and we were able to use them in conjunction with
operations being conducted by 2/25th US Regiment with good result.[8]

Former members of 1 Squadron, including the commander, have
recalled that from about mid year they conducted more and longer
reconnaissance patrols and less recce–ambushes, and at the end of
June Teague wrote in his Commander's Diary: 'SAS patrols are now
being employed on correct SAS tasks and are achieving better results'.
A study of the patrol reports, however, indicates only a slight change
in emphasis and effectiveness. For example, in April the squadron
conducted 29 recce–ambush, 1 surveillance and 2 ambush patrols. A
total of 68 enemy were sighted; there were 11 contacts, 9 enemy were
killed, possibly a further 12 were killed, and 3 more were wounded. In
May, the month before Henderson arrived, there were 19 reconnaissance, 4 recce–ambush and 1 ambush patrols. A total of 30 enemy
were sighted and in 6 contacts 3 VC were killed, possibly 5 more were
killed and 1 was wounded. In June there were 36 reconnaissance
patrols in which 25 enemy were sighted, and in 5 contacts 3 enemy
were killed, plus possibly 6 more. In July there were 20 reconnaissance, 1 ambush and 1 fighting patrols. A total of 17 enemy were
sighted, and in 3 contacts 4 VC were killed plus possibly 1 more. In
August there were 25 reconnaissance, 2 surveillance, 1 ambush and 1
fighting patrols. A total of 32 enemy were sighted, and in 5 contacts 4
VC were killed plus 1 possible. More than likely, the statistics do not
tell the full story; perhaps the patrols were indeed producing better
information—but while it can be seen that there was a decline in the
number of recce–ambushes, it is equally clear that there was less
enemy activity. Indeed at the end of July Teague wrote in his Commander's Diary: 'Due to current restrictions on 1 ATF TAOR there is
a distinct lack of SAS tasks'. At the end of August he again wrote: 'The

355

1 ATF restricted TAOR still exists and hence the continuing lack of proper SAS tasks'. There is certainly a hint of frustration in Teague's comments.

After almost five years of SAS operations the VC had become familiar with SAS insertion techniques. Some days before an insertion there would be a reconnaissance of the proposed area of operations by a light plane or reconnaissance helicopter. Then perhaps the day before the insertion or on the day itself there might be a further reconnaissance of the area, particularly of the proposed LZ, by the patrol commander. Finally, on the day of the insertion two or three gunships and two lift helicopters would swoop into the area. And there were only a limited number of suitable LZs in any area of operations. Not surprisingly, SAS patrols continued to be contacted soon after insertion. Sergeant Gebhart's contact on 7 March was the first instance during 1 Squadron's tour, but there were many others. For example, on 27 June Second Lieutenant Lloyd Behm's five-man patrol was inserted on an open, swampy LZ near the Song Rai river seven kilometres west of Xuyen Moc. They were engaged by automatic small arms fire from a group of perhaps ten VC while they were actually deplaning. The helicopter took off leaving three SAS soldiers on the ground and two in the aircraft. The SAS soldiers on the ground engaged the VC, possibly killing two, and supported by the light fire team the lift helicopter extracted the patrol ten minutes after their initial insertion.

Lieutenant Zot Simon was one patrol commander who had a large number of patrols detected by the VC on insertion. Indeed by 30 April he had had seven contacts in which he had accounted for two, and possibly two more, enemy killed.

Another patrol commander with a similar run of bad luck was Jack Gebhart. On 12 June he was contacted on insertion into an LZ nine kilometres west of Nui May Tao. On 2 July, with a six-man patrol, including the squadron signal officer, Second Lieutenant Brian Schwartz, as patrol signaller, he was inserted near the Bien Hoa border, north of Nui Thi Vai. They contacted one VC about an hour after insertion, but then were engaged by a group of about five VC and were extracted within the hour. His bad luck continued when he was inserted six kilometres south of Nui May Tao on 15 August. On deplaning they spotted a VC in a tree about 300 metres south of the LZ and moved south with the intention of capturing him. However, a group of six VC was located at the base of the tree and they engaged the patrol with small arms fire. In the ensuing firefight the VC in the tree fell and the patrol was quickly extracted. Some months later, on

10 November, his patrol was again contacted as they were inserted into an LZ in Binh Tuy Province, about five kilometres from the coast.[9]

In an effort to ensure that patrols were inserted safely in areas where there was a possibility of a contact the squadron introduced the practice of 'Cowboy Insertions'. In a Cowboy Insertion a stand-by patrol was inserted by a second 'slick' helicopter following the tasked patrol into the LZ. Both patrols exited from the LZ, travelled together for some five minutes from the LZ after which the stand-by patrol waited for another five minutes before making its way back to the LZ. The helicopters, which had been holding for their normal twenty minute period some five kilometres away, returned to the LZ and picked up the stand-by patrol. In the event of a contact on the insertion the firefight was controlled by the patrol commander of the tasked patrol. After the contact had been completed, provided there were no casualties in the tasked patrol, they replenished their ammunition from the stand-by patrol and continued with their mission after the stand-by patrol had been extracted in two helicopters, thus giving the appearance that all personnel had left the area.

It was easier said than done. For example, on 21 October Lieutenant Andrew Freemantle, who had joined 1 Squadron at the end of Mike Eddy's tour as operations officer, was deployed with a five-man patrol near the Binh Tuy border, nine kilometres south of Nui May Tao. Sergeant John Ward commanded the stand-by patrol. Fifteen minutes after insertion, while Ward was returning to the LZ, Freemantle's patrol was engaged by three VC. The SAS returned fire, striking one VC, and then withdrew to the LZ. The stand-by patrol, 35 metres to the west, also received a large volume of automatic fire, but being separated from the other patrol could not join in the firefight. The Bushrangers observed three enemy moving away from the contact area, and because of shortage of ammunition both patrols were withdrawn. The problem of enemy detection continued, and as late as 31 December the squadron commander wrote in his Commander's Diary: 'The enemy are becoming increasingly more aware of our methods of operations and there is a continuing requirement to vary SAS techniques to reduce the number of patrols being compromised and to keep the enemy guessing'.[10]

Of course when one patrol was detected on insertion it often enabled another patrol to be inserted without alerting the enemy, and on 2 July, while Gebhart's patrol was in contact on the LZ near the Bien Hoa border, Second Lieutenant Robin McBride's four-man reconnaissance patrol was operating three kilometres to the south. On 5 July they found an occupied squad-sized enemy camp, and for the next three days observed VC activity. On one occasion a VC patrol of five,

Second Lieutenant Robin McBride's patrol, Nui Dat, 1970. *From left:* Signaller Allan Keys, Lance-Corporal Barry Lansdown, Trooper John Frazer, Corporal Greg Mawkes and McBride. McBride celebrated his 22nd birthday in Vietnam, making him the youngest SAS officer to see active service either in Borneo or Vietnam. (PHOTO, R. MCBRIDE)

including one female, moved out of the camp with weapons at the 'ready'. At 1.30 pm on 8 July McBride's patrol was reinforced by Sergeant Allan Smith's six-man patrol, and McBride prepared to attack the camp. The attack was planned for the late afternoon of the following day, but as the SAS approached the camp they were detected by one VC who was engaged with a sustained burst of fire from an M60. Fire was returned by an unknown-sized group of VC. The patrol poured a large quantity of fire into the camp and then withdrew to allow a light fire team to continue to engage the enemy position. A secondary explosion was heard from the camp, and just before last light the patrol was extracted by helicopter. A few days later McBride's second-in-command, Trooper John (Zed) Frazer, and the signaller, Trooper Lawrie Sams, guided C Company 2 RAR back through a swamp to the enemy camp. The SAS had had difficulty fully reconnoitring the camp because of concern that they might leave traces of their movement in the swamp, and it turned out to be much larger than they had anticipated. C Company 2 RAR ended up with a substantial contact.

McBride's patrol near the Bien Hoa border was an example of the SAS's task of maintaining surveillance towards the edge of the Task Force TAOR while the battalions concentrated their operations around the more settled areas in the centre of the province. Another example of a reconnaissance patrol on the extremity of the TAOR was Lieutenant Zot Simon's four-man patrol in Long Khanh Province, 8 kilometres north west of Nui May Tao. On 30 July they sighted six VC following their tracks towards their LUP, and the Australians initiated a contact, killing at least two VC. The enemy continued their attack against the SAS patrol which eventually broke contact and was extracted later in the day.

While the patrols near the provincial border continued, by August, 1 Squadron was beginning to conduct patrols closer to the populated areas in support of the battalion operations. These patrols had the potential problem of identifying the enemy from among the civilians who often wandered out of their access areas. Robin McBride's four-man patrol, six kilometres south of Xuyen Moc on 1 August, was clearly outside the civilian access area. After sighting small groups of enemy carrying packs and moving south and west, they initiated an ambush with small arms on two VC. Although the VC were found to be carrying food, equipment for living in the field, and documents, they were unarmed. McBride commented later that 'the result horrified me and changed my attitude towards other operations'. But the Task Force intelligence officer assured him that the documents confirmed that the bodies were indeed those of VC and it was suspected that an enemy camp was in the vicinity. While McBride's actions were entirely justified, the incident underlined the problems of operating near civilian access areas. The danger was that an SAS soldier might hesitate to fire at an enemy and thus place himself and his patrol in jeopardy.

A little later Sergeant Arthur (George) Garvin's four-man patrol was inserted about six kilometres north of Xuyen Moc. On 22 August they heard voices of two people and when Garvin saw a figure raise what he thought was an SKS rifle he fired one single shot, while the remainder of the patrol suppressed the area where the other enemy was thought to be located. On advancing, the patrol realised that they had wounded a woman and had possibly engaged a group of civilians. They rendered first aid and arranged for two civilians, including the wounded woman, to be evacuated to Nui Dat for questioning. In the report of the patrol the squadron operations officer at that time, Robin McBride, commented: 'SAS recce patrols must rely on instinctive reaction in order to extricate themselves from contact with the enemy. By tasking patrols in areas where civilians may illegally wander, it places an unnecessary restriction on each member, and removes from the patrol what in the past has been a distinct advantage, SPEED and AGGRESSION'.

Corporal Hans Van Eldik was faced with a similar problem a few days later. As part of in-country training he was leading a patrol of which his troop commander, Zot Simon, was a member. The patrol was inserted about 5 kilometres north east of Binh Gia and soon after midday on 29 August they heard chopping and voices near a well-used track. Van Edlik and his scout, Trooper Gerry Vale, carried out a close reconnaissance, sighted a hutchie with a woman in a hammock, and brought the rest of the patrol forward in extended line to within five metres. With Garvin's recent experience on his mind, Van Edlik crept forward again to establish definitely whether the persons were civilians, as no weapons had been seen. When they were three metres from the woman they challenged her, but immediately two other persons appeared, one carrying an M1 carbine. The SAS engaged the area and two enemy fell while two more dragged a body into thick jungle. No bodies were found but two weapons and seven packs were recovered.

Sergeant Lawrie Fraser was critical of the decision to conduct in-country training by requiring junior NCOs to command patrols while accompanied by established patrol commanders. He believed that in war 'you always fielded your first eleven', and he thought that there could possibly be confusion over command relationships if quick decisions were required. On the other hand, it was argued that junior men had to be given the opportunity to develop their skills as commanders. Second Lieutenant McBride recalled that it was 'not a topic of great debate and was a matter for patrol commanders to decide when his patrol 2ic was ready for it'.

Corporal Greg Mawkes was one patrol second-in-command who was given the opportunity of commanding a patrol while accompanied by his normal patrol commander. The patrol demonstrated the pressure likely to be faced by a junior patrol commander when operating near a civilian access area. On 24 September the six-man patrol was inserted two kilometres east of Nui Dat Two into an area that was outside civilian access but had been traditionally farmed and used by the former inhabitants of the abandoned village of Long Tan, now living in Hoa Long and Dat Do. Mawkes later described the patrol in an article, and the following extract, although written in the third person, describes the pressures he was under:

> The commander of the SAS patrol was a corporal. Just 22 years old, he was a regular soldier in Vietnam for a second time. He had been given

360

Flexibility and frustration

the opportunity to lead this reconnaissance mission under assessment by the usual patrol commander. A good job would mean promotion and his own patrol.

Their task was to find the Viet Cong and report the location back to base by radio. A simple enough assignment, complicated only by the fact that the terrain to be searched bordered a civilian access area (a non-populated area subject to curfew). Boundaries and time restrictions, however, meant little to hungry peasants. Risking the consequences was accepted. As per the usual brief, the SAS men would be the only 'friendlies' in their designated area of operation. Anyone else was enemy.

Late on the second day, the patrol discovered an unoccupied VC camp. Following their orders, the soldiers began to search and sketch the fortifications. This was done with more haste than normal in the encroaching darkness. The still twilight created an eerie atmosphere. Each man could feel unseen eyes observing his every move—visualise hidden VC taking aim. The tension was evident on their sweating camouflaged faces. They estimated that the camp was in regular use and had been occupied recently. Quickly completing their task and concealing any sign of intrusion, they stealthily moved off into the dark.

Away from the camp the dense black jungle enveloped them. Silent movement became impossible. Although the corporal wanted to move further away, he knew it was dangerous to do so. He estimated that they were about 400 metres from the camp but had moved to within 500 metres of the civilian access area. It would have to do. On his whispered order the patrol halted and established an all-round defensive position. Safe in the knowledge that it was almost impossible for anyone to find five men in the jungle at night, they slept.

Just prior to first light each man awoke instinctively. As the surrounding blur became distinguishable they silently increased their perimeter to visual distance. The corporal remained in the centre from where he could see every man.

He had decided what action they would take that morning, but he wanted to discuss his proposal with his senior, now acting as second-in-command...As he turned to attract the sergeant's attention the sound of movement through the undergrowth could clearly be heard. His heart went into overdrive at the sight of the sergeant's left arm extended—fist clenched and thumb pointing down. The silent signal for enemy.

The noise was rapidly increasing and coming from the direction of the camp. Was it possible Charlie had discovered their incursion and was now tracking them? He glanced around and saw that the other three had instantly taken up fire positions.

The small trees and undergrowth appeared to be advancing on a 20 metre front. Glimpses of black clad figures could be seen through the foliage. The glint of metal caught his eye. The distinctive curve of an AK-47 magazine. Or was it?

361

Something wasn't right. The enemy he had fought for almost two years and come to respect was a professional. Almost as good as his own unit. If *he* was hunting down *his* most feared adversary would he be so careless? 'Christ!' he thought, 'What is it?'

He could see the sergeant's trigger finger slowly tightening. One shot and they would all open fire.

'Dung lai!' the corporal shouted in Vietnamese.

All movement ceased—the sudden silence unnerving. The sergeant looked over his shoulder and glared, his eyes querying, 'What now?' The corporal slowly stood up. 'Lai dai!' [come here] he yelled, beckoning whoever it was to come to him. Nothing happened.

'Lai dai! Lai dai!' he screamed.

A woman stepped forward. She was holding a cutting tool about the length of an AK-47 assault rifle. It had a curved blade. Another woman hesitantly appeared, two small children clinging to her legs. They began to cry as more women and children became visible. He counted fifteen as they cautiously moved towards him.

'Jesus!' he said aloud. 'Another second and...'

Helicopters were called and the women and children were carried to Nui Dat where their identification papers were checked and they were interrogated by the Task Force intelligence staff. Only half of them had identification cards but they were all subsequently released.[11]

The SAS operations near the more settled areas in August and September 1970 were principally in support of Operations CUNG CHUNG II (3 August–10 September) and CUNG CHUNG III (10–22 September) in which 7 RAR had the mission of denying enemy access to Xuyen Moc, Dat Do and Lang Phuoc Hai. As part of this operation, a five-man patrol commanded by Sergeant John Ward provided the infantry support for a section of APCs operating near the coast on the Phuoc Tuy–Binh Tuy border. On 6 September, while patrolling with APCs, they found a recently used camp which was destroyed by engineers, and the next day they located an enemy bunker system which was also destroyed.

They returned to a night defensive position, known as NDP Beverly, and remained behind in ambush when the APCs departed on the afternoon of 8 September. Soon after noon the next day three enemy approached the position and the patrol initiated an ambush, killing two VC. For the next four days they again operated with the APCs, and in his patrol report Ward observed that 'small infantry patrols could be well employed in this manner'. Indeed they could; while Ward had been successful, his task had hardly been one for which the SAS was trained. Realising this, Lieutenant-Colonel Grey, commanding officer of 7 RAR, arranged for the SAS to train one of his platoons in recce–ambush techniques.

These operations to deny access to the villages significantly disrupted the VC activities. For example, a Task Force intelligence report in November 1970 observed: 'Captured documents from district and village cadres referred constantly to shortages of food, lack of contact with the local population, allied harassment of courier routes [and], the prevalence of "ranger" [that is SAS] type activities'.[12]

Although Sergeant Lawrie Fraser had seen little action during the first six months of his second tour, his experiences in October showed that it was not necessary to sight the enemy or become involved in a contact to make a contribution to Australian success. On 2 October, while patrolling on the Bien Hoa border, north east of Nui Thi Vai, he noticed a section of pipe in the ground, realised that it was an air vent, and searched for and eventually found a tunnel complex. The next day they located 400 metres of tunnel containing large quantities of munitions plus two 90 mm recoilless rifles and five Bangalore torpedoes. A platoon from V Company 2 RAR/NZ was inserted to secure the area and a combat engineer team from 1 Field Squadron destroyed the complex.

On 25 October 8 RAR ceased operations to prepare to return to Australia. The battalion was not replaced and Brigadier Henderson now had to maintain operations in Phuoc Tuy with only two rather than three battalions. When warned of this development in September Teague had written in his Commander's Diary: 'The increase in size of 1 ATF TAOR and the withdrawal of one infantry battalion from extended jungle operations has increased the number of tasks for SAS'. But his optimism was short lived, and at the end of October he wrote that 'there is still a lack of proper tasks due to both the restricted TAOR and the conspicuous absence of enemy forces'. On 6 November he wrote: 'The decreasing level of enemy activity and capabilities in Phuoc Tuy Province during 1970 has led to a decreased employment of SAS patrols on reconnaissance tasks which are beyond the capacity of infantry battalions'.[13]

The problem outlined by Teague had a direct impact on the effectiveness of his squadron. There was a period of about twenty days when nine patrols were inserted across the border into Binh Tuy Province and several had contacts on or just after insertion. For example, Second Lieutenant McBride's patrol was inserted 16 kilometres south east of Nui May Tao on 3 November but was extracted almost immediately when it engaged an enemy group. Next day his patrol was inserted into a nearby LZ and again had to be extracted after killing one enemy.

McBride recalled:

We were deployed a long way from 1 ATF, with a long response time, with apparently no intention of deploying 1 ATF forces as a follow-up. 7 RAR, I believe, requested permission to enter the area for our operation in direct response to our pattern of contacts, yet were not allowed. In response to my question, voiced in some exasperation to my OC, as to why no one was being sent in, I was told that that may happen after we had identified the enemy. To us at the patrol level it seemed that sufficient information (albeit not hard intelligence) existed that something was up, if only because the LZs were obviously being watched and on at least one LZ a heavy machinegun was reportedly used against gunships...Our suspicion was that it was elements of 33 NVA Regiment which was relatively quiet at the time engaged in some 'rest in country'...The frustrating element and the sense of futility was that politically it was unlikely that Australian forces would be deployed outside their restricted AO within the province, let alone across the border of another, regardless of the results.

Political considerations aside, it seemed that we were being deployed into an increasingly hot area for no real purpose. I requested a ride in one of the gunships for a Jack Gebhardt patrol into this area because I was certain the patrol would be in contact shortly after insertion. That occurred some ten minutes after insertion and they were placed under considerable pressure before the gunships suppressed the area sufficiently for the patrol to be extracted.

In fact Gebhardt's patrol was inserted at 2.40 pm on 9 November and was extracted at 3.25 pm. They were engaged by four VC and possibly killed one.

McBride concluded that while the squadron 'maintained a presence in another area, when the remainder of the TF was on a very tight leash, it was a point of some disappointment'. Eventually, on 30 November Teague wrote in his commander's diary that there were 'now both suitable tasks and sufficient scope with terrain to adequately keep all SAS patrols occupied on normal SAS tasks'.

Teague was also faced with personnel problems as the pressure of over five years of operational service was now affecting the levels of training within the SAS. Towards the end of 1970 there was a shortage of personnel with patrol skills and a number of cases occurred when patrols were inserted with one member having to perform the tasks of both signaller and medic. In one instance the member was also second-in-command of the patrol. Fortunately, the squadron still retained a large measure of experience. For example, a survey of the squadron found that of 96 personnel, 59 had previous operational experience: 56 had previously served in Vietnam, 24 in Borneo, 8 in Malaya and one in Cyprus.

Although patrols developed close teamwork and tried to remain together, that was not always possible. This patrol, commanded by Second Lieutenant Robin McBride, comprised completely different personnel to that shown on page 358. *From left:* Corporal Graham Smith, Corporal Brian Kelly, McBride, Trooper Graham Wearne and Signaller Tony Jacques. (PHOTO, R. MCBRIDE)

Teague's problem in trying to arrange suitable tasks for his squadron was reflected in a memorandum prepared by Captain Graye Shatkey of 4 Troop NZSAS, on 26 October. He observed that since June his troop had been tasked primarily with reconnaissance tasks and had avoided contact. In the same memorandum he was critical of the Task Force commander for ordering quick fighting and ambush patrols at the expense of detailed preparation and briefing. He stated that his patrols had been deployed for periods of up to seven days while he thought that this period could be extended to up to fifteen days in the wet season. He went on to outline the SAS method of operations and concluded: 'That this report details an operational procedure that is followed without variation, only emphasises the fact that 4 Troop NZSAS is misemployed in its current role. The recce and ambush missions given to 4 Troop are well within the capabilities of the Recce Platoon, 1 RNZIR [Royal New Zealand Infantry Regiment] and could be successfully accomplished by a well-trained infantry section. To continue to employ NZSAS in the present manner will only cause the unit to lose sight of its Unconventional and Special Warfare role. The high degree of individual training is not fully utilized on present operations nor does 1 ATF make use of the specialist skills which the Troop possesses'.[14]

While Teague agreed with some of Shatkey's views, he believed that the New Zealand troop commander did not take into account the constraints under which the Task Force commander had to operate. While it was true that the SAS was not using its full range of skills, it had to be flexible enough to meet the requirements of the theatre commander. With an eye to the future, it was preferable for the SAS to obtain operational experience for as large a number of officers and soldiers as possible, even if not all tasks were completely SAS in nature.

It is difficult to know how much Shatkey's views influenced decision-makers in New Zealand. Although the New Zealand government had been planning for eventual withdrawal, on 5 February 1970 the head of the New Zealand defence liaison staff in Canberra told the Australian Department of Defence that as 'SAS patrols would be of increasing importance during [the] latter stages of 1 ATF withdrawal', they 'would not seek to withdraw [the] SAS Troop until Australia [was] prepared to release' them. Perhaps Shatkey's views about the value of SAS operations were heeded because in December 1970 the New Zealand government asked Australia if they had any objection to the withdrawal of the SAS troop. The Chief of the General Staff, Lieutenant-General Daly, said that he had no objection and the New Zealand withdrawal was announced soon afterwards.[15]

On 3 December Major Teague returned to Australia to attend the Army Staff College in 1971, and Major Geoff Chipman, commander of 2 Squadron which was due to relieve 1 Squadron in February, arrived to take command. Teague had been an outgoing and imaginative commander, but in the frustrating circumstances of 1970 it had been hard to please the expectations of many of his soldiers. Chipman would have no easier a task.

Despite Shatkey's reservations, 1 Squadron's operations in December 1970 and January 1971 showed that the SAS still had an important role to play. An example was Sergeant George Garvin's four-man patrol which was inserted eight kilometres south west of Nui May Tao on 27 November. On 30 November they located a camouflaged fish trap in a creek and set an ambush. Next day they ambushed three VC, killing two and wounding the other, who was winched out by helicopter and taken to Nui Dat. Perhaps more importantly, the patrol reported that every night they had heard heavy machine-gun and small arms fire and numerous explosions from the Nui May Tao feature. Of 27 patrols deployed during December, all except four were deployed on the approaches to the May Taos, with the patrols spread evenly between the north east corner of the province, and Binh Tuy

Province from the May Taos to the sea. In five contacts four enemy were killed, plus possibly three more, and two were wounded. In two other sightings, four enemy were observed. This was not a markedly better result than in other months, but the squadron persisted with its patrols in this area and the following month they paid off.

Acting on information from an earlier patrol, on 5 January Sergeant Lawrie Fraser and a twelve-man ambush patrol was inserted ten kilometres south east of Nui May Tao. On the morning of 11 January the patrol was in its LUP eating breakfast when Trooper Graeme Antonovich left the LUP for his 'morning constitutional'. Some patrol commanders insisted that patrol members should defecate within the LUP. Fraser was content to allow members of his patrol to move slightly out of the LUP as long as the other patrol members were warned. Suddenly, from Antonovich's direction, 20 metres from the LUP, came a series of shots, and then silence. Fraser called out to Antonovich, asking what had happened. There was no reply. Then Antonovich appeared still buttoning up his trousers. While squatting he had seen three VC approaching the position. Without moving, he had shot two and had wounded the third who had run off leaving a blood trail.

Unfortunately the patrol signaller was inexperienced, and before Fraser could stop him he had called for support from Nui Dat. Before long a fire team was overhead, and later the squadron commander would be critical of this waste of helicopter hours. Nevertheless, Fraser's patrol followed up the blood trail and within the hour had located an enemy camp abandoned not 30 minutes earlier. The fires were still hot and half-eaten food had been discarded. Late in the morning a bunker system was discovered 100 metres further on and, supported by fire from helicopter gunships, the patrol mounted an assault, clearing the bunkers with grenades. Enemy camp fires and discarded food indicated that the enemy had been in occupation up to ten minutes earlier. Demolitions were lowered to the patrol and the camp and bunker system were destroyed.

On 10 January, the day before Fraser's contact, Lieutenant Zot Simon had been inserted with a five-man patrol only 3 kilometres south of Nui May Tao. On 12 January they heard digging, and the next day the voices of from five to ten people and chopping sounds about 40 metres from their position. They withdrew south for 50 metres to report to Nui Dat, and while in their LUP they heard one enemy move to within 20 metres of their position and start chopping. He was unaware of the SAS's presence and Simon called for the support of a light fire team to attack what appeared to be an enemy camp under construction. When the helicopters arrived they could not establish communications and circled the patrol's position, compromising the patrol. All chopping stopped and no further sounds were heard. At 2.30 pm the patrol was winched out of the jungle while Bushrangers engaged the camp. Then half an hour later an airstrike went in against the camp. Meanwhile Simon's patrol had been transported to Xuyen Moc where they had married up with a US Ranger platoon, tasked to return to the area to assess bomb damage in the enemy camp. By the time they were back on the ground it was too dark to carry out the bomb damage assessment and both patrols were extracted.[16]

In one respect this had been a frustrating result, but it appeared as though Simon's patrol might have located an important VC camp, and three days later a twelve-man patrol commanded by Sergeant John Duncan was inserted to locate the camp. Trooper Gerry Vale, Simon's scout, accompanied the patrol to lead them back to the same spot. He successfully located the camp and Duncan planned a silent assault to be led by Sergeant John O'Keefe and Corporal Greg Mawkes, both with silenced Sterlings. At 1.15 pm on 19 January they entered the southern perimeter of the camp, and Mawkes saw one VC walking away to his flank as though he had not seen the Australians. In the centre of the camp were two VC. One had his back to the Australians and appeared to be drilling on a structure. Beyond him was another VC with his weapon stripped for cleaning on a table. Suddenly the VC who was walking away started running and said something in Vietnamese to the other two. The man with the drill looked around while the man with the stripped weapon looked up, and then with a look of horror looked down again at the pieces of his stripped weapon. Both jumped to their feet as O'Keefe opened fire. The rounds from the silenced Sterling struck home, but neither man stopped as they disappeared into the jungle.

Years later, looking back on the incident with the advantage of hindsight, Mawkes, who was then a major in the SAS Regiment, realised that the SAS had not been trained for an infantry style assault on an enemy camp, but fortunately the enemy did not put up a fight and scattered to the north, north east and north west. The patrol had located the VC BA Long Province Armoury Section workshop. In an area 75 metres by 50 metres they found five bunkers, a workshop, three kitchens, and four individual shelters. There was a large quantity of documents, three SKS rifles, an M16 and an M1 carbine. A stand-by patrol commanded by Sergeant John (Blue) Mulby was winched in with demolitions and the camp was destroyed. In the ensuing explosion the patrols were surrounded by a cloud of CS gas, and a later inspection showed that the VC had filled seven 81 mm mortar bombs with CS powder.

An LUP was established and next day the stand-by patrol was winched out with captured equipment. Meanwhile, a reconnaissance patrol had discovered another enemy camp a kilometre to the south and a platoon from the Australian Reinforcement Unit arrived in APCs to prepare the camp for demolition. On 21 January another enemy cache was discovered.

Patrols continued to be inserted deep into Binh Tuy Province, and on 23 January Lawrie Fraser's six-man patrol was inserted 13 kilometres south east of Nui May Tao. It was almost 45 kilometres from Nui Dat and was an illustration of the fact that enemy activity in the populated areas of Phuoc Tuy had declined substantially. Fraser established an ambush on a recently used track, and during the next two days small groups of VC moved past the ambush site. Fraser held his fire, hoping for a more worthwhile target. Then on 26 January three VC moved into the ambush. The third VC seemed to drop back, as though he had seen a Claymore, and Fraser initiated the ambush. Two VC were killed and the third could be heard crawling through the jungle moaning. He could not be located, and suspecting that a camp was nearby the patrol withdrew from the area. The patrol was extracted, and obtaining a resupply of Claymores was re-inserted about 1500 metres to the north west. They discovered a recently abandoned enemy camp and were extracted on 30 January. Fraser commented later that the nature of SAS patrols had changed considerably by this time. Initially they had avoided walking on tracks; now they actually walked on tracks hoping to run into the VC.[17]

1 Squadron had its last contact on 4 February when a New Zealand patrol killed two VC 7 kilometres north west of Thua Tich. By now 2 Squadron's advance party had arrived, and on 18 February 1 Squadron emplaned for Australia. Major Geoff Chipman, who had been in Vietnam since the end of November, summarised the problems of 1 Squadron's last month of operations in his Commander's Diary for January 1971.

> The decision to withdraw the NZ SAS Troop on the 20 February 1971 was announced on 27 January 1971. I am reluctant to lose the troop; however, if there had to be a reduction, under the circumstances it is best that it is the NZ troop. The time for the reduction is inopportune and will mean that only one patrol will be able to be deployed during the period 2 SAS Squadron is doing 'in country' training.

He continued with comments on SAS operations:

> Not only is there a requirement to vary methods of insertion of SAS patrols but there is also a requirement to vary SAS missions so that the enemy are unable to execute drills against a pattern of friendly activity. Associated with this is a morale and man-management aspect also. In

the situation that currently exists in the 1 ATF and Phuoc Tuy context I would like to see SAS patrols tasked to provide information directly required by infantry battalions in addition to HQ 1 ATF. I would like to see all SAS patrols issued with a particular aim in view, that justifies deployment of Special Forces on the mission, and which ultimately mean that there is likely to be some reactions to what the patrol reports. When SAS patrols locate an enemy installation or complex there is a limit to what the small group of men can do and it is necessary to get supporting troops in to assist in clearing and searching. I believe the principle should be that Special Forces having located each installation/complex, conventional follow up action is taken by conventional troops.

While 1 Squadron's tour had been frustrated by changing roles and a marked decline in VC activity, there is no doubt that it had been a successful operational year. In 52 contacts the squadron had inflicted 52 kills and 28 possibles on the enemy for the loss of no Australians. In 102 sightings it had observed over 300 enemy and had discovered several significant complexes. It had played an important role in helping to chase the VC Main Force units out of Phuoc Tuy Province and in severely restricting the activities of the VC local forces. Whether that favourable situation could be maintained would be tested during the following months.

21
Withdrawal from Vietnam

2 Squadron: February–October 1971

The prospect of Australian withdrawal from Phuoc Tuy Province presented the local VC commanders with a dilemma. Although no announcement had been made, by the end of 1970 it was obvious to all that Australia's involvement in the province was winding down and perhaps the combat troops would be withdrawn before the end of 1971. From the VC point of view, the quicker the Australians withdrew the better; but in the meantime what strategy should they follow? VC ralliers spoke to Australian intelligence staff of the many hardships enemy soldiers had to endure, and prisoners from enemy hospitals were a clear indication of the poor condition of many of the combat and even logistics units. Captured documents told of shortages of men, key cadre, food, medical supplies, ammunition and weapons; and told that villagers were demanding higher prices from the VC for food and other commodities normally obtained by VC 'Entry/Exit' and 'Forward Supply' organisations. Political re-orientation and indoctrination sessions were the order of the day in all VC units. Some units disappeared for months undergoing re-organisation and retraining. The enemy were suffering in almost every way and the success of the Australian pacification programme seemed assured.

But the VC could not afford to remain completely on the defensive waiting for the Australians to withdraw. If the VC were to mount an offensive, would that hurry the Australian withdrawal or delay it? There was a case for waiting patiently for the Australians to depart, but if the VC were to retain any credibility with the local population they had to be seen to be capable of taking the fight to the South Vietnamese government and its allies. The problem was that if they did go on the offensive they faced the risk of another defeat with a loss of credibility and, more importantly, of manpower and equipment, leading to a further deterioration in morale and effectiveness. There was only one order that the high command of Military Region 7 could give; the NVA and VC main force units had to re-enter Phuoc Tuy Province and challenge the Australians.

On 29 November 1970 part of D445 Battalion supported by elements of the disbanded D440 Battalion successfully attacked the district capital of Xuyen Moc. Nine days later a 7 RAR night defensive position was mortared, followed by a deliberate attack by elements of D445 Battalion. On the night of 30/31 December elements of 7 RAR ambushed a large part of D445 Battalion south of Xuyen Moc. In these incidents the enemy displayed classical VC tactics and fought with a tenacity thought long dead in Phuoc Tuy. Plans for a 1971 Tet offensive, however, were postponed due to Australian chance encounters with enemy reconnaissance parties and protective patrolling. The VC withdrew to lick their wounds and again the war went quiet in the province.

In this situation it was the task of the SAS, along with the other units of the Task Force, to detect when the VC again entered the province, and this mission was confirmed when Brigadier Bruce McDonald succeeded Bill Henderson as the Commander of 1 ATF on 28 February 1971. McDonald explained later that to prepare for future operations and to ensure the security of the Task Force the SAS was deployed to:

1. Gain essential intelligence on enemy movement and strength.
2. Check or confirm information of the enemy as gained from other sources.
3. Harass through ambush, and take full advantage of casual encounters to maximise enemy casualties.
4. Leave the enemy in no doubt who controlled the 'outer space' of the Task Force territorial limits as defined from time to time.

2 SAS Squadron arrived at Nui Dat on 26 February, and during March the squadron deployed 23 patrols. All but one of these patrols were deployed to the eastern sector of the province in an arc stretching from the South China Sea to the Long Khanh border. Only one patrol sighted the enemy but that sighting was significant. On 12 March a six-man patrol led by Sergeant Terry O'Farrell was inserted just east of Route 329, twelve kilometres north east of Xuyen Moc. That evening, while establishing a night LUP, they heard a large force of enemy moving north east along a track, 30 metres to the west of their LUP. A little later one enemy approached to within ten metres of the LUP, remained there for two or three minutes in a listening attitude, and then fired one shot. He was engaged by the patrol and a loud scream was heard. The patrol then withdrew from the area and was extracted

the next day. Four days later 3 RAR, which had relieved 7 RAR in February, contacted a company of D445 Battalion about 4 kilometres north west of Xuyen Moc. In an operation extending well into April, 3 RAR and elements of 2 RAR/NZ located and attacked a bunker system recently occupied by D445 Battalion. Most of the enemy battalion escaped the Australian cordon and began to withdraw from the province.

O'Farrell's patrol had successfully detected a group from D445 Battalion, but otherwise it had been a quiet month; as Major Geoff Chipman commented in his Commander's Diary at the end of March: 'The squadron is settling down and all patrols have completed or are in the process of completing their second patrol. Contact with the enemy to "Blood" new members is required. A variety of reconnaissance and fighting patrols have been tasked, but so far the enemy has been conspicuous by his absence in our AOs. The squadron has not been misemployed in my view. As a result of 1 ATF deployment, being based on getting maximum effect from the immediate operational situation, the following out-of-the-ordinary situations have arisen: patrols have had to be extracted earlier than planned to make way for battalions; patrols have been extended and resupplied; and [one] patrol AO was reduced in size twice due to battalion move-[ment], and had to be allotted and moved to another AO'.

The rapid redeployment of patrols placed the soldiers under increased pressure. Terry O'Farrell, on his second tour of Vietnam, recalled that 1971 was more arduous physically than 1968. A lot of time was spent in the jungle for no action. On one occasion he received a warning order for his next patrol as soon as he returned from his previous one.

In an effort to intercept the withdrawing VC a series of ambush patrols were deployed north east of Thua Tich and further east towards the Binh Tuy border. Sergeant Frank Cashmore led a ten-man ambush patrol which was inserted by APC into an area 1500 metres north east of Thua Tich, and at 7 am on 3 April they established a Claymore ambush on an east–west track. At 10.15 that morning one enemy was seen approaching the ambush position from the west, moving fast. The SAS flank protection group advised Cashmore by radio that there were two more enemy, 20 metres apart, moving about 45 metres behind the first man, and Cashmore let the first man through, waiting until the second man had reached the eastern edge of the killing area before initiating the Claymores. Both enemy were killed, and weapons and documents were recovered.

None of the other patrols sighted any enemy, but Sergeant Danny Wright's ten-man ambush patrol in an area 12 kilometres north east of Xuyen Moc had an additional task. It was here that Terry O'Farrell had detected an enemy party during the previous month, and a little later a patrol led by Second Lieutenant Brian Jones had reported finding a recently occupied VC bunker system. If the situation allowed, Wright's patrol was to be reinforced by Jones's patrol and an engineer party, and under Wright's direction the bunkers were to be destroyed. Wright was now on his third operational tour in Vietnam. After his tour with 2 Squadron in 1968, both he and Sergeant Dave Scheele had returned to Vietnam for a short tour with the AATTV before rejoining 2 Squadron towards the end of 1969.[1]

Wright's patrol was inserted by helicopter on 4 April and next day established an ambush position on a track. They remained there for three days without result, and then on 8 April Wright was ordered to move to an LZ to receive Jones's patrol which would guide him to the bunker system to be destroyed. Next day Wright located an old enemy camp, and the following morning Jones' six-man patrol plus four engineers married up with Wright's patrol. They were about to move off towards the area where Jones had reported the bunker system when they saw smoke rising from the area where Wright's patrol had found the old enemy camp the previous day. Wright sought clearance to assault the position and at 12.30 pm they began their attack. It transpired that the smoke was coming from a smouldering log, and there was no sign of enemy.

After a lunch break they moved off again with Jones leading, trying to locate the bunker system. They had gone about 300 metres when they came across an ox-cart track which appeared to have been used within the last seven days, and near the track was an abandoned ox-cart wheel. Wright distinctly recalled having seen the wheel during his first tour of Vietnam in 1968, and years later he commented that in his opinion that the few bunkers in the area had not been used for years; he believed that Jones had mistaken old bunkers for new ones. When he had seen the bunkers at the end of March Jones had been commanding only his second operational patrol. Whatever Wright's private doubts might have been, Jones told him that this was the western edge of the bunker system and that the bunkers themselves were 60 metres further on. Wright, who was in charge of the overall group, therefore deployed his troops in a defensive position while he went forward with Jones and Warrant Officer Tough, commander of the engineer party, to locate the bunkers. Before leaving the defensive position Wright briefed Sergeant Kim Pember, second-in-command of his own patrol, and Jones' second-in-command, a corporal, that he would be going forward on a reconnaissance.

The reconnaissance party moved off to the east but could not locate the bunkers. After about 50 metres Jones remarked, 'Someone has stolen the bunkers'. They moved a further 60 metres and then turned north east. Wright realised that they had moved in an arc and he reminded Jones that if they continued they would end up in front of the area where Jones' patrol was in a defensive position. Jones merely acknowleged Wright's warning and kept moving, first north and then north west. By now Jones was some distance ahead of Wright and he started to walk across an open area of long grass. Wright realised that he would have to stop Jones, and tried to gain his attention with a series of louded hisses and 'psts'. Jones kept walking. Wright said later that years of training and the experience of numerous operational patrols made him hesitate to yell out. He opened his mouth to call to Jones when there was a burst of fire from an M16 and the tall figure of Brian Jones collapsed. Wright yelled, 'cease fire' and ran forward.

As Wright had feared, Jones had wandered back towards the eastern flank of the defensive position where his own patrol was located. The patrol had been placed in a half moon with the corporal in the centre and the signaller, Private Jeff Mathews, on the left flank. The corporal thought that he had briefed each of the other four soldiers in the patrol, but it turned out that he had omitted to brief Mathews. The corporal recalled 'that after about 15 minutes Private Ken Freeman and I heard movement to the south. I took a kneeling position. The movement continued from south to north across our front. Finally when the movement had gone across our front, about twenty metres out, and was almost in front of Private Mathews, I identified the recce party. I instructed Private Freeman to brief Mathews'.

Mathews had also seen the movement. He remembered that the corporal and Freeman 'were about 15 metres to my right rear. After about five minutes I heard movement to my right about 50 metres away. I looked around to [the corporal] to see if he had identified the movement. His weapon was still in the shoulder ready to fire and he was looking to his front. Finally after looking at [the corporal] I looked back to my front to see, ten metres away, a weapon pointing at me from the scrub with the figure still completely concealed by the foliage. I had my safety catch on automatic when I first heard the movement. I aimed instinctively and pulled the trigger. My weapon had a stoppage after about five rounds. As soon as I had the stoppage I heard [the corporal] screaming out my name. I looked to the front and heard Sergeant Wright shout "You have just shot Jonesie". I put my weapon down, ran to the front and saw Second Lieutenant Jones lying there'.

Jones died at about the time he was being placed aboard the Dustoff chopper. Although inexperienced, he was a popular and capable officer. The accident was investigated by Captain Robin Letts, second-in-command of the squadron, and he reported that he could not find that the fatal wounding was caused or aggravated by misconduct or neglect by any persons. Second Lieutenant Jones had been over the ground before. He should not have led the recce party outside the Eastern flank protection group. However, it is easy to become disoriented in the jungle. Sergeant Wright as overall group commander should have stopped Jones much earlier. However, he cautioned him once. Jones had seen the positioning of the flank protection groups and seemed confident, and he had seen the ground before. [The corporal] does not appear to have briefed the patrol before placing them individually in the Eastern protection group. However, had he briefed them fully they would still have been suspicious of movement outside their position as he himself was. I find Private Mathews in no way blameworthy. He could do litle else in the circumstances. He did not know there was a recce party out, the vegetation and SAS camouflage made positive identification impossible. To have challenged at ten metres would have been suicide had it been an enemy.

Letts concluded: 'The fatal wounding was caused by disorientation on the part of Second Lieutenant Jones, the thickness of the vegetation where the wounding took place, and the build up and release of tension occasioned by the events of the morning'. The incident underlined again the problem faced by the SAS when operating in groups larger than a normal patrol.

It had been a discouraging month for 2 Squadron. In 23 patrols they had sighted three enemy and killed two, all during Cashmore's patrol on 3 April. And one Australian had been accidentally killed. Chipman's comments at the end of April provide a good summary:

The situation has not appreciably changed in the past month. Contacts are still required, but the enemy are very elusive; he is using existing tracks less and less and cutting his own tracks. Once he realises that 1 ATF troops are looking for him he becomes very careful in his daily routine and movement. He continues to move freely about the province from bunker system to bunker system. Further to last month's comments in relation to 1 ATF deployment being based on getting maximum effort from the immediate operational situation, the Comd 1 ATF has expressed the wish to deploy SAS patrols as quickly as within two hours notice under exceptional circumstance. This is being planned for and will depend considerably upon the RAAF in relation to the possibility of the need for rapelling and suspended extraction. Despite the closeness of operations within Phuoc Tuy Province it is reiterated that SAS patrols should be employed on reconnaissance beyond the capacity of infantry battalions and other supporting arms. Also close secure liaison with supporting arms at the highest level and as early as possible is essential.

Second Lieutenant Brian Jones' patrol, shortly before he was killed in April 1971. *Left to right:* Trooper Jeff Mathews, Corporal Ken Bowen, Jones, Troopers Stephan Dembowski and Graham (Percy) Beplate. (PHOTO, D. SCHEELE)

It was now obvious that the enemy had withdrawn from Phuoc Tuy Province and was capitalising on the fact that the allies did not generally operate in the border areas. The commander of 1 ATF, Brigadier McDonald, therefore obtained permission to extend the 1 ATF AO a further four kilometres north across the border into Long Khanh Province, and during May nine patrols were tasked to operate on and across the border west of the Courtenay rubber plantation. A further eight patrols were tasked to operate in the border area east of the Courtenay rubber plantation.

This extension to the area of operations brought immediate success. On 1 May Sergeant Danny Wright's five-man patrol was inserted just across the border, about sixteen kilometres east south east of the allied operational base at Black Horse. In the next six days the patrol sighted various enemy tracks and harbour positions and concluded that the VC had moved through the area in strength during the previous two weeks. Wright returned to the area, and on 15 May, four kilometres from the Courtenay rubber plantation and a kilometre south of the border, his patrol located an enemy bunker system which looked as though it had been constructed within the previous two weeks. Nearby

Patrol 22, 2 SAS Squadron, 1971. *Left to right:* Sergeant Danny Wright, Corporal Kerry (Curley) Clarke, Troopers Graham Wearne, Brian Jones and Lindsay Bennett. (PHOTO, D. WRIGHT)

they found another unoccupied enemy camp with bunkers, tables and three kitchens. Further patrols were inserted into the area and Sergeant Ray Swallow's five-man patrol from 26 May to 1 June was perhaps the most successful. They heard numerous enemy signal shots and sometimes bursts of automatic weapons. Three times on 30 May they heard enemy voices and the next day they heard mortars being fired. At one stage two males and a female came to within 15 metres of their LUP. Swallow believed that a nearby enemy radio station was operating on morse close to the patrol's frequency because the patrol had its radio set tuned to minimum volume and the enemy station still came through loudly.

The SAS patrols on the other side of Route 2 were even more successful. On 11 May Sergeant Adrian Blacker's five-man patrol was inserted just across the border about eight kilometres south west of Black Horse. On 12 May they killed one enemy and captured a quantity of documents, and they found signs that small parties of enemy had moved through the area during the previous two weeks. A few days later Sergeant Swallow and a six-man patrol including Major Chipman

378

were inserted about two kilometres closer to Black Horse. Chipman recalled that for reasons of command and morale he wished to share 'some of the patrol risks and discomfits with selected patrols'. For his part, Swallow was pleased to have the opportunity of demonstrating to Chipman (who said that he was under no misapprehensions anyway) that the problem of estimating when and how many enemy had used a track was more difficult in the jungle than it might seem from the operations room at Nui Dat. They mounted a number of ambushes, and on the night of 18 May established an OP near a track and detected groups of enemy moving in each direction using torches and signal shots. Next morning Chipman advised Swallow to ambush the track that night. Swallow replied that he did not like ambushing at night; he was the patrol commander and they would ambush the track during that day. The ambush was set up and Swallow had barely told everyone to get their heads down while he connected the Claymore leads when four VC approached along the track. All four VC were killed.

Sergeant Graham Brammer's five-man patrol was deployed to the same area and on 26 May located a newly constructed bunker. Although they heard digging, they could not conduct a close reconnaissance because of noisy undergrowth and open patches. The following afternoon they directed a light fire team onto the enemy position, and while the helicopter fire was keeping the enemy occupied another five-man patrol, commanded by Lieutenant Andrew Freemantle, was winched in to join Brammer. Within half an hour they began their assault but almost immediately came under fire from an automatic weapon. The firefight against an unknown number of AK47s and possibly an RPD lasted about twenty minutes until the enemy threw two grenades wounding Trooper Michael Crane in the left shoulder and right ankle. The patrol then conducted a tactical withdrawal using fire and movement and was extracted by winch. By now Blacker had been re-inserted in the area and that afternoon his patrol contacted a group of enemy withdrawing from the area of Brammer and Freemantle's contact and killed one. SAS patrols continued in this area and on the afternoon of 3 June, two kilometres to the west, a five-man patrol led by Sergeant Jeff Kidner contacted and killed another enemy. Three hours earlier they had sighted fifteen enemy moving east through the jungle.

The information from the SAS contributed substantially to the Task Force intelligence picture, and on 5 June the Task Force deployed on Operation OVERLORD against the positions, suspected to be occupied by 3/33rd NVA Regiment and D445 Battalion, east of Route 2. The newly arrived 4 RAR/NZ and a US battalion occupied blocking positions while 3 RAR, supported by tanks, swept through the area astride

Patrol 23, 2 SAS Squadron, Nui Dat 1971. *Left to right:* Corporal Hans Fleer, Troopers Ian Lawrence and Robert Kilsby, Sergeant Graham Brammer, and Trooper Hartley Smithwick. Fleer had been awarded the Distinguished Conduct Medal while serving with 6 RAR in 1969–1970. At the age of 18 he had earlier served in Vietnam with 4 RAR in 1968–1969.

the provincial border. The NVA Regiment was attacked in its bunker system and dispersed back into Long Khanh Province. D445 Battalion was also identified in the area and it too was dispersed.

Although Major Chipman could now feel more satisfied with his squadron's operations, he was beginning to face manning problems not experienced to the same extent by the other squadrons. As the Vietnam commitment had continued the number of soldiers being replaced for one reason or the other during each year had increased. For example, if a soldier had become sick in mid 1969 he would have been replaced by a reinforcement who in turn would have to be replaced in mid 1970 and again in mid 1971. In the meantime if another soldier had become sick in, say April 1970, he would have been replaced by a reinforcement who in turn would have had to be replaced in April 1971. The accumulative effect over five years meant that when 2 Squadron arrived in Vietnam in February 1971 it had to

Preparing for a patrol. From the left is Corporal Clive Mason (with M16 rifle with grenade launcher attached). Trooper Rod Swanson, Trooper Graham (Dixie) Lee (back to camera), Trooper Les Waring and Lieutenant Andy Freemantle. Both Freemantle and Mason had served with the British Army in Borneo. Freemantle returned to the British Army and is now a brigadier. Mason later served with the Rhodesian SAS and Selous Scouts and was killed in Mozambique in 1977.

accept 21 personnel from the previous squadron.

Furthermore, out of a squadron strength of 95 in February 1971, ten were National Servicemen. By 14 April, out of 95 personnel, 25 were National Servicemen.[2] While it was intended that National Servicemen should not be sent to Vietnam unless they had sufficient service remaining to allow them to complete one year of operations, many situations could arise whereby it was thought convenient to send a National Serviceman to Vietnam with less than one year's service remaining. On 29 April Chipman expressed his concerns to Task Force headquarters, pointing out that with 'SAS type operations it is highly desirable to have a patrol working as a team at all times'. He stated 'that specialist units such as SAS squadrons should change over as a unit' and he continued: 'By September 1971 it appears (without any reinforcement demands) [that] 2 Squadron will have replaced 40 out of the 114 in the squadron. Of this number 26 are national servicemen affecting 14 of the 15 patrols. The amount of training for such a limited return is out of all proportion and is demoralising to the dedicated professional regular soldiers in the squadron who must ultimately wear the "backlash". It is true to say that most of the national servicemen are quite good soldiers and that some of them may extend their tour of duty, but they have no intention of declaring their intentions until the last moment and only then if everything suits them. It is disappointing to see the SAS manning getting worse every year. There are outstanding soldiers in our small number and it is felt that they should be looked after outstandingly well'.

Officer manning was also a cause for concern. It took a month for Second Lieutenant Jones to be replaced, and when the new officer, Lieutenant Bob Brett, arrived, it was learned that he had already been nominated to attend a training course in Britain and would have to return to Australia before the end of August. Brett was a 23 year old Portsea officer who had been awarded a Military Cross while serving in Vietnam with 5 RAR in 1969. Clearly, as the end of the war drew near an attempt was being made to give as many officers as possible the opportuniy to gain operational experience with the SAS.

By 1971 the 9th US Division had departed, depriving the SAS of an opportunity of working with other long range reconnaissance troops, but fortunately the SAS still had its exchange programme with the US Navy SEAL teams operating in the Mekong Delta. While much of the SEAL training was similar to the SAS they had different tasks to the SAS. Operating out of Sea Anchor, on the coast in the Delta area, they had two main tasks during 1971; the neutralisation of the VC infrastructure, and the liberation of US and South Vietnamese prisoners being held in prisoner of war cages by the VC in remote areas of the Delta. The first task was conducted as part of the Phoenix Programme, while direct Australian involvement in Phoenix operations elsewhere had ceased in 1970. The second task was conducted with only limited success in the Delta, and could not be applied in Phuoc Tuy Province as no Australians were held prisoner by the VC.[3]

On one occasion an Australian patrol operating with SEALs in the Delta moved into a VC village to rescue some Vietnamese hostages. The hostages were not there, but a VC returnee with the patrol indicated the homes of several VC and these were set on fire. One house exploded as munitions were hidden under the floorboards. On another occasion six SAS, six SEALs and a group of Nung mercenaries raided a suspected VC prisoner-of-war camp in a mangrove swamp. They were led by a VC returnee through canals into a VC machine shop complex. The camp was held by a VC caretaker group and it was attacked by the SAS–SEAL team. One Nung mercenary grabbed a VC flag and was killed by a booby trap.

While the SEAL methods of operation were quite different from the SAS the attachments provided valuable experience and established a relationship with the US organisation which has endured to the present day. When the exchange programme terminated in September 1971 the Commander US Naval Forces, Vietnam, wrote to Brigadier McDonald: 'I wish to extend my personal congratulations to the Special Air Service Squadrons for the superb examples of professionalism displayed by those who participated in this worthwhile endeavour. Additionally, on behalf of the US Navy SEAL Teams, I take great pleasure in expressing the esteem in which the Australian SAS is held. The interchange of tactical knowledge and operational techniques has provided the basis for mutual respect and admiration between the two organizations as well as many long-lasting friendships. The camaraderie established serves as ample evidence of the unique affinity which has evolved between all connected with the program'.

Following Operation OVERLORD, enemy activity declined and 2 Squadron spent the remainder of June and most of July trying to locate D445 Battalion and reconnoitring several parts of the province that had received little attention in recent months. An eight-man patrol commanded by Lieutenant Freemantle located a bunker system in the southern foothills of the May Taos and marked it for an airstrike. The airstrike missed; the enemy hastily evacuated under cover of darkness and rain, and the patrol recovered miscellaneous equipment the following day.

Most signs of enemy activity were found across the border in Long Khanh Province, west of Route 2. In mid June Lieutenant Freemantle's patrol in the area six kilometres south west of Black Horse saw little sign but heard sufficient signal shots to pinpoint what they thought were two enemy bunker systems further to the west. On 28 June a nine-man patrol commanded by Lieutenant Brett was inserted into the same area and found two recently used bunker systems. While investigating the second bunker system they were detected by the enemy and the ensuing firefight continued for ten minutes before the enemy withdrew. Airstrikes were directed onto the enemy position.

In late June and throughout most of July 4 RAR operated against the VC 274 Regiment in the north west of the province. Then at the end of July, using information collected by the SAS, the Task Force mounted another hammer and anvil operation, Operation IRON FOX, against the positions located astride the provincial border west of Route 2. This time 3 RAR was the blocking force while 4 RAR pinned

Sergeant Paul Richards,
checking his map while on
patrol in 1971.
(PHOTO, L. ALVER)

elements of 274 Regiment, now reinforced with a large number of NVA, in a bunker system, and inflicted considerable casualties before the enemy escaped.

While operating as a western cut-off group for Operation IRON FOX a ten-man patrol commanded by Sergeant Paul Richards located signs indicating the presence of an occupied enemy camp about ten kilometres south west of Black Horse. Endeavouring to confirm the camp, the patrol was seen by a sentry who was in the process of aiming his AK47 at a patrol member. He was shot before he could fire and a blood trail was followed for 30 metres in the direction of the camp. Headquarters was contacted and the patrol was ordered to direct a gunship strike onto the camp and then withdraw to allow artillery to engage the camp area. The following day A Company 3 RAR with tanks and a guide from the SAS patrol swept through the area, destroying the camp which turned out to be the C24 convalescent company of 274 Regiment, complete with operating theatre and sixteen bunkers. The camp showed signs of having been well hit by the gunships, but the enemy had escaped, probably during the night. Operation IRON FOX had completely disrupted the efforts of 274 Regiment to set up operational bases on the north west provincial border.

Again it had been shown that when operating close to battalions in a fluid Task Force operation patrols had to be very flexible. On occasions patrols were inserted without completing the normal preparation cycle, and patrols had also had been extracted rapidly to allow the battalions to manoeuvre. The SAS did not like these situations, but they were inevitable when operating in the restricted area of Phuoc Tuy Province.

While the SAS was now putting its main effort in the north west of the province it maintained surveillance over the eastern area, and on 27 July Sergeant O'Farrell's eight-man patrol was inserted five kilometres south east of Nui May Tao. The following morning they sighted five enemy at a range of ten metres. Two were washing in a creek. An OP was established and O'Farrell reported a suspected camp. On orders from Nui Dat, they set up a time-delay device but the airstrike was delayed and O'Farrell and Corporal Clive Mason had to crawl back into the occupied camp to reset the time-delay device. At 5 pm on 29 July an airstrike went in, followed by RAAF gunships which struck the camp successfully. Next morning the SAS entered the camp and recovered documents showing that the enemy were from C3 Company D445 Battalion. In his patrol report O'Farrell wrote: 'Too long a delay elapsed between the detonation of the delay device and the time until the airstrike went in. The airstrike was 100 metres to the east of the camp and left the bunkers intact. In [my] opinion...the enemy had been in the area for some time and felt quite secure. [I] recommend that the area be left alone for a few days, then another airstrike be conducted'. Chipman later recalled that O'Farrell's 'patrol was a fine effort and in my view was information wasted by an air strike. This information justified a battalion operation against the camp but at this stage...the least possible "friendly casualty action" appeared to be the one selected by the Task Force'. Patrols deployed to the area during August found that the camps had not been re-occupied.

Meanwhile, the patrols continued across the border in Long Khanh because, as Sergeant Frank Cashmore said later, 'we had hunted out our province and there were no VC left'. During a reconnaissance flight ten kilometres south west of Black Horse, near where A Company 3 RAR had located the enemy convalescent camp, he saw what appeared to be a well-used track and sought permission to lead an ambush patrol to that area. On 12 August his ten-man ambush patrol was inserted and before last light they located a north–south track showing signs of recent movement in both directions. They withdrew to an LUP for the night, awoke at 4.00 am and cooked a hot breakfast in pouring rain. They then returned to the track and soon after first light set up a Claymore ambush with an M60 machine-gun on each flank.

Within an hour enemy were sighted moving south along the track. They were moving tactically with two scouts armed with AK47s 20 metres in front of the main body. When the first man reached the left or southern flank of the killing zone Sergeant Brian Blake, second-in-command of the patrol, sprang the ambush. The first four enemy were killed instantly, and after the M60s had raked the area Cashmore gave the order to cease fire. Corporal Graham Smith, manning the M60 on the right flank, yelled back that there were more of them, and continued firing. For some inexplicable reason Cashmore then ordered the remainder of his patrol to swing around in extended line and to assault the enemy. During his first tour of Vietnam in 1968 Cashmore had felt that he 'was operating in Charlie's backyard'. But in 1971 he believed that he was operating in his 'own backyard'. He was supremely confident, until then he had seen few enemy, and he was not going to let this group escape. With both M60s firing they advanced on the VC over broken ground; however within a few minutes Cashmore's radio operator started yelling that he would get them all killed. Cashmore told him to 'shut up', but when Smith quietly stated that this was not the sort of job the SAS was supposed to do Cashmore called a halt to the attack, having wounded at least three more enemy.

Meanwhile Major Chipman was heading towards the contact site in a helicopter and when he arrived over the area Cashmore told him that the VC had headed north. By about 8.20 am a heavy fire team was operating along the possible escape routes and between 8.30 and 11.30 am four airstrikes were directed onto the area. Artillery then continued the bombardment as it was thought that because of the early time of contact an enemy camp could exist within 1000 metres of the contact area. By now Cashmore's patrol had recovered four packs, two transistor radios, five weapons and a quantity of documents, and had been extracted to Nui Dat.

Although Cashmore himself put the enemy's strength at over 100, the official report stated that his patrol had engaged about thirty VC, and the captured documents indicated that they were members of 3 Battalion 274 Regiment heading towards an important conference. One of the dead was probably the assistant political officer of the battalion, and another, possibly from the battalion rear services staff, was carrying 30 000 piastres.[4]

As a reward, Cashmore's patrol was informed that it could spend a few days at the Peter Badcoe Club at Vung Tau. Meanwhile, the intelligence staff examined the documents and Brigadier McDonald ordered the SAS to return to the area. The squadron was now in the midst of a change-over of some 25 personnel. As Chipman wrote at the time 'the replacements are mostly national servicemen who did not complete the normal training cycle in Australia for Vietnam and who are therefore very much learners at this stage'. There was no alternative other than to cancel Cashmore's patrol's trip to Vung Tau, and on 15 August they were re-inserted into the area. At about the same time another ambush patrol, commanded by Sergeant Dave Scheele, was inserted nearby.

For two days Cashmore's patrol lay in ambush without result. They then patrolled north and observed the effect of the airstrikes on the withdrawing VC. There were marks on rotten logs where the VC had obviously tried to escape the bombs, and an NVA sandal was prominently displayed on a log at a junction, with the toe pointing to the east. They returned to the site of their earlier ambush and while eating lunch some 60 metres from the ambush area Trooper Bill Nisbett commented on the bad smell. Investigating further, they found that behind the log on which they were leaning was a dead VC, obviously one of the casualties from the ambush. A polished brown leather satchel was attached across his shoulder and he was wearing a gold Bulova watch. While Nisbett went for the watch Cashmore cut off the satchel. It contained 120 000 piastres and six maps with arrows. Despite Cashmore's advice to Nui Dat that he had 'vital information' he was directed to remain in the jungle. While in ambush on 19 August the patrol signaller received a perplexing message which read MAC DEC HOM. He took it to Cashmore who informed him and the rest of the patrol that it was a pre-arranged message from the squadron intelligence sergeant indicating that the Prime Minister, McMahon, had announced that the Task Force was to be withdrawn from Vietnam before Christmas. The ambush tasks continued until 21 August by which time, according to the patrol report, they were 'extremely tired'.

The dead VC was the assistant chief of staff of 274 Regiment, and the documents provided an extensive and up-to-date picture of 274 Regiment's condition and intentions. The regiment was moving into position to attack the populated areas in northern Phuoc Tuy to disrupt the South Vietnamese elections due to begin on 29 August, and 4 RAR was deployed to counter this threat.

While Cashmore was operating near the site of his successful ambush, Dave Scheele's patrol was active about 3 kilometres to the south. On 17 August they located an enemy camp, saw two VC and established an OP for 24 hours. Continuing their patrol they found an unoccupied but recently used bunker system and established an ambush position. On 20 August one VC walked into the ambush area and was killed. The patrol was extracted later that day and re-inserted on 23 August to assess the effect of airstrikes on the two camps they had reported. The airstrikes had missed, the artillery had been more accurate, but no damage had been done. While communicating with Nui Dat three VC approached the patrol and in the ensuing contact one was killed and a quantity of documents was recovered.

The news that 2 Squadron was to cease operations on 6 October and return to Australia on 10 October was welcome but it was tempered by a rather startling signal from Perth on 18 September that the squadron

387

was to be disbanded on return to Australia. As Chipman wrote in his commander's diary: 'This had had a natural adverse effect on the morale of the unit as the future of every man was at stake. In an attempt to overcome the lack of information I informed every man in the unit that he would be kept informed when further information came to hand and that the disbandment was probably related to manning problems which when overcome would result in the re-raising of 2 SAS Squadron'. He was to not know that it would be over ten years before the squadron would be re-raised. A week later Chipman received a signal from Perth stating that all members of the unit would be absorbed into the regiment. The commanding officer, Lieutenant-Colonel Clark, informed Chipman that the disbandment was 'out of the blue' and that the regiment had not been consulted.

Operationally, the last month in Vietnam was quiet. Twelve reconnaissance and fifteen ambush patrols were mounted but there were no sightings or contacts. The SAS did not operate in an area where it could have detected that 33 NVA Regiment had re-entered the province, and it was not involved in Operation IVANHOE, the Task Force's last operation in which 4 RAR was engaged in heavy contacts with the North Vietnamese north of Binh Gia.

On 18 August Second Lieutenant Richard Gurney, a 22 year old Portsea officer, had replaced Second Lieutenant Lawson-Baker as squadron signals officer. And the next day Second Lieutenant Lindsay Hansch arrived as a replacement for Lieutenant Brett who had returned to Australia to begin a training course overseas. Aged 24, Hansch had served as a lance-corporal with 5 RAR in 1966 and 1967 and had graduated from Portsea in December 1969. He was to have the honour of commanding the SAS's last operational patrol. On 1 October his ten-man fighting patrol was inserted about seven kilometres east of the Courtenay rubber plantation with the mission of intercepting enemy withdrawing from the area of Operation IVANHOE which concluded on 2 October. There was no sign of enemy activity within the previous week. At 9.24 am on 5 October the helicopter extracted them from an LZ two kilometres south of the Long Khanh border.

By a strange coincidence the last operational patrol had been commanded by an officer who had been serving, admittedly as an infantry rather than as an SAS soldier, in Phuoc Tuy Province when 3 Squadron had begun operations in 1966, five years earlier. Meanwhile, serving with 4 RAR in Vietnam in 1971 were Captains Peter Schuman and Trevor Roderick, two of the troop commanders from 1966. Major Geoff Chipman also spanned the beginning and the end of the SAS in Vietnam, having been second-in-command of 3 Squadron in 1966.

388

From this advantageous position his final comments in his Commander's Diary are worth stating in full:

> This final narrative would not be complete without paying tribute to the magnificent assistance received by 2 SAS Squadron on their second operational tour of duty in South Vietnam. Firstly to the commander and staff of 1 ATF who have at all times been extremely helpful in all matters. Secondly to the officers and airmen of 9 Squadron RAAF who have consistently closely supported the squadron both on and off operations. Without doubt over the years in Vietnam there has developed a unique special relationship between 9 Squadron and SAS at all levels which I believe has been firmly established for all time. Thirdly to the officers and men of A Squadron 3 Cavalry Regiment who have inserted and extracted numerous SAS patrols when it has been neither possible nor ideal for the RAAF to do so. It is indeed most comforting to have had their firepower and mobility at our disposal. Fourthly the pilots of 161 (Independent) Reconnaissance Flight who have consistently, speedily and professionally flown in support of SAS patrols that have needed ground air communications, regardless of the time of the day or night. Finally to all the other units that we have had the pleasure of being associated with on operations at varying times throughout the tour.
>
> In conclusion I would like to say that in my opinion the SAS within 1 ATF in South Vietnam have at all times been a most valuable asset for any commander; in fact all Task Force commanders have had a division's allocation of SAS. However, the scope for employment associated with the extensive skills and training of the SAS trooper justify and permit more advanced employment of this specialist unit when committed to operations in the future.

These sentiments were supported by almost all those involved with the SAS in Vietnam.

Brigadier McDonald was highly satisfied with the work of 2 Squadron during 1971. In his view the deep patrolling by the SAS had deterred the enemy from entering Phuoc Tuy Province, and he wrote later:

> No large operation to bring the enemy to battle was undertaken until such time as all available intelligence was fully analysed and tested through targetting operations (albeit this was a continuing process). The role of the SAS in this collection was of dominant importance. I could rely on the accuracy of their reports to either confirm information or add to the overview which had been built up of the enemy and his habits.
>
> Perhaps I used the SAS unremittingly but again I demanded the same operational capacity of the two battalions.

Despite the fact that the world was told that the Task Force would be withdrawn by the end of 1971, I saw my task as ensuring that no pre-condition would or could exist to prejudice the Task Force throughout the period of my command.

On the day of withdrawal from Phuoc Tuy more than half the Task Force was on operations within the province. The SAS was withdrawn simultaneously with all those troops covering the withdrawal.

On 7 October the advance party of 2 Squadron left Vietnam, and on the morning of 10 October the main body flew to Vung Tau by Caribou before boarding the C130 for Australia. During seven months of operations they had conducted 100 reconnaissance and 67 fighting patrols. Although they had killed only sixteen VC, they had nonetheless made a substantial contribution to the effectiveness of the Task Force. Except for short periods, the VC were absent from the province waiting for the Australians to withdraw. In three of the four major incursions into the province they had been detected by the SAS in time for the remainder of the Task Force to mount operations against them.

The return of 2 Squadron to Australia concluded six and a half years of operational service in Borneo and Vietnam. While it is hard to arrive at exact figures, because accounting methods changed from year to year, it appears that in 298 contacts in Vietnam the SAS inflicted 492 kills, 106 possible kills, 47 wounded, 10 possibly wounded and captured 11 prisoners. A total of 5366 enemy had been sighted in 801 sightings. The Australian and New Zealand casualties in Vietnam were one killed in action, one died of wounds, one killed in a grenade accident, two accidentally shot while on patrol, one missing during a rope extraction, and one died of illness. Twenty-eight SAS soldiers received wounds serious enough to be notified.

The types of patrols conducted in Borneo and Vietnam by Australian SAS squadrons are shown in the following table:

Type of patrol	Vietnam		Borneo	Total	
	Aust	NZ	Aust	Aust	NZ
Reconnaissance	678	76	38	716	76
Recce–ambush	238	47	7	245	47
Ambush	131	6	3	134	6
Fighting	86	1		86	1
Surveillance	35		21	56	
Special	7		3	10	
Psy Ops			14	14	
	1175	130	86	1261	130

Obviously some squadrons described a patrol with an ambush mission as a fighting patrol while other squadrons decribed it as an ambush patrol.[5]

These figures underline several aspects of SAS operations. Firstly, that they were devoted substantially to reconnaissance and surveillance—a fact that would no doubt have pleased the purists. Secondly, when required, the SAS could undertake offensive action. And the extremely favourable ratio of enemy to friendly casualties attests to the effectiveness of SAS training and combat expertise. Thirdly, there were a very large number of patrols conducted by only a relatively small number of men. During a period of just over five years a total of some 580 SAS soldiers served in Vietnam. There was therefore ample opportunity to build up a considerable store of experience, enabling the SAS to refine and further develop its patrol techniques.

This operational experience was used to determine the most effective size for an SAS patrol. For example, out of 1391 missions, 197 were conducted by four-man patrols, 732 by five men and 263 by six men. Only nine patrols were over twenty men. Patrol missions extended in time from a few minutes (when patrols were surprised on an LZ) to over 80 days. But 728 patrols (over half) lasted from four to seven days. The patrols in Borneo were generally longer with the majority lasting for about fourteen days.

From the time of its formation in 1964 until the withdrawal from Vietnam in 1971 the SAS Regiment had directed almost all of its energy towards preparing for operations; first in Borneo and then in Vietnam. The operations had imposed a tremendous burden on the training resources of the new regiment. While there were periods of frustration, for most soldiers it had been a time of exciting professional development. The SAS had met the challenge of over six years of active service and had established its reputation. But the frustrations of peacetime soldiering were to present perhaps an even bigger challenge. Its long term reputation would eventually rest on how it met this new challenge.

22
The defence of Australia: 1972–1981

The end of Australia's commitment to the Vietnam war brought mixed feelings to the members of the SAS. Most were regular soldiers and many had completed two, some even three operational tours of Vietnam. For seven years the regiment had operated on a war footing and training had been urgent and realistic. The operations were a challenging test of the SAS soldiers' professionalism, courage and initiative. It would be no easy task to return to peacetime soldiering.

Yet the SAS needed a rest. It had become increasingly difficult to provide trained soldiers for the operational squadrons, and frequent absences from home had put a strain on the family life of many soldiers.

Vietnam had also brought frustrations. While SAS officers and senior NCOs had reservations about the effectiveness of the roles they had been directed to undertake in Vietnam, they nonetheless were proud of their efforts. It was disappointing to learn that the rest of the Army did not always seem to share the view that they had performed well in Vietnam. The fact was that the SAS had still not been accepted fully as part of the Australian Army. And obviously it would not be accepted fully until it was clear to all that it had a role to play in the defence of the nation.

Contrary to popular belief, the formulation of a new Australian defence policy was underway well before before the announcement in 1971 of Australia's withdrawal from Vietnam. There were a number of contributing factors. Following the ousting of President Sukarno in 1966 Indonesia was no longer seen as a threat, and the requirement to station ground troops in Malaysia had become less persuasive.

The defence of Australia

Britain's decision to withdraw from east of Suez demanded a re-appraisal of Australia's policy in the Malaysia–Singapore area. And in 1969 President Nixon stated at Guam that in future the United States would expect its allies to provide substantially for their own home defence.

But it took some time for Australia to determine a new defence policy, and in the meantime the SAS Regiment had to continue to support its operational squadron in Vietnam. Nevertheless, senior officers in the SAS were aware that they would have to change the regiment's training to take account of the new strategic environment. As described in Chapter Seventeen, in 1970 Lieutenant-Colonel Clark had advocated increased attention be given to long range vehicle pa-trolling in Western Australia, and the SAS had indeed begun this training. However, it was still not clear whether Australia would con-tinue with forward defence or withdraw into a form of continental defence, described by some journalists as 'Fortress Australia'. Until this broader picture was drawn it was difficult to define a role for the SAS.

A new role for the SAS was one of the major problems facing Lieutenant-Colonel Ian McFarlane when he succeeded Lawrie Clark as commanding officer on 26 January 1972. Aged 38, McFarlane had graduated from Duntroon in December 1954 and had served in vari-ous infantry units before joining the SAS Company as a platoon com-mander in July 1962. He was later second-in-command of the company. In March 1964 he had been posted to 1 RAR and had commanded a rifle company during the battalion's first tour of Vietnam in 1965 and 1966. Before his appointment as commanding officer he had been an instructor at the Army Staff College.

By the time McFarlane assumed command the SAS commitment to Vietnam had ended. 2 Squadron had returned, had been disbanded, and its members had been absorbed into other squadrons, particularly the Training Squadron which, as described in Chapter Seventeen, had been formed towards the end of 1971. The raising of specialist wings within the Training Squadron swung the emphasis back to specialist training. This period also saw the re-introduction of troop specialis-ation into free fall, water operations, vehicle-mounted and climbing.

McFarlane recalled that when he assumed command the 'SAS was a dirty word in the army'. Officers contemplating joining the SAS were often warned that such a posting would curtail their military career. No-one at Army Headquarters seemed interested in the regiment and Headquarters Western Command seemed intent on finding fault with the unit. The lack of direction was compounded by the decision in

393

Lieutenant-Colonel I. D.
McFarlane, Commanding
Officer of the SAS
Regiment from 26 January
1972 to 9 December 1973.
Despite the lack of support
from Army Headquarters
he developed long range
desert patrols and the
Unconventional Warfare
Wing.
(PHOTO, I. MCFARLANE)

1973 to re-organise the Army on functional lines, and the SAS was to
come under command of the newly formed Field Force Command in
Sydney.

One result of the lack of direction was that McFarlane had to
initiate and control his own regimental exercises. North West Cape
and Onslow was the setting for the first regimental-sized exercise in
April and May 1972. After years of jungle warfare the exercise in open,
arid terrain was a new experience for many soldiers. Squadrons practised vehicle-mounted patrols as well as the more traditional foot
patrols. Regimental headquarters was located at Learmonth with 1
Squadron at Exmouth racecourse and 3 Squadron at Onslow. Freak
winds and rocky ground caused a number of serious injuries during
parachuting activities.

By the end of 1972, the regiment was still lacking direction from Army Headquarters, as noted in an article in the *Army* newspaper on 14 December 1972. 'SAS on parade, but what of the future?' was the newspaper's headline, and the article continued: 'When soldiers of the Special Air Service Regiment go on parade today for the Beating of the Retreat ceremony there will be a question mark in the public mind. Now that the regiment has ended its service in Vietnam—what is its main role?' In the article McFarlane was quoted as stating that the regiment would continue carrying out exercises which would prepare it for any situation. This might mean going to New Guinea, the Kimberleys, the Gibson Desert or some other remote spot. Referring to an exercise in the Gibson Desert he added: 'It is harsh country and poses particular problems to soldiering. We learn the best way of overcoming these, and at the same time raise the general standard. It is essential that we should know what conditions are like in these remote areas'. However, McFarlane also commented that with the end of operations in Vietnam the SAS could concentrate on specialist training.

The formation of Training Squadron with its specialist wings; waterborne operations, parachuting, reinforcement, climbing, demolition and operational research, meant that the regiment took on many of the characteristics of a special warfare school. The development of these skills enabled each troop in the two remaining sabre squadrons to be given specialist responsibilities. Thus before long there were water operations, climbing, vehicle-mounted and free fall troops. Brigadier Rod Curtis, then a squadron commander, recalled that there was 'renewed interest in vehicle mounted operations and the development of the Long Range Patrol Vehicle (LRPV) including patrol and troop procedures. Leading this development was one of my troop commanders, Lindsay Hansch, and his troop sergeant, Sergeant Garvin. Together with McFarlane's support, they were responsible for modifying in-service landrovers and converting them to LRPV. These modifications were initially undertaken without Canberra's authorisation or support for which the Regiment and the Commanding Officer was severely criticised, but the initiative taken was necessary to provide new direction for the Regiment, and Hansch's well-considered work in establishing a long range vehicle mounted capacity has stood the test of time'.

The lack of support from Army Headquarters was also shown by the difficulty experienced in gaining Military Board approval in December 1972 for six members of the regiment to participate as an Army sponsored regimental team in the Australian National Free Fall Championships. And sponsorship did not mean that money was available to support the SAS team.[1]

In the early 1970s, following the end of the Vietnam commitment, the SAS concentrated on long range patrols across north west Australia. This Series 2 Land Rover has been set up as a Long Range Patrol Vehicle.

The emphasis on operations in remote areas of Western Australia continued with Exercise APPIAN WAY in July and August 1973. In the largest special forces exercise held in Australia to that time 900 men from the SAS Regiment, 1 and 2 Commando Companies, a rifle company and elements of HQ 8 RAR, and Airfield Defence Guards were deployed to an area 200 kilometres north west of Derby. The exercise was designed to test special force units, RAN patrol boat and RAAF Caribou and Iroquois crews in specialist techniques, including the conduct of raids and surveillance in tropical waters and rugged subtropical terrain. The exercise was the first practical test for the regiment's vehicle-mounted troop with its newly modified Land Rovers. The SAS also began experimenting with tactical motorcycle patrols.

The Chief of the General Staff, Lieutenant-General Sir Mervyn Brogan, visited the exercise and McFarlane wanted to demonstrate the modified Land Rovers, the motorcycles, parachute drops and camouflage techniques. Unfortunately, a Land Rover ran over a camouflaged soldier, one of the air drops went astray, and generally the demonstration was unconvincing.

To support future operations in northern Western Australia Warrant Officer Roy Weir was given the task of researching survival methods in this remote area. His work predated that of the 'Bush Tucker Man', Major Les Hiddins, whose documentary series was shown on ABC television in 1988. With an Aboriginal elder, Sam Woolagoodja, Weir toured through 4800 kilometres in the Kimberley Ranges learning Aboriginal methods of living off the land.

While Exercise APPIAN WAY provided many valuable lessons, it was clear that a role for the SAS in Australia had still not been resolved, and this was revealed in a newspaper interview with Lieutenant–Colonel McFarlane in October 1973. He was quoted as saying that it was 'most likely that the SAS would not fight on Australian soil. Their primary task [was] to operate in patrols of as few as five men, isolated behind enemy lines'. The article noted that: 'The SAS did this in Malaysia and Vietnam to telling effect—if any SAS man cares to tell, which he invariably doesn't. "Most of our men don't talk much", the colonel admitted laconically'.[2]

During APPIAN WAY a number of local people had complained of strange boats landing in remote localities and McFarlane wrote a paper suggesting that a Citizen Military Force unit be raised from local people, regardless of age or fitness, to watch the coast and to undertake some training for guerilla warfare. Nothing further was heard of this suggestion.

In another paper, circulated within the Army, McFarlane outlined the capabilities of the SAS. While arguing for the proper use of the SAS he stated that the SAS had been misused by the Task Force commanders in Vietnam; rather than conducting offensive operations, they should have concentrated on clandestine reconnaissance and the direction of airstrikes onto enemy positions. These papers did not win

McFarlane many friends at Army Headquarters. Indeed he recalled one staff officer, during a trip to Canberra, telling him that there was no place for the SAS in the Army. It would have to be disbanded.

In August 1973 McFarlane forwarded to Army Headquarters another paper entitled 'The Employment of Special Forces in the Defence of Continental Australia'. It was one of the first attempts to define a role for the SAS in the new strategic environment that had developed after the withdrawal from Vietnam and the election of the Whitlam government in December 1972. McFarlane argued that SAS skills could be used to counter urban guerillas and ransom threats, and should be employed on long and medium range reconnaissance. They could provide advice to the police on methods of dealing with terrorists, they should familiarise themselves with remote geographic areas including the coastline, and could provide information on civil resources.[3]

Although McFarlane's paper was not accepted as a blueprint for SAS training, he decided to form an Unconventional Warfare Wing within Training Squadron and directed Sergeants Greg Mawkes and Paddy Bacskai to conduct a two-week special warfare course. It became obvious that an officer was needed to command the wing and Captain Tony Tonna returned from the Jungle Warfare Centre at Canungra in Queensland. So little was known about unconventional warfare that when he received his posting to the 'UW' Wing Tonna thought that he had been posted to an underwater operations wing.

Tonna and his two sergeants set out to learn all that they could about unconventional warfare and special operations, and early in 1973 McFarlane decided that rather than a two-week course the wing would put ten SAS NCOs (sergeants and corporals) through a series of unconventional warfare courses for about one year and thereby form a special mission team, somewhat like the A Teams in the US Special Forces. The courses included basic unconventional warfare, tactics, survival, intelligence, foreign weapons, demolitions, sniping and close-quarter fighting. Although quiet and sad faced, Tonna was a dedicated soldier and an excellent trainer, and in due course he was producing highly capable special mission teams, each of ten to twelve men.

This development again focused attention on the old problem of differentiating between SAS and Commando operations. It will be recalled from Chapter Seventeen that in 1969 there had been discussions at Army Headquarters as to the possible amalgamation of the SAS and the Commandos into a Special Action Force Group. The roles of both the SAS Regiment and the Commandos covered unconventional warfare, including guerilla warfare, the organisation of rescue, evasion and escape and sabotage. Since Australian doctrine for unconventional warfare had not been developed, various US pamphlets were obtained for use by the SAS and the Commandos. McFarlane arranged a conference with the commanders of the Commando companies to try to co-ordinate SAS and Commando activities, but firmer direction was needed from Army Headquarters.

While some officers believed that there was little difference between the roles of the SAS and the Commandos, others were adamant that the SAS was designed for long range reconnaissance in small teams who generally avoided contact, while the Commandos had the role of conducting raids in groups ranging from a few men up to more than 100 men behind enemy lines. The case for the Commandos was put in an article entitled 'The Case for Special Forces' by Major Philip Bennett in the *Australian Army Journal* in 1964. Bennett later reached the rank of general as Chief of the Defence Force. In the late 1950s he had undergone Commando training in Britain and had then commanded 2 Commando Company. He argued that the Commandos had to become part of the regular force. Their tasks would include reconnaissance, harassing, destruction raids and raising local anti-guerilla forces, and they would form the basis of an efficient intelligence network. 'If it means teaching underhand tactics such as assassination and spying', he concluded, 'let us teach them. The next war will not be fought in accordance with agreed rules'.[4]

The Vietnam war changed the perception that the SAS was concerned only with reconnaissance. Many members of the SAS had served with the Training Team in Vietnam and had seen that the potential for special operations went well beyond that displayed by the SAS in Phuoc Tuy. Operations included raising and training indigenous units, particularly as irregular forces, raiding by small strike forces, psychological operations and recovery of prisoners or equipment. These operations were, in the main, conducted by US Army Special Forces, but were also conducted by the CIA and by US Navy SEAL teams, with whom the SAS had had an exchange programme. In 1969 a paper published by the Rand Corporation advocated the use of small strike teams behind enemy lines. The paper stated that strike teams in Vietnam included 'among other, the two USMC [US Marine Corps] reconnaissance battalions in I Corps, the hundreds of US Army LRRPs in I, II and III Corps, the Australian SAS in III Corps, several Special Forces detachments working in the central Highlands with the Vietnamese Civilian Irregular Defense Groups, the Vietnamese Provincial Reconnaissance Units and the US Navy SEAL platoons in III and IV Corps'.[5]

The paper did not mention the involvement of the CIA, but in recent years numerous published sources have shown that the CIA played a prominent role in unconventional warfare in Vietnam. The history of the Australian Army Training Team in Vietnam published in 1984 revealed that Australian Army officers and NCOs were seconded to the CIA both to raise indigenous forces and to conduct operations against the VC infrastructure, described loosely as the Phoenix Programme. Captain Barry Petersen, who operated on behalf of the CIA in the Central Highlands with the Montagnards, has described his experiences in his book *Tiger Men* published in 1988. One of his assistants was an SAS warrant officer, Danny Neville. A leading figure in the establishment in 1965 of People's Action Teams under the CIA umbrella was Captain Ian Teague who later commanded 1 SAS Squadron in Phuoc Tuy. Undoubtedly, the experience gained by the SAS in the Training Team led to the development in the Unconventional Warfare Wing of skills which until that time had been seen as the province of the Commandos.

The SAS interest in operations in northern Western Australia and in unconventional operations continued when Lieutenant-Colonel Neville Smethurst assumed command on 10 December 1973. After a year, the Unconventional Warfare Wing had prepared two special mission teams and was able to form an unconventional warfare troop. Smethurst decided to retain the troop and began training another. Before the end of 1974 B and C Troops in 1 Squadron had been designated unconventional warfare troops.

But before long Smethurst had become concerned that some members of the Unconventional Warfare Wing were too melodramatic about their activities. Years later, as a major-general and Land Commander, Australia, he recalled that it 'was all too secretive for my liking. On one occasion an NCO even tried to slip me a training programme while I was having a drink in the Sergeants' Mess'. When Smethurst asked why the programme was not being forwarded through the squadron commander he was told that it was too secret for him to see it. Smethurst decided to put this training 'on the right plane' and changed the title of the wing to Guerilla Warfare Wing. Most of the skills and courses were retained, but the training was directed more towards preparing the SAS to conduct irregular warfare behind enemy lines rather than clandestine operations in a para-military environment. Furthermore, Smethurst directed that all unit personnel could undergo various selected elements of the guerilla warfare training cycle, thereby demystifying these activities. Meanwhile, B and C Troops were converted to water-operations and vehicle-mounted operations respectively. Initially it was suggested to the squadron commander that he raise a mule troop from one of the former guerilla warfare troops.

Smethurst had an additional purpose in disbanding the guerilla warfare troops. If the SAS was to survive in a period when the Army was contracting, it had to be seen to be an integral part of the Regular Army, not some special, secretive separate unit. Smethurst was well equipped to institute these changes. Aged 38, he had graduated from Duntroon in 1956 and had served in the SAS company from 1958 to 1960. For his second year in the company he had been Adjutant and Quartermaster. But for the next fourteen years he had had no direct contact with the SAS. As a company commander he had had two operational tours in Borneo and had been second-in-command of 7 RAR in Vietnam. He therefore returned to the regiment as somewhat of an outsider, but as the GOC of Field Force Command, Major-General Don Dunstan, had told him, the regiment 'needed sorting out'. This comment was somewhat unfair to McFarlane who had battled lack of interest of Army Headquarters and widespread antipathy towards the SAS. Perhaps Smethurst would receive greater support from higher formations than McFarlane had received.

The first problem, as McFarlane had also realised, was the quality of the officers. In the post-Vietnam period the regiment had attracted, or had been sent, a number of officers who were not suitable for the SAS and needed to be removed. In addition, Smethurst was concerned that the young lieutenants in the regiment did not have the maturity or experience to command older, more experienced and mature soldiers who were not afraid to give their views. After posting to the unit, young officers were required to undergo the basic selection course with other non-officer SAS applicants, but there was no guarantee that they would pass. Even if they did, it did not provide a good basis for later command within the regiment. Although many officers appeared suitable on paper, they often failed to qualify either through lack of physical fitness, lack of individual fortitude and motivation, being temperamentally unsuitable to a special forces environment, or being psychologically unable to cope with some specialist skills such as parachuting or diving. Failure to qualify had a damaging effect on the individual's morale, and often the officer had already moved his family and possessions to Perth.

To help remedy this shortcoming, in August 1974 the first SAS Officers' Selection Course was conducted. The idea was that officers would be sought initially from infantry units and later from all corps. If suitable they would undergo the Officers' Selection Course, which involved, among other things, thinking and planning under pressure. After completing the course they would be posted to the regiment once a vacancy occurred, and could go on to attend the basic SAS selection and training course in which they would learn patrol techniques. Having already completed the Officers' Selection Course it was almost certain that they would do well and take a leadership role during the

Major-General Neville
Smethurst, AO, MBE,
Commanding Officer of
the SAS regiment, 10
December 1973–6 January
1976. One of his major
innovations was the
establishment of the
Officers' Selection Course.

basic selection course. Furthermore, the conduct of the Officers' Selection Course required the assistance of NCOs and soldiers from across the entire regiment, not just from Training Squadron, and these NCOs and soldiers witnessed at first hand the gruelling, demanding and at times exhausting series of tests the candidates were subjected to. There could no longer be any doubts among the soldiers about the quality of their officers. The introduction of the course was to have a profound effect on the quality of future officers of the SAS.[6]

In addition to introducing the Officers' Selection Course, Smethurst proposed that troop commanders be promoted from lieutenant to captain. It took a little longer to win approval for this change to the establishment, but before the end of 1975 all troop commanders were captains.

Smethurst was also concerned about the NCOs in the regiment. A number were so ingrained in their old ways that they could not adapt their thinking to the post-Vietnam requirements. Many more had been in the regiment for ten or even fifteen years. They were good men but were blocking promotion within the unit. Smethurst believed that they needed to move out into the rest of the Army where they would show the Army that SAS NCOs were of the highest quality and would help attract good quality applicants back to the regiment. In addition, he would therefore be able to promote some of the good young soldiers who were becoming dispirited at the lack of opportunities within the unit.

Finally, Smethurst identified a number of potential RSMs who needed to gain experience away from the SAS. They could then come back and exert strong control over the extremely powerful Sergeants' Mess. Even more than an infantry battalion, the sergeants formed the backbone of the unit; in an infantry platoon there were one officer and one sergeant while in an SAS troop there were one officer and four sergeants. And the sergeants, as patrol commanders, were commanders in their own right, usually reporting directly to squadron headquarters when on operations. A later commanding officer thought that Smethurst's greatest contribution to the regiment was in setting 'the foundation for the transfer of *power* from the Sergeants' Mess to the Officers' Mess'.

Smethurst recalled that he removed fifteen senior NCOs from the regiment: 'I was not very popular in the Sergeants' Mess, but I knew that it had to be done'. His RSM, Lawrie Fraser, observed that Smethurst 'came in to get rid of the dead wood and was particularly ruthless. He was highly professional and taught me how to be an RSM'. Major Reg Beesley, who had served many years in the regiment and was to return as commanding officer several years later, recalled that before Smethurst had taken command he had discussed with him the need to replace the men who were 'burnt out or getting old'. It was a hard task but someone had to do it.

On top of these worries, the regiment, along with most other Army units, became the subject of severe manpower restrictions, and the authorised establishment was set at 272 personnel, not including 152 Signal Squadron. Suddenly, without prior discussion, Army Office deleted the Training Squadron from the establishment. After considerable argument the squadron was restored with a slightly reduced establishment and with its name changed to Support Squadron. The implication was that there was no role for a training squadron within a field force unit but there was a role for a support squadron.

Until the end of the Vietnam War the SAS had had few visits from allied SAS-type units, but during the early 1970s valuable individual exchange programmes were developed, particularly with the Special Boat Squadron (SBS) of the Royal Marine Commandos and the US Navy Seals. Troops from the British and New Zealand SAS, the SBS, the Malaysian Special Service Regiment, and even the Indonesian RPKAD visited Swanbourne. In March and April 1975 1 Squadron, commanded by Major Bill Hindson, exercised with 22 SAS, NZSAS and 28th Brigade in Malaysia, and B Squadron 22 SAS joined the regiment on Exercise WINDFALL near the newly acquired training area at Yampi Sound.

From the British visit the Australians discovered two important facts. Firstly, that their standards were as high as those of the British SAS; and secondly that the British had begun counter-terrorist (CT) training. The British visitors included Brigadier John Simpson, commander of the British SAS Group, and Lieutenant-Colonel Tony Jeapes, the commanding officer of 22 SAS Regiment. Simpson gave a graphic lecture to the regiment on the British SAS, highlighting their CT activity, and the Australians wished they could get on with this exciting development, even though Simpson warned that the CT role was a great detractor from the unit's main role.

To some members of the Australian SAS, especially those who had been in the Guerilla Warfare Wing, this new form of training warranted further investigation. Lieutenant Greg Mawkes, by now commissioned with the idea that he would succeed Tony Tonna as senior instructor of the Guerilla Warfare Wing, resolved to try to send two NCOs to Britain on the exchange programme known as LONG LOOK, where they would make every effort to learn all they could about CT training; this was to pay dividends in the future.

SAS training also proved valuable in quite an unexpected quarter. At the beginning of 1974 the Australian Red Cross had sent a medical team to work in a drought-stricken area of Ethiopia, but in late February a grave crisis developed when the Ethiopian armed forces mutinied and demanded pay rises and other concessions. The Ethiopian Foreign Minister even stated that he could not guarantee the safety of delegates to the Council of Ministers of the Organisation of African Unity, then meeting in Addis Ababa. By this time the Red Cross was considering replacing its first medical team with a second, and it asked the SAS if it could send one or two members to accompany the second team, which consisted of Dr David Dallas and three women—two nurses and a nutritionist. The request was forwarded to Canberra but permission was not granted. To Smethurst it seemed both a worthy cause and an opportunity for good training, and he asked the Red Cross whether it could provide the salary for one of his soldiers if he granted him leave without pay. The Red Cross agreed.

Volunteers were soon forthcoming and Trooper Harold Marten was selected. Aged 27, Marten had enlisted in the Army in 1967 and had served as an infantryman in Vietnam. When he had joined the SAS in 1971 he had reverted from the rank of corporal to trooper. Since then he had completed a number of courses, including medic and unconventional warfare. His tasks with the medical team included driver,

SAS Trooper Harry Marten in Ethiopia in May 1974. He had been included as part of an Australian Red Cross team after concern was expressed about their safety. The SAS wings can be seen on his right sleeve. (PHOTO, H. MARTEN)

mechanic and medical assistant, but as he wrote to his squadron commander before leaving Australia on 14 April: 'I personally have been given additional tasks that include finance, communicating with the government and assuring a good working relationship with the military. My only concern is having large sums of money on my person; I have been told that this is to pay workers, interpreters, truck and camel drivers etc. For this reason alone I will carry a current service revolver and I will be fully responsible for that accountable item...I am fully aware of the political and sensitive job ahead of me and I will always uphold the honour of our regiment and understand my duty to the Australian government'.

The team operated for three months in the province of Wollo near the town of Dessi. Some understanding of their experiences can be gained from a report written by Marten towards the end of May:

All of us are following the military revolt very closely on our radio. Although we are situated three hundred miles [480 kilometres] from the capital, there is a garrison at Dessi, which is our only link with the outside world...Our actual position is eighty miles [130 kilometres] north of Dessi and about ten miles [16 kilometres] inland off [the] main road.

I suppose you would like to hear about the work we are doing. We operate as a mobile medical team and initially German Luftwaffe helicopters moved us from place to place. This gave us access to areas never before visited by white people. David and I worked out a system of drop sites and by this method we airlifted grain and milkpowder at least twice per day as well as the medical staff. Problems we did have were roofs blowing away and the locals running away from the helicopters. Eventually, my being able to speak German let me acquire a ground to air radio and thus we operated efficiently. On the other hand for the last nine days we have operated with our vehicles. I dare say this will be the case for the remainder of our stay here.

Marten was right, for the helicopters had been withdrawn by the German government.
He continued:

Currently we are operating in the field or at one of the five clinics we have set up. Our clinics are at Tis-A-Bilima, Wurgessa, Girana and Gafra, the latter being a two-day mule trip through the mountains. Wurgessa and Girana are accessible by four-wheel drive. The weekly roster is worked so that one of the nurses remains at Tis while David and I work with one of the girls. Hence the girls are never on their own.

When arriving at the site we try to organise the crowd into groups according to their complaints. After this we try to get them to form lines; doing this is always very frustrating because all the people are constantly pushing, and when you have three or four thousand it is a constant battle. On many occasions we have just refused to treat the people until they get themselves organised. Once this backfired on us and they rushed us grabbing everything we had. The reason being that Emperor Haile Selassie had told them that he was giving them all the things we had, and therefore we were denying them what was rightfully theirs. We were careful after that. When we do get them organised we operate very efficiently. Many will receive an injection and then move to another line. Subsequently they receive another. This is very hard to control as they all look alike. They believe that the more tablets and needles they receive the better chance they have to get better. Some come back three days in a row for the same complaint. Many have only minor complaints while others have complaints we cannot treat. It distresses David terribly when we cannot help them, especially when some have walked for several days to get to the clinic. We do not like to send them away with nothing, so aspirins and vitamin tablets have to do.

On many occasions I have taken mule trips into the mountains with clothing from OXFAM. On one of these trips a few weeks ago I had ten mules loaded with clothes. Long before I arrived at the village the locals had already known I was coming. Very soon I was in the middle of at least eight hundred people who all wanted clothing, so I approached the chief and took refuge under the biggest hut in the village. Meanwhile outside the crowd were chanting. The walls of the hut started to fall

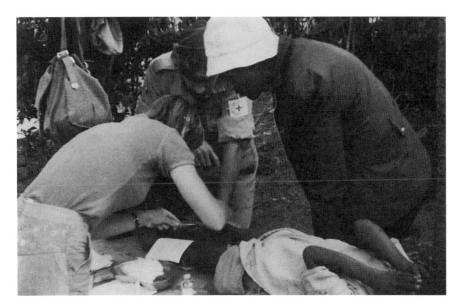

SAS Trooper Harry Marten and a nursing sister, Mrs Doreen Kops, suturing an injured Ethiopian man, May 1974. (PHOTO, H. MARTEN)

down. I was somewhat disturbed at what was going to happen. To end this story I eventually had families come through one at a time and fitted them out. They all wanted jumpers because of the bright colours, even though they would have been terribly hot. One thing I could not stop was the fact that they they would change back into their native clothes and come through again.

During his three months in Ethiopia Marten delivered four babies and undertook one amputation. He was highly commended for his 'outstanding service' by the Chairman of the Australian Red Cross Society.

Despite the wide range of interesting activities taking place within the unit, the search for a role in the new strategic environment continued to dominate thinking within the regiment. The Guerilla Warfare Wing promoted the idea that Australia should plan on conducting guerilla war against an invader. In July 1975 Captain David Mason-Jones, who although no longer with the regiment, had recently been operations officer of 1 Squadron which had two guerilla warfare troops, published an article in the *Pacific Defence Reporter*. He argued that a threat to Australia might develop at short notice and that the Army needed to prepare doctrine for guerilla war and develop plans to train the population, even going as far as to prepare television programmes on specific topics such as how to use weapons. Towards

SAS Trooper Harry
Marten and guide leading
a mule team loaded with
clothing and medicine into
the remote Danakil area of
Ethiopia, May 1974.
(PHOTO, H. MARTEN)

the end of 1977 Captain Tony Tonna, who for four years had been
senior instructor of the Guerilla Warfare Wing, presented similar ideas
in an article in the *Defence Force Journal.* He concluded: We have
never needed guerilla warfare in the past as the prospect of Continental Defence has been a remote possibility. Its relevance to the present
cannot and must not be ignored'.[7]

These ideas did not win wide-spread approval throughout the Army,
nor for that matter within the SAS. Many soldiers resented the elitist
and secretive nature of the Guerilla Warfare Wing which seemed to
absorb a large slice of resources. Other specialist troops, such as climbing, water operations, parachuting and vehicle-mounted operations,
felt that they were more relevant to the possible role of harassing
operations, including troop-sized raids, across northern Australia.
From the beginning the SAS had looked upon the Long Range Desert
Groups from the Western Desert in the Second World War as a model
for operations in Australia. The SAS had exercised in the north west of
Australia for many years, and they knew the problems of operating
there.

Exercises in these remote areas continued throughout 1975, but
Smethurst's tour as commanding officer concluded in January 1976.
He had completed much of the hard and unpopular work in increasing
the efficiency of the unit when many members might otherwise have

rested on their laurels after Vietnam. His lasting achievement was the introduction of the Officers' Selection Course, which has endured and has ensured a consistently high standard of officer in the regiment. His successor could build on this strong foundation.

One important facet of Smethurst's command was his open door policy for the unit. Invitations were actively distributed to senior Army officers to visit the regiment, either at Swanbourne or on exercise and to receive detailed unit briefings. This approach was in concert with the desire of the GOC Field Force Command, Major-General Dunstan, to promote the regiment, and to demonstrate his regard for the regiment he directed Smethurst to provide him with an SAS officer to be his ADC.

That Smethurst's reforms might have ruffled a few feathers among the older SAS soldiers was demonstrated when one recently discharged soldier, who had never served on operations with the Australian SAS, wrote to the Minister for Defence to complain that there was a decline in SAS morale, the unit was conducting the wrong sort of training, and that a special forces directorate should be established in Canberra. The Chief of the General Staff, Lieutenant-General A. L. MacDonald, provided a detailed reply to the Minister which is reproduced here because it provides an informed view of the state of the SAS in mid 1976:

> The training undertaken by the Regiment is related to Defence needs as a whole and, because of this, programmes and syllabi are reviewed periodically. Some years ago, when Australian strategy was to adopt a forward defence posture, its training concentrated on activities likely to be required of it in counter-insurgency situations on the Asian mainland. Subsequently, as a shift in strategic thinking occurred, the thrust was varied to take into account, for example, long range operations in an Australian environment.
>
> The personal characteristics of the officer or soldier who joins the Regiment, which has tended to consider itself to be elite with its emphasis on independent small scale operations and individuality, usually means that he is more inclined than his confreres to probe and question the merits of decisions by persons outside the Regiment. Accordingly, despite the fact that, so far as is reasonably possible, the underlying reasons for changes are made known to the Regiment, they are not always understood or appreciated by all of its members.
>
> There has been no lessening in the standards of entry for soldiers joining the Regiment; moreover, in recent years the standard of applicant has risen. Selection board procedures and the basic SASR training course have improved and, not withstanding the better standard of applicants, the average pass rate has remained at slightly under 50 per cent over the years 1971 to 1976.

I am quite satisfied with the standard of officers. They are required to undergo a rigorous selection course and then receive the same basic training as the soldiers. I might add that the greatest number of officer volunteers on record has applied to appear before the next selection board, which hardly supports the allegations of decreasing morale in the Regiment.

Because of changes in the likely employment of the Regiment in war, the range and number of specialist courses conducted for its members have also come under scrutiny...Training standards continue to be rigorous, and high qualification rates at courses reflect the standard of student attending them.

I do not believe that creation of a special forces directorate...is necessary at this time. Direct comparison with the British Army is not valid, if for no other reasons than that there is a vast difference in the strengths of the two Armies and a need in the Australian Army to avoid a disproportionate number of officers serving in staff appointments.

Changes which have taken place in the Regiment have been initiated to enable it better to perform its likely functions in war. The standard of its men—both officers and soldiers—is as good as, if not better than, before. I am satisfied that its morale continues at a high level although obviously, there is at least one member who, for reasons which do not bear close examination, cannot be regarded as a contributor to it.[8]

On 7 January 1976 Lieutenant-Colonel Mike Jeffery assumed command of the regiment. Aged 39, he had last served in the SAS in 1966. After graduating from Duntroon in December 1958 he had been posted to 17 National Service Battalion at Swanbourne and towards the end of 1959 was transferred to the SAS Company. He remained with the SAS until November 1962, during which time he ran some of the early selection courses. After service in 2 RAR and 3 RAR in Malaya he had been ADC to the Chief of the General Staff until he was posted from July 1965 to March 1966 as operations officer at SAS Headquarters Far East on Labuan. He returned to the SAS Regiment as adjutant and in December went to Papua New Guinea as a company commander in the PIR. Later he was a company commander in 8 RAR during its tour of Vietnam and was awarded the Military Cross. After attending the British Army Staff College at Camberley he became commanding officer of 2 PIR.

By the time Jeffery took command of the SAS Regiment he was already an experienced commanding officer, and he had had wide SAS and infantry experience. Tall, clear thinking, confident and politically aware, he had the fortunate task of building on the base provided by Neville Smethurst. Like Smethurst, his ideas did not always win the approval of pockets of old soldiers within the regiment, but he was determined to give the SAS a relevant role in the defence of Australia.

Despite the fact that the regiment was no longer on active service, tough realistic training continued, and the danger of this training was emphasised on 4 March 1976 when Corporal Doug Abbott of 152 Signal Squadron was killed while parachuting at Cunderin after a mid air collision with Lieutenant John Wright (US Navy SEALs). Wright sustained serious injuries. Four months later, on 22 June Corporal Lawrence (Burma) Mealin was accidentally killed during a night raiding exercise when he fell down a 12-metre coal storage pit at the South Fremantle power station. In an attempt to rescue Mealin, Captain Allen Valentine descended the shaft in total darkness—for this he was awarded a Commendation by the Chief of the General Staff.

The Guerilla Warfare Wing continued to conduct its courses but with less emphasis than before. Greg Mawkes sent Warrant Officer Dave Scheele and Sergeant Michael (Red) Webb to Britain on Exercise LONG LOOK, during which, on their own initiative, they completed the 22 SAS Regiment bodyguard course and became close-quarter battle instructors. Lieutenant-Colonel Jeffery was not impressed, telling them that there was no pressing requirement for those sort of skills in the SAS and that they were not to tell anyone what they had done. As Jeffery said later, Mawkes and his men saw the natural progression from close-quarter battle to a CT role faster than he did as he had his mind focused on developing the surveillance role.

Initially the regiment under Jeffery continued developing its capacity to conduct raids against an enemy force that might have landed in northern Western Australia. In the view of Colonel Chris Roberts, then a squadron commander, 'by the end of 1976 the Regiment had developed a very good capacity in infiltrating into an area of operations and conducting troop size raids with a good chance of success, provided the targets were not well-defended by well-trained combat troops. We had also demonstrated the viability of taking aerial resupply in the area without being detected. What had not been properly evaluated was the extraction of the force from the target area'.

The regiment was not to have the chance of developing the concept further. In October 1976 Captain Terry Mellington took a small party to Shoalwater Bay in Queensland for the Australian Defence Force's big exercise KANGAROO II with the task of detecting the landing of the Orange Force. Despite all the RAAF and RAN assets available to seek out the Orange Force invasion fleet, the air and electronic defences of the Orange Force amphibious group were too good. Mellington's party, located on a very small island (more of a rock outcrop), was the first Blue Force asset to identify and accurately locate the Orange

Major–General P. M. Jeffery, AO, MC, Commanding Officer of the SAS Regiment, 7 January 1976–22 October 1977. He was instrumental in developing the surveillance concept for northern Australia, and as Director Special Action Forces prepared the plan for the development of the counter-terrorist capability.

Force fleet. However, they did more than that. They correctly identified the principal vessels, and their locations were so accurately reported that Blue Force was able to 'inflict damage' on the fleet with 'stand-off' missiles. Eventually Orange Force put a naval landing party on Mellington's small and rocky haven but failed to locate the patrol, despite moving close to it. The SAS performance was highly commended by the Blue Force commander, and it demonstrated that the SAS did have a viable role to play in the defence of Australia. Despite all the technological assets available for reconnaissance and surveillance, the man on the ground was needed to confirm or deny the readings from electronic assets.

Politically aware, Mike Jeffery saw the opportunity to acquire a role that was acceptable to the Army hierachy and which fitted the defence thinking at the time, and under his direction the regiment began to develop its surveillance role in northern Australia. In November 1975 the *Army Journal* had published an article describing the operations of the North Australian Observer Unit during the Second World War. Commanded by an anthropologist, Major, later Professor, W. E. Stanner, the unit had the task of surveillance in northern Australia. A few years later, early in 1979, the role of the unit was elaborated in an article in the *Defence Force Journal.*[9]

After two squadrons exercised in the Pilbara region in May 1977, Jeffery's concept of surveillance in northern Australia was tested in a

Field Force Command exercise, LONG VIGIL, in the Northern Territory in July and August. An enemy force of over two divisions, represented by 1 RAR and a Commando company, was landed by RAN ships and RAAF aircraft both to the west and east of Darwin, and advanced on the strategic airfield of Tindal near Katherine, 500 kilometres inland. The SAS mission was to provide information to a corps headquarters located at Tindal on enemy amphibious lodgements, to continue reporting the enemy build-up and breakout, and to impose maximum delay on the enemy once they advanced, particularly by directing airstrikes. The SAS was not informed of the likely landing areas, and more than 200 SAS soldiers were deployed in 22 patrols across 66 000 square kilometres of the north west. Several SAS patrols were detected and two were captured, but it was a highly successful exercise; the invaders were detected and SAS headquarters was kept fully informed. The link between LONG VIGIL and the North Australian Observer Unit in the Second World War was underlined when Professor Stanner was invited to visit the regiment during the exercise.

One patrol was positioned to observe an airfield near the South Alligator River which SAS headquarters had picked as a likely landing site for the invading force. 'We suddenly found ourselves sitting right in the middle of the "enemy" build up, complete with a couple of long range Land Rovers', said the patrol leader, 'but we decided to sit it out'. They sat it out for three days and nights, 'living like dogs on survival biscuits and onions'. A member of the patrol said: 'After the first day it became a challenge; we expected at any minute to get caught. One patrol came within 20 metres of us. The only time we could have a bit of a scratch or a cough was when an aircraft landed with more troops...and that was about every two hours'. Another added: 'We must have been well concealed because on the last night we were there a dingo came within 10 metres of us, and they are pretty hard to fool'. The patrol finally slipped away undetected after passing valuable information on the enemy build-up back to Tindal.[10]

The SAS post-exercise report concluded: 'Notwithstanding the overall success of the surveillance concept as practised by SASR, there is no doubt that the task of coastal and depth surveillance would be at least well handled by territorial/reserve units who could be raised, trained and equipped to operate in their home areas. Such a force could also provide the basis of a guerilla organisation if required and could takeover the functions of coastwatcher and military reporting tasks. A paper further developing this theme is being prepared by SASR'. LONG VIGIL had been an outstanding exercise, and staff of Field Force Command were impressed with the capabilities of the SAS.[11]

While LONG VIGIL was underway the SAS was abruptly reminded that at any time it might be deployed on active operations. On the

The wreckage of the Australian helicopter that crashed in remote, rugged terrain in Irian Jaya while assisting survey operations in July 1977. Two patrols led by Captain Allen Valentine were extracted from Exercise LONG VIGIL in the Northern Territory, and several members were winched in to the crash site to provide first aid to the injured survivors. The pilot had been killed.

afternoon of Friday 29 July news was received in Canberra that an RAAF Iroquois helicopter operating in support of an Australian survey operation had crashed through 60 metres of jungle canopy in mountainous terrain in Irian Jaya. The pilot was killed and four passengers were injured, two seriously. Next morning an RAAF Hercules aircraft collected from Darwin an eleven-man SAS party led by Captain Allen Valentine and continued north to Irian Jaya. Arriving at Wamena the rescue party found that a second RAAF Iroquois had located the crash site. But because of the high altitude only one rescuer could be winched down per sortie. Initially the SAS medical officer, Captain John McLean, and Warrant Officer Frank Cashmore were winched down, with Corporal A. J. Bowen being landed at a clearing about four kilometres from the site. The SAS members administered first aid to the injured persons who were then winched out and brought to Wamena. There was extraordinary co-operation from the Indonesian authorities who waived all diplomatic clearances, permitted the SAS to operate in Indonesian territory, and handed complete control of the operation to Australian authorities.[12]

The crash was in one of the most remote and rugged areas in Irian Jaya, and it would take some time for Indonesian troops to reach the area. In the meantime there was concern that members of the OPM, the Irian Jaya independence movement, might capture valuable equipment from the crash site. Consequently, the second SAS patrol, led by Captain Jim Wallace, was inserted to protect the wreckage until the Indonesians arrived. In difficult, rainy weather Wallace's party of eleven SAS soldiers, six engineers and two RAAF aircraft fitters assisted the Indonesians to move into the area. After several attempts were abandoned due to poor weather, the dead body was recovered on 6 August. The operation was completed on 8 August.

Exercise LONG VIGIL provided tremendous impetus to the development of the SAS. During July the new Chief of the General Staff, Lieutenant-General Donald Dunstan, visited Swanbourne and LONG VIGIL, and on return to Canberra instructed his staff to investigate the formation of an Army Reserve component within the SAS to use the expertise of the ex-SASR soldiers living in Western Australia. He wanted to know why SAS soldiers were not being recruited from corps other than infantry.[13] He asked whether a study had been done to investigate whether Army Office should have a directorate to co-ordinate special forces aspects. And he asked: 'What is the present situation with regard to the agreed roles and tasks of the SASR?'[14]

One of the officers consulted by Dunstan was Colonel Smethurst, then serving in Army Office. He replied that 'the seemingly never ending discussion on the role and tasks of SASR and Commando Companies is fundamentally the result of a failure to create a proper command and control organisation for special forces by Army Office and Defence. The failure to achieve this organisation is perhaps based on a strong feeling that special forces are unnecessary. They may well be. If they are considered to be necessary the vital aspects of command and control in peace and war must be considered with their roles and tasks'.

Dunstan considered this paper, and directed that a special action force section consisting of a lieutenant-colonel and a major be established in the Directorate General of Operations and Plans. Lieutenant-Colonel Jeffery, who was already SO1 Joint and Special Warfare in that Directorate General, was posted to head the section. Work was also accelerated in Field Force Command on preparing a concept of operations for the SAS. At the end of 1978 an instruction was issued stating that the roles of the SAS Regiment were: 'to conduct long range reconnaissance and surveillance, often by deep penetration; to harass and disrupt the enemy in depth; and as a secondary role, to provide officers and senior NCOs who have been trained for guerilla warfare cadre postings'. The primary role had not changed for many years, but the secondary role was new. The instruction spelt out the organisation and concept of operations in detail and provided clear guidance for the training of the regiment. But by no means was it the end of discussion about the roles and tasks of the SAS.[15]

In Canberra Mike Jeffery, by now promoted to colonel as Director Special Action Forces, privately and in his own time continued work on the concept for surveillance in northern Australia and in March/April 1980 published an article in the *Defence Force Journal* entitled

'Initial Thoughts on an Australian Land Surveillance Force'. His concept was 'to raise on a phased basis an Army reserve territorial surveillance force of around 3500 men which would provide seven Regiments of three operational squadrons, each containing twenty patrols of six men. The SAS would play a major role in raising and training the new units and preferably would command the regiments and the squadrons. The regiments would be placed under command of the Military Districts with technical control exercised by the Director Special Action Forces in Canberra.[16]

The delay between LONG VIGIL in 1977 and the completion of Jeffery's paper in mid 1979 can be attributed to the extra workload caused by the introduction of a counter-terrorist force in 1978, the development of which will be discussed in the next chapter. But Jeffery's paper appeared at an opportune time, for in December 1979 Soviet troops advanced into Afghanistan and in the period of increased tension the Minister for Defence, Jim Killen, authorised the expansion of the Army Reserve to a force of 30 000 personnel. In June 1981 the North-West Mobile Force (Norforce) was established at Darwin, commanded by Lieutenant-Colonel John George, who had served as a captain with the SAS in the late 1960s. In 1983 he was succeeded as commanding officer of Norforce by Lieutenant-Colonel Doug Gibbons, who had served as second-in-command and officer commanding an SAS sabre squadron in the early 1970s. Other regular officers and senior NCOs came from the SAS, and Norforce was trained in SAS techniques. In 1982 the 5th Independent Rifle Company was formed in the Pilbara region of Western Australia, and in 1985 it became a regional force surveillance unit, the Pilbara Regiment. Its commanding officer was a former SAS officer, Lieutenant-Colonel Bruce Wallis. In 1985 another regional force surveillance unit, the 51st Battalion, the Far North Queensland Regiment, was formed at Cairns. The first commanding officer, Lieutenant-Colonel Kel Ryan, had served as a major in the SAS in the late 1970s.

Thus in the period of ten years since the withdrawal from Vietnam, the SAS had developed and refined its concept for operations within Australia and had been accepted as an integral part of the Australian Defence Force structure. The acceptance could be attributed to two factors. The first was the decision to introduce the Officers' Selection Course. As the standard of officers improved they took a closer grip on the unit and the Sergeants' Mess was forced to surrender a little of its previous influence. NCOs were sent to non-SAS postings within the Army and the knowledge of SAS capabilities and professionalism became more wide-spread.

The second factor was the SAS's substantial contribution to the concept and the formation of the regional surveillance units. Surveillance represented an acceptable and viable role within the concept of 'continental defence' being discussed at the time. Furthermore, the SAS was seen as being part of the conventional battle; the new role did not represent the mystical 'special operations' that were viewed with the skepticism and hostility that had confronted McFarlane, Smethurst and Jeffery. It was a role that conventional military thinkers could understand and with which they felt comfortable, and it thrust the regiment squarely into the strategic sphere of employment.

Although the new regional surveillance units played a key role in Australian defence, they still could not completely cover the huge areas of Australia. Undoubtedly, in times of tension they would need to be supplemented by the SAS which needed to maintain and further develop its expertise. But the regional surveillance units, which were in effect on an operational footing, released the SAS from immediate operations and enabled it to support Regular Army exercises and to reassess whether surveillance ought to be a main role, or whether special operations should be given more emphasis. The SAS's concern to step back from the surveillance role was shown by the fact that whereas the early commanders of the surveillance units were all SAS officers, within a few years none of them were.

SAS training for special operations continued, and involved exercises with other special force units. For example, between September and November 1979 twenty members of the regiment were in New Zealand for a special warfare exercise with the NZSAS, US Special Forces, SEALs, 22 SAS Regiment and the SBS. In 1974 and 1979 there were exercises with Philippine forces; in 1979 the Philippines Special Warfare Brigade (now called the Home Defence Forces Group) exercised in Australia.

In 1981 a twelve-man team commanded by the officer commanding 3 Squadron, Major Reg Swarbrick, exercised in the Philippines where it practised sea-to-shore insertion, including submarine entry and exit, use of special underwater swimming vehicles and covert beach reconnaissance. The aim of the exercise was to infiltrate onto an island at the head of Subic Bay, locate and assassinate an 'enemy' leader and destroy an 'enemy' communications centre. During the exercise, on 26 February 1981, a US special warfare Hercules crashed into the South China Sea soon after taking off from Subic Bay naval base. Only one of the 24 passengers survived. Those killed included Americans, Filipinos, New Zealanders and three Australians from the SASR, Sergeants Ewan Miller and Murray Tonkin, and Signaller Gregory Fry.[16]

With the decision in 1981 to form regional surveillance units the planning for the defence of Australia was not over. Indeed it was just beginning; but it marked the end of a phase in SAS development, and established a clear requirement for SAS skills and expertise. Furthermore, it was fortunate that the regional force surveillance units were able to relieve the SAS of this operational commitment because the threat of international terrorism was creating another, more immediate demand on SAS professionalism.

23
Counter-terrorism: 1978–1988

At 12.40 am on the morning of Monday 13 February 1978, shortly after a warning from an anonymous caller, an explosion shattered the still night air at the George Street entrance to the Hilton Hotel in Sydney. It also shattered any illusion Australians might have had about their remoteness from the threat of international terrorism. Two persons were killed, another died later and six more were injured. It was not the first terrorist incident in Australia, but this explosion was to have far reaching implications. The Commonwealth Heads of Government Regional Meeting, said to be the biggest gathering of overseas heads of government in Australia, was due to begin the following day, and already several visiting heads of government had arrived at this entrance to the hotel. The Indian Prime Minister, Morarji Desai, was staying in the hotel, and there was press speculation that the attack had been carried out by the Ananda Marga religious sect.

The Commonwealth heads of government were due to travel to Bowral, south of Sydney, on 14 February, and the Australian government called out troops from the Army's 1st Task Force in Sydney to protect the route to Bowral. The conference concluded without any further incident, and on 23 February the Prime Minister, Malcolm Fraser, announced in Parliament a range of measures to counter terrorism in Australia. He explained that since it had come to office at the end of 1975 his government had acted to improve the effectiveness of liaison and co-ordination between various Commonwealth and State agencies and police authorities. A Protective Services Coordination Centre had been established in the Department of Protective Services and contingency plans had been developed. He proposed to set up a

Standing Advisory Committee on Commonwealth State Cooperation for Protection against Violence (SAC–PAV) and he stated that CT training would be stepped up within the police forces. He announced that Sir Robert Mark, the recently retired Commissioner of the London Metropolitan Police, would visit Australia to advise on CT measures. Mr Justice Hope, then in the midst of a Royal Commission on Intelligence and Security, would also examine protective security.

Among a range of tasks, Hope was instructed to examine the 'relationship between the Defence Force and civilian authorities in the matter of civilian security', but there was no indication of what preparations were being instituted within the Australian Defence Force. Hope's Report was not available until 1979, but in the meantime Mark reported to the government that: 'the close quarter battle is a task for the most experienced soldiery, not for the police, whose role should be that of containment until military aid arrives. The higher degree of training, more sophisticated weaponry, experience and fitness of specialist troops is likely to reduce rather than increase the possible loss of life in a close quarter battle'.[1]

Clearly there was a role for the Australian Defence Force. In fact, throughout 1977, well before the attack in Sydney, there had been lengthy discussions between government departments in Canberra over Defence Force Aid to the Civil Power in CT operations. The Minister for Defence had already approved specific roles and tasks for the Defence Force and the Army had proposed to form 'response forces' in each Regular Army Task Force and in the 5th Field Force Group in Perth. Initially, each 'response force' would consist of an infantry platoon, an engineer section and an Explosive Ordnance Disposal (EOD) team. Training of the response forces began in November 1977, and the SAS conducted a special course for 24 men towards the end of that month.

While the SAS already possessed a certain degree of expertise in close-quarter battle, it was not yet able to provide the specific training needed by the response forces. Consequently, in December 1977 it was decided that the Commandant of the Infantry Centre, Colonel John Essex-Clark, should proceed overseas to study counter-terrorism. This tour was still being planned when the Hilton bombing took place, and there was an acceleration of activity within the Department of Defence. One Army proposal was that a team from 22 SAS Regiment, which already possessed a substantial reputation for its CT assault capability, should visit Australia to train selected members of the Australian SAS in CT procedures and combat techniques. In putting forward the proposal the Chief of the General Staff, Lieutenant-General Donald Dunstan, wrote: 'Although the threat of international terrorist operations in Australia is assessed as being low it is accepted that the situation can change rapidly, as was graphically illustrated by the Hilton bombing. It is possible that an incident requiring siege/assault techniques could occur at any time with little or no warning. Should such an incident take place before a properly trained response force is available, the lives of any hostages taken will be in additional jeopardy. The requirement is therefore for us to develop an effective counter terrorist response force as soon as possible against the contingency that it will be called out in aid to the civil power'.

The Army also proposed that Colonel Essex-Clark continue with his overseas tour to gather 'the information needed for Army to determine the best course of action in training for, and planning the conduct of, counter terrorism operations'. Dunstan knew that forming a CT force would be a lengthy process, as he explained to the Chief of the Defence Force Staff and the Secretary of the Department of Defence: 'Given the planning and preparation required to agree on and then establish an adequate training facility, it is unlikely that we can develop a fully effective counter terrorist response force before 1979 unless we resort to other means'.[2]

Colonel Essex-Clark was absent from Australia from 30 March to 6 May 1978, visiting Britain, the Netherlands, the Federal Republic of Germany and the United States, and on his return reported that there were three critical requirements: the selection and grouping of personnel; the selection and acquisition of equipment; and special training facilities. In his report, known as the IRONBARK Report, he concluded that:

a special counter terrorist force was required to rescue hostages in the siege/assault situation...The force must be able to react rapidly. Command of the force must be at the level at which the decision is made to use it, to prevent delay. Information on terrorist organisations and methods of operation is essential for our own training and research and it is the key to lowering the risk. The force must acquire, and maintain, high standards of proficiency. This means continuous training on a wide range of weapons and facilities. Special training facilities must be constructed to develop and maintain essential skills...There is a need for overseas training assistance...Security is essential for success. Information about the organisation and equipment of the force enables terrorists to gauge its tactics and capabilities. Any warning that the force is about to be committed to the assault endangers the lives of both hostages and members of the force. On the other hand the fact that the force exists may act as a deterrent.[3]

Colonel Essex-Clark provided useful and detailed information which greatly assisted planning in Army Office. But his suggestion that the CT force should be based at the Infantry Centre at Singleton under his command did not win general approval. Rather, the Director-General of Operations and Plans, Brigadier Ron Grey, argued that the SAS was 'quite capable of training a "pagoda" type team to meet any demands made upon the Army. The area of greatest difficulty...is whether reaction forces should be established in more than one location, that is, elsewhere than at Swanbourne. I believe it is possible to have a reaction force at Swanbourne, backed up by very small recce elements in the commando companies which could react in time to a threatened or actual terrorist situation. Unfortunately this can never be given as a guarantee. It must be remembered that an Army response force is essentially a back-up to other law enforcement agencies, not the initial or indeed primary means of dealing with a terrorist situation'.[4]

The Army operations staff, particularly the SO1 (Joint and Special Warfare), Lieutenant-Colonel Mike Jeffery, continued work on Essex-Clark's proposals, and on 1 August 1978 the plan to form an Army CT force was forwarded to the Chief of the Defence Force Staff (CDFS) and the Secretary. The plan envisaged the formation of a special unit skilled in the selective use of force and capable of rapid deployment to any part of Australia. This unit was provisionally called the Tactical Assault Group (TAG), and it was proposed that it be formed within the SASR. It was expected that the commanding officer of the SASR would also command the TAG. He needed to be a relatively senior officer of mature judgement, capable of assessing the nature of the problem, making a sound plan and giving balanced advice. The actual assault force would be commanded by a captain, and in turn the force would include a number of assault teams, each commanded by a sergeant. Rules of engagement figured prominently in the Army's plan.

It was envisaged that once activated, the commander of the TAG would be placed directly under the command of the CDFS who would therefore require expert advice on TAG capabilities and limitations to be available, both on a day-by-day basis and in the event of an incident. There would also be a need for liaison with other Commonwealth and State law enforcement agencies on matters relating to the TAG and for the conduct of appropriate exercises. This need was best satisfied by establishing a Director of Special Action Forces at Army Office who, in addition to these important responsibilities, would look after the policy, planning and co-ordination of the activities of the Special Action Forces, which included the SAS, the Commando companies and their associated signals squadrons. He would also develop policy concerning the acquisition of special equipment and would perform the function of Head of Corps for all Special Action Forces. Special Action Forces were defined as Army units which performed operational roles that were outside the scope of conventional forces.

The conclusions of the Army proposal were: 'The Army counter terrorist assault force, provisionally named TAG, must be specially trained and equipped and organised in order to perform its role with minimum risk of failure. It must have proper legal cover agreed at the highest level. The SASR comprises selected men with the skills and training most suited to employment in the TAG'. However, the Army plan noted some drawbacks: 'the manpower dedicated to this task will reduce the capability of the SASR in its long range reconnaissance and surveillance roles...Army is unable to meet the financial commitment from existing allocations without detriment to normal Army objectives and activities'.[5]

For the next six months the Army proposal was refined as it was scrutinised by other Defence and government agencies. Eventually, on 22 February 1979 Colonel Essex-Clark and Lieutenant-Colonel Jeffery met with the Minister for Defence, Jim Killen, to discuss CT developments. Mr Killen raised the question of the likely degradation in the performance of present SAS roles and tasks if it took on the CT responsibility. He was informed that with the approved manning level there would be considerable degradation in capability that could only be overcome through an increase in manpower. Should such an increase come from Army resources it would mean a reduction in Army capabilities elsewhere. The Minister did not comment on this point, but when told that the cost of raising a TAG would be $3.1 million, intimated that he was confident of finding extra money for defence 'because of the situation in Iran, Vietnam and China'.

After the interview Lieutenant-Colonel Jeffery reported his impressions to the Army Chief of Operations, Major-General Ron Grey: 'I am of the view that the Minister has already made up his mind to have a Defence Force CT capability based on SASR. I feel he will support the Cabinet Submission as written including the provision of additional funds, although he may insist that they come from an increase to the defence vote about to be announced by the PM. I feel he would personally support an increase in manpower ceiling for the Army for CT purposes if we wished to inject this into the draft submission'.[6]

Jeffery's assessment was right, and on 3 May 1979 the government gave approval for the establishment of 'a specialised and dedicated counter terrorist assault team'. The team was to be available to the Commonwealth to deal, where authorised, with high risk terrorist incidents and be made available to States and Territories for the same purpose. State and Territory governments were informed of the decision and were invited to participate in the development of arrangements for co-ordinating action.

As foreshadowed in the original plan, a proposal was submitted to form a new Directorate of Special Action Forces (SAF), and in July approval was received. Mike Jeffery was promoted to colonel and became the first Director, Special Action Forces (DSAF).

The possibility of a TAG being located on a rotation basis in the Eastern States raised a further command problem as there might be an unnecessary delay if, in an emergency, the commanding officer of the SASR had to travel from Perth. Furthermore, a senior officer would be required in the Eastern States to look after the TAG's special training, liaison with government departments and administrative requirements. It was therefore decided to raise the headquarters of the 1st Commando Regiment at Randwick in Sydney and to increase considerably the Regular Army manpower of the newly formed regiment. This new headquarters filled a long felt need to provide proper command facilities for the two Commando companies and the associated 126 Signal Squadron in respect of training, exercises and support to outside agencies. The Army Reserve Commando companies were located in Sydney and Melbourne; both were commanded by Regular Army officers and had a substantial Regular Army cadre, most of whom had served with the SAS.

Considering the role of the commanding officer of the 1st Commando Regiment in CT operations, and the role of the SAS in the development of the regional surveillance units, described in the previous chapter, the decision to form a Directorate of Special Action Forces was commonsense. Furthermore, it underlined the eventual acceptance by the Army of the integral role of Special Action Forces in the Army's organisation.

Although government approval to set up a CT assault team was not given until May 1979, the SAS was already well underway with CT training. However, counter-terrorism was hardly in the forefront of Lieutenant-Colonel Reg Beesley's mind when he assumed command of the regiment on 23 October 1977. Exercise LONG VIGIL had only recently concluded and the development of a surveillance force in northern Australia appeared to have priority. Beesley also had his own ideas about the development of the unit. In particular he wanted to improve the basic SAS skills, and to this end he introduced a patrol course. He was assisted by his RSM, John Sheehan, whom he had first met as a lance-corporal in the SAS Company in 1961. Sheehan had served with the SAS in Borneo and had been wounded with the Training Team in Vietnam. He was one of the group of NCOs whom Smethurst had identified as potential RSMs and had posted out of the unit to gain experience.

It did not take long before Beesley found himself immersed in CT training. Indeed, barely a month after he assumed command the SAS

Colonel R. P. Beesley, AM, Commanding Officer of the SAS Regiment, 23 October 1977–21 December 1979. The first commanding officer to have previously served both as a troop or platoon commander and as a squadron commander, he had the initial responsibility of developing the counter-terrorist capability within the Regiment.

was required to provide training for the 5th Field Force Group response force. It was then that he discovered the extent of the training that had been carried out by Captain Greg Mawkes, Warrant Officer Dave Scheele and Sergeant Red Webb in the Guerilla Warfare Wing. The Hilton bomb incident in February 1978 made it clear that CT training would gain in importance and Beesley changed the title of the wing to Tactics Wing to reflect its new role.

When Mawkes was given nine months to develop a training system to produce a CT assault team, he soon realised that despite the fact that all members of the SAS had been specially selected and trained, not all were suitable for this new task. In effect, Mawkes had a free hand in selecting his team, and naturally he tended to select men who had previously successfully completed unconventional or guerilla warfare courses. While the Australians learned much from British SAS experts, they did not accept the British approach uncritically. The Australians also developed their own techniques which they believed were superior.

Until the government approved the raising of the TAG, however, little additional money was available and training was severely hampered by lack of equipment and appropriate weapons. For example, soldiers wore green coveralls, green Army issue flak jackets and gas masks borrowed from the Western Australian police. They carried F1 sub-machine-guns and 9 mm pistols. A 'killing house' was labouriously built into a sandhill behind the Swanbourne classification range using thousands of sandbags, wood and galvanised iron. When finished it

425

was a very basic structure with only two rooms and an open roof. Under the system at the time, Tactics Wing trained personnel in CT assault techniques and then they returned to their parent squadron and troop, from which they could be recalled in case of emergency. The trained personnel in the squadrons plus personnel from Tactics Wing formed an interim team of roughly troop size. It suited the government to be told that the SAS had developed a CT capability, but in reality that capability was quite rudimentary.

In May 1978, less than three months after the Hilton bombing, The *Australian* newspaper carried a two-part series on the SAS. 'Australia's last resort in the event of terrorist attack is the army's Special Air Service Regiment', stated the article, which went on to explain that the SAS was preparing for the task with typical thoroughness. It had upgraded its unarmed and close-quarter training. Men who were highly specialised in moving unseen and fighting from cover were 'now tearing into abandoned buildings, learning how to fight and move quickly indoors. They are emphasising accuracy in close-quarter rapid fire, the silent techniques of taking a man with a garotte or a fighting knife or boot. They are practising assaults on dummy aeroplanes. They are training their men in the intricacies of high-rise buildings, emphasising the points of access such as lift shafts and air conditioning ducts. They are doing a lot of things the SAS, with its attitude to security, will not discuss'. According to the article, the SAS was 'touchy about the job and, given the political and public ramifications of a military force moving into a civilian role in a democratic society, it is understandable'. Lieutenant-Colonel Beesley was quoted as saying: 'We will be called in only—and I mean only—when every other avenue has been exhausted, when all the efforts of the police forces have failed. We wouldn't have it any other way'.[7]

To some extent, all this was a smokescreen. A year later, one month after the government authorised the establishment of the TAG, the Army Chief of Operations, Major-General Grey, reported that the TAG was 'not expected to be fully operational until 18 April 1980 because of equipment procurement, facilities construction and training requirement leadtimes'. If an urgent operational requirement developed before then, a partially trained TAG could be deployed at 24 hours' notice using standard issue weaponry, ammunition and military communications. However, any decision to commit the TAG before the completion of its training cycle had to be made with the full knowledge of the risks involved, particularly against practised and dedicated terrorists. The regiment owes much to Grey's decision not to hurry the introduction of the CT capability as it meant that the war role side of the unit was not completely decimated.

Lieutenant-Colonel Beesley finally received authorisation to begin raising the TAG on 31 August 1979. According to the directive from the Chief of the General Staff, the tasks of the TAG included: the neutralisation, including capture, of terrorist groups, which might include snipers, hijackers, kidnappers, bombers or assassins; the neutralisation of aircraft or ships; the recovery of hostages and property held by terrorists; and the recovery of buildings and installations occupied by terrorists. The strength of the assault team was not to exceed three officers and 26 soldiers, and the codeword GAUNTLET was to be used when referring to the TAG. The TAG headquarters was to consist of the commanding officer of the SASR, his operations officer and two signallers. The assault force headquarters was to comprise a captain, a signaller and a signaller/driver. There were to be three assault teams each of three men, and a sniper team of ten men.

Although Captain Greg Mawkes continued training the personnel for the TAG, Beesley realised that the TAG needed an appropriate command structure, and the task of supervising the formation of the TAG was given to Major Dan McDaniel, who assumed command of 1 Squadron which was providing the troops for the TAG. McDaniel had initially joined the Army as a National Serviceman and had served as an infantry platoon commander in Vietnam in 1971. He had been posted to the SASR to conduct operational research and once there, had been given the opportunity of completing the Officers' Selection Course. As the officer responsible for operational research he had already been involved in the efforts to purchase new equipment for the TAG. His SSM was that capable, long-serving SAS soldier, Warrant Officer Frank Sykes, and the key position of SQMS went to Staff Sergeant Joe Van Droffelaar, who had to ensure that the wide range of equipment needed for CT operations was transported and available at short notice anywhere in Australia. It was not appreciated that CT operations would require the use of large quantities of stores and equipment. Van Droffelaar started with a staff of three, but it eventually grew to eight.

Up to this time, A Troop had been a free fall troop, B Troop had concentrated on water operations, and C Troop had been vehicle-mounted. With the decision to form the TAG on a permanent basis, B Troop became the TAG and C Troop was absorbed into A Troop which continued with training for the SAS's normal roles and tasks. The first commander of the TAG was Captain Martin Hamilton-Smith, who, with Sergeant Leigh Alver, received additional CT training with 22 SAS Regiment in Britain in late 1979. The first TAG, known as GAUNTLET 1, began training in March 1980 in special training facilities constructed at Swanbourne, and it became fully operational in May 1980. While conducting their final exercise at Laverton

RAAF base, Victoria, in May 1980 they watched on television 22 SAS Regiment undertaking its famous assault on the Iranian Embassy in London. It provided excellent motivation for the soldiers and a timely reminder of the purpose behind their training. Brigadier Rod Curtis later commented that 'the training of the first GAUNTLET team and the excellent standards it achieved were due to [the] dedication, hard work and professionalism' of Captain Greg Mawkes. 'GAUNTLET 1 set the standard for other teams to follow.'[8]

Having completed its training within the SASR, GAUNTLET 1 extended its training with the police forces of the various States to resolve problems of command and control and to familiarise themselves with local police and governmental procedures. Once the TAG was deployed the OC 1 Squadron had no command function. The commanding officer of the regiment moved to the scene of the incident where he was in direct contact with the local police commander. If the TAG was required to undertake an assault, the commanding officer gave his orders directly to the troop commander.

The task of identifying, requesting, purchasing and introducing the myriad assortment and quantities of equipment to enable GAUNTLET 1 to reach operational effectiveness in such a short time was a tremendous effort by the higher staff of the SASR and the Directorate of Special Action Forces. In retrospect, there were few errors in selection of equipment and most of the gear purchased in this initial period, or updated versions of it, is still in use today. The equipment included the black uniforms, seen by everyone when 22 SAS assaulted the Iranian Embassy, the gas masks, the Heckler and Koch sub-machine-guns and the special grenades.

In December 1979, some months before the first TAG became operational, Lieutenant-Colonel Rod Curtis succeeded Beesley as commanding officer. The initial work to establish the TAG had been one of Beesley's major achievements, and the new commanding officer had the task of continuing that work. Aged 39, Curtis had graduated from Duntroon in December 1963 and had served as a platoon commander with 4 RAR in Borneo during which he earned the Military Cross. He was second-in-command of a company and intelligence officer with 9 RAR in Vietnam in 1969 and had joined the SAS as a sabre squadron commander in 1972. In 1974 and 1975 he was an instructor at the British Army's School of Infantry; in 1976 he attended the Australian Army Staff College, and before his appointment back to the SAS was chief instructor at the Royal Military College, Duntroon.

By the time Curtis assumed command, it was already clear that the CT commitment was becoming a major drain on the resources of the regiment. Members of the TAG were kept at a high state of readiness and it was estimated that after about one year their effectiveness

Brigadier R. G. Curtis, AM, MC, Commanding Officer of the SAS Regiment, 22 December 1979–22 December 1982. Under his command the regiment developed the offshore assault capability.

would start to decline. There would therefore be a requirement to train a new TAG each year. In addition, in July 1980 a directive was received from Canberra that the regiment was to develop a capability to retake an offshore oil platform in Bass Strait if it was seized by terrorists. The codeword BURSA was to be used when referring to offshore operations. The first Offshore Installations Assault Group (OAG), to be known as NULLAH 1, was to be operational by October 1980 and was to continue through to the end of 1981. Thus by the latter months of 1980 the regiment had one troop 'on line' as GAUNT-LET 1 and two more troops in training. The first inter-state deployment of GAUNTLET 1 took place towards the end of 1980 on Exercise CAMPBELL PARK in Canberra.

Curtis wrote later that '1980 was an extremely busy and challenging year; a national CT capability was established to counter onshore terrorist incidents; interim CT training facilities were completed at Swanbourne and work started in preparing submissions for more permanent facilities; other accommodation and works programmes were completed to house the TAG and its equipment, weapons and vehicles; an offshore CT capability was worked up; and specialist weapons, equipment and vehicles procured'.

429

SAS: Phantoms of War

Not only were the tactical skills and techniques of the team developed, but internal command arrangements had to be established and practised. Liaison and training with State and Federal Police Forces had to be co-ordinated, and procedures, in concert with the National Anti-Terrorist Plan, agreed and practised. Of particular concern and sensitivity were the legal provisions to call out the team in support of the civil authorities (police), and the essential control mechanisms which needed to be set in place and understood before the TAG could be employed.

Many of these and other matters were discussed at the SAC–PAV Command and Management Conference, co-ordinated by the Protective Services Coordination Centre and conducted at Campbell Barracks, Swanbourne in June 1980. The conference was chaired by the Commissioner of the Federal Police, Sir Colin Woods, and examined the co-ordination and control arrangements between various Commonwealth and State government departments and agencies involved in combating terrorism. It also provided an early opportunity for the regiment to demonstrate to delegates the standard of the TAG capabilities to counter high risk terrorist incidents. As a marketing exercise, it was both necessary and highly successful.

The CT capability was a mixed blessing for the regiment. On the one hand it provided realistic training and an actual operational commitment. The regiment received weapons, equipment and facilities that until then it had not been able to acquire, and it gave the regiment a great sense of purpose. On the other hand, it had a highly disruptive effect. The soldiers not involved in the CT training resented the attention being given to CT training, and the commander of 3 Squadron found it difficult to maintain the integrity of his patrols as he was continually required to provide reinforcements to the CT troops. Obviously the 'war role' training of the regiment was suffering from the manpower drain which the emphasis on CT training required. This problem was highlighted by an article in the *West Australian* newspaper in 1981.

Since 1979 the Special Air Service Regiment in Australia has had to find about 40 of its best men, at any one time, for a force tied in with the Commonwealth Police. This in a unit which always had recruiting problems anyway...Distracting units from their wartime tasks can be seen as a triumph for the spectre of subversion any way. With 40 men permanently allotted to the Federal anti-terrorist squad, a SAS regimental commander has to be satisfied if he can put about half a regiment...into the field in its proper role of intelligence gathering and raiding.[9]

430

If it is assumed that a TAG consisted of 40 men, that there were to be two operational groups, the TAG and the OAG, and that two more had to be training to relieve the operational teams, it can be seen that at certain times up to 160 men could be involved in CT activities, not counting training, stores and other administrative staff.

There were a number of solutions to this manpower problem. The first was to seek assistance from the Royal Australian Navy's Clearance Diving Teams, and seventeen Naval personnel were placed under operational control of the commanding officer of the SASR from 4 August 1980. There was some heartburn about the fact that the Naval personnel had not completed the SAS selection course. This view failed to take account of the operational need for additional manpower, and their use was accepted as they were employed for a specific purpose. In fact, all RAN clearance divers were required to complete a purpose designed selection course, the first of these being conducted by the Commando Regiment on behalf of the SAS. Subsequent courses were conducted by the SAS.

The second solution was a further rationalisation of personnel within the regiment. There were various employments, such as driver, storeman, cook or clerk, which did not have to be performed by 'badged', or qualified SAS soldiers, and the use of non-badged soldiers released trained SAS soldiers for operational tasks.

But these measures were only a stopgap. Ultimately the SAS could only handle the CT task and maintain its 'war role' expertise by obtaining more soldiers. Eventually, towards the end of 1981 authorisation was received to reraise 2 Squadron, but the difficulty was to find the additional soldiers to man the new squadron. The British Army, with a strength of 165 900 regular soldiers, had trouble maintaining an SAS Regiment that was only marginally bigger than the expanded Australian SASR. The Australian Regular Army had a strength of only 32 000 personnel.

There were some observers both within and outside the SAS who doubted whether the third sabre squadron could be manned, and indeed it took a number of years before it reached its full strength. The regiment had always had an active recruiting campaign, but now it had the clear and open support of the Chief of the General Staff. It was a telling commentary on the quality of men in the Australian Army and the high regard of the SAS among commanders and staff in the conventional units, that sufficient high quality applicants were eventually found. Standards were not lowered but the unit's operational commitment attracted an increased number of applicants. The Australian SAS had always had a slightly different recruiting policy to the British SAS, where there was a tendency to seek men who appeared already to have all the attainments of an SAS soldier. In Australia the selection teams

431

looked for men with the right physical and psychological attributes who, with training, could develop the SAS skills. In other words, the Australian SAS sought men of character and with demonstrated potential to serve in the regiment, and was willing to spend a little longer on training them once they completed the Selection Course. But it still took almost five years to provide sufficient men to form three sabre squadrons.

In September 1980 A Troop began CT training so that it could relieve B Troop at the end of the year. CT training was highly demanding, and the standards of marksmanship required for CT operations called for speed of action and accuracy unprecedented in other areas of military training. Consequently, the programme of close quarter battle training was also unprecedented in intensity, intending to prepare individuals to be able to identify targets instinctively in the shortest possible time. The training was described by a reporter from The *Australian* newspaper who observed one session in 1984. 'I saw members of 1 SAS Squadron', wrote Ken Brass, 'men trained not to blink if a hand-grenade explodes beside them, in ferocious action in a series of precision and extremely dangerous exercises. Even through the suffocating smoke and the staccato thump of sub-machineguns you could almost smell the adrenalin'. To maintain standards the soldiers had to shoot at least once a week in close quarter battle, and sometimes a soldier might get too close to explosives.[10] As Ken Brass wrote, 'Here the living casualties are the "blokes" whose hands or fingers have been blown off and the "guys" deafened by the sound and fury of some counter-terrorist measures. For the SAS doesn't fake it. All the bullets are real, the explosives potentially deadly. With almost every shot at Swanbourne someone's life is at risk. You can't pretend in this sort of work and when you're playing for real things can go wrong'.[11]

Things did go wrong on 10 October 1980 at the indoor range known as the killing house. The firing practice was a torch shoot in which individuals armed with a sub-machine-gun and a pistol entered a darkened room, illuminated targets with a torch fixed to their sub-machine gun, and engaged those targets identified as terrorists. Each student would enter the darkened room, and then return to an outer room before being recalled to the darkened room for another practice. While waiting in the outer room the students were encouraged to rehearse their actions in what was known as a dry practice. During one of these dry practices one of the soldiers momentarily lost concentration and shot Lance-Corporal Peter Williamson in the head. He was dead by the time he arrived at Royal Perth Hospital. The Army conducted a high level court of enquiry which concluded 'that it was an instinctive action that occasioned [the soldier to fire his weapon] *not* carelessness, neglect, intent or misconduct'. It was only the second full-scale CT course conducted at Swanbourne, and the Court of Inquiry directed that additional safety measures be instituted to 'minimise as far as possible the likelihood of a further such serious incident, yet retain the realism needed in this type of training to ensure operational success'.

While GAUNTLET 2, commanded by Captain Bob Quodling, was preparing to relieve GAUNTLET 1, the first OAG, NULLAH 1, commanded by Captain Graham Ferguson, was training for a role which was perhaps even more demanding than that of the TAG. Whereas the members of the TAG had to be familiar with high rise buildings, passenger aircraft, buses and power stations, the OAG had to learn the intricacies of offshore oil and gas rigs. It is not generally appreciated that one of the chief problems on an isolated oil rig is one of orientation. If a soldier were to burst out of a door onto the outside of an oil rig on an overcast day or at night he would be presented with an expanse of sea in every direction with no indication of north. Furthermore there were the obvious problems of actually getting the OAG onto the oil rig without alerting the terrorists. Finding an oil rig at night in difficult seas would be only the first of many problems. Major McDaniel commented later that the 'risk management of Captain Ferguson was superb'. By November 1980 NULLAH 1 was 'on line', and at the same time GAUNTLET 2 relieved GAUNTLET 1.

The establishment of the OAG underlined one considerable difference between the British and Australian SAS. Since its early days the Australian SAS had developed and maintained a substantial water operations capability. That capability had been reinforced by the existing active exchange programme with the Royal Marine Commando SBS, which was responsible for British water operations. There was also an exchange programme with the US Navy SEALs. In Britain there had always been competition between the SAS and the SBS, and there was the possibility of a breakdown in command and control if both organisations were involved in an operation. For example, in February 1987 the London *Times* claimed that in the Falklands there were accidental gun battles between SAS and SBS patrols which cost the life of at least one SBS soldier. Because of secrecy one unit was often completely unaware of what the other was doing. The article stated that a joint special forces headquarters was being established to increase co-operation between the two British units.

There was no such difficulty in Australia where both onshore and offshore capabilities were conducted by the same unit. Nevertheless, there were initial problems to be resolved over the conduct of offshore operations. The original idea was that the OAG would be manned and trained by the Navy at the SASR, thus following the United Kingdom

A Troop, 1 SAS Squadron 1984. Sergeant Maurice (Spud) Murphy, Trooper Peter Cape and Corporal Michael (Dinger) Bell, in the holding area preparing for an oil platform assault in Bass Strait.

example where the SBS was responsible for offshore operations. However, the Navy could not provide sufficient qualified and suitable men, and there was much common training between the TAG and the OAG. Eventually the SASR assumed full responsibility for both commitments, although Naval Clearance Diving Team members were still posted to the OAG.

The integration of members of the Naval Clearance Diving Team was initially not without difficulties. These men sometimes lacked the maturity and experience of their SAS counterparts, yet might be required to parachute into the Bass Strait at night. To their great credit the clearance drivers responded positively to the challenge of the OAG and performed their tasks to a very high standard. As one CT squadron commander wrote later, in undertaking the SAS selection course and the basic parachute course 'they were being tested in the completely foreign environment of the field soldier. That as many as did passed is a great credit to the Branch and the Service'.

A further problem concerned the higher command of offshore operations which invariably would involve RAN ships and helicopters. At first the Deputy Fleet Commander was the joint force commander, but with his headquarters in Sydney and his lack of expertise in special forces operations, he was hardly the best man to advise the local police commander or to direct the offshore assault. Following the first complete and realistic exercise in Bass Strait to test both the command and control arrangements and the OAG, it was accepted that the commanding officer of the SASR should be the joint force commander for any assault on an offshore oil or gas platform.

By early 1981 the OAG was being referred to as the OAT, the Offshore Assault Team. The commitment of personnel had grown, and almost 100 members of the SASR (with attached naval personnel) were dedicated to, or had primary CT roles. The total Australian Defence Force commitment was larger, involving RAN and RAAF crews that might be required for specific operations. The size of this commitment was of some concern to the Chief of the General Staff, but when it was pointed out to him that the Australian CT assault force was smaller than those of the United States, Britain, Germany, France, Thailand and Indonesia, and also had both onshore and offshore capabilities, which in many countries were carried out by different groups, he was satisfied.

In December 1981 Major Jim Wallace succeeded Dan McDaniel as commander of 1 Squadron and continued to develop the CT capability. A major deployment was to Brisbane in September 1982 for the Commonwealth Games. The assault teams kept a low profile but were familiar with the layout and procedures at all the Games venues. By this time the CT squadron consisted of GAUNTLET 3 and NULLAH 2.[12]

The inherent danger of offshore operations was underlined on 16 April 1982 when Trooper David O'Callaghan was drowned while diving in Bass Strait. O'Callaghan and another diver were swimming underwater at night using a closed-circuit breathing apparatus at a depth of about five metres. Suddenly O'Callaghan appeared on the surface in great distress. Apparently he had not breathed out sufficiently while coming to the surface and the pressure had severely damaged his lungs. He was dead by the time he had been flown by helicopter to the Sale base hospital. The exact reason why he had surfaced quickly was not clear. Perhaps he had been frightened by some sort of animal life in the water, such as a seal. At 21 years of age he was inexperienced and had only recently completed the water operations course, but all safety procedures had been followed.

As CT exercises and training proceeded, it became apparent that neither the TAG nor the OAT would be large enough in a real terrorist incident, and if either group was required to conduct an assault the other group would be needed as a back-up. In particular, in preparing for the Commonwealth Games in Brisbane it became obvious that two troops might need to assault simultaneously. Despite the fact that initially the squadron commander was not in the chain of command, he was now deployed with his small headquarters to co-ordinate both groups and to command the assault force.

The insertion of the squadron commander in the chain of command changed the command arrangements, which by 1981 were as follows. The DSAF, now Colonel Reg Beesley, acted as the special action forces adviser to the Chief of the Defence Force Staff for CT operations, and in time of an incident moved to the Crisis Policy Centre in Canberra.[13] The commanding officer of the SAS was located in the Police Operations Centre alongside the Police Commander. The commanding officer of the Commando Regiment moved to the Police Operations Centre on the basis that, being on the east coast, in most instances he could get there more quickly than the commanding officer of the SAS. The commanding officer of the Commandos was to establish initial liaison with the police; start preparing for the requirements of the CT force; brief the commanding officer of the SAS when he arrived; and then act as a relief for the commanding officer of the SAS. The commander of the CT squadron was located at the Police Forward Command Post with the dual tasks of commanding his squadron and maintaining liaison with the Police Forward Commander. That is, the two tactical commanders were co-located.

While the CT capability was being refined the normal role and tasks of the SASR languished. The raising of 2 Squadron did not help much because initially there were insufficient soldiers to man the squadron. Manpower for the squadron was to be phased in over three years at the rate of 30 men per year, and at this rate it would take until 1984 for the squadron to be fully manned.

The new squadron commander, Major Paul Robottom, has left some vivid impressions of these early days.

In January 1982, apart from a title and names on paper, 2 SAS Squadron essentially did not exist. There was no headquarters building, no desks, no chairs, no phone, no stationery and only a few headquarters personnel...2 Squadron started its new life as a squadron (-) [sic] in all areas. The squadron headquarters was reduced in manning and there were only two troops, E and F. Another annoying feature of our re-raising was that key positions such as operations officer, a troop commander and the SSM were absent on courses...The fact that the squadron was manned almost entirely by troops who had only recently come off the reinforcement cycle meant that they had a good level of specialist skills but little or no experience of their application within an SAS squadron.

It was not until the end of 1982 that the squadron was moderately efficient as an SAS squadron. Robottom's frustration was shown by his later comment that there appeared to be a 'lack of interest in the war roles side of the SASR shown by Regimental headquarters'.

Curtis was adamant that he did not ignore the problems of 2 Squadron. 'I was aware at the time of the frustrations felt by the war role squadron and by Major Robottom in 1982. But it must be remembered that priority of effort had to be given to establishing a sound CT capability. A task not to be taken lightly when one considers the Regiment's CT force is in the final analysis, a National asset of last resort for the Government to counter high risk terrorist incidents.'

Curtis was instrumental in having established the SAS Regiment Army Reserve component with the aim of utilising, on a part-time basis, the specialist SAS skills of NCOs leaving the regiment and the Regular Army. The Reserve component provided an additional source of manpower which could be called upon for instructors. The scheme quickly became effective and assisted in relieving manpower pressures in the sabre squadrons, enabling them to get on with training.

On top of these manpower concerns, Rod Curtis was faced with some lack of clarity in higher headquarters about the role of the SAS. For example, in September 1981 he wrote to Headquarters Field Force Command to complain that his roles appeared to be restricted to surveillance and reconnaissance. Parts of his letter deserve quoting at length:

> there still remains no clear perception as to how SASR should be employed operationally in time of war. This has occurred for a number of reasons. Firstly, there is no authoritative document which specifies a concept for SAF operations including command and control in war. Secondly, officers and NCOs returning to SASR from postings or detachments to SAF units overseas bring back with them their newly gained skills from either 22 SAS, SBS, US SEAL or USASF [US Army Special Forces] and inject into SASR training their recently acquired knowledge. This information flow can have many advantages but it has also been responsible at least in part for SASR departing from primary skills applicable to its wartime roles. As a consequence, SASR is now covering the broad span of SAF skills and techniques. This has resulted in veneer specialisation and degradation of primary war skills.
>
> Perceptions regarding the employment of SASR have unfortunately also been coloured by the unit's operational employment in South Vietnam (SVN). Sabre squadrons operating in SVN were primarily employed on tactical reconnaissance tasks with harassment being a secondary task. This latter task was also tactical, and was achieved essentially by ambush. These tasks, in the main, could have been undertaken by specialist sub-units from infantry battalions of the RAR.

Ironically, the AATTV which was raised for service in SVN had as one of its tasks the responsibility for training irregular forces. A task which SASR was and is ideally suited for...By definition SAF units conduct tasks which are not normal or are outside the scope of conventional forces. These tasks will be essentially strategic in nature but operations of a tactical nature may be undertaken (as in the case of SVN) where specific SAF skills are required.

Curtis went on to argue that the roles of the SAS should be: operations behind enemy lines; acquisition of strategic information and harassment and disruption of the enemy; and unconventional warfare.[14]

As to be expected, the commanding officer of the 1st Commando Regiment, Lieutenant-Colonel Peter McDougall, a long-serving SAS officer, would not remain silent on this issue, and he argued that the Commandos could not be ignored in any discussion of special operations and unconventional warfare. A draft Army Office instruction outlining the roles of the Commandos and the SAS was still being discussed when Curtis concluded his tour as commanding officer in December 1982.

Curtis had been an approachable and popular commanding officer who had the common touch with the soldiers. He had done a magnificent job in developing the CT capability within the regiment, and the fact that he had not ignored the 'war role' side of the regiment was underlined by the gruesome statistic that during his time in command he lost six soldiers killed in training incidents, two in CT training and four in 'war role' training.[15]

While there was an In Memorium Board for members of the regiment killed in action, there was nothing to remember those killed in training accidents. Curtis felt strongly that it was necessary to do something about the latter, and he had another In Memorium Board built and located in the Assembly Hall. Furthermore, he directed his RSM, Warrant Officer Blue Mulby, to have constructed a simple memorial stone, comprising a large rock set on a concrete base, with a brass plate bearing the names of all members killed in service with the regiment in war and peace. He directed that the memorial stone be erected on the lawn opposite regimental headquarters, close to and visible to all entering and leaving the regiment by the front gate. The Rock, as it was soon known, was dedicated on Anzac Day 1982 by the unit chaplain.

In December 1982 Lieutenant-Colonel Chris Roberts assumed command. Just a month short of his 38th birthday, he was returning for his third posting with the regiment. After his operational tour of Vietnam in 1969–1970 he had been adjutant of 5 RAR, instructor at the Officer Cadet School, Portsea, had obtained a Bachelor of Arts

Colonel C. A. M. Roberts, AM, Commanding Officer of the SAS Regiment, 22 December 1982–30 June 1985. By the time of his command additional resources were becoming available to enable the third sabre (operational) squadron to be built up and for war roles training to be strengthened.

degree at the University of Western Australia, and in 1976 and 1977 had commanded 1 SAS Squadron. He had then attended the Australian Army Staff College, had been brigade major of the 1st Task Force and had been a staff officer in the Directorate of Special Action Forces where he was responsible for preparing the draft Army Office staff instruction on employment of special action forces.

On assuming command Roberts was very impressed by the high quality of work that had been done in developing the CT capability, but thought that through no fault of his predecessor the war role side of the regiment had been degraded. When the CT capability had first been raised, Reg Beesley had been conscious that the system, employed in Vietnam, of rotating squadrons rather than individual personnel, had been highly successful. The disadvantage was that at the beginning of each tour the level of effectiveness was not as high as it could have been, but this was outweighed by the good esprit de corps which led to a higher level effectiveness after only a short period of time. With these thoughts in mind he decided that the TAG would be formed from a troop which would serve in that role for a year before being replaced completely by a new troop. The problem was that before 2 Squadron was re-raised, the only source of a new troop was 3 Squadron, which therefore found it extremely difficult to maintain continuous and worthwhile training.

In an effort to rectify this problem Major Wallace suggested to Lieutenant-Colonel Curtis that the troop replacement system be

Warrant Officer Class One
M. J. Ruffin, OAM,
Regimental Sergeant Major
of the SAS Regiment, 18
October 1982–10
December 1985. He gave
great support to
Lieutenant-Colonel Roberts
during the difficult period
when the regiment was
expanding from two to
three sabre squadrons.

changed to one of trickle reinforcement. Curtis was about to hand over to Roberts and he asked Wallace to present his proposals to both lieutenant-colonels. Roberts agreed to the proposal and ordered the new system to be instituted, with the understanding that it was not to be a permanent solution.

From mid 1983 the two assault troops in 1 Squadron remained static and the individual personnel were changed at regular intervals. By this time Tactics Wing had moved to the Eastern States to run special warfare courses for the Commando Regiment and a Training Cell was set up in 1 Squadron to conduct four CT courses each year. As the SSM of 1 Squadron wrote: 'It may have tortured Training Cell, but at least it greatly reduced the burden on the roles and tasks squadrons'. Major Duncan Lewis, who succeeded Paul Robottom as commander of 2 Squadron, recalled that the trickle system 'was a marked improvement on the previous system where virtually whole troops would be moved. The trickle system was not, however, viewed by the war roles squadrons as a final solution. We all wanted to rotate the CT role, but the fluctuating levels of CT training that would result were, at that time, considered unacceptable. So we persisted, often losing our best patrols to 1 Squadron at the rate of one every three months...Troops returning from 1 Squadron were often tired and jaded after the intense levels of training activity they experienced while on-line'. Despite its drawbacks, the trickle system nonetheless had the virtue of enabling Roberts to restore the war role capability of the regiment.

Of course Roberts had resources not available to Curtis. He was able to build on the work of his predecessor, and reaped the benefit of the solid recruiting campaign to gain more soldiers for the regiment. His methods were not always popular. Tough and determined, he pushed through measures against pockets of strong resistance from conservative elements across all sections of the regiment. For example, his efforts to move the Rock and establish a dignified and more fitting memorial attracted great opposition. Curtis, who had initiated the Rock, was not a supporter of moving it because it had been dedicated and accepted by members of the unit. In recent years, however, most SAS soldiers have accepted the new position and enhanced surroundings of the Rock, and indeed look upon it with a degree of pride, perhaps even affection. How perceptions change with time! But it gave emphasis to Sniffy Roberts' nickname, 'Rock', which had started as a derogatory term for his 'gung ho' approach in Vietnam but now referred to his somewhat determined and uncompromising attitude on certain issues.

The war role side of the regiment continued to improve, and in October 1985, 2 Squadron flew direct from Australia and parachuted into the Dobodura area of Papua New Guinea to begin the first SAS exercise in that country since 1970. 3 Squadron followed the next year. The war roles were emphasised further when, after agitation from the regiment, one of the two war role squadrons was earmarked to support the Army's Operational Deployment Force which was based on the 3rd Brigade at Townsville.

Although Roberts put great emphasis on developing the war roles, he never lost sight of his first priority, the CT commitment, and this capability continued to improve. Soon after he took command he rationalised the names of the CT force. The CT force (1 SAS Squadron, 1 Signal Troop and the commanding officers's command group) became the TAG. GAUNTLET became A Troop; NULLAH became B Troop; and the terms OAG, GAUNTLET and NULLAH were dispensed with.

Another change concerned the command structure. It was found that the CT squadron commander at the Police Forward Command Post had too much work to do. He had to plan the assault and if, as was likely, members of both teams were involved, he had to be prepared to command the assault. Yet he had to keep abreast of developments in the terrorist incident, advise the police Forward Commander on SAS capabilities, arrange assistance from other Defence agencies, such as the RAAF or the RAN, and keep the commanding officer at

The Rock, the SAS memorial inside the entrance to Campbell Barracks, Swanbourne, commemorates all those who died in service of the regiment in war and peace.

the Police Operations Centre informed. On the other hand the commanding officer did not have a particularly onerous task but was hampered by his lack of feel for the incident. The solution was for the commanding officer to move to the Police Forward Command Post with his small headquarters of an intelligence officer, operations officer and communicators, thus releasing the squadron commander to concentrate on mounting the actual assault. The commanding officer of the Commando Regiment remained at the Police Operations Centre. The change was implemented formally in early 1984.

By the time Roberts relinquished command in July 1985 the regiment was beginning to settle down after the effort of raising and developing the CT capability. As the new commanding officer, Lieutenant-Colonel Terry Nolan, commented, he could not have taken over from a better man than Roberts. Many of the hard and unpopular decisions had been taken, some old soldiers had left the regiment, the CT capability had been refined, 2 Squadron had been successfully re-raised, the war role had gained its proper importance, the manning

Colonel T. J. Nolan AM,
Commanding Officer of
the SAS Regiment, 1 July
1985–7 January 1988. He
was the first commanding
officer to have graduated
from the Officer Cadet
School, Portsea, and the
last with operational
service from Vietnam.

level of the regiment had improved, and new equipment, ordered in the late 1970s, was starting to appear.

Aged 40, Nolan had not served with the regiment since he had been second-in-command of a squadron in 1970, but he had gained wide experience in infantry, having served for a total of four and a half years in 5 RAR and 5/7 RAR. He had been an instructor at both the Officer Cadet School, Portsea, and at the Royal Military College, Duntroon, and had been a Standardisation Representative in London with a secondary duty as SAS liaison officer between the Directorate of Special Action Forces and British SAS headquarters. Before his appointment to command the SASR he was a staff officer in the Directorate of Special Action Forces.

Nolan presided over a period of substantial consolidation in the regiment. In April 1986 the Minister for Defence, Kim Beazley, announced that new facilities for CT training were to be constructed at Swanbourne, at the Bindoon training area, and at Gin Gin airfield. Work was to begin in June 1986 and $22 million was to be spent over the next three years. As early as November 1978 the SASR had initiated a paper outlining the facilities required for CT training, and by 1986 the interim facilities had deteriorated considerably. Mr Beazley stated that the new facilities at Swanbourne would 'include an indoor and outdoor close quarter battle range complex, an outdoor sniper range and an urban mock-up of a scene. At Bindoon...a special urban complex for counter-terrorist training [was to] be built as well as a sniper "plunging" range (enabling vertical firing training not just horizontal firing as now), and an administration and site services area. At Gin Gin, an aircraft mock-up [was to] be constructed'.[16]

One of Nolan's early concerns was the deterioration in the offshore capability which, due to trade union problems on the Bass Strait oil platforms, had not been exercised properly for over a year before he assumed command. The union difficulties were resolved in May 1986 and training recommenced. The offshore capability continued to be restricted by the aging Wessex helicopter fleet, which was not replaced by Sea King helicopters until December 1986.

Another concern was the trickle system of reinforcement for the CT squadron, the disadvantages of which had been recognised by Roberts when he had introduced the system. Having maintained a CT capability for six years, there was a wider spread of CT experience throughout the regiment, and Nolan realised that the situation now allowed him to change the system and to replace 1 Squadron at the end of 1987 with 2 Squadron as the CT squadron. It became the first squadron other than 1 Squadron to hold this role. The squadron commander, Major Mike Hindmarsh, discovered the meaning of the word 'attrition' during the work-up training towards the end of 1987. In one six-week period the squadron suffered the following incidents: a sergeant lost a leg in a motorcycle accident; a sergeant lacerated his hand when he caught it in the propeller of an outboard motor while conducting diving training; a sergeant broke a wrist while diving in Bass Strait; a sergeant lost the forward sight of an eye when hit by grenade fragments while training overseas with US forces; a corporal smashed his left forearm while training in the Method of Entry House; a trooper shot off the tip of a finger; a trooper severed a tendon in one of his fingers; and a trooper broke a leg while parachuting.

It is difficult to gauge how good the SAS CT capability really is. Senior SAS officers who have seen CT forces in other countries believe that the Australian force is as good if not better than any other force. In 1987 a book titled *The Rescuers—The World's Top Anti-Terrorist Units* by Leroy Thompson, attempted to rank the world's best CT units. Thompson claimed to have asked friends in each of the different groups to rank their top ten. His top ten in order were: Britain's SAS, West Germany's *Grenzschutzgruppe 9*, France's *Gendarmerie d'Intervention de la Gendarmerie Nationale*; Australia's SAS, Israel's *Sayaret Matkai*; the US military's Delta Force; Spain's *Grupo Especial de Operaciones*; the US Federal Bureau of Investigation's hostage rescue team; Italy's *Group Interventional Speciale*; and Holland's Royal Dutch Marines.[17] Unlike many of the other forces, the Australian SAS has never been tested in a live situation. According to a newspaper report in 1984, the SAS had been placed on a Special Counter Terrorist Alert—one step short of Full Alert—on three occasions in the previous two years. The first was in December 1982 with the bombing of the Israeli consulate in Sydney; the second was in January 1983 when a threat was made to blow up a TAA aircraft; and the third was in August 1983 after the Australian Security Intelligence Organisation (ASIO) reported that Armenian factions were preparing to strike at Australian targets.[18]

The SAS continued to develop its CT capability, and in August 1988, for the first time, the CT squadron exercised on the North Rankin A offshore gas platform off North West Cape. The soldiers found that the threat from sharks and sea snakes was not as great as they had been led to believe.

The CT requirement still places a substantial burden on the SASR but it has now been placed in a better perspective. The war role of the regiment has again received the attention it deserves. Furthermore, the development of CT skills has meant that if the regiment is required to exercise its war role skills, they will probably be executed with just a little more expertise—an expertise honed in the very real world of counter-terrorism.

24
Phantoms of the operations: 1987–1999

By the beginning of 1987 the SAS Regiment had accumulated a decade of counter-terrorist experience, with its CT squadron being held continuously at a high state of readiness. But while the regiment had continued to develop its war roles, there seemed little chance of their being used. The following twelve years, however, were to see momentous changes in the regiment's roles and capabilities and, importantly, a return to overseas operations. As might be expected, most of the SAS's activities were within larger operations conducted by the Australian Defence Force (ADF) and were undertaken with very little publicity. No longer just the phantoms of the jungle, as in Borneo and Vietnam, they were now truly the phantoms of the operations, which were likely to take place in a range of environments, from jungle to desert and urban areas. But gradually the SAS was to take a more prominent role, and there were several reasons for this development.

The end of the Cold War in 1989 brought fundamental change to the strategic environment, resulting in increasing numbers of peace operations under United Nations auspices. Within Australia, as the Vietnam War receded from popular memory it became more acceptable to deploy forces overseas again. Special Forces frequently offered the government an attractive, cost-effective and discreet alternative to the deployment of larger military units, particularly in regional countries where Special Forces are held in high esteem, and they brought particular skills that were needed in sensitive political situations where multiple government agencies were involved. The SAS adapted to meet these new demands, in many cases anticipating the demands with new capabilities already introduced.

In December 1997 Colonel Don Higgins, Commander Special Forces,

446

reflected on these changes. He had joined the regiment in 1978 when there 'were few overseas training activities and operational deployments were non-existent'. By the time he returned as commanding officer in 1993 he found

> a unit with a wide range of regional training activities as well as on-going CT functions. Peace operations were trickling in although we could never seem to get enough to please everyone. The unit contribution to the UN missions in Cambodia, Somalia, Rwanda and Iraq were largely positive experiences and our contingents were widely praised.

By the end of 1997 he noted that 'as well as very demanding peacetime activities the unit has coped with a modest but constant string of operations and some unique challenges'.[1] The period covered by this chapter saw a series of small deployments and activities which, individually, were often insignificant, but when taken together represent a major development in the history of the Australian SAS.

During the 1970s and 1980s it had become increasingly difficult to maintain operational experience and readiness under peacetime conditions. After withdrawing from Vietnam in October 1971 the SAS had turned its attention to the direct defence of Australia and had developed concepts for operating across the north of the continent. These developments led to the formation between 1981 and 1985 of three regional force surveillance units based in northern Western Australia, the Northern Territory and North Queensland. Eventually these new, mainly Army Reserve units took over most of the surveillance operations across this area, thus relieving the SAS Regiment for other duties.

During the late 1970s and the 1980s the SAS Regiment concentrated on developing its CT capability. The exploits in Borneo and Vietnam still dominated the collective memory, even though most of the experienced Vietnam-era soldiers had left the unit. The regiment continued to build on this wartime experience, developing both tactical concepts and individual skills for surveillance and raiding, but there was uncertainty about the situations in which these capabilities might be used.

Perhaps more than any other military unit, the capability of the SAS relies on the skill, initiative and character of its individual soldiers. Adventurous activities such as mountaineering, however, provide the means of fostering and sustaining vital initiative and spirit. The climbing wing was formed in 1971 with Sergeant Paul (Yogi) Richards as the first senior instructor, and by 1975 nine soldiers were qualified as cliff leaders class 1. In 1976 the climbing wing took over the regiment's survival training and became known as the climbing/survival wing. When K Troop was designated as a climbing troop, one of its

447

early commanders was Captain Pat Cullinan, an unassuming but enterprising young officer.

Even at this early stage SAS climbers began planning an attempt on Mount Everest, and in November 1978 fifteen soldiers, led by Captain Greg Nance, climbed Mount Cook in New Zealand in preparation for an attempt on the difficult Mount Gauri Shankar (7213 metres) in the Himalayas. The attempt, led by Cullinan and conducted under the auspices of the Australian Army Alpine Association from 19 February to 26 April 1980, was the first Australian expedition to climb in the Nepalese Himalayas. Dogged by bad weather, the expedition failed to reach the peak but gained valuable experience. On one occasion Paul Richards and a sherpa were buried in snow for 24 hours. On another occasion Warrant Officer Barry Young fell almost 400 metres down a gully. A sherpa, seeing a pack fly past, moved out onto the slope and grabbed Young as he came past.

SAS members continued climbing around the world. Captains Cullinan and Tom Moylan climbed Mount Kilimanjaro in Tanzania, while Captain Jim Truscott climbed Ganesh IV in the Himalayas (during this expedition his climbing partner was killed in an avalanche). Captain Tim McOwan scaled the Matterhorn in Europe, and in 1986 ten members of K Troop reached the summit of Mount Victoria in Papua New Guinea.

Cullinan booked Everest for 1988, and to prepare for the ascent he led a series of climbing expeditions, including one to Broad Peak (8047 metres) in Pakistan from June to September 1986. The difficulty of climbing in this area is demonstrated by the experiences of nine expeditions on K2 (8611 metres) during the summer of 1986 while the Australians were nearby on Broad Peak. A total of 27 climbers reached the summit of K2 but thirteen of them died, seven after reaching the summit. Above 8000 metres climbers are faced with rapid physical and possible mental decline. Only a third of the oxygen that is available in the atmosphere at sea level leads to a degree of strain on the heart and lungs that makes even rest of little worth. Climbers who spend even a couple of nights above 8000 metres run the risk of severe physical damage.

Pat Cullinan reached the summit of Broad Peak late one afternoon and while descending met a German climber approaching the summit by himself. Realising that the German was disoriented, Pat persuaded him to return down the mountain. They had not travelled far when night fell and together they bivouacked above 8000 metres. Next morning the German was too exhausted to move. At about 11 am observers, looking through binoculars from the base camp, saw Pat slowly dragging the German along a ridge in his sleeping bag. A rescue expedition was mounted and late that afternoon the party reached Pat and the German. Cullinan had kept the German alive by himself for 24 hours. By now a storm was developing and it took a further six days for the party to descend from the mountain. The German was hospitalised for six months and undoubtedly would have perished during the first night had Cullinan not supported him. In that environment it took all of a man's determination and strength to keep himself alive, let alone look after another disoriented, irrational and uncooperative climber. Disorientation and belligerence are symptoms of hypothermia. For his courage and determination Cullinan was awarded the Australian Star of Courage.

Discovering that the Australian Alpine Climbing Club had also booked Everest for 1988, the Army Alpine Association joined with the civilian club to form the Australian Bicentenary Everest Expedition. They decided to attempt the ascent without the help of high altitude porters. While this self-imposed handicap reduced the chances of ultimate success, it meant that if there were success, it would be due solely to Australian efforts. Of the eighteen climbers and eight Australian support staff, Majors Cullinan, Jim Truscott and Rick Moor were ex-SAS officers (Cullinan was in the 1st Commando Regiment), and Sergeant Norm Crookston was a serving SAS soldier. Only three climbers eventually reached the summit, after the first attempt was defeated by deep snow. Major Peter Lambert, a member of the expedition, described the success:

> Two weeks later [on 25 May 1988], in a magnificent display of endurance and skill, Major Pat Cullinan and Paul Baynes [a civilian] reached the summit. Although in improved conditions, it still took them six hours to reach our high point and a full 25 hours for the round trip back to Camp Four. Both climbers' oxygen ran out on the way down . . . Pat thought the Col was Charlotte's Pass and Paul thought he saw an android on the Col. Such are the effects of high altitude oxygen deprivation! More incredible was that they climbed after spending six days at 8000 metres waiting for good weather—the longest anyone has spent at that debilitating altitude.[2]

It was the first successful expedition to climb Everest from Nepal without high altitude porters.[3] Continuing the SAS's climbing activities, Ric Moor led an expedition to Mount McKinley in North America in 1987. In 1990 Jim Truscott climbed Mount Aconcagua in South America; he made the third Australian ascent of Puncak Jayawijaya in Irian Jaya in 1991 and the first Australian ascent of Nanda Devi East in India in 1994. In 1991 Lieutenant-Colonel John Trevivian led an expedition to Mount Kilimanjaro. In May 1992 eleven SAS members, led by Sergeant Peter Hooper, reached the summit of Mount McKinley. The other members were Captains Richard Campbell and Martin Skin, Corporals Gary Boylan, Ralph Folie and Leon Kouts, Lance-Corporals Mick Aiton and Dick Tracey, and Troopers Brian Hale, John Sullivan, Mark Forester and Stuart Young. Other members continued climbing in the greater mountain ranges of the world. In the absence of active service the expeditions provided the challenges needed by the young men of the regiment. As Cullinan wrote about his ascent of Everest: 'I think that when you've served in SAS for any period of time, what you would like to do once, even just once in your life, is put your skills to one real test, and I mean a real test.'⁴

The fact that Cullinan had not served in Vietnam was an indication that the SAS was not living in the past, but was adapting to a new strategic environment. When Lieutenant Colonel Jim Wallace, aged 36, assumed command of the regiment on 8 January 1988, for the first time the SAS had a commander with no war service. Having graduated from Duntroon in 1973, he had served as both a troop and squadron commander in the regiment. He had also been a UN observer in Syria and Lebanon in 1980 and had attended the British Staff College at Camberley. Considering the way the strategic environment was developing in the late 1980s, perhaps this United Nations experience was as relevant as any other might have been.

Throughout the 1970s and 1980s the SAS had continued to train for overseas operations, including special operations, guerilla warfare and small-scale offensive operations, although the government's defence policy seemed to make such operations unlikely. This policy did not change substantially when the government released its Defence White Paper, *The Defence of Australia*, in March 1987, in which it was argued that low-level threats to Australia could develop relatively quickly, and that Australia's military forces had to be structured for the defence of Australia and its direct interests. The ADF also had to be able to respond to attacks in the area of direct military interest, defined as 'Australia, its territories and proximate ocean areas, Indonesia, Papua New Guinea, New Zealand and other nearby countries of the South-West Pacific'.⁵ The ADF's future development, however, focused on the continental defence of Australia and gave no priority to deploying forces from Australia, even into the area of direct military interest.

Two months later the comfortable assumption that no forces would be deployed overseas was challenged by a military coup in Fiji. Faced with the likelihood of a breakdown of civil order, the Australian government directed the ADF to prepare to evacuate Australian citizens. In Operation MORRIS DANCE several Australian ships, with a company of infantry from 1 RAR embarked, deployed to the vicinity of Fiji. A squadron from the SAS Regiment was placed on readiness for deploy-

Lieutenant-Colonel J. J. A. Wallace, AM, Commanding Officer of the SAS Regiment, 8 January 1988–16 December 1990. He had previously served as both a troop and squadron commander in the regiment.

ment. By the end of May the situation had stabilised and the Australian forces began to withdraw from Fiji.

One year later a similar situation arose when it appeared that civil disturbance in Vanuatu might place the lives of Australian citizens there in danger. Forces were not deployed, but the Fiji and Vanuatu incidents underlined the need to prepare plans and develop command structures to facilitate operations to protect or evacuate Australian citizens. The SAS Regiment was ideally suited for these types of operations.

The end of the Cold War in 1989 gave added impetus to preparing the SAS for unexpected contingencies. Some in the regiment began to recognise a need to change its emphasis and perspective from a strict delineation between CT or 'black' roles and war or 'green' roles to a more generic challenge—an offshore recovery operation with what seemed to be a black role in a green setting. The regiment would have to become more multi-skilled.

In February 1990 the Directorate of Special Action Forces in Army Office became Headquarters Special Forces and the Director, Colonel Chris Roberts, became the first Commander Special Forces. He was given responsibility for both the SAS Regiment and for the Army Reserve 1st Commando Regiment. The latter regiment always had a high percentage of Regular staff, most of them coming from the SAS Regiment. The Commander Special Forces and his headquarters continued to be located in Army Office and came under the Chief of the General Staff for training, administration and development, but reported directly to the Chief of the Defence Force for operational deployments, including CT operations. This was a substantial step forward in the command and control of Special Forces and was in line with the view that the SAS needed to be considered as a strategic asset, even if a joint command structure to encompass the Navy and the Air Force was not developed.

Other countries had already seen the need to place special operations on a joint footing. For example, in October 1986 the US Congress directed the Joint Chiefs of Staff to set up Special Operations Command, to consist of all US special operations forces with the exception of Naval Special Warfare Groups. The new command was to include nine US Army Special Forces battalions, three Ranger battalions, four psychological operations battalions, one civil affairs battalion, six US Navy SEAL teams, three US Air Force (USAF) special operations wings, one special operations aviation brigade, 165 dedicated USAF fixed-wing aircraft and helicopters, and 89 dedicated Navy boats and other craft. General James A. Lindsay, the first Commander-in-Chief, established his headquarters at MacDill Air Force Base in Tampa, Florida.

The US Department of Defense bitterly resisted the formation of the new command. But, as one analyst wrote, it was important that the President understand what was available to him in times of crisis.

452

The line that runs from the Commander-in-Chief and the operators he is actually commanding has to be made as short and elegant as possible . . . As national assets, the President should know what these forces look like, what they do, how they train, and what they need, and he must have confidence in the character and judgement of the men who lead them.[6]

The value of Special Forces had been shown in the 1982 Falklands War, in which the British SAS and SBS had played crucial roles. It was demonstrated again in the 1991 Gulf War, in which the Commander-in-Chief Central Command, General Norman Schwarzkopf, had six major subordinate forces for operations—the US 3rd Army, Joint Forces Command (Arab forces), Central Command Marines, Central Command Air Force, Central Command Navy and Central Command Special Operations. The latter included the US 5th Special Forces Group, the 75th Ranger Regiment, USAF Special Operations Forces, US Navy SEAL Forces and British SAS and SBS units. These Special Forces conducted reconnaissance in Iraq and directed attacks on Iraqi SCUD launcher sites. Four members of the SAS lost their lives behind enemy lines, and several members of SAS patrols in Iraq have since written books about their experiences.

Considering the large numbers of Special Forces units deployed to the Gulf, the Allies did not need Australian SAS elements. In any case, the Australian government was reluctant to place any forces on the ground, or to deploy forces that might have incurred casualties. Australia's contribution was restricted to naval forces, a small staff of intelligence analysts and several personnel on exchange with British forces. Lieutenant-Commander John Griffith, the commander of the RAN clearance diving team that did extensive and difficult work around Kuwait City, was a former member of the SAS Regiment. On the initiative of the regiment, two troops of the Australian SAS were placed on standby in case they were deployed, but primarily the war merely reminded the SAS that they had to be prepared for any such operation. Despite years of training SAS still had little capacity to conduct operations overseas, and it was not until the major ADF Exercise KANGAROO 1989 that an SAS element was deployed as part of the 'enemy' force to undertake clandestine operations.

Although the SAS was not employed in the Gulf War, during succeeding years SAS soldiers were deployed on several overseas operations under United Nations auspices, thereby increasing the concentration of operational experience in the regiment. The soldiers were not deployed in their SAS role, but rather as individual medics or signallers. Personnel with those skills were deployed from many regiments in the Army, and the SAS soldiers formed only a small proportion of them. They proved, however, to be valuable members of the

Lieutenant-Colonel D. E. Lewis, Commanding Officer of the SAS Regiment, 17 December 1990–4 January 1993. During his command members of the regiment began to be deployed on a series of UN peacekeeping operations.

deployed teams because of their high level of training, multiple skills, resourcefulness and initiative. Within each SAS patrol at least one member is trained as a medic and another as a radio operator. Additional medical expertise is found in the regiment's medical centre, so that when an SAS squadron is deployed on operations or on exercises Royal Australian Army Medical Corps members are deployed to provide additional medical support. The regiment's medical centre staff provide continuation training and trade testing of patrol medics.

Because the SAS operates in small groups over long distances, communications are crucial, and 152 Signal Squadron is an integral part of the regiment. It consists of Royal Australian Corps of Signals personnel and provides multi-band communications and deployable secure networked information systems for the SAS squadrons and regimental headquarters. Although each SAS patrol has its own radio operator, some operators from 152 Signal Squadron have the necessary training to allow them to fill in as patrol members.

By the time the first SAS personnel were deployed overseas with the UN the commanding officer was Lieutenant-Colonel Duncan Lewis, who had assumed command on 17 December 1990 during the build-up for the Gulf War. Aged 37, he had graduated from Duntroon in 1975 and joined the regiment in 1977. During 1982 he was a UN observer in the Middle East and was in Lebanon during the Israeli invasion. Between 1983 and 1985 he was commander of the SAS Base Squadron, second-in-command of the regiment and commander of 2 SAS Squadron.

The first UN deployment to involve the SAS was between 16 May and 16 June 1991 when 72 Army and three RAAF medical, dental, engineering and logistic personnel were deployed as part of Operation HABITAT to Turkey and northern Iraq to assist Kurdish refugees who had been driven north by Iraqi forces. The SAS provided several medics.

The second operation also grew out of the Gulf War. Following the end of the war the UN Special Commission (UNSCOM) was established to oversee the destruction of Iraq's weapons of mass destruction. Between 1991 and 2000, as part of Operation BLAZER, about 125 Australian personnel served with various UNSCOM inspection missions. They usually deployed to Bahrain to join up with their assigned UNSCOM team (ten to twenty personnel), which then deployed into Iraq on chartered UN aircraft to conduct their assigned tasks for periods of one to two weeks. SAS medics were deployed as part of some teams; they were particularly valuable because of their skills with communications, navigation and driving.

The third operation was the deployment in September 1991 of a 45-man contingent of radio operators and drivers to the Western Sahara to provide communications for a UN force monitoring the ceasefire between Moroccan forces and Western Sahara independence fighters. It was hoped this ceasefire would lead to a referendum. The SAS provided several signallers from 152 Signal Squadron. The Australian radio operators deployed to each of the observation sites lived in extremely trying conditions, at times operating in the desert without a break for a month and working up to 15 hours per day. Five contingents were deployed, each for six months, before the force was withdrawn in May 1994 with no decision concerning the referendum in sight. Lieutenant-Colonel Gary Barnes, a former SAS officer, commanded one of the contingents.

The commitment of signallers to the Western Sahara corresponded with an even larger commitment of communicators to Cambodia. During 1991 the United Nations passed two resolutions aimed at producing a neutral environment in which the four warring armies in Cambodia would disarm and demobilise and so allow free and fair elections. The first resolution established the UN Advance Mission in Cambodia and in October 1991 Australia contributed a 65-strong communications unit, while the details for the deployment of a larger force were determined. In March 1992 the UN Transitional Authority was established. Australia provided the Force Communications Unit, with 488 personnel drawn mainly from the 2nd Signal Regiment (plus RAN and RAAF communicators); this force included 40 New Zealand personnel. Australia also provided fourteen staff for the Force Headquarters.

Seeking signals operators from across the ADF, it was natural that the Army should draw on the highly trained operators of 152 Signal Squadron. The first contingent that landed at Phnom Penh in November 1991 included eight SAS members—Captain Peter Bartu (who became ADC to the force commander), Sergeant Tony Wills, Corporals Carlos von Bishoffhausen, John Gallarello, Gavin Hillman, Roger Morish and Mick Ryan, and Signaller Fraser McKenzie. The signallers spent a couple of weeks in Phnom Penh before being deployed in three-man communications teams to support military liaison officers scattered across the country with the various warring factions. Inevitably, the SAS signallers found themselves working in the more remote locations, imposing a heavy responsibility on relatively junior personnel, often in delicate and at times dangerous situations. Some signallers had seven-month tours, others were there for thirteen months. The dispersal of the Australian signallers in 56 locations enabled the UN force commander, Lieutenant-General John Sanderson (an Australian officer), to maintain a strong grip on the force. As Sanderson later explained, 'the Australian contingent was the glue that held the mission together'.[7]

The Cambodia operation was still continuing when, on 5 January 1993, Lieutenant-Colonel Don Higgins assumed command of the regiment. Aged 38, he had been in the same Duntroon class as Lewis and had joined the regiment in 1978. He had commanded a troop, had been a staff officer in the headquarters of the British special forces group in London, and had been both the regimental operations officer and the commander of 3 SAS Squadron.

Higgins's main memory of this period is the constant preparation of contingency plans in case members of Australian forces deployed overseas came under threat. This was particularly important in Cambodia, where elections were held during 1993; several SAS teams visited the country as part of this planning process. By this time the commanding officer of the Force Communications Unit was Lieutenant-Colonel Martin Studdert, a former SAS signals officer.

The UN operations in Turkey, Iraq, the Western Sahara and Cambodia gave active service to selected SAS soldiers between 1991 and 1993, but they had not been required to operate as discrete SAS teams. In Somalia, in 1994, however, the SAS was to undertake its first operational deployment (as a formed force element) since the Vietnam War 23 years earlier. It was to be a relatively low-key deployment, but one

Lieutenant-Colonel D. G. Higgins, Commanding Officer of the SAS Regiment, 5 January 1993–8 December 1994. His command was marked by an increasing tempo of operations, including the deployment to Somalia in 1994.

that demanded a high degree of professionalism, discipline and dedication.

Australia's involvement began in October 1992 when the first members of an ADF movement control unit arrived in the Somali capital, Mogadishu. Civil war had led to mass starvation and anarchy, and UN forces were being deployed to protect the delivery of humanitarian assistance. The following January, in Operation SOLACE, an Australian battalion group based on the 1st Battalion, The Royal Australian Regiment (1 RAR), arrived to undertake security operations around the town of Baidoa. SAS members were not deployed on this operation. When this force returned to Australia in May 1993 the movement control unit remained, to be joined by ADF air traffic controllers as part of UN Operation in Somalia II (UNOSOM II). This contingent experienced tense periods, especially on 3 October 1993 when two US Black Hawk helicopters were shot down in Mogadishu and US Rangers and Delta Force members were surrounded, losing eighteen killed in a fourteen-hour battle. This incident is graphically described in Mark Bowden's detailed account, *Black Hawk Down*.[8] After dead US servicemen were dragged through the streets of Mogadishu President Clinton decided to withdraw US forces, the last leaving in March 1994.

A security patrol by the SAS detachment in Somalia passes a Nepalese Ferret reconnaissance vehicle in the airfield compound at Mogadishu in May 1994.

As other countries began withdrawing from Somalia, 3 SAS Squadron was warned in February 1994 of a possible deployment to protect the Australian Services Contingent (ASC), which by then numbered about 67 personnel.[9] On the advice of Major Greg de Somer, commander of 3 SAS Squadron, Lieutenant-Colonel Higgins appointed Sergeant Gary Kingston to command a ten-man team, to be known as J Troop 3 Squadron. Members were drawn from throughout 3 Squadron and included specialists in mounted operations, free-fall, water operations and communications.[10] On 31 March a two-man advance party led by Sergeant Malcolm Wood departed for Somalia and the remainder of the team departed on a civilian flight on 14 April.

The team's main task was providing security for the ASC, which was split between two locations—a camp at the airfield and the Australian Embassy building (Anzac House) in Mogadishu, three kilometres away. The team's duties included VIP protection and protection for ASC personnel on high-risk tasks, providing quick response teams for extraction of ASC personnel under threat, conducting weapons training and PT for the contingent, developing evacuation and security plans, route reconnaissance, and providing medical and signals support. The contingent was equipped with Steyr rifles, Minimi light machine guns, M79 grenade launchers, pistols and Heckler & Koch

Members of the Australian Services Contingent security group—the Gerbils—before their first route reconnaissance to Baledogle, Somalia, in May 1994. They are pictured with their two M113 APCs.

sub-machineguns. They also had night-vision equipment and body armour. In addition to Toyota Land Cruisers and Datsun utilities, the UN issued the security team with two M113 armoured personnel carriers (APCs). Team members had undertaken training in Perth with APCs from the 10th Light Horse Regiment, and one soldier, Lance-Corporal Alan Reid, had previously driven APCs in 5/7 RAR (Mech), the Army's mechanised infantry battalion.

The SAS team named its camp at the airfield Camp Gerbil, after the mouse-like rodents that roamed the hills, and team members were soon nicknamed 'the Gerbils'. This was clearly a dangerous mission. On 19 April, three days after the team arrived, two Nepalese soldiers and one Pakistani were killed and the Nepalese compound received rocket fire. Fighting between Somali clans continued over the next few days and the Australians occupied bunkers to gain protection from rocket and heavy machinegun fire. In this threatening environment the SAS team took the security task seriously. When in May 1994 the main body of ASC IV arrived to replace ASC III, one member of ASC IV gave a vivid description of being met by the SAS security team at the Mogadishu airfield:

There were some military personnel dressed in flak jackets waiting to greet us at the hangar, but it's the others that caught my eye.

They were dressed in full combat gear, all carried the standard Steyr (with some 'minor adjustments'), wore frag jackets which bulged with goodies and big boy toys. Most of them wore dark ski goggles, fitted to black stone guards that the Motocross riders wear to protect themselves from the dust and stone chips. They positioned themselves around the aircraft in such a manner that at any given moment should an attack occur there were three overlapping arcs of fire that could have been returned.[11]

Sergeant Kingston later summed up the problem of operating in an area where many Somalis were carrying weapons:

J Troop were therefore in a situation in that each time they carried out a task on the streets they came face to face with all types of armed men. With no identified enemy they relied heavily on their own training in identifying hostile action and the use of appropriate counter measures. This was to become a weekly occurrence with other contingents often taking casualties. On many occasions diplomacy prevailed and the Troop took a step backwards if they were completely out gunned and had no back up.[12]

Many UNOSOM contingents were not prepared to give support to members of any other national force than their own, and the Australian SAS team often found itself filling the breach. An example of this occurred on 21 May 1994, when a civilian Canadian helicopter went down 20 kilometres north of Mogadishu and an armed crowd threatened the crew. In response to tardy reaction from the Pakistani quick reaction force, the commander of ASC III, Lieutenant-Colonel Brian Millen, ordered Kingston to secure the helicopter. Eight Gerbils flew in to the crash scene on another Canadian helicopter with the doors closed. When the doors opened to reveal the heavily armed Australians the Somali gunmen disappeared very quickly. The Gerbils secured the area until further help arrived. When two SAS soldiers rendered medical aid to several Somali children the tension eased.

With their numerous tasks J Troop often found themselves with insufficient personnel. But two members of the SAS Regiment's 152 Signal Squadron—Corporals John Gallarello and Wade Hughes—were part of ASC IV, and with the approval of ASC IV's commanding officer, Lieutenant-Colonel Stuart Ellis (a former commander of 3 SAS Squadron) they boosted J Troop to twelve personnel when required.

Although there were no direct attacks on the Australians, there was considerable fighting between rival militia gangs and occasional attacks on other UN personnel. In late June the security team was confined to Anzac House for two days because of fighting taking place between it and the airfield where the APCs were stationed. There were insufficient security team members at the airfield to drive the APCs to Anzac House. The security team finally departed Anzac House in soft-skinned vehicles; when they approached parties of militia exchanging fire, the militia ceased firing to allow the Australians to pass.

The danger to vehicle convoys was shown when eight to ten Somali gunmen ambushed a three-vehicle UN convoy on 18 July. The UN force included three Italian officers, a New Zealand warrant officer and a protection party of nine Malaysians. The driver of one vehicle was killed in the ambush. Although some UN personnel returned fire, within minutes the patrol commander surrendered to the gunmen with other members of the patrol not having fired a shot. The UN members were kicked and punched and another one was killed. The New Zealand warrant officer was nearly killed, but was saved by the intervention of another Somali. The UN team was released after an approach to the Somali clan leaders, Farah Aideed and Ali Mahdi.[13]

The Australians were determined not to be similarly humiliated, as demonstrated on 16 August. The Australian movement control unit had been given the task of driving two UN dual-cab vehicles from the old port to the airfield for air transportation. The Pakistan contingent was unable to meet a request to escort the vehicles. The convoy consisted of an APC, the two dual-cab vehicles and another APC. Members of the security team drove the APCs and several rode in the back of the dual-cabs. The convoy passed a number of technicals (utilities or light trucks with mounted weapons) and militia vehicles without incident.

As the convoy approached the airport a Somali utility approached it from the opposite direction, with a gunman standing in the rear pointing his weapon (an AK47 variant) at the first dual-cab. Lance-Corporal Gary Porter, standing on the tray of the second dual-cab, raised his Minimi light machinegun to cover the gunman. At this instant the gunman aimed his weapon at Porter, who fired a burst instinctively. The gunman and another Somali were killed. The convoy proceeded to the airport without stopping.

The bodies were later identified as those of a Somali policeman and a Somali UNOSOM employee, but at the time they had no visible identification. The SAS soldier's actions were completely within the rules of engagement that allowed him to fire when someone aimed a weapon at him. As Ellis wrote soon after, 'the shooting was an instinctive reaction to a life threatening situation . . . the fact remains that the weapon was pointed directly at the vehicles. I am 100 per cent confident the soldiers acted correctly'.[14] Lance-Corporal Reid, driving the leading APC, was adamant that he had seen at least three weapons in the Somali vehicle, although only one was later found. Some Somali witnesses travelling with the gunman later agreed that he had acted unwisely.

The incident demonstrated that SAS soldiers would not hesitate to use deadly force while operating strictly within the rules of engagement. It also demonstrated the high degree of skill possessed by the SAS soldier in firing accurately with a short burst from a vehicle moving at more than 30 kilometres per hour against a target in another vehicle moving at a similar speed. While the remainder of the SAS team congratulated Porter for his professionalism, some other members of the ASC were less complimentary, being fearful that the action would evoke a reaction from the Somali militia. The Gerbils were disappointed at this attitude, which they saw as a lack of professionalism from fellow military personnel. Members of the dead men's families sent a note to the Australians demanding a satisfactory reply or 'we will take immediate action', but they took no further action.[15] Although there were rumours that the warlords had vowed to avenge the deaths, thereafter the Somalis were wary of challenging the Australians. Kingston noted a 'marked increase of Somali weapon discipline on the streets', with Somalis openly turning the weapons away from Australian convoys.[16]

The Australian government decided not to replace ASC IV at the end of its six-month tour, and the main body of the contingent, including J Troop, departed Mogadishu on 23 November 1994 after a tour of seven months and one week.[17] The most noticeable feature of the operation was the professionalism of the SAS team—on every occasion, they conducted their tasks dressed in helmets and protective vests and were well armed. Neither they, nor the Australians they were protecting, were ever attacked by Somali gunmen. By contrast, during the same period other much larger national forces were frequently attacked and took casualties. The Somalia tour was, however, a tactical deployment that might have been undertaken by a well-trained infantry platoon, and it was not a model for future Special Forces operations.

Before the Gerbils arrived back in Australia, a second group of SAS soldiers had been deployed on another operation. In August 1994 the first contingent of Australians deployed to Rwanda to provide medical support for the UN Assistance Mission in Rwanda (UNAMIR II). Its secondary role, utilising any spare capacity, was to provide humanitarian relief to the citizens of Rwanda, where a civil war had resulted in hundreds of thousands of deaths and similarly large numbers of refugees. The Australian contingent included a medical company, an infantry company group (to provide protection) and a logistics company. The medical company numbered 84 personnel, including seventeen from the RAAF and seven from the RAN, with the remainder from the Army; of these, nine came from the SAS Regiment: Major Lindsay Bridgeford (medical officer), Corporals Adam Kestel, Wayne Tait and Mal Vaughan, Lance-Corporal Brian Hale, Troopers Scott

SAS Regiment medical assistants Corporal Dominic Boyle and Trooper George Taulelei survey the Kibeho Camp, Rwanda, in 1995 with members of the Zambian Battalion.

Carnie, Darren Oldham and Todd Ridley, and Signaller Travis Standen. The SAS members were either medical assistants from the SAS Regiment medical centre, or patrol medics.

In February 1995 a second Australian contingent replaced the first. Its SAS members included: Warrant Officer Class Two Rodrick Scott, Corporals Dominic Boyle, Paul Jordan and Shane White, and Troopers Jonathon Church and George Taulelei. During April 1995 forces of the Rwanda Patriotic Army—mainly from the Tutsi tribe—carried out a massacre of Hutu internally displaced persons (IDP) at the Kibeho refugee camp, killing perhaps 4000 and injuring about 600. In a highly dangerous situation a seven-person Australian medical team under the command of Captain Carol Vaughan-Evans struggled to give medical aid to the injured refugees while protection was provided by two sections of Australian infantrymen. Corporal Paul Jordan wrote:

There were lots of mothers who had been killed with their babies still strapped to their back. We went around and freed the ones that we saw. There were a great many children sitting on piles of rubbish . . . there was so much rubbish around and quite often some would move under

463

SAS: Phantoms of War

Trooper Jonathon Church, a medical assistant with the second Australian contingent to Rwanda in 1995, carrying an injured child. He was killed in the Black Hawk accident in 1996.

your feet, exposing a baby who had been crushed. I counted 20 such babies, but I could not turn over every piece of rubbish.

As Warrant Officer Scott commented: 'It is hoped that our soldiers never witness such obscenities ever again.'[18] The massacre was an extremely distressing episode for the Australian infantrymen who, under the UN rules of engagement, were not permitted to open fire to protect the refugees. They could only return fire if they or the medical teams were attacked.

After the incident an Australian Army psychology team was deployed to Rwanda to counsel the Australians present at the massacre. Vaughan-Evans (later to be an SAS Regiment medical officer) and Scott, along with two infantrymen, were awarded the Medal of Gallantry for their actions during the Kibeho massacre. They were Australia's first decorations for gallantry awarded since the Vietnam War and the first under the Australian rather than the British honours system. The second Australian contingent returned to Australia in August 1995.

The UN operations in Iraq, Cambodia, Western Sahara, Somalia and Rwanda indicated that the SAS had to be ready to face a variety of

464

operational situations. But by late 1994 the numbers of SAS soldiers deployed overseas had declined considerably from those in previous years. Some in the regiment were disappointed that it had not been given a more important role in Somalia, in particular. It seemed that some senior military officers and government officials did not have full confidence in the regiment's capabilities. Meanwhile, in November 1994 the government released its Defence White Paper, *Defending Australia,* which announced that the ADF would be involved in 'increasing engagement' with the South-East Asia region. This, along with the overseas deployments during the preceding five years and constant contingency planning, led Colonel Wallace, Commander Special Forces, to direct the SAS Regiment's commanding officer, Lieutenant-Colonel Michael Silverstone, to begin an extensive review of the SAS's capacity to undertake likely new tasks. The outcome was to result in improvements to capabilities, but some of the developments were to be interrupted by the Black Hawk accident of June 1996 that placed the regiment under greater scrutiny.

Silverstone assumed command in December 1994. Aged 39, he had graduated from the Officer Cadet School, Portsea, in 1974 and had had long experience in the regiment. He had spent eighteen months training in Britain with 22 SAS Regiment and the Special Boat Squadron and had been commander of J Troop when three members were killed in a C-130 crash in the Philippines in 1981. Later he was commander of 1 Squadron.

Silverstone set up a team of majors (led by the Operations Officer, Major Keith Perriman) to undertake a wideranging review of SAS tasks and organisation, realising that if the SAS was to respond to demands for a more unconventional approach some old ideas might have to be abandoned. Lieutenant-Colonel Higgins had already obtained a language instructor to improve language skills within the regiment. Silverstone took this further, looking at ways to improve the education of patrol commanders and to place adventurous activities, especially those designed to develop mental agility, on a firmer basis. Trials were carried out to see if the sabre squadrons could be reorganised, with their three troops focusing on operations in particular countries rather than on the traditional water, vehicle and air operations. Already exchange and training programmes had been established with Special Forces elements and police counter-hijack units in regional countries. Some regional countries conducted CT training in Australia, concentrating on countering aircraft and offshore platform hijacking. The SAS was at the forefront of implementing Australia's foreign policy in actively engaging with the ASEAN countries (Thailand, Malaysia, Philippines, Indonesia, Singapore and Brunei) at this time. The one area of disappointment was in Papua New

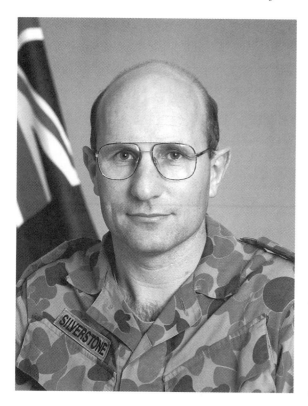

Lieutenant-Colonel
M. J. W. Silverstone, CSC,
Commanding Officer of the
SAS Regiment 9 December
1994–12 January 1997.
He oversaw a wide-ranging
review of SAS tasks and
organisation, but his
command was
overshadowed by the
Black Hawk accident in
June 1996.

Guinea, where the SAS was prohibited from training, following the outbreak of conflict in Bougainville in 1989.

Throughout this period of relatively minor, but demanding, overseas deployments, the SAS continued training for its prime tasks—conducting 'strategic and operational intelligence gathering, small-scale offensive operations and specialist recovery operations'. Through the SASTRAIN and later SASQUAL programmes, the regiment introduced rigorous individual and collective training methodologies which ensured that each squadron could be reinforced directly from elsewhere in the regiment. With the maturing of the CT training regime there was a substantial increase in the standard of the SAS trooper in terms of knowledge, skill and attitude.

As well as exercises with overseas armies, the SAS took part in major Australian exercises, including the KANGAROO exercises in 1992 and 1995 held across northern Australia. KANGAROO 95 saw the introduction of new command arrangements, based on the US Joint Special Operations Task Force (JSOTF) model. During CT exercises the commanding officer of the regiment had usually deployed with a small headquarters to the Police Forward Command Post, and the various contingency plans for overseas recovery operations had also demonstrated the need for a higher-level headquarters. Further, the complexity of operational planning, combined with an unrelenting tempo, necessitated the establishment of a command element to design and plan operations, thereby permitting the tactical commander (the squadron commander) and his small staff to get on and conduct all tactical planning and operations. For KANGAROO 95, Silverstone established a small tactical headquarters, known as the Special Operations Command and Control Element (SOCCE), at the headquarters of Northern Command in Darwin. Silverstone became the special operations adviser to the Commander Northern Command, who commanded the exercise 'Blue Force'. The Blue special operations forces included 1 SAS Squadron and elements from the 1st Commando Regiment. Meanwhile 2 SAS Squadron operated as part of the enemy 'Orange Force' and effectively applied JSOTF doctrine. The officer commanding 2 Squadron, Major Rick Burr, set up his own SOCCE and provided special forces advice to the Orange Force commander. The concept of deploying a 'green' SOCCE for warlike operations and exercises and a 'black' SOCCE for CT activities was established. The regiment therefore had to develop the capacity to staff both SOCCEs simultaneously. The regiment's second-in-command took on added responsibilities and later (in 1998) became the regiment's executive officer.

When a journalist visited 1 SAS Squadron at RAAF Base Tindal near Katherine during KANGAROO 95, he found frustration that the regiment had not been deployed on operations and entitled his article 'Warriors Without a War'.[19] During the exercise squadron members carried out surveillance and raiding, using long-range patrol vehicles and the Army's Black Hawk helicopters.

In 1989 the Army had taken over the operation of the newly arrived Black Hawk helicopters from the RAAF, and the 5th Aviation Regiment, based in Townsville, began developing a close rapport with the SAS Regiment. Some pilots trained with US special operations aviation units. Many RAAF pilots transferred to the Army to continue flying helicopters, but the acquisition of this new capability placed a heavy burden on the Army's training and maintenance resources. By the time of Exercise KANGAROO 95 the 5th Aviation Regiment had reached a low level of aircraft availability that was resulting in a degradation in aircrew skills. At the same time the SAS was demanding more force projection skills from their helicopter support, for the recovery role in addition to domestic CT. During 1995 Silverstone formally noted his concern about the level of training in the 5th Aviation Regiment.

Since 1990 the SAS Regiment's CT squadron and the 5th Aviation Regiment had carried out two exercises per year involving the insertion

467

The SAS Regiment's counter-terrorist team conducting training with a Black Hawk helicopter from the Townsville-based 5th Aviation Regiment in a training area near Townsville.

of an SAS force directly onto a target by helicopter. Conscious of concerns about the low level of pilot training, the 5th Aviation Regiment undertook an extensive work-up period before it conducted a similar exercise with 1 SAS Squadron in the High Range training area near Townsville during June 1996. In the evening of 12 June six helicopters were approaching a 'defended area', the pilots wearing night-vision goggles. The three leading helicopters flying abreast were to deliver their SAS soldiers directly onto the target; a fourth helicopter travelling behind carried further troopers. Fire support was to be provided by helicopters flying on each flank. It was the sixth time the helicopters had flown towards the target during that day.

As the helicopters approached the target the left forward aircraft veered sideways into the centre aircraft. One helicopter crashed to the ground and twelve personnel on board died. The other helicopter managed a crash landing but burst into flames. Six on board died. SAS soldiers on the ground as exercise staff rushed to pull their mates from the burning wreckage. Immediate first aid was applied to the injured soldiers and the surviving helicopters began ferrying the injured to Townsville. Army pilot Captain David Burke, who displayed outstanding calmness and professionalism in landing the damaged Black Hawk, was amazed at the 'incredible courage' of those who risked their lives to save the men who lay in the wreckage:

As soon as we were literally on the deck, the safety staff on the ground were there, they had medics there, they were organising, they were pulling people out, there were people going into the flames of aircraft to pull them out.

There were rounds going off, there was ammunition flying in the air, there were explosions in the back of the aircraft going off, and these men, both SAS and air crew were going into the flames and cutting people out and bringing them out.

Burke gave special mention to the Squadron Sergeant-Major, Warrant Officer Class Two Peter Green, who took charge of the emergency 'with remarkable professionalism'.[20]

The SAS members killed in the crash were: Captain Timothy Stevens, Sergeant Hugh Ellis, Corporals Mihran Avedissian, Michael Bird, Andrew Constantinidis, Darren Smith and Brett Tombs, Lance-Corporals Gordon Callow, David Frost, Glen Hagan, David Johnstone and Darren Oldham, Troopers Timothy McDonald and Jonathon Church, and Signaller Hendrick Peeters. Three members of the 5th Aviation Regiment—Captains John Berrigan and Kelvin Hales, and Corporal Michael Baker—were also killed. Corporal Gerry Bampton was left a paraplegic as a result of the accident and, ably supported by his wife, fought a tough battle to obtain appropriate compensation.

The following year thirteen SAS members were recognised for their actions in the aftermath of the crash. Corporals Dominic Boyle and Gary Proctor were awarded the Star of Courage, Captain James Ryan, Sergeant Mick Williams and Lance-Corporal Brian Morriss were awarded the Bravery Medal, Sergeants Bob McCabe and Nick Coenen received Commendations for Brave Conduct, Corporal Rob Cousins, Lance-Corporals Craig Naumann and George Taulelei received commendations from the Chief of the Defence Force and Corporals Steven Frerk and Jamie Sharpe received commendations from the Chief of Army. Boyle was aboard Black Hawk 2 and sustained a fractured elbow when it crashed. Despite this, according to his citation, he 'made numerous attempts to save those trapped, to free the bodies of those killed in the wreck of the burning fuselage and to quell the fire with extinguishers'. Then on the way to hospital he performed heart resuscitation on another injured airman. Proctor received a broken coccyx in the crash, but 'placed the lives and welfare of his fellow soldiers above his own pain'.

The sudden deaths came as a devastating blow to the regiment even though it had suffered deaths in training since its inception. The Special Airborne Services Trust Fund was established to hold and

SAS: Phantoms of War

disperse unsolicited funds donated to support the families of those involved. By the end of September 1996 the fund had reached almost $160 000. By 2001 the figure had grown substantially and continues to support the families affected by the accident.

Following the disaster there was press speculation that it had severely degraded Australia's CT capability. But the regiment had a greater capacity to regenerate that capability than was generally appreciated, and Silverstone directed the execution of the standing plan to establish the standby CT capability. Within 24 hours the national CT response capability was re-established, and 1 SAS Squadron soon resumed its CT role. The SAS soldiers understood that they were always likely to suffer casualties during training—it was a consequence of training 'on the edge'. They took pride in their capacity to rebound from such accidents and even though the numbers of deaths was far in excess than had been previously experienced they were determined to get on with their business.

During the preceding four years the regiment had lost four other members in training accidents. Corporal Stephen Daley was killed on 12 November 1991 when a Pilatus Porter crashed shortly after take-off near Nowra. He was attending a jump master free-fall course. Sergeant Paul Kench was killed while diving in Bass Strait on 9 December 1992. An outboard motor propeller struck him shortly after he entered the water for a diving task. Corporal Gordon Holland died of natural causes while training at Swanbourne on 21 May 1993. And finally, Signaller Lachlan Martin was killed while parachuting at Bindoon on 14 May 1996, barely a month before the Black Hawk accident. He was taking part in a rehearsal jump before a visit by the Defence Force Remuneration Tribunal.[21] The regiment had therefore developed well-tested procedures for dealing with the aftermath of accidents, although no accident had previously matched the scale of the Black Hawk crash.

The damaging fallout from the accident related not so much to problems within the regiment as to perceptions among members of the SAS community that the Army hierarchy unfairly blamed both regiments for the accident rather than looking for root causes. The Army's Board of Inquiry, headed by the chief of staff at Land Headquarters, Brigadier Paul O'Sullivan, recommended charges be laid against Major Bob Hunter, officer commanding 1 SAS Squadron, his operations officer, Captain Sean Bellis, and Major Chris Jameson of the 5th Aviation Regiment. In September 1997 the convening authority for the charges, Major-General John Kingston (Support Commander Army) dropped the charges against Hunter and Bellis, and in December dropped the charges against Jameson. By then Jameson had left the Regular Army and was critical of the way the matter had been handled. In a press interview he said that he was disgusted by the actions of

470

some senior army officers and in particular by the performance of one who, on national television, laid blame for the accident with those in charge on the night. 'The hierarchy has been able to damage reputations with relative impunity and the accused have had no opportunity to clear their names,' he said. 'I would have been more than happy to stand my ground so the real faults could have been exposed. They have slunk out of it and left a sour taste in everyone's mouth.'[22] Hunter and Bellis were censured for their actions and although these censures were later rescinded they left the Army. Most in the SAS still believe that they were unfairly blamed for the accident.

Colonel Don Higgins, Commander Special Forces at the time of the Black Hawk accident, saw a direct connection between the accident and a decision made soon afterward that his headquarters be placed under the Land Commander in Sydney. Previously, ever since its establishment in 1990, Headquarters Special Forces had come directly under the Chief of Army in Canberra. The decision to remove Special Forces from the direct control of the Chief of Army was the first step in establishing new command arrangements that were to have a fundamental effect on the conduct of operations by the SAS Regiment and on the development of Special Forces units in the ADF.

Whether there was any reason for Higgins to see any connection between the Black Hawk accident and the timing of the decision, it was a natural outcome of new command arrangements announced by the Chief of the Defence Force (CDF), General John Baker, in January 1996, six months before the accident. A new headquarters—Headquarters Australian Theatre—would be established in Sydney, and a new commander—Commander Australian Theatre—would be appointed, answerable directly to the CDF, for the command of all Australian joint operations. The existing Maritime, Land and Air Headquarters would become components of Headquarters Australian Theatre. The three Headquarters Commanders would become Maritime, Land and Air Component Commanders for joint operations, although they would remain responsible to their own service chiefs for raising, training and sustaining their respective forces.

Headquarters Special Forces was to move to Sydney and become the Special Operations Component of Headquarters Australian Theatre. The Commander Special Forces would therefore be responsible to the Land Commander for the raising, training and sustaining of Special Forces units, but as Special Operations Component Commander would answer directly to the Commander Australian Theatre and on occasions directly to the CDF for operations.[23] The Commander Special Forces remained responsible directly to the CDF for the conduct of CT operations. Headquarters Australian Theatre became operational in January 1997, located in a rented building at Potts Point in Sydney.

Headquarters Special Operations, as it was now known, did not move to Sydney until January 1998 when Brigadier Philip McNamara succeeded Colonel Higgins as commander. McNamara had been a squadron commander in the SAS Regiment and commanding officer of the 1st Commando Regiment. His headquarters was located in the Garden Island dockyard, only 100 metres from Headquarters Australian Theatre. He took command of the newly formed Special Forces Group. Higgins' hard work and persistence had been instrumental in the development of the Special Force Group.

During 1995 the government had authorised the raising of the 4th Battalion, The Royal Australian Regiment (4 RAR), and in 1997, as part of a programme to restructure the Army, the government authorised the conversion of 4 RAR to become a Regular commando battalion, with the role of large-scale raiding and amphibious operations. The Special Forces Group would therefore consist of the SAS Regiment, the 1st Commando Regiment (Army Reserve), 4 RAR (Commando) and 126 Commando Signal Squadron (previously part of the 1st Commando Regiment). It was planned that 4 RAR, based at Holsworthy in Sydney, would eventually take over the SAS's CT role, enabling the SAS Regiment to concentrate on its traditional roles of strategic reconnaissance and surveillance and unconventional warfare. By 2000, however, 4 RAR had formed only two commando companies and it was decided that the very demanding CT role should remain with the SAS Regiment, as the most suitable unit to conduct these activities.

Some Special Forces officers criticised the move of Headquarters Special Operations to Sydney, claiming that it isolated the Commander Special Forces from the ADF's strategic-level headquarters in Canberra, that the SAS Regiment was a strategic asset and that the Commander Special Forces should be available to provide the CDF with advice about the use of Special Forces and the coordination of inter-departmental special operations. Others recognised that, while the Commander Special Forces was only a brigadier, he had been elevated to the level of a component commander on a par with the Maritime, Land and Air Component Commanders within Headquarters Australian Theatre. When the Directorate of Special Action Forces had been formed in the late 1970s it had a staff of four officers. Headquarters Special Forces was formed in 1990 with a staff of fifteen, and Headquarters Special Operations opened in Sydney in 1998 with a staff of 42 personnel.

These new command arrangements were matched at lower levels. The Headquarters of the Army's 1st Division at Enoggera in Brisbane became the Deployable Joint Force Headquarters (DJFHQ). The commanding officer of either the SAS Regiment or the commando regiments became the Special Operations Component Commander in

Lieutenant-Colonel M. S. Hindmarsh, Commanding Officer of the SAS Regiment, 13 January 1997–7 January 1999. He commanded the Anzac special operations force in Kuwait in 1998.

DJFHQ when DJFHQ undertook contingency planning or conducted command-post exercises and operations such as in East Timor. This development was to prove important as the pace of contingency planning and operations increased.

By this time the SAS Regiment had a new commanding officer, Lieutenant-Colonel Mike Hindmarsh, who took over from Silverstone in January 1997. Aged 40, he had graduated from Duntroon in 1978 and had served as a troop and squadron commander in the regiment. He had also served in the headquarters of the British Special Forces and in Headquarters Special Forces in Canberra

The existence of Special Operations Component Commanders at both Headquarters Australian Theatre and DJFHQ ensured that the SAS was considered during the planning phase of the many operations that were now taking place. For example, in July 1997, at a time of high tension in Papua New Guinea following the controversy concerning the employment of Sandline mercenaries, plans were developed for Australian troops to support the evacuation of Australian nationals from Port Moresby.

Also in July 1997, RAAF C-130 aircraft airlifted 455 Australians and other nationals out of Cambodia during a period of civil unrest caused by a coup in that country. Major Bob Hunter made a significant contribution to the ADF Mission Assistance Group that had deployed to Phnom Penh. A small SAS team commanded by Captain Jim Ryan deployed in a C-130 to Butterworth at very short notice, should *in-extremis* evacuation operations be required, while Major Truscott also deployed to Butterworth, in case a liaison officer should be required with the US ground evacuation operation, which was planned to be conducted from across the Thai border.

In October 1997 a truce was agreed in the long-running rebellion in Bougainville, and the following month Australians arrived in Bougainville as part of the New Zealand-led Truce Monitoring Group (TMG). Major Hunter joined the commander's reconnaissance party and then continued on in the headquarters. Major Richard Campbell later relieved him. In April 1998 Australia took over leadership of the renamed Peace Monitoring Group (PMG). Numerous SAS officers, NCOs and troopers served in the TMG and PMG over the next four years as members of the headquarters and monitoring teams.

Not all operations were in foreign countries. In October and November 1997 small JTFs, including RAAF C-130 aircraft and RAN ships, apprehended two vessels fishing illegally for Patagonian toothfish in Australia's exclusive economic zone around Heard and Macdonald Islands in the Southern Ocean. Meanwhile ADF elements brought food to drought-stricken areas of Papua New Guinea. Major Tony John served in the headquarters of the JTF based in Port Moresby with the mission of providing drought relief on Operation PLES DRAI, while SAS medics deployed in the field in this operation. Three SAS troopers cross-trained as linguists and medics deployed as part of the JTF to Vanimo, Operation SHADDOCK, to assist in the post-tsunami disaster. While the SAS was not necessarily involved in all these operations, they illustrate the tempo of planning then taking place in Headquarters Australian Theatre.

After several years of planning for operations near to Australia it was significant, in terms of the need to be ready for any contingency, that the SAS's largest overseas deployment since Vietnam—to Kuwait in February 1998—was to an area well beyond Australia's region. During the latter months of 1997 Iraq had denied entry to the UN Special Commission inspection teams, and the United States and its allies had begun preparing to conduct air attacks against Iraq. On 10 February 1998 the Australian government announced that it had responded to a request from the US to participate in an international coalition should military action against Iraq become necessary. Australia would commit one SAS squadron (actually an SAS task group), two RAAF Boeing 707 tanker aircraft to provide air-to-air refuelling, and specialist personnel to provide back-up medical, technical and intelligence support. The force to be deployed to Kuwait would number between 200 and 250 people.[24] The SAS's main role, as part of the US-led multinational force Operation DESERT THUNDER, would be the recovery of US and British pilots shot down inside Iraq.

The government's decision to deploy the SAS was a great boost to the morale of the regiment, which received the warning order for Operation POLLARD on Friday 6 February; by Sunday the force was ready for deployment. The US forces in Kuwait would come under the US Central Command based in Florida, and by the second week of February Australia had two special operations liaison officers in Florida. US Special Forces had begun to deploy to Kuwait in January.

Applying the recently developed concept for the deployment of Special Forces, the Commander Australian Theatre decided to deploy a SOCCE headed by Lieutenant-Colonel Hindmarsh. Accompanied by his logistics officer, Major Brad Rickerby, Hindmarsh arrived in Kuwait on 15 February to form a Combined Special Operations Task Force (CSOTF) with units from the US Special Operations Command Central Command (SOCCENT). The main body of the contingent came from 1 SAS Squadron, but as the commander of 1 SAS Squadron was occupied elsewhere, Major Mark Smethurst (actually the officer commanding 3 SAS Squadron) commanded it. Smethurst deployed with A and B Troops and a troop from 152 Signal Squadron. This force of 110 personnel departed from Perth on an RAAF Boeing 707 at 6.30 pm on 17 February. The Prime Minister, John Howard, and the Opposition leader, Kim Beazley, were at Swanbourne to wish them 'Godspeed, a safe return and a very successful mission'.[25]

Meanwhile, the New Zealand government had decided to send an SAS troop, which became the third SAS troop (D Troop) in the 1 SAS Squadron group when it arrived in Kuwait on 23 February. The New Zealand troop was completely integrated into the squadron. Hindmarsh was appointed commander of the Anzac Special Operations Force (ANZAC SOF), and commander of the Anzac component of the headquarters of the CSOTF–Kuwait. Despite years of operations in Borneo and Vietnam in the 1960s and early 1970s, this was the first time that an SAS tactical headquarters had been deployed on operations outside Australia and beyond the confines of domestic CT and other precautionary deployments. Hindmarsh's intention was that the Anzac force would become the 'force of choice' for the Commander CSOTF–Kuwait and be employed on missions which optimised SAS capabilities. The squadron was located with the Americans at a large Kuwaiti air base known as Ali Al Salem. The regimental SOCCE was initially located at Camp Doha in Kuwait City, then it too moved to Ali Al Salem.

If the allied forces were to go into combat the CSOTF would come under either a Commander Coalition Task Force–Kuwait or the Commander-in-Chief Central Command. In either case, the Commander Australian Theatre foresaw the need for a liaison officer with a Special Forces background to ensure that SAS interests were protected and the Australian authorities were kept abreast of all developments within the area of operations. Lieutenant-Colonel Tim McOwan, a former SAS officer, was withdrawn from the Joint Services Staff College overnight and deployed almost immediately as the Commander Australian Theatre's liaison officer to the Coalition Task Force–Kuwait. He was accompanied by Major Grant Walsh, a former troop and squadron commander in the SAS Regiment, whose task was to assist the Australian Department of Foreign Affairs' representative to the Kuwaiti government. McOwan's responsibilities broadened through necessity to become the de facto Australian National Command Element, as he had to represent the interests of the RAAF 707 squadron as well as the SAS squadron.

Under the directive from the CDF, the SAS was to conduct military operations without undue risk to Australian lives, operate with coalition partners to demonstrate the ADF's professionalism and competence, and demonstrate an ability to integrate into a multinational force. The force was to be logistically self-sufficient during transit and for the first seven days, and then was to be supported by US forces. The force's primary task was to undertake combat search and rescue (CSAR) missions into Iraq to find and rescue coalition personnel, especially downed aircrew. Second, in cooperation with Kuwaiti and other coalition forces, it was to undertake surveillance and response operations on the Iraq–Kuwait border to provide warning and protection against Iraqi attacks. Third, it was to form a quick response force to respond to small-scale Iraqi raids. (These second and third tasks were never agreed to by the US-led coalition nor did the Kuwaiti government ever endorse them.) And finally, it had to be prepared to undertake surveillance and response operations against Iraqi sea-mining activities. (This task was never confirmed.)

Within three days of their arrival the SAS had worked up and deployed a quick reaction force capability. At the same time, B Troop (water operations) under Captain Sean Bellis began training with US forces, including US helicopters, for the CSAR task. This training culminated after three weeks in a full-scale operational rehearsal with the Americans. Meanwhile A Troop (vehicle and free-fall), under Captain John Van Der Klooster, began familiarisation patrols along the Kuwait border using the long-range patrol vehicles that had been flown

Phantoms of the operations

Members of 1 SAS Squadron carrying out a rehearsal for a combat search and rescue mission with a US helicopter in Kuwait in February 1998.

in by a US C-5 aircraft from Perth. The three troops in 1 Squadron rotated through the CSAR and vehicle-mounted activities.

Two weeks after the Australians arrived in Kuwait the UN Secretary General, Kofi Annan, struck an arrangement with the Iraqi leader, Saddam Hussein; the planned air attacks were put in abeyance and the US multinational force became Operation DESERT SPRING. On 9 April Lieutenant-Colonel Rowan Tink, chief of staff at Headquarters Special Operations in Sydney, arrived in Kuwait to take over from Hindmarsh as commander of the Anzac Special Operations Force. Hindmarsh was required back in Australia where Headquarters Australian Theatre was busy planning new operations. The Asian economic crisis had led to instability in Indonesia, and in May, following violent demonstrations against the Indonesian government, SAS force elements and RAAF C-130s were deployed to Darwin to be prepared to evacuate Australians from Indonesia. Meanwhile, the Australians in Kuwait continued training with the Americans and Kuwaitis until 1 May, when the force was reduced from a squadron to a troop. At that stage Major Dan Fortune (who had replaced Smethurst as squadron commander) succeeded Tink as force commander. On 10 June the remaining troop returned to Australia; the rear party left on 19 June. An ADF liaison officer remained in Kuwait with the Combined Joint Task Force on a three-month rotation.

1 SAS Squadron members on exercise in Kuwait in March 1998 with their long-range patrol vehicles. The value of these vehicles was the cause of many debates within the regiment.

Although the Australians were not involved in active service in Kuwait, the operation had provided extremely valuable experience. This was summed up in a briefing given by Hindmarsh to the Minister for Defence and CDF during their visit in April. 'It is important,' he said, 'to understand that we easily fitted into this mission. In comparison to the other US Special Operations units here, we are the only unit with a tier one, or national CT response force background.' He acknowledged that they lagged behind the Americans in technology, particularly that affecting command and control and intelligence, and there was the danger that they might become a liability to the command and control of coalition special operations. In comparison with the Americans, however, the Australians also had advantages:

We have a different culture, more reliant upon an animal cunning and individual initiative approach borne from years of having to make do with what we've got. Indeed our culture still allows us to prepare for a range of options for which the US doesn't even plan for any more.

If there is one overarching lesson to come from our experience of working with the US it is the need to blend rather than replace our basic soldierly common sense and nous with their application of technology; the danger is of course in taking the technology route, as we must do, we become lazy and lose over time those fundamental soldierly skills which have traditionally set us apart from the rest.

Hindmarsh concluded that while his force had not 'fired a shot in anger' the deployment had been 'a tremendous vote of confidence for me and my men that you committed us to the operation'.[26]

The operation confirmed some weaknesses in the SAS's capabilities. For example, the Steyr rifle was unsuitable for some aspects of special operations. It could not be fitted with the array of sensors that the SAS needed to do their job by night and in all environments. As a result, the regiment purchased 40 US M4 carbines that could be fitted with additional items, such as 40mm grenade launchers and torches, in a modular manner. The SAS also fielded the new long-range patrol radio (LRPR) for the first time in Kuwait, taking a quantum leap forward in high-frequency radio communications after many years of working with interim radios and infantry radios not suited to special operations.

The role of the long-range patrol vehicles in Kuwait was also subject to many debates within the regiment. In the years before Operation POLLARD some SAS members had argued that the vehicles should be handed over to the regional force surveillance units as they would be the only units operating in the Australian outback. Others thought that the vehicles should be retained, but that vehicle-mounted operations should no longer be considered an SAS specialisation. Obviously there was no role for these vehicles in conducting the principal CSAR mission deep in Iraqi territory, but they might have been used in low-threat areas or in tasks of short duration. The argument about the use of high-signature patrol vehicles would again arise in East Timor.

The most important outcome of Operation POLLARD was the recognition that the SAS could deploy a squadron-sized task group at short notice across the world, and that the government was willing to order such a deployment. The ADF also learned not to do so again without deployable secure communications and intelligence information systems. The next deployment would be much closer to home—to East Timor.

When Hindmarsh was recalled from Kuwait it was clear that Australian defence planners were considering how the SAS would be used in any operations that might result from the unstable situation in Indonesia. Already the CDF, Admiral Barrie, had publicly discussed the difficulty of evacuating 10 000 Australians working in Indonesia. In January 1999 Lieutenant-Colonel Tim McOwan succeeded Hindmarsh, and his first year in command promised to be as interesting as any experienced by his predecessors in the previous 27 years. In January 1999 the Indonesian President, B. J. Habibie, said that he was willing to grant East Timor independence if it rejected an offer of autonomy. Teams from the UN Office for the Coordination of Humanitarian Relief began to be deployed to East Timor, where Indonesian military-backed militia groups stepped up violence against those favouring independence. Whereas previously the ADF had been preparing to evacuate Australians from Java, McOwan's first task was to prepare for operations that might eventuate in East Timor as the province moved towards voting for autonomy or independence.

Aged 41, McOwan had graduated from Duntroon in 1980, had been a troop and squadron commander in the regiment, had served with the headquarters of British SAS, and had been the SO1 Plans and the chief of staff in Headquarters Special Operations. During 1995 he had been commander of the Australian contingent and acting chief of staff of the Multinational Force and Observers in Sinai, and in 1998 was sent to Kuwait as the Commander Australian Theatre's liaison officer in the Coalition Task Force headquarters.

McOwan took command of a unit that had been involved in a multitude of minor deployments and planning activities over the previous eight years. During that time there were probably three times as many movements of SAS elements to forward bases (in preparation for operations) as there were actual operational deployments, and the unit had prepared perhaps five times as many plans for operations that had not eventuated.

The Operations Officer, Major Jim Truscott, summed up some of this activity in an article in the SAS Association magazine *Rendezvous* in December 1998. He noted that the new headquarters building had been in operation for a year, during which time the operations and intelligence staff had achieved closer working relations by occupying the same office, and some of the signals operators had also moved into the building. This staff had previously spent three months in an air-conditioned tent in Kuwait and 'another month in an air-conditioned warehouse in northern Australia'. As well as commenting on the substantial operational and developmental activity, Truscott noted another significant event in the life of the regiment—in mid-1998 Language School West had been put on a formal basis in renovated premises. Finally, Truscott noted the establishment of the Special Forces Wide Area Network, and that it was 'now possible for Trooper Bloggs to send a secret email to Theatre Headquarters from the comfort of the Troop office'. The infusion, he said, 'of other, much needed technology this year continues to make us the force of choice for many ADF responses'.[27] Moving quickly, in the next edition of the magazine McOwan reminded members and past members of the regiment of the importance of maintaining security during a time when the unit was 'going through a period of unprecedented technological and organisational change . . . accompanied by heightened operational activity'.[28]

As the likelihood of operations increased regimental headquarters began planning in earnest, submitting peace operations plans as early as February. During April 1999 there were reports from East Timor of the killing of seventeen people in one incident, 25 in another incident in a church yard at Liquica, and 21 in the capital, Dili. On 27 April President Habibie agreed to a UN 'civilian police force' supervising a vote on the autonomy package. On 11 June the UN Security Council passed Resolution 1246 establishing the UN Assistance Mission East Timor (UNAMET) to monitor the fairness of the political environment and to take responsibility for activities relating to registration and voting in the ballot to be held on 8 August. Under Operation FABER, the ADF provided six military liaison officers to the UNAMET military observer group. Australian civilians and police officers fulfilled other roles within UNAMET.

Meanwhile the Australian government directed the ADF to step up its readiness as a reaction to events in Indonesia. On 11 March 1999 Defence Minister John Moore announced that the Army's 1st Brigade, based near Darwin, would be brought to a level of preparedness that would allow it to be deployed at 28 days' notice. This was the same level of preparedness as the 3rd Brigade in Townsville and, as Moore said, the ADF therefore would be able to 'deliver forces of up to two brigade or task force size groups with associated naval and air units.

This is the most significant level of force readiness for two decades'. It was, he said, 'a prudent and necessary measure which gives Australia maximum flexibility to respond to contingencies [in the region, including East Timor] at short notice'.[29]

The SAS was also brought to a higher state of readiness. In January 1999 Major James McMahon assumed command of 3 SAS Squadron and began preparing for possible operations in East Timor. The squadron was completing its first year as the regiment's 'contingency' squadron and after Exercise JUPITER in the early months of the year, it began its second year as the contingency squadron. At McOwan's direction, during May the squadron undertook a month of intensive language training, with one half learning Bahasa Indonesia and the other half Tetum, the native language of East Timor.

McMahon was an experienced SAS officer. Aged 34, he had started his military career as a soldier in 5/7 RAR before attending the New Zealand Officer Cadet School in an exchange programme. After serving as a platoon commander in 3 RAR he had commanded the water operations troop in 3 SAS Squadron, served in the Operational Support Squadron and, after postings in Canberra, served with the British Special Forces headquarters. At the end of April 1999 he visited DJFHQ at Enoggera where the commander, Major-General Cosgrove, and his staff were planning Operation BRANCARD, the evacuation of Australians from Indonesia. McMahon represented his commanding officer as the Special Operations Component Commander. As the officer who would probably command the SAS on the ground, McMahon brought a valuable perspective.

In view of the continuing violence from the East Timor militias, on 28 July the United Nations decided to delay the vote for independence or autonomy until 30 August. Political campaigning began on 15 August and the final pro-autonomy rally on 26 August left five pro-independence supporters dead. That day Moore announced that 'as a matter of routine precaution' he had directed the ADF 'to assist in the evacuation of personnel from East Timor, should that be required'.[30] The SAS Regiment was to be at the heart of this operation. The phantoms of the operations were about to emerge from the shadows.

25
The force of choice:
1999–2000

In 1999–2000 the SAS Regiment conducted its most significant operations since the Vietnam War. During the preceding decade the SAS had become a vital part of the ADF, conducting many small-scale operations and deployments both near Australia and farther afield. But in 1999–2000 the emphasis changed. In three major ADF operations—the evacuation of Australian citizens and others from East Timor (Operation SPITFIRE) in 1999, the peace enforcement operation in East Timor (Operation STABILISE) in 1999–2000 and the support to the Sydney 2000 Olympic Games (Operation GOLD)—the SAS played a dominant role. Each of these operations suited the particular capabilities of the SAS, but they also illustrated the fact that for many situations the SAS had finally become the ADF's force of choice. This created new dilemmas. If the regiment was indeed to be the force of choice, then it might have to take on operations that some purists might earlier have seen as outside its normal field. And inevitably, if the regiment were to play central roles in operations that attracted a high level of public interest, its part in them could not remain concealed for long. In the case of Operation GOLD there was good reason for publicising some of the regiment's capabilities as part of a deterrent to possible terrorist activities. In a two-year period the regiment amassed a level of operational experience that had not been achieved for almost 30 years, and provided a firm base for its further development over the years to come.

The operation to evacuate Australian and UN personnel from East Timor, Operation SPITFIRE, had been expected for some months and various plans had been developed but, as always, the actual order to deploy came unexpectedly. On the morning of Thursday 26 August

1999, the commanding officer of the SAS Regiment, Lieutenant-Colonel Tim McOwan, was travelling north from Bangkok, Thailand. In the same vehicle were Brigadier Philip McNamara, Commander Special Forces, and several other senior officers from Australia's Special Forces Group, taking part in the triennial visit to Special Forces organisations in South-East Asia. Suddenly the officers' mobile phones started ringing. McOwan answered to hear Lieutenant-Colonel Gus Gilmore, the operations officer in Headquarters Special Operations, telling him to return to his unit immediately. McOwan had been due to deliver a lecture to the Thai National Security Command forces, but gave his lecture notes to Lieutenant-Colonel Neil Thompson, commanding officer of 4 RAR (Commando), and with a military escort returned to his hotel in Bangkok. That night he was on the plane to Sydney.

That same day Major James McMahon was completing his five-week Intermediate Operations Course at the Army Promotion Training Centre at Canungra, south of Brisbane. The course was due to conclude that day and McMahon had already completed all the formal course work. He received a similar phone call, and at 5.30 am the next day left Canungra for Brisbane airport.

McOwan arrived back in Perth in the morning of 27 August to find the unit preparing to deploy, using well-practised procedures. McMahon appeared about midday and did not even have time to return home; he met his wife briefly in the officers' mess. Most of the SAS soldiers had learned that they were to be deployed when they arrived at work that morning, and only had time to phone their wives and loved ones to tell them they would not be coming home that evening. That afternoon McOwan, members of the headquarters staff, McMahon and the 3 SAS Squadron group travelled to RAAF Base Pearce north-west of Perth, where they boarded C-130 aircraft. They arrived in the evening at RAAF Base Tindal, near Katherine in the Northern Territory.

The Tindal airstrip has existed since the Second World War, but the new RAAF base was carved out of the Northern Territory scrub in the late 1980s as the forward operating base for a squadron of F/A-18 Hornet fighters. The main defence base in the Northern Territory, it can accommodate additional aircraft and large numbers of troops. The aircraft ordnance loading areas are protected from the sun and from view by individual camouflaged hangars, while each area has earthen walls to confine any blast that might be caused by enemy action or accident. The permanent RAAF staff live in a nearby new housing estate or in comfortable messes, while visiting aircrew can be accommodated in semi-underground bunkers. Just south of Katherine, the RAAF base is away from prying eyes and, unlike defence establishments down

484

south, uniformed personnel rather than contractors carry out most work on the base.

At Tindal, 3 SAS Squadron was housed in a large shed, with McOwan and his headquarters located closer to the base headquarters. McOwan was appointed commander of a Joint Task Force (JTF 504) with the task of preparing to evacuate Australian and UN personnel from East Timor. His force included Black Hawk helicopters from the 5th Aviation Regiment at Townsville, Hercules C-130 and Caribou DHC4 transport aircraft from the RAAF's Air Lift Group at Richmond, a rifle company from 1 RAR in Townsville, and a rifle company from 3 RAR in Sydney. Although the commander of the RAAF aircraft element, Group-Captain Grahame Carroll, was one rank higher than McOwan, the Commander Australian Theatre, Air Vice-Marshal Robert Treloar, appointed McOwan as joint task force commander with Carroll as his deputy. McOwan was selected as commander because of the possible need to conduct a recovery operation and found Carroll very helpful and cooperative. For particular special operations McOwan would report to the Commander Special Operations, Brigadier McNamara, while for a straightforward evacuation operation he would come under the Commander DJFHQ, Major-General Cosgrove. Of course, Treloar was superior in rank to both Cosgrove and McNamara, but it was recognised that because of the political nature of the operation the CDF, Admiral Barrie, would probably exercise control through his Strategic Command staff in Canberra.[1]

The ADF is only able to operate with these overlapping command relationships because of improved communications. All major headquarters across the continent are linked by secure computers that allow for the rapid dissemination of information and the circulation of plans. From the beginning, McOwan realised that once any evacuation operation began Cosgrove would probably exercise overall command. As directed by McNamara, Lieutenant-Colonel Jeremy Logan, commanding officer of the 1st Commando Regiment, assisted by the adjutant of the SAS Regiment (Captain David Tonna), went to Cosgrove's headquarters as the special operations liaison officers.

Following the practice developed in KANGAROO 95 and employed in Kuwait, McOwan deployed to Tindal with his 'green' SOCCE and additional operations, intelligence and communications staff. There was also a small logistic element known as the LOGPAK that included the unit medical officer, Major Carol Vaughan-Evans.

The strength of the SAS is its flexibility and the high level of skills of its individual soldiers. This enabled McOwan to plan for a range of different types of operations and to hold forces in readiness for them. While McOwan's joint task force might have been deployed in an operation of limited scope, it was insufficient for a large-scale

SAS: Phantoms of War

evacuation of Australian citizens, UN personnel and East Timor refugees—a type of mission described by the ADF as a Services Protected Evacuation. In this case the mission was to be passed to the 3rd Brigade—the high-readiness brigade—based in Townsville.

While the operations officer, Major Jim Truscott, and his staff struggled with numerous and ever-changing contingency plans, 3 Squadron under McMahon began training for operations. The squadron included a small command staff, headed by the executive officer, Captain Chris Johns, and several members of 152 Signal Squadron. There were three troops: Captain Gavin Keating commanded I Troop (water operations), Captain Carl Marning commanded K Troop (free-fall) while Captain Jon Hawkins had L Troop (mobility). Each troop is able to deploy four five-man patrols, but also has to be able to operate as a troop. For its surveillance operations the SAS usually deploys five-man patrols, although CT operations are based on the employment of larger groups. McOwan realised that operations in East Timor might require the entire squadron, and at his direction McMahon began practising squadron-level operations, using both C-130 aircraft and Black Hawk helicopters.

On 30 August, nearly 99 per cent of voters in East Timor voted on the question of independence or autonomy. But on polling day two East Timorese poll workers employed by the United Nations were killed by pro-autonomy militia, and violence over the following days caused further deaths. On 4 September the UN Secretary-General, Kofi Annan, announced that 78.5 per cent of East Timorese voters had decided in favour of independence. The following night hundreds of East Timorese scaled the wall around the UNAMET compound in Dili to escape the militia bands roaming the streets. The head of UNAMET, Ian Martin, asked the Australian government to assist with the evacuation of his non-essential UN personnel and the refugees within his compound, and the Indonesian government agreed to allow Australian aircraft to land at Dili and Baucau.

The joint task force at Tindal was ready, and on 6 September RAAF C-130s began their flights to Dili. On board were parties of SAS soldiers who had been rehearsing their procedures with the RAAF crews. Their tasks included security for the aircraft on the ground, marshalling the UN staff and refugees at the airport, checking that they were not carrying weapons and escorting them to the aircraft. The Indonesians had guaranteed security and the Australians were indicating that they trusted the Indonesian Armed Forces (TNI) with their safety.

It was, however, a highly dangerous situation. McOwan, in command of the evacuation with 46 SAS soldiers, was on the first of five C-130 flights into Dili, and recalled seeing a city ablaze as the plane

486

Members of the SAS security party escorting East Timorese refugees to a C-130 aircraft on Komoro airfield during Operation SPITFIRE.

circled. Convoys of 20 to 30 trucks were heading west out of Dili, carrying looted goods and furniture and many East Timorese. Once the Australians landed, automatic fire could be heard in close proximity. Indonesian air force Special Forces troops and troops from a local TNI battalion ringed the airport. The TNI troops were in fire positions with their weapons pointing inwards towards the airstrip, not outwards towards the militia threat. At the airport the Australian defence attaché from Jakarta, Brigadier Jim Molan, and the army attaché, Colonel Ken Brownrigg, worked hard to calm the Indonesian forces, who were understandably upset at the sight of the Australian soldiers on their soil. Brownrigg and other embassy staff also helped journalists to move from the city, past roving militia bands to the airport. By the end of the day the Australian aircraft had evacuated about 300 people. Next day McOwan was back in Dili with 28 soldiers to continue the evacuation.

Meanwhile, Major McMahon with eight SAS soldiers landed with two C-130s at Baucau. UNAMET staff began to board the aircraft, but when East Timorese refugees arrived a local militia band insisted on screening each person. McMahon was dismayed by the absolute terror displayed by the refugees when the militia singled them out. The TNI troops were taking no action, so the SAS soldiers stood between the militia and the refugees to protect them. Fortunately, after negotiations by Brigadier Molan, who had journeyed from Dili, the militia agreed to allow the refugees to be carried to Dili in a UN Puma helicopter. At this stage an SAS officer, Major John Gould, working as a UN military observer in the UNAMET compound, had the distressing task of telling the East Timorese refugees there that they could not be evacuated. (They were evacuated the next day.)

The Australians now discovered that the East Timorese religious leader, Bishop Belo, was at the airfield, and the SAS soldiers began escorting him to an aircraft. They had proceeded about halfway before the militia leader realised what was happening and surrounded the party with about 30 militiamen and 40 TNI soldiers. Weapons were raised and the TNI commander told McMahon that he could not prevent his men shooting because of their 'high emotional involvement'. McMahon recalled later that as the tension arose, 'I really thought—we could lose our lives here—the whole lot, including the plane.' Molan persuaded the militia to allow the aircraft to depart with Belo, but before take-off a militia vehicle drove onto the airfield to block the runway. After more negotiation it was agreed that Belo could fly to Dili. As previously planned, however, the aircraft flew directly to Darwin. The SAS soldiers on the ground waiting to depart on the second aircraft saw shots being fired at Belo's aircraft.

Responding belatedly to the violence and destruction across East Timor, Indonesia now placed the province under martial law, and on 8 September the United Nations announced a total withdrawal from East Timor. Meanwhile, in Jakarta, protesters attacked the Australian Embassy and several Indonesian officials talked of attacking Australian forces if they arrived in East Timor. Militia leaders vowed to kill any Australians in the province.

The evacuations continued for another week, often involving careful negotiation. Each day the aircraft left Tindal with their SAS security teams, flew to Dili, collected their cargoes of refugees and flew them to Darwin. Then it was back to Tindal for debriefing and planning for the next day's operation. By the last flight on 14 September they had evacuated over 2700 personnel.

Operation SPITFIRE was a demanding operation for the SAS soldiers, who reflected later that it was just as dangerous as the subsequent more extensive operations in East Timor, most particularly as the rules of engagement were very constrained. It had been a delicate mission in which the soldiers had had to rely on the TNI for their security in a volatile environment. The SAS brought particular skills to this task including weapons discipline, training as medics, an ability to speak Bahasa Indonesia and Tetum, excellent radio and satellite communications, and experience in operating in small groups. Their most important attributes, however, were discipline and an ability to defuse sensitive situations by developing rapport with individual TNI soldiers. By the end of SPITFIRE nearly every member of 3 Squadron had visited East Timor, many on up to six occasions. They had become familiar with Dili airport and had seen the surrounding countryside from the air. It was valuable preparation for Operation STABILISE.

As soon as the SAS had arrived at Tindal the Joint Task Force staff had begun planning for further contingencies, planning that continued while SPITFIRE was under way. The planners, who included officers and NCOs from all the units involved, had to contribute to Headquarters Special Operations and DJFHQ planning on how the SAS would be used if a larger force were to be deployed to East Timor. The DJFHQ had already begun considering the deployment and maintenance of the Australian force, which was given the name Operation WARDEN. The DJFHQ sub-contracted planning for the deployment of the land force (apart from specialist and logistics elements) to the 3rd Brigade, which had already prepared its plans for the services-protected evacuation of Australian and UN personnel from East Timor, expecting a deployment for up to 96 hours.

For Operation WARDEN the plan had to be changed from one based on a short-term limited deployment to a large-scale longer-term commitment. The commander of the 3rd Brigade, Brigadier Mark Evans, was told to prepare for a commitment of from 40 to 60 days but, given the time constraints placed upon him, his plan could really only be an expanded version of the evacuation plan. Planning could not begin in earnest until 12 September, when President Habibie announced that Indonesia would accept an international peacekeeping force. Staff then worked throughout the night preparing plans to secure a point of entry. The logistic support was relatively light, reflecting the early plans for a very short-term deployment, and the belief that large numbers of combat troops had to be deployed in the first few days to seize the initiative from the militia.

On 14 September, the last day of Operation SPITFIRE, the UN abandoned its compound in Dili to the vengeance of the militias and warned that thousands of refugees were close to starvation. That day the Australian government accepted the UN Secretary-General's request to lead and manage a coalition force. Next day the International Force East Timor (INTERFET) was established by UN Security Council Resolution 1264, with the mandate of restoring peace and security in East Timor, protecting and supporting UNAMET in carrying out its task and, within force capabilities, facilitating humanitarian assistance operations.

Cosgrove was appointed Commander INTERFET, and the necessary

Australian and overseas forces were assigned to his command. The UN-authorised coalition operation, as distinct from the Australian operation, was known as Operation STABILISE, and DJFHQ became Headquarters INTERFET. Within the headquarters, Commodore James Stapleton, Commodore Flotillas, became the INTERFET Naval Component Commander. Brigadier Evans was the Land Component Commander; the land forces would initially include his brigade group and a company of British Gurkhas that arrived in Darwin on 17 September. Air Commodore Roxley McLennan, commander of the RAAF's Air Lift Group, became the Air Component Commander; he would assign missions to aircraft from several countries that had been provided for transport tasks between Darwin, Tindal and East Timor. He also commanded the RAAF Caribou transport aircraft and Army Black Hawk helicopters based in East Timor, and the RAAF combat force support units operating the two airfields at Dili and Baucau.

Special Forces figured prominently in INTERFET's plans to secure the points of entry and later to conduct operations throughout the province. As Cosgrove's Special Operations Component Commander, Lieutenant-Colonel McOwan would deploy with a small tactical headquarters to command Special Forces operations. The coalition Special Forces were given the title of Response Force—a term that was more acceptable to both the international community and the United Nations. The Response Force was built around the SAS force at Tindal, including McMahon's 3 SAS Squadron group; while at Tindal they were joined by elements of Allied Special Forces, allowing 3 Squadron to deploy 20 patrols. The Response Force numbered about 200 personnel.

Cosgrove planned to conduct Operation STABILISE in four phases. The first phase (gaining control) involved securing the points of entry at Dili and Baucau and the initial lodgement of INTERFET. In phase two (consolidation) INTERFET would establish a secure environment to enable UNAMET to recommence operations and to facilitate humanitarian assistance. During this time the force was to reach its full strength. Phase three (transition) involved the transfer of responsibility to the UN Transitional Administration East Timor (UNTAET) peace-keeping force. Phase four would be withdrawal. Cosgrove identified a number of essential tasks. The first was to establish liaison with all the key organisations, ranging from the TNI and the East Timor independence groups to the UN and the non-government organisations (NGOs). He had to secure the points of entry, protect and support UNAMET, protect and sustain his own force, create a secure environment to protect the East Timorese people, and facilitate NGO humanitarian assistance operations.

On 12 September McOwan flew to Brisbane to brief Cosgrove on his ideas for the Response Force concept of operations. He explained that the Response Force was structured to provide discrete elements offering survivability, sustainability, linguistic, communication, medical and protection capabilities. It would be structured as self-contained force elements able to operate in remote areas with limited external support and with a focus on providing 'situational awareness'. Situational awareness is military jargon for intelligence gathering and, in the case of the SAS is normally achieved through reconnaissance and surveillance.

In the first phase the Response Force would secure the air and sea points of entry to enable the 'follow on force' to arrive securely. It would support Cosgrove's command element and establish initial liaison with the TNI.

The Response Force would, however, come into its own in the consolidation phase when it would concentrate on situational awareness. This would include establishing contact with all the 'stakeholders', primarily the non-TNI elements, although contact would also be made with TNI units until effective military observer groups could be deployed. Liaison and Communication Teams (LCTs), equipped with secure communications, would deploy to all major militia leaders, the National Council of Timorese Resistance (CNRT) leaders, the FALINTIL guerilla cantonments and camps, and the large camps of internally displaced personnel (IDPs). (FALINTIL was the Portuguese acronym for the Armed Forces for the National Liberation of East Timor.) In a brief to Cosgrove, McOwan described the role of the LCTs as providing the 'commander's telescope' and 'ground truthing'. As the media would be reporting every activity, it was crucial for Cosgrove to receive immediate and accurate information about the situation throughout the area of operations; also, any factional complaints would require a rapid response. The Response Force aimed 'to provide a trusted channel directly responsive to the commander', using voice communications from critical areas at decisive times. In addition to this liaison and communications role, the Response Force had to retain the capacity to conduct raids (known as ADD—apprehension, disarmament and detention—operations), to provide close personnel protection for senior officers and dignitaries, to provide a military search and rescue capability, and to undertake surveillance and reconnaissance operations.

In considering the possible threat posed by the militia forces if they failed to cooperate, the Response Force staff were aware that the militias relied on support from both covert operatives working among the population and from the TNI. The Response Force's first task would therefore be to sever the connections between these three elements. A similar task would be necessary with the FALINTIL guerillas if they did not cooperate; they would have to be separated from their collaborators and the general population.

On 18 September nine warships from the international force set sail from Darwin with the landing planned for 20 September. On 19 September Major-General Cosgrove and a small staff visited Dili for discussions with the Indonesian commander, Major-General Kiki Syahnakri. An SAS patrol accompanied Cosgrove as the escort team. When Cosgrove departed from Dili he left behind a liaison officer, Lieutenant-Colonel Roger Joy, and the SAS patrol. Joy explained to Colonel Brownrigg that the INTERFET plan envisaged landing some of the Response Force in Black Hawk helicopters. Brownrigg was concerned that the Indonesians might react adversely to the helicopters and Joy therefore radioed Cosgrove who decided to postpone the Black Hawks' arrival by 24 hours.

Back at Darwin that evening Cosgrove telephoned McOwan at Tindal with his orders concerning the helicopters. McOwan had planned to split his force between the C-130s and the Black Hawks, with the C-130s following the Black Hawks into Komoro airport standing off the northern coast, 'at call'. The C-130s were to bring in fuel bladders to refuel the Black Hawks and give the Response Force immediate tactical mobility and observation during the first day of the operation. The Response Force staff had to quickly change their plans. Between 8 pm and 3 am they repacked their load pallets and rescheduled their lists of personnel on the first five C-130s. The operation was launched 30 minutes after the last aircraft was loaded. Senior Response Force officers believed that the change in plans was not warranted; the tactical integrity of the Response Force had been disrupted and its full capability was being carried on only the five C-130s, reliant upon the airfield being unobstructed. As for upsetting the Indonesians, one officer observed that by 20 September INTERFET ships carrying helicopters were offshore from Dili, and the additional helicopters could hardly have upset them even more.

Nonetheless, the decision was made. Cosgrove's instinct to trust the Indonesians proved to be justified by the relatively smooth operations on the first day and their subsequent cooperation. In the morning of 20 September the C-130s landed unobstructed at Komoro airfield, Dili, with the main elements of the Response Force. The Australian soldiers had visited Komoro during SPITFIRE and, as many of them spoke Bahasa Indonesia, they quickly made contact with the TNI soldiers around the airport. McOwan was soon able to assure Cosgrove that the airport was secure and further aircraft began landing with troops from 2 RAR. K and L Troops, securing the Komoro airfield, guided the 2 RAR troops to their positions.

Meanwhile Major McMahon, accompanied by Allied Special Forces

Captain Carl Marning, commander of K Troop 3 SAS Squadron, meeting with TNI soldiers soon after arrival in Dili during Operation STABILISE.

elements and some of I Troop, set off to secure the seaport, travelling in four Land Rovers brought in on the C-130s. Komoro is on the western edge of Dili and the seaport is about ten kilometres away in central Dili. The port was like a scene from Hades. The air was full of acrid smoke from the huge piles of burning rubbish burning throughout the compound. Amidst this surreal and disturbing scene were thousands of people—the sheer numbers inside the small compound were staggering. Many were armed with machetes and seemed unfriendly. In the heat, the stench of humanity in close confines without sanitary support was overpowering. A company of 2 RAR arrived to extend the security of the port before HMAS *Jervis Bay* and HMAS *Tobruk* arrived the following day with 3 RAR and a squadron of 2 Cavalry Regiment.

Once K and L Troops had handed over the airfield to 2 RAR they began their next task, securing the heliport in Dili. The TNI provided several vehicles to transport the Response Force elements and INTERFET tactical headquarters to the heliport. INTERFET had wanted to secure the heliport as quickly as possible to provide a base for the Black Hawks on the first day and although the helicopters were now to arrive on the second day, the heliport still needed to be secured. McOwan was determined to base his Response Force at the heliport so that he could best achieve Cosgrove's directive of maintaining a personnel recovery capability at short notice to move throughout the operation.

Meanwhile, elements of L Troop went to the UN compound in Dili to take possession of a dozen UN Land Rover Discovery vehicles. The SAS had intentionally deployed to Tindal without their long-range patrol vehicles and with only two standard Land Rovers. At Tindal they had been issued with four more standard Land Rovers. While in Darwin, McOwan had asked the head of the UN mission if the Australians could use the UN vehicles left behind in Dili. The UN official had replied that he fully expected the vehicles to be destroyed but that if they were still operable the Australians were welcome to use them. Finding the vehicles usable, but having no keys, the SAS soldiers hot-wired them, and shared their use with INTERFET headquarters. Eventually the Response Force had six Land Rover Discovery vehicles, three Toyota Hi-Ace type vehicles, and one Hi-Ace ambulance.

Next day the Black Hawk helicopters arrived and the Response Force could now constitute an immediate reaction force. Cosgrove required two helicopters always to be ready to move at a moment's notice, and McOwan usually had at least two patrols ready for immediate deployment. One of the Response Force's earliest tasks was to lift a patrol to rescue a British journalist being attacked by the militia.

Members of an SAS patrol entering the jungle town of Dare, fifteen kilometres south of Dili, soon after their arrival in East Timor. They were welcomed by cheering refugees.

By the evening of 21 September the Response Force at the heliport numbered 152, with a further sixteen members of the LOGPAK in Darwin to join them shortly. A few days later HQ INTERFET moved to the premises of the Dili Public Library, several kilometres away, as did McOwan's Response Force headquarters. McMahon and his Response Force squadron group remained at the heliport.

Even before the 3rd Brigade's battalions took control of Dili, the Response Force conducted reconnaissance patrols by vehicles beyond the city and sought to establish contact with FALINTIL. On 21 September L Troop and elements of I Troop wound their way through the hills fifteen kilometres south of Dili to the jungle town of Dare, where they were welcomed by cheering refugees. Next day they were joined by K Troop, escorting UNAMET staff carrying out a humanitarian assessment. Response Force had received information that there were about 37 000 IDPs in the Dare area. L Troop, which detained eleven personnel on 23 September, remained at Dare until 28 September when it became clear that it was a CNRT base, not a FALINTIL one.

While L and K Troops were at Dare, on 22 September the Head of UNAMET, Ian Martin, accompanied by McMahon and I Troop, flew by helicopter to Uamori, on a high plateau in the middle of the jungle 70 kilometres east-south-east of Dili (an area called the 'Lost World' by FALINTIL). McMahon met the FALINTIL commander, Taur Matan Ruak, who had been leading the independence fight for the previous nine years while Xanana Gusmao was held in prison in Indonesia. Ruak agreed that the Response Force's Liaison and Communication Teams (LCTs) could be deployed with his forces in their cantonments. I Troop remained at FALINTIL headquarters and in the Group 2 and 3 Cantonment at Uamori until 27 September, when it was reduced to two patrols. It had been reduced further to a patrol-sized LCT by 1 October, when it relocated to Atelari.

A few days after McMahon's visit to Uamori, Major Truscott was deployed as a liaison officer with Taur Matan Ruak, with whom he established a close relationship. The militia commanders had made a major mistake in trying to oppose INTERFET. If they had cooperated with INTERFET in the same manner as FALINTIL, Cosgrove would have been obliged to deploy LCTs with them also, which perhaps would have given them some legitimacy. INTERFET's task with FALINTIL was firstly to gather situational awareness across the area of operations, and secondly to demonstrate that it could provide security throughout the province and thereby persuade the guerillas to remain confined to their cantonments in the mountains and not take action against the militia. If FALINTIL had left their cantonments and begun to fight openly with the militia, Indonesia could then have presented the problems in East Timor as a civil war.

An SAS patrol talking to a senior CNRT officer.

Response Force moved quickly to increase its contacts with FALINTIL. On 27 September a troop was deployed to Odolgomo with Cantonment 4-South. On 3 October this was reduced to an LCT and moved to Bobonaro where the LCT carried out medical clinics in late October, ceasing operation on 18 November. An LCT was deployed to Ermera on 2 October with Cantonment 4-North and remained there until 17 November. Other teams were deployed to Group 1 Cantonment at Atelari, Remixio and Aileu. Eventually, between 15 and 20 November, the FALINTIL fighters were concentrated at Aileu after Groups 2 and 3 had been temporarily located at Remixio. All LCTs were then based at Alieu. The relocation of the guerillas was a huge logistic operation for Response Force and was undertaken with a wide range of INTERFET vehicles and local modes of transport. By using limited resources INTERFET demonstrated its impartiality, and the SAS officers, NCOs and soldiers did an excellent job in difficult circumstances. Warrant Officer Bob McCabe and Sergeant Bill Maher proved outstanding LCT leaders.[2]

The Response Force LCTs performed an absolutely vital task. They supplied a flow of intelligence about the militia and alerted INTEREFT authorities to war crimes sites. By providing communications between FALINTIL locations and giving FALINTIL commanders confidence that INTERFET forces were gaining control

The force of choice

The SAS squadron provided close personnel protection for a range of visiting officials and East Timorese leaders. The most demanding task was protecting the East Timorese leader Xanana Gusmao, shown here with the FALANTIL commander, Taur Matan Ruak.

of the province, they ensured that FALINTIL remained out of action and contained within their cantonments.

The provision and maintenance of the LCTs posed a considerable drain on Response Force resources, but the task reaped crucial benefits. Another demand on personnel was the provision of close protection personnel for visiting officials and East Timorese leaders. The first task was the protection of the Head of UNAMET, Ian Martin, who arrived at the beginning of the INTERFET deployment. Other tasks included protection of the US Deputy Secretary of State during his visit on 25 September, the protection of Bishop Belo on his return to East Timor, and protection for the Australian Defence Minister and the New Zealand Chief of Defence Force during visits in October. The most demanding task was that of protecting Xanana Gusmao when he returned to the province on 22 October. Initially two troops were assigned, but the group was later reduced to a patrol. The task ceased on 15 November, but the Response Force continued to provide a liaison officer with Gusmao's party.

Although the Response Force gained valuable information through the deployment of LCTs with FALINTIL, there was no substitute for gaining intelligence by deploying Response Force patrols in the

497

traditional SAS role of reconnaissance and surveillance. Some of the earliest reconnaissance activities consisted merely of driving Land Rovers to selected points near Dili.

On 24 September Lieutenant-Colonel McOwan, accompanied by his RSM, Warrant Officer Greg Jack, and another soldier, carried out a headquarters reconnaissance to the small village of Tibar, west of Dili. While there the villagers showed McOwan some bodies in a grave. Next day Jack led a party, including the SAS doctor, Major Vaughan-Evans, several military police and a small security party from Response Force, back to identify the bodies and record the graves.

Later that day McOwan visited the party and stayed for five or ten minutes. Just as he was getting into his vehicle to depart there was a burst of gunfire from the nearby Tibar village. In an instant, between 70 and 80 East Timorese fled the village and disappeared into the scrub to hide. The villagers who had been showing the graves to the Australians immediately looked to McOwan and his men, expecting them to act. Taking Warrant Officer Jack, Major Andrew Shaw (the SAS intelligence officer) and three SAS soldiers, McOwan and his small force drove slowly down the road towards Tibar where they dismounted and moved carefully through the village. Suddenly a vehicle with three occupants drove into the centre of the village from the other direction; the SAS party stopped the vehicle at gunpoint. One of the occupants, a uniformed Indonesian policeman with a handgun, immediately jumped from the vehicle and disappeared into the scrub. INTERFET rules of engagement prevented troops from firing at fleeing individuals, even though they might be armed. The patrol captured the other two occupants, one of whom had an SKS rifle and grenades, fixed plasticuffs around their wrists and ankles and placed them beside the road, then continued clearing the village as they were unsure what lay ahead.

Within a few minutes three trucks, each with 20 or 30 TNI soldiers (or militias, as they were wearing an assortment of clothes), drove into the village, stopping short of the other vehicle. The soldiers dismounted and, with much yelling, pointed their rifles at the Australians. Jack, Shaw and one SAS soldier pulled back to a hut on one side of the road, while McOwan and two other soldiers also withdrew quickly, their exit route being blocked by the brick wall of a coffee warehouse compound. They were no more than ten metres from the first truckload of soldiers, all of whom were pointing their weapons at the Australians. McOwan told the Indonesians not to fire, but took the precaution of moving the safety catch of his M4 rifle to fire, and pointed back. The other Australians did likewise. An Indonesian lieutenant moved forward, cut the plasticuffs off the prisoners and gave one of them his SKS rifle. The latter immediately ordered the TNI to kill the Australians. The TNI took no further action, however. They returned to their trucks, someone climbed into the original Indonesian vehicle, and they all departed.

The small SAS force withdrew into the coffee warehouse compound, consolidated and radioed for assistance from the Immediate Reaction Force. When the Black Hawk helicopters arrived they reported that the TNI had formed an assault line and were advancing on Tibar from the north. The TNI burned a few houses; then, with the Black Hawks still circling with SAS troops, mounted their trucks and disappeared en route for West Timor. For the small group it had been an unnerving experience.

In the afternoon of 27 September Truscott, at FALINTIL headquarters at Uamori, informed McOwan by secure radio that a large militia band had looted and burned the coastal town of Lautem in the far east of East Timor and was heading east towards the ancient port of Com. More than 2000 East Timorese were on the wharf at Com awaiting forced deportation to Indonesia. Truscott reported that he had heard that FALINTIL guerillas had already clashed with the militia, killing three, and now planned to rescue the displaced persons at Com. Such an action would have inflamed the security situation.

Over the previous days McOwan had proposed several raiding operations to Cosgrove, recognising the need to reassure FALINTIL of INTERFET's resolve in dealing with the militia. In each case the INTERFET Commander had withheld approval; his first priority was to secure Dili and he still lacked intelligence about what was happening throughout the rest of the area of operations. McOwan now asked if he could mount an immediate raid on Com; this time Cosgrove ordered him to proceed. McOwan had two major problems—one, dusk was approaching (and the aviators had been ordered not to fly their Black Hawks at night) and two, the distance of the task would be on the limits of their endurance. He phoned a warning to the heliport and drove there to brief McMahon. The available Black Hawks of the 5th Aviation Regiment immediately began start-up procedures. Barely 20 minutes after the phone call, three Black Hawks took off with McMahon, Captain Marning's K Troop and an Allied Special Forces element.

Just on last light the Black Hawks delivered the SAS force near the coast several kilometres west of Com. Each man was equipped with night-vision goggles and individual radios, making it relatively easy to advance in the dark along the coastal strip towards Com. Meanwhile the helicopters carried out a reconnaissance down the coast, also using night-vision goggles, reporting by radio to McMahon. They refuelled on HMAS *Adelaide*, stationed offshore, before returning to Dili.

Within a few hours the SAS reached Com and found the displaced East Timorese in a compound beside the wharf. No militiamen were identified, but from the attitude of the refugees, there were evidently some militiamen among them. Assessing that to enter the compound might lead to a dangerous and possible bloody confrontation, McMahon left word that if the militia did not come out of the compound he would move in and apprehend them. McMahon was anxious to clear a small rocky hill overlooking the town, so about midnight he withdrew his force to the hill and settled into a secure perimeter to await the militia reaction. Soon a group left the compound, carrying only machetes and swords. McMahon suspected that this was a decoy group of refugees, forced by the militia to depart with a few weapons in the hope that the Australians would then leave. He therefore gave another ultimatum to the group in the compound and withdrew again to his perimeter.

The SAS had been on the hill for a barely an hour when through their night-vision goggles they detected a group of between 20 and 30 armed men moving out of Com towards a truck waiting on the road. Directed by McMahon, the SAS moved quietly down to the hill. When the group was about 20 paces from the SAS force, several SAS members shone torches onto them, ordering them to raise their hands. Fortunately the militia group obeyed. If they had tried to return fire there is little doubt many would have been killed, as each SAS soldier had picked his target with the assistance of his night-vision goggles and his laser designator.

The group, numbering 24, proved to belong to a militia band that had previously clashed with FALINTIL, as several members were already wounded. Fourteen automatic weapons as well as some 30 sharp-edged weapons were collected. At first light the Black Hawks arrived and the entire force, with the captives, returned to Dili. The displaced persons in the compound at Com were then free to return home.

It had been an outstandingly successful operation, conducted at very short notice and involving the deployment of almost 50 men at a distance of over 170 kilometres into an unknown area, arriving at last light. The speed of reaction and the flexibility that came from each member being equipped with individual radios was a legacy of the SAS Regiment's experience with counter-terrorist operations. These superb communications, and continual training and exercising in compressed planning cycles, enabled McMahon to direct his men onto the militia party in the dark without the need for extensive orders. He could also emphasise the rules of engagement so that no shots were fired. The incident demonstrated the extreme discipline and competence of each SAS soldier. The Allied Special Forces element operated effectively with the Australian SAS, demonstrating the excellent inter-operability between them.

The operation also gave Cosgrove the confidence to use the Response Force in further raids at some distance from Dili. There was one sour note, however. The Response Force was required to pass the detainees—found to consist of sixteen militia members and eight farmers—to INTERFET military police for detention. A little later Headquarters Response Force received information that one of the detainees was of particular interest. They then learned that the military police, under the policy of release after questioning unless there was a more serious prima facie case to answer, had released the entire group. This caused considerable disquiet among the FALINTIL, who were able to tell the SAS of it before they learned of it from within INTERFET.

By late September, with the 3rd Brigade in control of Dili and the surrounding area, Cosgrove began considering how to deploy his battalions to the other parts of the province. He had been unable to deploy them earlier because of logistic constraints. Response Force led the way with the insertion of two patrols into the Ossu area on 26 September and the deployment of a vehicle patrol to Manatuto on 27 September, both from Uamori and with FALINTIL guides. The Manatuto patrol recovered a small child who had lost a leg from an airdrop of food. The recovery of sick and injured Timorese was to happen many more times.

On 29 September two patrols were inserted into the Balibo area to act as a pathfinder group for 2 RAR, which was to seize the town two days later. Sergeant Brett McCosker commanded one patrol of six men. Receiving twelve hours' notice, he was told that the patrol could last for between seven and 21 days and that he had to carry sufficient rations for the entire period. Just after last light on 29 September they rappelled from a Black Hawk helicopter into an area of tangled lantana bushes, then struggled for four and a half hours to cover a distance of 340 metres to reach a hill from which they could observe Balibo. Next morning they could see vehicles moving in and out of the village. Corporal Andrew Cameron commanded the other patrol of five men. During the insertion into thick scrub the patrol radio operator landed heavily, broke his leg, and was immediately evacuated. The four remaining men had to cover about four kilometres until they reached high ground from where they could observe Balibo through a telescope. Both patrols had difficulties with their radio communications, and their orders to prepare a landing zone in Balibo for 2 RAR in the morning on 1 October were received late. McCosker's patrol made a forced march of two and half hours to secure the landing zone just before 2 RAR arrived. Cameron's patrol covered two kilometres

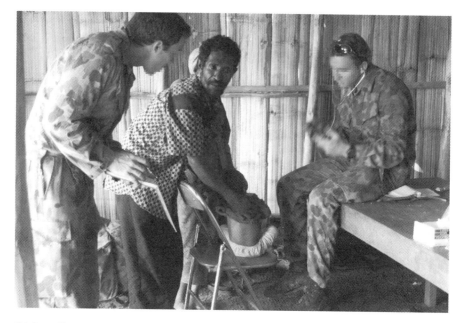

SAS medics provided assistance to the East Timorese population at every opportunity, especially in the more remote areas.

through difficult terrain and arrived at one end of Balibo just as the helicopters landed at the other.[3] These early patrols enabled the head-quarters staff to reassess subsequent movement rates.

Two Response Force patrols undertook similar pathfinder tasks for the deployment of 2 RAR elements to Cova on 1 October and Batugade on 2 October. For the pathfinder mission before 3 RAR secured the border town of Maliana, Lieutenant-Colonel McOwan planned a different operation. Information from FALINTIL indicated that the militia controlled the town. McOwan hoped to replicate the success of the Com raid with a rapid descent on Maliana with eight patrols, with three more patrols providing a cut-off force and another kept in reserve. The raid would use twelve of the Response Force's 20 patrols, but additional patrols were to be formed out of squadron headquarters. Insufficient Black Hawks were available to mount the operation as planned, thus the assault forces had to be inserted progressively at some distance from the town. McOwan accompanied the raid although Major McMahon retained tactical command. When the Response Force patrols entered the town on 4 October they were greeted enthusiastically on the soccer field. The militia had apparently departed two weeks earlier, and the FALINTIL information was either out of date or had been incorrectly assessed. Two patrols remained in the town until relieved by 3 RAR. Although there was no militia contact the raid provided Response Force with valuable experience in conducting a raid against a town on converging axes. The experience would soon be put to good use.

The Black Hawk helicopters enabled the Response Force to mount operations quickly and then return to base. As soon as the troops returned to Dili from Maliana they began preparing for the next operation. On 29 September a reconnaissance helicopter carrying Response Force personnel had received fire while flying over Suai airstrip, near the south-west border. FALINTIL intelligence had consistently reported that the main militia supply route was from Atambua, in West Timor, along the road though Suai, and then further in East Timor. The area contained the militia group Laksuar Merah Putih and possibly other militia organisations. FALINTIL was deploying forces into the Suai area to prevent the militia from burning rice crops, and had also deployed a patrol into the Zumalai area to prevent the militia from forcing displaced persons to move across the border into West Timor. On the night of 3 October, Response Force inserted two patrols into the Suai area and they were in position the following morning. That night they reported shooting in Suai and the next day, 5 October, reported that some rice crops were burning and that several trucks were moving between Suai and the East Timor/West Timor border.

In response to FALINTIL concerns about militia activity in the Ainaro Regency south of Dili, bordering on the south coast, McOwan sought and received Cosgrove's agreement to Response Force undertaking a vehicle patrol through this area. In the morning of 5 October L Troop, commanded by Captain Jon Hawkins, and including an Allied Special Forces patrol, headed south with two Land Rovers, two Light Armoured Vehicles (ASLAVs) provided and driven by the 2nd Cavalry Regiment, a Unimog truck and two motorbikes. They drove through Aileu, in Aileu Regency, and then through Maubisse, Ainaro, Cassa and Zumalai, in Ainaro Regency. Although most members of the vehicle patrol came from L Troop, it also included Corporal Mark Hogno from K Troop. Hogno had injured his ankle and had been on restricted duty. A friend from L Troop had recognised Hogno's disappointment at not being able to take part in operations and arranged for him to join the vehicle patrol, as it would not involve a great deal of walking. The patrol did not encounter any militia, but provided much confidence to the villagers along the way. As the vehicles passed through the town of Cassa, Hogno was heard to exclaim, 'I don't think anyone is ever going to fire a shot in this whole deployment!'

Meanwhile, acting on FALINTIL intelligence, Response Force, with approval from Headquarters INTERFET, was planning a raid on Suai, described as a squadron-level 'brush and block' operation. The task

Vehicles from the patrol that crossed East Timor from Dili to Suai in early October 1999. Corporal Mark Hogno, in the closest vehicle, is speaking to the commander of L Troop, Captain Jon Hawkins.

group included 3 SAS Squadron headquarters, I Troop (four patrols), a patrol from Headquarters Response Force, a patrol from the Media Support Unit (including serving or former members of the SAS), the two K Troop patrols inserted on the night of 3 October and already observing Suai, L Troop (advancing by road from Ainaro Regency), the Immediate Reaction Force on standby in Dili, and four Black Hawks. HMAS *Adelaide* was located ten to fifteen nautical miles offshore from Suai. On board was an adviser on Caribou operations (as it was hoped to open Suai airfield for Caribou operations) and a Clearance Diving Team to survey the beach for a later amphibious landing.

At 8.50 am on 6 October Black Hawk helicopters landed west of Suai and Response Force troops moved quickly into blocking positions. At the same time, more helicopters landed north of Suai where squadron headquarters and I Troop, commanded by Captain Keating, began the 'brushing' operations towards Suai. By that time L Troop, advancing by road three kilometres east of Suai, had encountered a militia group—two members of the group were now fleeing towards Suai on motorbikes. Suai town spreads over several kilometres and shots were heard as the 'brushing' force swept from north to south through the town. A truck and a bus tried to rush the Response Force roadblock, commanded by Captain Marning, south-west of the town and the SAS soldiers fired at the truck's tyres and engine. Six suspected militia members were wounded, two seriously, and the drivers of both vehicles were detained. Eventually the Response Force had 115 detainees. Aided by a local person, the Response Force carried out a rudimentary and quick sorting of the detainees to determine who among them might be of intelligence value. Ten detainees, thought to be important militia members, were despatched by helicopter to Dili along with the wounded militiamen.

McOwan, who was present during most of the operation, directed McMahon to transport the remaining detainees to the West Timor border, where they would be released to proceed on foot. About 5 pm Captain Hawkins and his troopers, in two Land Rovers and the Unimog, were given the task of escorting the detainees in seven civilian trucks. About four or five kilometres west of Suai the trucks moved away from the escorting vehicles, which were suddenly ambushed by a burst of automatic fire from their left front at a range of about 50 metres. The driver of the second Land Rover, Corporal Hogno, was shot in the shoulder and only the swift action of his front-seat passenger, Corporal Phillip Beresford, in grabbing the steering wheel prevented the vehicle from crashing. Trooper Ron Juric, manning the machinegun in the first Land Rover, was struck in the right forearm and left leg. The vehicles immediately stopped and the Response Force soldiers spread out in the scrub along the right hand side of the road, returning fire.

While Captain Hawkins reported to Major McMahon by radio, several other soldiers, including Trooper Brook Burgess, ran back to the vehicles, under fire, to recover equipment and the patrol medical kit so that they could treat their wounded comrades. McMahon immediately joined the two ASLAVs that had remained in Suai and drove to the ambush site, taking between five and ten minutes to arrive. Once there the ASLAVs engaged the area, while Response Force troops swept along the southern side of the road for about 100 metres. They found two dead militiamen and several weapons. As the assault party returned to assist with the treatment of the wounded soldiers they were engaged with fire from the northern side of the road. Heavy chain-gun fire from the ASLAVs and small arms fire from the Response Force caused the enemy fire to stop. At last light the two wounded Australians were flown by helicopter to Dili. They were evacuated to Darwin the following day.[4]

Back in Suai, where Response Force members came across the first evidence of the massacre in the half-built cathedral, McMahon occupied a defensive position on orders from INTERFET Headquarters.

505

After a few days the force was reduced to two patrols; they were still there when the headquarters of the 3rd Brigade landed near Suai in Australia's first amphibious operation since the Second World War.

While the majority of Response Force was preparing for the Suai operation, two patrols were inserted near the village of Alto Lebos, close to the East Timor border. After several days, one of the patrols was extracted while the other, consisting of Allied Special Forces members, remained in an observation position. In the morning of 9 October they saw a group of between sixteen and 20 people at a distance of two kilometres. Later in the morning a smaller group approached to within seven metres of the observation post, firing indiscriminately. In the early afternoon several militia groups appeared, one group firing shots that came within metres of the observation post. When the militia pointed the barrels of their weapons directly at the observation post the Response Force patrol opened fire, killing one. The remaining militia started running towards the border as the Immediate Reaction Force, led by Captain Hawkins, arrived by helicopter. From the air they could see 20 armed persons with green camouflaged fatigues, about one kilometre from the incident, but under the rules of engagement could not fire at them. One dead militiaman wearing olive green trousers and shirt was recovered and taken to Dili.

Response Force now planned a raid on Aidaba Salala, near the East Timor border, as intelligence had indicated considerable militia activity there. Response Force wanted to initiate the raid without a reconnaissance, but the village was in the 2 RAR area of operations and brigade headquarters would not approve the raid. Eventually Response Force was permitted to conduct a preliminary reconnaissance provided the patrol was inserted from the north-east of the village, and on 13 October a six-man patrol, commanded by Sergeant Steve Oddy, was inserted. Aged 32, Oddy was an experienced patrol commander with twelve years' service in the regiment. Over the next three days the patrol moved south-west towards their target, spending a whole day in a village with up to 500 people, without being detected. Early in the morning of 16 October, approaching Aidaba Salala, the patrol crossed the dry bed of the Moto Meuculi Creek and prepared to establish an observation post. The creek bed was about ten metres wide with high banks of two to three metres. The area was covered with scattered scrub, long grass and lantana. It had been reported that the militia had used the creek bed as their route in and out of West Timor.

Oddy was instructing his second scout, Lance-Corporal Keith Fennell, to clear an area overlooking the creek bed, when a group of five or six militia appeared in the creek, the leading man moving stealthily with his weapon to his shoulder. They were wearing uniforms with webbing. Fennell sank quietly to the ground, but the

An SAS reconnaissance patrol being extracted by a Black Hawk helicopter near the East Timor/West Timor border.

leading man had seen him and brought his weapon to the aim position. Fennell immediately engaged his opponents, firing half a magazine and striking three of them. Oddy stepped alongside him and fired several shots from his 40mm grenade launcher at the escaping militia.

There was no further noise or movement, so Oddy ordered Fennell and the patrol medic/machinegunner to recross the creek to obtain a better field of view. Meanwhile Oddy reported by radio to Response Force headquarters (although his initial message was not received). A few minutes later the medic saw two militiamen moving up the creek in the same direction as the previous party, obviously intending to attack the patrol. He shot and killed one of them and fired further rounds into their general area. Oddy recalled the two men from across the creek bed and went into an all-round defensive position, each patrol member being assigned an arc of responsibility.

Before long several patrol members reported movement towards their position. It seemed that the militia group had left the creek bed and was trying to encircle the Australians. From the movement of the undergrowth it looked as though up to five militiamen were approaching in extended line. The patrol second-in-command opened fire at a distance of about ten metres and believed that he struck two militiamen. Other attackers then fired wildly towards the Australians, who

507

SAS: Phantoms of War

returned fire with their M4 carbines, 40mm grenade launchers and hand grenades. The militia party, now thought to number 20, pressed the attack, with one of them giving orders for using fire and movement. Facing a numerically superior force willing to accept casualties, Oddy decided to break contact. Leaving their large packs, the Australians quietly withdrew across the creek bed. After 300 metres they saw two Immediate Reaction Force Black Hawk helicopters. The patrol discharged a smoke grenade; one of the helicopters landed in an open area and the patrol scrambled aboard. The helicopters then flew back to the contact site, where they could see no sign of activity. Spotting an open area near the contact site which had not been evident to the Australians when they were in contact, the helicopter landed, three patrol members alighted, recovered their packs, collected several weapons near one of the dead bodies, and quickly departed. Six hours later a company from 2 RAR swept through the area but found nothing. Reports from villagers suggested that the militia party had suffered five killed and three wounded; the official figure was four killed.

The series of contacts had lasted for about one and half hours, during which the Australian patrol had displayed excellent fire discipline. One patrol member did not fire at all as no enemy appeared in his arc of responsibility. In total the patrol fired about 200 rounds from the Minimi and only 67 rounds from M4 carbines.[5]

The Response Force continued mounting raids at diverse locations, sometimes with FALINTIL guides. On 11 October two troops raided Hato Hudo and Cassa, but no militia were detained. On 16 October two troops raided Fatomean and Fohorem, again without detaining any militia. On 19 October a troop went to Atauro Island to demonstrate INTERFET's capacity to operate offshore, to investigate the situation and to reassure the public on the island. On 20 October the immediate Reaction Force went to Salale, where it was reported that militiamen were forcing people onto trucks to take them across the border.

Meanwhile, General Cosgrove was under pressure to send forces to the Oecussi (or Ambino) enclave where there were reports of continuing militia violence and destruction. On 14 October a Response Force patrol assisted the Navy to land an East Timor youth (from Oecussi) on a beach in the Oecussi enclave. This person reported back to INTERFET headquarters for several days by satellite telephone, via the Remexio LCT, on militia activity in the enclave.[6] Cosgrove then appointed Lieutenant-Colonel Mike Crane, commanding officer of the 4th Field Regiment, as commander of Ambino Force and directed him to plan an operation to secure the enclave. Major McMahon was ordered to undertake a reconnaissance of the enclave using a troop from Response Force, a platoon of Gurkhas and a mechanised platoon from 5/7 RAR.

In the morning of 22 October McMahon and L Troop flew into the Oecussi enclave in Black Hawk helicopters. They immediately encountered a group of between 30 and 40 militia, with an assortment of primitive weapons, who surrendered when they saw the force arrayed against them. An hour later the remainder of McMahon's small force arrived on a Navy landing craft. McMahon then quickly deployed his patrols by helicopter to the main villages in the enclave. On 25 October Crane arrived with a larger force and took command. L Troop remained in the enclave until 17 November, manning an overt observation post at Bobometo near the Indonesian border.

With the deployment to the Oecussi enclave, INTERFET had secured most key areas of the province and about 80 per cent of East Timor had been returned to peace and stability. On 19 October the Indonesian government recognised East Timor's independence, and on 21 October the independence leader, Xanana Gusmao, returned to the province. On 31 October the TNI forces withdrew from East Timor, and on 18 November planning began for the transition from INTERFET to the United Nations Administration East Timor (UNTAET).

Throughout this period Response Force had worked extremely hard. In its first 40 days it conducted 56 missions and spent almost 80 per cent of its available time on tasks, leaving only 20 per cent of its time for rest and battle preparation. Towards the end, troops were actually falling asleep in the doorways of helicopters while deploying on operations. Although the Response Force soldiers were very tired, both General Cosgrove and Lieutenant-Colonel McOwan were determined that the 'best soldiers in the Australian Defence Force' should take the lead in INTERFET's operations. As McOwan had hoped, the Response Force had become Cosgrove's force of choice. None of this implied that the infantry battalions were not fulfilling vital and difficult tasks. Indeed, it was because the infantry battalions were undertaking the main security missions that the Response Force could be kept 'in reserve' for tasks at short notice.

Operation STABILISE had seen the SAS conducting the full range of operations—from liaison with and direct support to guerillas, including intelligence collection, to close personal protection, medical assistance, reconnaissance, surveillance, raids and immediate reaction. While the raids received some publicity, the most important work was with the guerillas and the extensive surveillance patrols. These relied on the training, initiative and discipline of NCOs and soldiers who received little acknowledgment for their excellent work. The operation also revealed some weaknesses in equipment. The PINTAIL radio was found to be unreliable for Response Force operations. As in Kuwait, the soldiers mostly preferred the M4 carbine over the Steyr rifle. The Australian camouflaged uniform was too heavy for the hot oppressive

Lieutenant-Colonel T. J. McOwan, DSC, CSM, Commanding Officer of the SAS Regiment, 8 January 1999–5 January 2001, with his Regimental Sergeant-Major, Warrant Officer Class One Greg Jack, greeting an East Timorese civilian during the Remembrance Day service in Dili on 11 December 1999.

climate. During the operation some SAS members had pressed McOwan to deploy the long-range patrol vehicles from Swanbourne; he was adamantly opposed to using them, as their unique configuration would have immediately drawn attention to SAS operations. Moreover, he was well aware of the vulnerability of vehicle-mounted troops throughout East Timor and was determined to employ vehicles only when no other option existed.

On 22 November McOwan handed over command of Response Force to Lieutenant-Colonel Neil Thompson, recently commanding officer of 4 RAR (Commando), and returned to Australia to participate in a pre-Olympic national level anti-terrorist exercise involving Cabinet.[7]

During December, 1 SAS Squadron replaced 3 SAS Squadron in East Timor. Later 3 SAS Squadron was awarded the Meritorious Unit Citation for sustained outstanding service in East Timor. 1 SAS Squadron, commanded by Major Jim Phillips, also performed exceptionally well in East Timor.[8] By this stage the Allied Special Forces elements had departed. The militia threat had not diminished, however, and a high state of readiness continued well into 2000 in the western border region and in the Oecussi enclave, where there were three contacts on 17 January in which at least one militiaman was killed. SAS liaison officers were placed at the West Force headquarters and with the battalions to ensure coordination between the units and the SAS surveillance patrols, which bore the brunt of border surveillance in difficult country. The Response Force continued to maintain the Immediate Reaction Force and to provide support to FALINTIL. Progressively, moving east to west, INTERFET handed over the sectors to UNTAET, until UNTAET took over completely on 22 February 2000. With the withdrawal of INTERFET, 1 SAS Squadron also completed its tour in East Timor.

The East Timor deployment placed a heavy burden on the SAS Regiment. With 3 SAS Squadron deployed from August 1999 onwards, 1 Squadron had to prepare to relieve it, while 2 Squadron continued training for the Sydney Olympics and remained the on-line CT squadron. But there were still officers and NCOs in training and operational support appointments that were not directly involved in operations, who continued to pursue the types of adventurous activities and skills that had maintained the SAS and had attracted the regiment's members in the past. For example, during 1999 Trooper Bjorn Aikman solo-climbed the extremely remote Tikora Peak in Irian Jaya in difficult alpine conditions. Captain Matt Shepley climbed the world's sixth highest mountain, Cho Oyu (8201 metres) on the border of Nepal and Tibet, with an Army Alpine expedition in May 2000. Another such activity was the expedition to climb Mawson Peak on Heard Island. At 2745 metres, Mawson, also an active volcano, is Australia's highest mountain. It had been climbed only twice previously—the last time in 1983. Heard Island is an Australian territory located in sub-Antarctic waters; it is uninhabited and as a world heritage area can be entered only with special permission.

The expedition was led by Captain Robb Clifton and included Captains Tim Curtis, Stuart Davies and Matt Rogerson, all members of the SAS Regiment. Clifton and Davies had climbed in Peru in 1996, while Rogerson had climbed several peaks in the Himalayas. The expedition had taken a year and a half to organise. On 20 November 1999 the party flew from Perth to Mauritius and on 3 December joined the Austral Fisheries vessel *Southern Champion,* spending most of December assisting with deep-sea fishing. Landing on Heard Island on 1 January, they established a base camp and several intermediate camps higher on the mountain. Temperatures on the mountain dropped to $-15°C$ with wind gusts estimated at up to 80 knots. Of their sixteen days on the island, only two gave a clear view of the summit. At times

Captains Robb Clifton, Tim Curtis, Stuart Davies and Matt Rogerson near the crest of Mount Mawson, Heard Island, 10 January 2000.

Members of the party that climbed Mount Mawson in January 2000 examine Corinthian Bay, Heard Island, in a Zodiac craft.

The force of choice

they climbed in whiteout conditions and light rain, and also had to use ropes in an alpine-style ascent. Their clothes were wet, and froze as soon as they left their tents. Finally, on 10 January, with glowing red lava flowing from the volcano, they reached the summit. They were collected by *Southern Champion* on 16 January and arrived at Albany on 25 January. The expedition report noted: 'The ascent of Mawson peak was a challenging and at times arduous undertaking. The team was required to operate in extreme isolation while conducting activities involving significant risk. Self-reliance, teamwork and sound risk management were essential for a safe and successful activity.'[9]

This was a world-class ascent. The establishment of a new route on such a remote mountain, involving 3000 metres of climbing with absolutely no support for rescue, is rarely attempted in the modern era of mountaineering.

The SAS Regiment's successful involvement in the security operation for the Sydney Olympic Games in September 2000 was the culmination of many years of training and preparation. The requirement to provide CT support for the Olympic Games was identified in 1993 when it was announced that Sydney had won its bid to stage the Games. Although not related specifically to the Olympics, the modernisation of the regiment's CT capability began when Colonel Jim Wallace, Commander Special Forces 1994–1995, initiated Project Bluefin to purchase the latest equipment. Australian CT staff visited the Atlanta Olympic Games in 1996 to witness the US CT operation.

Over the following years the SAS Regiment began developing techniques that might be needed in Sydney. As usual, each of the regiment's squadrons rotated through a year as the on-line CT squadron, each squadron contributing to the development of new techniques. As described in the previous chapter, 1 Squadron was practising CT procedures with Black Hawk helicopters when the helicopters crashed in June 1996.

As part of the preparations, the SAS purchased expensive equipment to counter possible chemical, biological or radiological threats, and began extensive training with it. The SAS also began developing techniques for the clandestine boarding of a moving ship at sea in the dark. This capability, known as Ship Underway Recovery (SUR), was begun by 2 Squadron in 1999 and further refined by it in 2000.

For the duration of the Olympic Games, one of the CT squadrons would have to be based in Sydney along with elements from regimental headquarters, and this required the selection and preparation of a staging area. The Black Hawk helicopters were an integral part of the CT plan, and they were to be based at Luscombe airfield at Holsworthy Barracks on the southern outskirts of Sydney. Next to the airfield,

An SAS CT team practises clearing a ship at night during training for Ship Underway Recovery operations in preparation for the Sydney Olympic Games. The photograph was taken through night-vision equipment.

living accommodation and office accommodation were constructed, as well as specialist shooting ranges and galleries approximating those at Swanbourne. The Naval base HMAS *Waterhen*, on the shores of Sydney Harbour, was identified as the base for water operations, while another forward operating base was established for launching an SUR operation out to sea.

As preparations progressed it became clear that the regiment would have to raise a second CT capability based on another squadron. With the Tactical Assault Group (TAG) in Sydney to deal with incidents in Sydney and Canberra, a second TAG would be needed for incidents elsewhere, possibly overseas, or for a second incident in Sydney. In January 1999 the CDF, Admiral Barrie, wrote to the Defence Minister seeking permission to raise a second TAG, advising that this would take a minimum of about eighteen months, mainly due to the time taken to purchase additional equipment. The cost, which was approved, was for an additional $15.135 million.

By early 2000 the command arrangements had been firmly established. The ADF's support to the Olympic Games, Operation GOLD, was conducted by two joint task forces (JTF). The first of these, JTF 112, commanded by Brigadier Gary Byles, was responsible for a wide range

The force of choice

Special recovery training at Swanbourne.

of support, including transport and general security; it reported to the Land Commander Australia, although later it reported directly to the Commander Australian Theatre. The second task force, JTF 114, was commanded by Brigadier Phillip McNamara, Commander Special Forces; he was responsible for the provision of CT support and reported directly to the Commander Australian Theatre. McNamara set up headquarters JTF 114 at Headquarters Special Operations. The principal CT capability was provided by JTF 643, the ADF TAG, commanded by the commanding officer of the SAS Regiment, Lieutenant-Colonel McOwan. The TAG's CT capability was provided by 2 SAS Squadron, commanded by Major Dan Fortune, and had the capability to conduct CT operations both on land and at sea. It had its own communications elements and its own liaison staff. JTF 643 included Black Hawk helicopters from the 5th Aviation Regiment and a response company, provided by 4 RAR (Commando). The latter would be used for tasks such as providing a cordon around an incident site.

The second TAG, JTF 644, was based on 3 SAS Squadron and was located at Swanbourne. After returning to Swanbourne from East Timor, 3 Squadron members spent several months undertaking courses around the barracks and improving their individual skills. They then

An SAS CT team descends by rope from a Black Hawk helicopter to practise an assault on the 'methods of entry' house at Swanbourne.

began preparing to take on the task as the back-up CT squadron for the Sydney Olympic Games. A second SOCCE was established at Swanbourne, commanded by the regiment's training officer, Major Grant Walsh. In addition to the two TAGs, McNamara also had access to RAAF aircraft, RAN ships and helicopters and the 1st Joint Incident Response Unit, an Army engineer unit with special training in chemical, biological and radiological response.

The most important work in Operation GOLD took place in the six months preceding the Olympic Games. In March 2000 the first TAG deployed to Randwick Barracks, Sydney, for three and half weeks of training. In May it returned to Sydney and was based at Luscombe Field for a series of five major exercises in and around the city, including an SUR operation to board the Navy ship HMAS *Manoora*, and a siege hostage exercise at the Olympic baseball stadium. All command elements were involved, and at one stage the Prime Minister, John Howard, took part. To deter possible terrorists, a deliberate effort was taken to show aspects of the CT forces training. The second TAG was raised at Swanbourne in July and in August the first TAG returned to Luscombe Field. JTF 643 was formally raised on 13 August and took part in the national anti-terrorist exercise during that month.

Fortunately there were no terrorist incidents during the Olympic Games or during the following Paralympic Games, for which the TAG was also on standby. Nonetheless, the security effort during this period was a major achievement, winning praise from allied officers who observed it in operation. In his post-operation report, Lieutenant-Colonel McOwan observed: 'In comparison to other operations involving SAS task groups in the last four years, this operation was notable for the substantial array of supporting mechanisms and a highly developed command and control architecture.'[10] The New South Wales Commissioner of Police, Peter Ryan, wrote:

> Joint Task Forces 112 and 114 performed their roles in a most professional and effective manner that has attracted international accolades . . . Each and every member of the ADF who contributed to providing the effective security template for the Games should take with them the pride of having done a challenging task well, and the thanks and recognition of the NSW Police Service for their efforts.[11]

By 2000 the SAS Regiment had become firmly established as the ADF's force of choice in handling difficult and delicate situations. This had been strengthened by the formation of Headquarters Special Operations and the establishment of the Special Forces Group in 1997–1998. Further, the government's Defence White Paper released in December 2000 included a commitment to bring 4 RAR (Commando) to full strength and to maintain the SAS Regiment at high readiness at its current strength of 'around 700 personnel'.

The SAS Regiment's successes during its history, and particularly in recent times, should not obscure the fact that for many years it struggled to gain recognition in a generally unfavourable environment. That it achieved these successes is a tribute to the dedication of the men who endeavoured to maintain the regiment at a viable strength while not compromising on standards. The value of the regiment as a force of choice rests squarely on the character, skill and resolve of its individual members. This is demonstrated in the way SAS officers and soldiers have prospered, both in their military careers and in civilian life. Compared with the remainder of the Army, a very high percentage of SAS soldiers has reached commissioned rank. Many former SAS soldiers have become successful businessmen, bringing to their civilian careers the qualities of careful planning, determination and flair nurtured within the regiment.

The strength of the SAS lies also in its family nature. It includes not only former SAS soldiers serving elsewhere in the Army or now in civilian life, but also the wives who have endured months of separation from husbands who invariably put the regiment first. As Warrant Officer Danny Wright explained it one night in the Sergeants' Mess, it is 'not a foot-stamping regiment but a way of life'.

The present-day SAS soldier is equal to the traditions of the regiment. With one exception, the old Borneo and Vietnam hands are no longer serving, although there are a few in the Army Reserve component who continue to pass on their experience to the new soldiers. The regiment also now has a large reservoir of soldiers with recent operational experience. But the men of the SAS will not rest on their laurels. They are a self-critical group. They have little concern for rank or class. Only merit counts. The accumulation of 40 years' experience in the regiment, complemented by the training, motivation and character of the soldiers, provides the ADF with its force of choice for dangerous and delicate missions. The East Timor operation and the Sydney Olympic Games dominated the SAS Regiment's activities in 1999 and 2000. But the regiment is already looking to the future. It will have to adapt to face new challenges, but its history indicates that if any unit can adapt to change it is the Special Air Service.

Notes

Preface

1 *Australian Joint Service Publication, (JSP (AS) 101). Joint Service Staff Duties Glossary* Department of Defence, 1984

Chapter 1

1 This chapter is based substantially on 2 SAS Sqn Ops 12/68, March 1968, SASR records; interview, F. C. Cashmore to author, 17 June 1987; an account written by Warrant Officer D. H. Wright, 1987; Wright to author, 29 June 1988; and A. Blacker to author, 2 July 1988.
2 Colonel John Jessup, the US Senior Provincial Adviser at the time, has confirmed that there was no concern about French sensibilities.
3 The 1 ATF Operational Log recorded that when the helicopter arrived over the area at 1.05pm fire was received from two enemy.
4 Cashmore was mentioned in despatches for his command of his patrol during this mission.

Chapter 2

1 Definitions for terms such as special operations and special forces are given in the preface.
2 M. R. D. Foot *SOE in France* London: HMSO, 1966, pp. 8–9; Ben Pimlott (ed) *The Second World War Diary of Hugh Dalton 1940–1945* London: Jonathan Cape, 1986, p. 62
3 DCGS to Minister for the Army, 8 July 1941 item 38/401/172, MP 729/6, Australian Achives; Lieutenant-General Sir Vernon Sturdee 'Foreword' in Bernard Callinan, *Independent Company, The Australian Army in Portuguese Timor 1941–43*, Melbourne: Heinemann, 1953, p. xiii
4 F. Spencer Chapman *The Jungle is Neutral* London: Reprint Society, 1950, p. 19
5 Chapman *The Jungle is Neutral* p. 19
6 Otto Heilbrunn *Warfare in the Enemy's Rear* London: George Allen and Unwin, 1963, p. 71
7 Memorandum by Colonel David Stirling DSO, OBE, on the Origins of the Special Air Service Regiment
8 D. M. Horner, *High Command, Australia and Allied Strategy, 1939–1945*, Sydney: Allen and Unwin, 1982, p. 143
9 C. A. Willoughby (comp) *Operations of the Allied Intelligence Bureau, GHQ, SWPA*, Tokyo: GHQ, Far East Command, 1948, p. 7
10 Horner, *High Command*, p. 236–237
11 Although some operations are mentioned in the Australian official histories, there is no comprehensive published history of AIB operations in the Second World War. Ronald McKie's book *The Heroes*, gives an excellent account of JAYWICK and RIMAU, but Myriam S. Amor *Operation Rimau, 11 September to 10 October 1944, What Went Wrong?* Department of Defence Historical Monograph No 76, 1988, and Charles Cruickshank *SOE in the Far East* Oxford: Oxford University Press, 1983 indicate that there was more to the story than McKie was able to discover. Colonel Allison W. Ind in *Spy Ring Pacific*

provides a popular history of the AIB. The problems of one SRD party in Borneo are told by Tom Harrison in *World Within*. *Ring of Fire* by Dick Horton, published in 1983, claims to provide 'for the first time' an 'authentic account of clandestine operations launched from Australia on the Japanese occupied islands' during the Second World War, but no sources or references are given, and the book deals with only a few operations. The British official history of SOE in the Far East makes only passing reference to Australian operations.

12 Gavin Long *The Final Campaigns* Canberra: Australian War Memorial, 1963, p. 621

13 Calvert to COs of SAS Regiments, 12 October 1945

14 Philip Warner *The Special Air Service* London: Kimber pp. 183–4. See also Brigadier J. M. Calvert 'The Survival of the Special Air Service After the 1939–45 War' *Mars and Minerva* Vol 1 No 7 June 1969

15 Correspondence in SASR records

16 ibid.

17 T. B. Millar 'Australian Defence, 1945–1965' in Gordon Greenwood and Norman Harper (eds) *Australia in World Affairs 1961–1965* Melbourne: Cheshire, 1968

18 Annual Report, Airborne Platoon, Royal Australian Regiment, 26 June 1953, item 240/2/27, MP 927, Australian Archives, Melbourne

19 From 1 December 1957 the company became 1 Infantry Battalion (Commando), City of Sydney's Own Regiment, but it still had a Commando company organisation.

20 Department of Defence 'Key Elements in the Triennial Reviews of Strategic Guidance since 1945' April 1986

21 Board Minute on Supplement No. 3 to Military Board Agendum No. 43/1956, 30 October 1956, item 41/441/18, CRS A6059

22 Commonwealth Parliamentary Debates, Vol H of R 14, p. 753; *Sydney Morning Herald* 9 April 1957

23 DMT to DSD, 4 June 1957, item 240/1/688, MP 927

24 DCGS to BGS (B), 20 June 1957, item 240/1/688, MP 927

Chapter 3

1 DCGS to N, E, S Comd, 1 April 1957, item 41/441/176, CRS A6059

2 DSD to DPA, 10 June 1957; DPA to Commands, 9 July 1957, item 41/441/176, CRS A6059

3 The unit was without a quartermaster sergeant for the first three months or so, and Wade was appointed acting quartermaster.

4 D Inf to DMO and DMT, 24 October 1957; DMO to DSD, 7 May 1958, item 60/441/74, CRS A6059

5 Len Eyles' wife, Norma, also played an important role and was very much the mother of the first company.

6 Tim Martin, 'Cherry Beret' in *People*, 20 August 1958, pp. 27–29

7 Report on Exercise GRAND SLAM, 21–27 May 1959, item 9323/4/261, MP927

8 'The Special Air Service Company' *Australian Army Journal* June 1960

9 Major Lewis L. Millet 'Recondo Patrol of Opportunity' reprinted in the *Australian Army Journal* October 1960

10 Major L. G. Clark MC 'SAS Recondo Training' *Australian Army Journal* August 1961

11 ibid.

Notes

12 Coleman was awarded the British Empire Medal for his efforts to save Smith.
13 Major L. G. Clark 'Organisation for Guerilla Warfare' *Australian Army Journal* May 1961
14 Another SAS soldier, Sergeant D. J. Neville, was a reserve for the Team.

Chapter 4

1 Military Board Minute 1/1962 12 October 1962
2 Correspondence in item 281–1–68, CRS A1945
3 Report on Exercise LONG HOP 26 March 1963 file 217/2/5, SASR records
4 Woodhouse spent five days in Australia and saw the SAS on exercise at Collie.
5 Note by Head of ANZAM Secretariat, 7 November 1963, item 211/A/2, AWM 121
6 Lieutenant-Colonel J. W. Norrie, 'Proposed use of elements of SAS Coy in operations South East Asia', 10 December 1963, SASR records
7 Woodhouse to Garland 2 January 1964, SASR records
8 Peacetime restrictions meant that sections numbered eight rather than the nine men allowed by the establishment.
9 Cables, Australian High Commissioner, Kuala Lumpur, to Department of External Affairs, 6, 15 April 1964, item 211/A/3, AWM 121
10 Peter Dickens *SAS, The Jungle Frontier* London: Arms and Armour, 1983, p. 119. Douglas-Home to Menzies, 10 April 1964, Defence Committee Minute 18/64, 13 April 1964, item 211/A/3, AWM 121. Defence Committee Minute 26/1964, 30 April 1964, Statement by Minister for Defence in House of Representatives, 16 April 1964, item 211/A/4, AWM 121
11 David Hawkins *The Defence of Malaysia and Singapore* London: RUSI, 1972, p. 24. See also Gregory Pemberton, *All the Way, Australia's road to Vietnam* Sydney: Allen and Unwin, 1987, p. 241
12 Army Headquarters to Western Command, 3 December 1964, item 4/4/1, AWM 121
13 On return to 42 Commando's camp at Kota Tinggi after Exercise LIGTAS, Skardon met Lieutenant-Colonel Woodhouse who invited him to join 22 SAS in Borneo. Skardon's attachment was approved provided that he did not cross into Indonesian territory.
14 Dickens *The Jungle Frontier* p. 116
15 For a description of the incident see Dickens *The Jungle Frontier* pp. 114–118
16 DMO & P to Western Command, 21 January 1965, item 4/4/1, AWM 121
17 Correspondence in item 211/A/6, AWM 121
18 Signal, AHQ to Western Command, 16 February 1965, item 119/1/19, SASR records

Chapter 5

1 Originally the NZ half squadron was designated to work with 1 SAS Squadron and they undertook their in-country training with 1 Squadron, but for a number of reasons they were eventually deployed to Sarawak.
2 Directive to Officer Commanding 1st Special Air Service Squadron, 17 February 1965, SASR records
3 The Golden Rules are quoted in Tom Pocock, *Fighting General, the Public and Private Campaigns of General Sir Walter Walker*, London: Collins, 1973, p. 197.
4 Report on Operations in Borneo, 1st Australian SAS Squadron, 17 August 1965, SASR records

5   1/65 Operational Notes for 1 Aust SAS Sqn Operations in Borneo, 12 March 1965, SASR records
6   John MacKinnon *Borneo* Amsterdam: Time-Life Books, 1975, p. 45

Chapter 6

1   E. D. Smith 'The Confrontation in Borneo, Part II' *Army Quarterly* Vol 106 No 1 January 1976, p. 35
2   For a discussion on these operations see D. M. Horner (ed) *Duty First, A History of the Royal Australian Regiment* forthcoming, 1990
3   Garland to Tuzo, 13 April 1965, SASR records
4   CGS to General Jolly, 12 April 1965, SASR records
5   Pocock, *Fighting General* p. 206
6   Report on Operations in Borneo 1 Aust SAS Sqn, 17 August 1965, item 234/4/16, SASR records
7   For a description of the Gurkhas' attack see E. D. Smith *East of Katmandu, The Story of the 7th Duke of Edinburgh's Own Gurkha Rifles, Volume II, 1948–1973* London: Leo Cooper, 1976, pp. 143–145
8   Dickens *The Jungle Frontier* pp. 125, 126

Chapter 7

1   He was awarded the BEM for his work in Vietnam in 1962 and 1963.
2   The account of this incident is based substantially on a typescript written by B. W. Littler, supplemented by information from SAS Ops 90/65/7, Patrol Report—Patrol 12, 24 May–3 June 1965, 11 June 1965, SASR records, and interview with Weir, 18 June 1987.
3   It was not part of operational procedure for Swanbourne to monitor operational radio traffic and the reception of the message was pure coincidence. Swanbourne could take no action.
4   Smith, *East of Katmandu*, p. 143

Chapter 8

1   Ops/90/80, 23 July 1965, SASR records
2   Dickens, *The Jungle Frontier* pp. 204, 205
3   Report on Operations in Borneo, 1 Aust SAS Sqn, 17 Aug 1965, and comments by CO SASR, 17 September 1965, SASR records

Chapter 9

1   'Official History of the Operations and Administration of Special Operations Australia', CRS A3270, Volume II
2   DMO & P to HQ W. Comd, 9 March 1965, item 9/C/8, Part 1, AWM 121
3   CO SASR to HQ W. Comd, 7 April 1965, item 9/C/8, Part 1, AWM 121
4   AHQ To W. Comd, 19 October 1965, item 334/R1462/1, CRS A3688
5   CO SASR to HQ W. Comd, 23 April 1966, SASR records
6   Report of Senior Commanders' Conference, Military Board records; CO SASR to HQ W. Comd, 9 February 1965, Brathwaite papers
7   Robert O'Neill *Australia in the Korean War 1950–1953, Volume II, Combat Operations* Canberra: Australian War Memorial and Australian Government Publishing Service, 1985, pp. 196–197
8   McNeill *The Team* p. 42

Notes

9 2 SAS Monthly Report, January 1966, SASR records
10 'Australian SAS Regiement' in *Mars and Minerva* Vol 2 No 5 December 1966, p. 30
11 Dickens. *The Jungle Frontier* p. 222
12 This operation is described in Dickens, *The Jungle Frontier* pp. 218–225
13 Smith *East at Katmandu* p. 151
14 Hughes to Brathwaite, 4 March 1966, SASR records

Chapter 10

1 The commander of D Company was awarded the only DSO to be given to a company commander during Confrontation.
2 Ayling was mentioned in despatches for this incident.
3 About this time one patrol captured an Indonesian operation order. The staff duties were perfect and quickly Major Hughes turned to the last page. The signature was that of Lt Col Sarwo Edhie RPKAD, an Indonesian student with him at Staff College.
4 Dickens *The Jungle Frontier* pp. 161–166
5 Young was mentioned in despatches for this and other work in Borneo.
6 Smith *East of Katmandu* p. 152
7 For a description of the contact see D. M. Horner *Duty First, The Royal Australian Regiment in War and Peace,* Sydney: Allen & Unwin, 1990.
8 Hughes protested at the decision to place his men 'under command' preferring 'operational control', but the brigadier had the final say.
9 2 SAS Sqn War Diary
10 Foreword to Smith *East of Katmandu* p. xvi

Chapter 11

1 DCGS to Chairman, Chiefs of Staff Committee, September 1965, item 1, CRS A6837. Colonel David Jackson, Commander, Australian Army Force Vietnam, recalled that he thought that General Westmoreland had asked Wilton for an SAS squadron: 'he often spoke of them and would have used them in medium and long range recce patrols—even outside Vietnam I suspect—given half a chance'.
2 Brathwaite to Commander W. Comd, 4 April 1966, SASR records
3 Roderick was accompanied by Sergeant John Sheehan. En route home from Borneo while staging through Singapore volunteers had been sought for a two week attachment to 1 RAR. The time was stretched to six weeks and as a result Roderick was 'nearly divorced'.
4 Report Ex TRAIIM NAU, 20 July 1966, SASR records. Murphy to Brathwaite, 11 July 1966, Brathwaite Papers
5 Frank Frost *Australia's War in Vietnam* Sydney: Allen & Unwin, 1987, p. 38
6 6 RAR deployed by helicopter from Vung Tau to Nui Dat on 14 June. Lex McAulay *The Battle of Long Tan* Melbourne: Hutchinson, 1986, p. 8
7 Anonymous 'The Third's First Step' *Excalibur, The Journal of The Special Air Service Regiment, Australia* Vol 1 No 1 December 1973
8 Murphy to Brathwaite, 22 June 1966, Brathwaite papers
9 One other consideration was the limitation of the UH 1B Iroquois. As the commander of No 9 Squadron RAAF wrote in July 1966: The helicopter needed two door gunners. 'This extra load, together with a requirement for the aircraft to have a reasonable climb out performace, limits the aircraft's load carrying capability to four fully equipped troops'.

10 Murphy to Brathwaite, 22 June 1966, Brathwaite papers
11 'Brief for Maj Gen K. Mackay MBE, A Special Air Service Squadron' 10 July 1966, 3 SAS Sqn War Diary, SASR records
12 Murphy to Brathwaite, 11 July 1966, Brathwaite papers
13 Schuman was awarded the Military Cross for his boldness and initiative in this action.
14 Murphy to Brathwaite 11 August 1966, Brathwaite papers
15 Nolan's patrol was the first insertion by the RAAF; the patrol was winched in from a height of 4 metres.
16 This was the first time No 9 Squadron inserted an SAS patrol by landing a helicopter. No 9 Squadron RAAF Unit History Sheet, 23 July 1966

Chapter 12

1 Address to CGS Exercise, 1971
2 Murphy to Brathwaite, 11 August 1966, Brathwaite papers
3 Davies lost an eye as a result of the mortar attack. The Task Force received a total of eighteen wounded in the attack.
4 No 9 Squadron RAAF Unit History Sheet, 21 August 1966, RAAF Historical Section
5 Murphy to DCGS, 28 July 1969, item 3 CRS A6838
6 HQ 1 ATF Combat Operations After Action Report, Operation CANBERRA, 27 December 1966 in the 1 ATF Commander's Diary states that B Company found that the approaches were heavily booby trapped. But the installation report of the area made by C Company reported no booby traps.
7 The No 9 Sqn Unit History Sheet for 14 December 1966 noted: 'This was the first planned assault the squadron had undertaken with so many aircraft'. There had been seven transport and two armed helicopters.
8 In December Lieutenant Ian Gay had replaced Second Lieutenant Schuman as commander of J Troop. Schuman had become ill and returned to Australia.
9 Murphy was awarded the Military Cross for his command of 3 Squadron in Vietnam. Sergeant Tonna was awarded the Military Medal. Thorburn and Urquhart were mentioned in despatches.

Chapter 13

1 Brief for Commander Western Command, 31 January 1967, SASR records. Brathwaite to Phil (Bennett?) 8 February 1967, item 841-R22-12, A3688
2 1 SAS Sqn War Diary
3 George Odgers, *Mission Vietnam, Royal Australian Air Force Operations, 1964–1972* Canberra; AGPS, 1974, p. 40
4 Odgers *Mission Vietnam* p. 42
5 Squadron Leader Jim Cox's Distinguished Flying Cross was awarded on 17 June 1968.
6 For this action and earlier patrols Hindson was awarded the Military Cross. De Grussa was mentioned in despatches.
7 Quarterly Summary Report, Implementation of the Combined Campaign Plan, HQ 1 ATF, 8 October 1967, 1 ATF Commander's Diary
8 No 9 Squadron Commanding Officer's Monthly Report, November 1967, RAAF Historical Records
9 Fraser to Bureau Chief Canberra, and Editors, 7 February 1968, SASR records
10 Baines was killed on the 13th day of his 13th month in Vietnam, 13 days before the squadron left for Australia.

11 Burnett was mentioned in despatches for his command of 1 squadron in 1987.

Chapter 14

1 MacDonald to CGS, 28 June 1968, item 'CGS to COMAFV, Period 1 Jan 67 to 28 Feb 70', AWM 98
2 Vincent to CGS, 18 December 1967, Vincent Papers
3 Vincent to Wilton, 10 January 1968, Vincent Papers
4 Assistant Defence Attaché Saigon, Detailed Report No 36, The Special Air Service Regiment in Vietnam, 6 March 1968, SASR records
5 AAORG Memorandum M36, 'Analysis of Australian Army Contacts in South Vietnam 1967–68', July 1969, prepared by R. G. Henderson. During the study period of 30 months there was a period of six months when no reports were submitted.
6 DCGS to CGS 31 March 1967, item]2[CRS A6837
7 Danilenko to CO SASR, 7, 27 December 1967; Danilenko to COMAFV, 6 December 1967, SASR records
8 McNeill, *The Team* p. 329. Danilenko was posthumously mentioned in despatches.
9 Robin Strathdee, 'The Supermen of Vietnam', cutting dated 7 August 1969 in item 9-C-8 part 1, AWM 121
10 'TR', 'The Australian Special Air Service', in A. Cameron (ed) *The Australian Almanac*, Sydney: Angus & Robertson, 1985, p. 219
11 Captain A. W. Freemantle 'Patrol Lessons in Vietnam' *Army Journal* February 1974
12 No 9 Squadron RAAF Unit History Sheet, July 1966. Appendix A to Form A50, July 1966
13 Freemantle, 'Patrol Lessons in Vietnam'. Freemantle served in Borneo with the British Army.
14 Freemantle, 'Patrol Lessons in Vietnam'

Chapter 15

1 Brathwaite believed that acclimatisation and toughening was the main purpose of the exercise but Wade persuaded the new commanding officer, Eyles, to allow tactical training.
2 Simpson was awarded the Military Cross for this action.
3 No 9 Squadron Commanding Officer's Report, June 1968
4 Letter, CO No 9 Squadron to CO 118 Assault Helicopter Company, 8 July 1968, No 9 Squadron Unit History Sheet
5 Captain Wischusen accompanied this patrol.
6 1 RAR Operation Order 8/68, 9 July 1968, 1 RAR Commander's Diary
7 Lobb was awarded the Military Medal for this action and for the action on 17 June 1968. The patrol report says 'APC tracks', but Lobb claims that the tracks were too wide to be APC tracks. James eventually reached the rank of major-general as Director General Army Health Services and Surgeon General of the Australian Defence Force.
8 No 9 Squadron RAAF Unit History Sheet
9 One soldier injured his back, and for future extractions the rope was clipped to the shoulder strap to keep the soldier upright during the extraction.
10 For a further discussion of the training wing see McNeill *The Team* pp. 210–214
11 The balloon was provided by the US Advanced Research Projects Agency.

12 In the ambush on 23 October Lance-Corporal B. P. Kelly received a fractured pelvis when a tree was blown over by one of the Claymores.
13 Wade was absent on R and R when the barbecue was conducted. In fact the Task Force commander had vetoed the planned celebration and officially the event was just a normal social function.

Chapter 16

1 Cullen was mentioned in despatches for his skill as a scout and for his aggressive yet calm action under fire in this and other patrols. He subsequently attended the Officer Cadet School at Portsea and later saw service in the SAS Regiment as a captain and major.
2 Wade claims that the message was 'no extraction today—move to night LUP'.
3 Pember was mentioned in despatches for this and other actions.
4 Stewart was mentioned in despatches for 'his courage, judgement and determination to succeed' in obtaining 'extremely valuable information on enemy movement under difficult conditions'.
5 In fact according to the official investigation three or four rounds were fired.
6 Major Wade was mentioned in despatches for his command of 2 Squadron during 1968.

Chapter 17

1 Lieutenant Colonel T. J. Nolan cannot recall any 'experienced' NCOs as students on the first course. The young officers on the course included Roberts, Howlett, Nolan, Fitzpatrick and Ison.
2 W. D. Baker, *'Dare to Win', The story of the New Zealand Special Air Service*, Melbourne: Lothian, 1987, p. 87. Apparently no one had considered the question of command and control for operations.
3 For a description of his experiences in Vietnam see McNeill, *The Team*, pp. 378–385
4 Grafton was a member of 2 SAS Squadron. He had recently returned from service in Vietnam with the AATTV and needed to requalify as a parachutist.
5 DCGS to AG, QMG, MGO, Secretary, 21 November 1969, item folder 1, CRS A6837
6 CGS No 80/1970, CGS to DSD, 16 July 1970, item 5, CRS A6835; D. M. Horner *Australian Higher Command in the Vietnam War* Canberra: SDSC, 1986, pp. 66–67
7 Memo by DDMO, 10 November 1969, item 9/C/10, AWM 121

Chapter 18

1 Dodd was later RSM of the Parachute Training School and since January 1988, a captain in the SAS Regiment.
2 Odgers, *Mission Vietnam* p. 89. Robinson was awarded the Distinguished Conduct medal for this action plus two other actions in 1969.
3 The Vietnam Digest for 17–24 May 1969 recorded that on 18 May the American FSB Husky, near Xuyen Loc, was attacked by 33 NVA Regiment. Twenty-six enemy bodies were found. The same day NVA troops attacked HQ 18 ARVN Division and a US calvary unit 3 kilometres east of Xuyen Loc. There were 30 enemy KIA and 8 PW. 1 ATF Commander's Diary May 1969.
4 L. D. Johnson *The History of 6 RAR-NZ (ANZAC) Battalion, Volume Two, 1967-1970* published by the battalion, Singapore, 1972, pp. 47–53

Notes

Chapter 19

1 Quoted in Frost *Australia's War in Vietnam* pp. 131–132

2 Standen was later commissioned, and as a major commanded 152 Signal Squadron.

3 Initially the ropes were of equal length, but they had different 'stretch' factors and were carrying unequal weights.

4 Van Droffelaar was awarded the Military Medal for this and other actions in 1969.

5 Roberts was awarded an immediate Distinguished Conduct Medal for this action.

6 Johnson *The History of 6 RAR-NZ (Anzac) Battalion, Volume Two, 1967–1970* p. 107

7 Phillips was mentioned in despatches for these and other actions.

8 The 1 ATF Commander's Diary shows that Weir visited Saigon between 6 and 8 January and visited 3 Squadron on 14 January. The 1 ATF operations log for 1043 hours 9 January reads: 'Patrol 19 has requested extraction. No decision yet'.

9 Beesley was mentioned in despatches for his command of 3 Squadron in 1969.

Chapter 20

1 During March the squadron deployed 12 reconnaissance, 19 recce–ambush, 2 ambush and one surveillance patrols. A total of 58 enemy were sighted in 16 sightings. In 6 contacts 3 enemy were killed plus 3 possibles.

2 A. Clunies-Ross (ed) *The Grey Eight in Vietnam, The History of Eight Battalion, The Royal Australian Regiment, November 1969–November 1970* Brisbane: published by the battalion, 1971, pp. 49, 80

3 Odgers *Mission Vietnam* pp. 125–126. Teague wrote in a Training Information memorandum on 22 April 1970: 'Rope extractions are not acceptable (from RAAF); however training for these should continue as it is felt that with improved harnesses and a change in RAAF personnel this type of extraction will again be acceptable'. SASR records

4 Clunies-Ross *The Grey Eight* p. 81

5 1 SAS Sqn Training Information memorandum, 22 April 1970, SASR records

6 7th Battalion The Royal Australian Regiment, Notes on Operations, Vietnam, 1970–1971

7 7 RAR Notes on Operations

8 The 1 ATF Commander's Diary indicated that Henderson visited 1 Squadron on 6 June 1970.

9 Gebhart was awarded the Military Medal for successfully commanding fourteen patrols, which involved several contacts, during 1970.

10 1 SAS Sqn Commander's Diary, SASR records

11 Paper from G. Mawkes, reproduced with permission, 9 January 1989. 1 ATF G Ops Log records that the civilians were 2000 metres outside civilian access. The 1 ATF Intsum No 270/70, 27 September 1970 states that the Div Int Unit had declared that all the civilians were from the Dat Do area with no knowledge of the VC.

12 Frost *Australia's War in Vietnam* p. 38

13 1 SAS Sqn Commander's Diary; Training Information memorandum, 6 November 1970, SASR records

14 Memorandum Prepared by 4 Troop NZSAS, 26 October 1970, SASR records

15 Head, NZDLS to Director Joint Staff, 5 February 1970; Daly to HNZDLS, 23 December 1970, item 161/D/1, AWM 121

16 Simon was mentioned in despatches for this and other successful patrols during 1970.
17 Fraser was mentioned in despatches for this and other successful patrols during 1970, the second time he had been mentioned in despatches.

Chapter 21

1 Scheele had been awarded the Distinguished Conduct Medal for his work with the Training Team in 1969.
2 These figures refer to SAS personnel and do not include Signals Corps personnel.
3 For Australia's role in the Phoenix Programme see McNeill *The Team* Chapter Fourteen.
4 Cashmore was mentioned in despatches for this and other patrols in 1971. It was the second time he had been mentioned in despatches.
5 SASR R852–1–2, 23 June 1972, SASR records

Chapter 22

1 Military Board Minute 528/1972, 14 December 1972, Military Board records
2 'There's something about a soldier (especially if he is an SAS man)' in *West Australian* 11 October 1973
3 CO SASR to DOP (JW) 4 August 1973, item 9/C/10, AWM 121
4 Major P. H. Bennett 'The Case for Special Forces' *Australian Army Journal* October 1964
5 F. J. West *The Strike Teams: Tactical Performance and Strategic Potential* Santa Monica, CA, The Rand Corporation, 1969, p. 3
6 Major-General D. B. Dunstan, GOC Field Force Command to Army Office, 4 June 1974, item 9/C/8 part 3, AWM 121
7 Captain David Mason-Jones 'Fighting the unforeseen war' *Pacific Defence Reporter* July 1975. Captain E. L. A. Tonna 'The Face of Reality—War Within Our Means' *Defence Force Journal* November/December 1977
8 Chief of the General Staff to Minister, 20 October 1976, box 2, CRS A6835, Australian Archives
9 Captain A. Vane 'Defence of Continental Australia—1942' *Army Journal* November 1975. Captain A. Vane 'The Surveillance of Northern Australia—its history. The Story of Stanner's Bush Commando, 1942' *Defence Force Journal* January/February 1979
10 Captain Kevan Wolfe 'SAS—The Quiet Professionals' *Triad* No 9, 1977
11 SASR Tactical Post Exercise Report, Exercise LONG VIGIL, 13 September 1977, SASR records
12 *Sydney Morning Herald, Canberra Times, Australian,* 1 August 1977
13 The first officers to enter the SAS Regiment in 1978 under the new 'all-corps' concept were Captains John Trevivian, RAEME, and Andrew Leahy, RAA. Since then the majority of corps have been represented in the regiment as officers or soldiers.
14 CGS to DCOPS, 8 July 1977, COPS, 3 August 1977, CPERS, 3 August 1977, box 3, CRS A6835, Australian Archives
15 Field Force Command Provisional Concept of Operations—SASR, 5 December 1978, SASR records
16 Melbourne *Sun* 27 February 1981; *Daily Telegraph* 28 February 1981; Adelaide *Advertiser* 27 February 1981. Yang Pei-shu 'The Australian and New Zealand Special Air Services' *Pacific Defence Forum* Summer 1988

Notes

Chapter 23

1 Sir Robert Mark, *Report to the Minister for Administrative Services on the Organisation of Police Resources in the Commonwealth Area and Other Related Matters*, Australian Government Publishing Service, Canberra, 1978, para 29

2 CGS No 280/1978, CGS to CDFS and Secretary, Department of Defence, 28 April 1978, item 1978/30, CRS A6839, Australian Archives

3 CGS No 365/1978, CGS to Minister for Defence, 25 May 1978, item 1979/47, CRS A6839

4 DGOP 896/78, DGOP-A to COPS 9 May 1978, item 1978/30, CRS A6839

5 CGS No 543/1978, CGS to CDFS and Secretary, 1 August 1978, item 1979/47, CRS A6839

6 COPS-A 83/79, COPS-A to CGS, 26 February 1979, item 1979/47, CRS A6839

7 Philip Cornford, 'The SAS, Our Ultimate Weapon', The *Weekend Australian* 6–7 May 1978

8 Details of CT training are from Major R. J. Tink to author 28 October 1988, and notes prepared for him by WO2 P. D. Lutley, Curtis to author, 16 December 1988.

9 Alan Hale, 'Paramilitarism and Paranoia', The *West Australian* 20 March 1981

10 The article understated the frequency of shooting. Soldiers were and still are required to shoot for a full day at least twice and preferably three days a week.

11 Ken Brass, 'Inside the SAS', The *Weekend Australian* 3–4 November 1984

12 Major Wallace was appointed a member of the Order of Australia for his work in developing the CT capability and for his command of 1 Squadron.

13 In 1981 Mike Jeffery had been promoted to brigadier to be Head of Protective Service Coordination Centre in the Department of the Special Minister of State.

14 CO SASR to HQ FF Comd, 15 September 1981, SASR records

15 In addition to the three soldiers killed in the Philippines in 1981 (Chapter 22, p. 417), on 24 July 1982 Lance-Corporal Peter Rawlings was killed in a parachuting incident while taking part in a skydiving display in Perth.

16 *Commonwealth Record* 14–20 April 1986

17 Mark Thompson, 'White Knights in Terror Fight', Melbourne *Sun*, 7 February 1987

18 Brass 'Inside the SAS'

Chapter 24

1 Colonel Don Higgins, 'Foreword', in *Rendezvous: Journal of the Australian Special Air Service Association*, Vol 14, December 1997, p. 3

2 *Canberra Times*, 2 July 1988. Another civilian climber reached the summit on 28 May.

3 Cullinan, Truscott and Crookston were awarded the Medal of the Order of Australia for their part in the expedition. Cullinan was later second-in-command of the SAS Regiment and Truscott was later its operations officer.

4 Quoted in M. J. Malone (ed.), *SAS: A Pictorial History of the Australian Special Air Service 1957–1997*, Imprimatur Books, Claremont, WA, 1998, p. 312

5 *The Defence of Australia 1987*, presented to Parliament by the Minister for Defence, the Honourable Kim C. Beazley, MP, Australian Government Publishing Service, Canberra, 1987, p. 2

6 Noel Koch, 'Special Operations Forces: Tidying Up the Lines', *Armed Forces Journal International*, October 1988

7 Lindsay Murdoch, 'The Peacemaker: Our Diplomatic Coup in Cambodia', *Sydney Morning Herald*, 2 October 1993, Spectrum p. 3A

8 Mark Bowden, *Black Hawk Down: A Story of Modern War*, Atlantic Monthly Press, New York, 1999

9 Colonel Phil McNamara, Colonel Operations at Land Headquarters, and Colonel Jim Wallace, Commander Special Forces, exerted influence to have this security task allocated to the SASR.

10 The team included Sergeants Kingston and Woods, Corporals V. J. Hartley, P. B. Sullivan and G. R. Thomas, Lance-Corporals G. K. Porter and A. A. Reid, and Troopers A. W. Cameron, N. R. Thompson and P. J. Vardenaga.

11 *Somalia: UNOSOM II, Australian Services Contingent IV, 19 May–22 November 1994*, privately published by *Australian and NZ Defender Magazine*, Brisbane, 1995, p. 13

12 Gary Kingston, 'J Troop, 3 SAS Sqn, Somalia, April–November 1994', in M. J. Malone (ed.), *SAS: A Pictorial History of the Australian Special Air Service 1957–1997*, Imprimatur Books, Claremont, WA, 1998, p. 69

13 Details of this incident are found in post-activity report in the ASC IV Security Group Log Book (from WO1 G. Kingston).

14 *Somalia: UNOSOM II*, 'J Troop, 3 SAS Sqn, Somalia, April–November 1994', p. 39

15 'Notice of Burial Period', 19 August 1994, copy held by a member of the SAS team

16 Kingston, op cit, p. 170

17 Kingston was awarded the Conspicuous Service Medal for his leadership in Somalia.

18 Warrant Officer 2 Rod Scott, 'Rwanda—An African Disaster: An Overview of Peacekeeping and SASR Involvement', in M. J. Malone (ed.), *SAS: A Pictorial History of the Australian Special Air Service 1957–1997*, p. 275

19 Roy Eccleston, 'Warriors Without a War', *Weekend Australian Magazine*, 21 October 1995

20 Leisa Scott, 'Heroes Emerge in the Darkest Hours', *Weekend Australian*, 15–16 June 1996, p. 6. Green was awarded the Conspicuous Service Cross for his part in the accident

21 The seriousness of SAS training persuaded the Remuneration Tribunal to increase the Special Forces allowance substantially, which contributed markedly to improved retention in the regiment in the late 1990s.

22 Ian McPhedran, 'Last of Black Hawk Tragedy Charges is Dropped', *Canberra Times*, 2 December 1997, p. 3

23 It is a quirk of terminology that the Commander Special Forces commands the Special Forces Group but is the Special Operations Component Commander in Headquarters Australian Theatre.

24 Headquarters Special Operations, through its close proximity to Headquarters Australian Theatre, was influential in having this task allocated to the SASR.

25 Matt Price, 'Farewell with Prayers and Dread', *Australian*, 18 February 1998, p. 1

26 Brief by COSASR for Defence Minister and CDF, 16–18 April 1998, SASR file 11-41-12(2)(A)(5)

27 Taipan, 'The Men Behind the Wire', *Rendezvous*, Volume 16, December 1998, p. 23

28 Letter to the editor, *Rendezvous*, Volume 17, June 1999, p. 4
29 Defence media release MIN 067/99, 11 March 1999
30 Defence media release MIN 246/99, 26 August 1999

Chapter 25

1 In the absence of Brigadier McNamara, the chief of staff at Headquarters Special Operations, Lieutenant-Colonel Rowan Tink, and his staff, had conducted most of the higher-level planning with Air Vice-Marshal Treloar and the staff of Headquarters Australian Theatre.
2 WO2 McCabe was awarded the Medal of the Order of Australia for his work in East Timor. Sergeant Maher was awarded a Commendation for Distinguished Service.
3 Corporal Cameron was awarded a Commendation for Distinguished Service for his service in East Timor.
4 Trooper Burgess was awarded the Commendation for Gallantry, and Corporal Beresford the CDF's Commendation, for this action. Captain Hawkins was awarded the Distinguished Service Medal for his leadership of L Troop in East Timor.
5 Sergeant Oddy was awarded the Medal of Gallantry, and Lance Corporal Fennell the CDF's Commendation, for this action.
6 Michael Ware, 'The Boy in Thongs who became a Secret Agent for Australia', *Daily Telegraph*, 11 December 1999, p. 5
7 Lieutenant-Colonel McOwan was awarded the Distinguished Service Cross for command and leadership of Response Force. Major McMahon was awarded the Distinguished Service Medal for his leadership of 3 SAS Squadron.
8 Major Phillips was awarded a CDF Commendation for his service in East Timor.
9 Report of the Army Alpine Association 1999/2000, Heard Island Expedition
10 Operation Gold Phase Three, SASR Post Op Report 3 October 2000
11 P. J. Ryan, 'Security Chiefs Write Open Letter to Defence', *Defence Information Bulletin*, October 2000, p. 4

Bibliography

Unpublished Records

1. Records of the SAS Regiment

a. Commander's Diaries and Patrol Reports
b. Administrative files and Post Exercise reports
c. Miscellaneous documents concerning the history of the Regiment
d. 2 SAS Squadron records

2. Australian Archives

a. *Canberra.*
   CA 36 Department of the Army (1939–1974)
   CRS A3688, Correspondence files, multiple number series,
   1962–1974
   CRS A6059, Correspondence files, multiple number series
   1956–1962
   CA 46 Department of Defence [III] Central Office
   CRS A3270, Vol 11, Official History of the Operations and Admini-
   stration of Special Operations Australia
   CRS A1945, Correspondence files, multiple number system,
   1957–1966
   CRS A6835, Outward Correspondence files of the CGS
   CRS A6837, Outward Correspondence files of the V and DCGS
   CRS A6839, Policy and working files of the Office of the CGS,
   1973–

b. *Melbourne*
   CA 36 Department of the Army (1939–1974)
   MP 927, General correspondence, 1952–1962

3. Australian War Memorial

AWM 98, Headquarters Australian Force Vietnam (Saigon) records
AWM 121, Army Office Operations Branch records
AWM 181, Herbicide series, Vietnam

4. Department of Defence (Army Office)

Military Board Minutes
Files of the Directorate of Special Action Forces

5. Department of Defence (Air Force Office)

RAAF Unit History Sheets

Bibliography

Books

Amor, Myriam S., *Operation Rimau, 11 September to 10 October 1944, What Went Wrong?*, Department of Defence Historical Monograph No 76, 1988

Baker, W.D., *'Dare to Win' The Story of the New Zealand Special Air Service*, Lothian, Melbourne, 1987

Barnett, Frank R., Tover, B. Hugh, and Shultz, Richard H. (eds), *Special Operations in US Strategy*, National Defense University, Washington, 1984

Battle, M.R., *The Year of the Tigers, The Second Tour of the 5th Battalion, The Royal Australian Regiment in South Vietnam 1969–1970*, published by the battalion, Sydney, 1970

Bottrell, Arthur, *Cameos of Commandos*, Specialty, Adelaide, 1971

Bowden, Mark, *Black Hawk Down: A Story of Modern War*, Atlantic Monthly Press, New York, 1999

Breen, Bob, *First to Fight*, Allen & Unwin, Sydney, 1988

—— *Mission Accomplished, East Timor, The Australian Defence Force Participation in the International Force East Timor (INTERFET)*, Allen & Unwin, Sydney, 2001

Burstall, Terry, *The Soldiers' Story, The Battle at Xa Long Tan, Vietnam, 18 August 1966*, University of Queensland Press, St Lucia, 1986

Callinan, Bernard, *Independent Company, The Australian Army in Portuguese Timor 1941–43*, Heinemann, Melbourne, 1953

Chapman, F. Spencer, *The Jungle is Neutral*, Reprint Society, London, 1950

Clarke, C.J. (ed.), *Yours Faithfully, A Record of Service of the 3rd Battalion, The Royal Australian Regiment in Australia and South Vietnam 16 February 1969—16 October 1971*, published by the battalion, Sydney, 1972

Clunies-Ross, A. (ed.), *The Grey Eight in Vietnam, The History of Eight Battalion The Royal Australian Regiment November 1969—November 1970*, published by the battalion, Brisbane, 1971

Connor, Ken, *Ghost Force: The Secret History of the SAS*, Weidenfeld & Nicolson, London, 1998

Cowles, Virginia, *The Phantom Major*, Collins, London, 1958

Curran, Jim, *K2, Triumph and Tragedy*, Hodder and Stoughton, London, 1987

Cruickshank, Charles, *SOE in the Far East*, Oxford University Press, Oxford, 1963

Dexter, David, *The New Guinea Offensives*, Australian War Memorial, Canberra, 1961

Dickens, Peter, *SAS, The Jungle Frontier*, Arms and Armour, London, 1983

Feldt, E.A., *The Coast Watchers*, Oxford University Press, London, 1946

Foot, M.R.D., *SOE in France*, HMSO, London, 1966

Frost, Frank, *Australia's War in Vietnam*, Allen & Unwin, Sydney, 1987

Geraghty, Tony, *Who Dares Wins, The Special Air Service, 1950 to the Falklands*, Arms and Armour Press, London, 1983

Gill, G. Hermon, *Royal Australian Navy, 1939–1942*, Australian War Memorial, Canberra, 1957

Hall, R.J.G., *The Australian Light Horse*, W.D. Joynt, Blackburn, Vic., 1968

Hawkins, David, *The Defence of Malaysia and Singapore*, RUSI, London, 1972

Heilbrunn, Otto, *Warfare in the Enemy's Rear*, George Allen & Unwin, London, 1963

Hope, R.M., *Protective Security Review Report*, Australian Government Publishing Service, Canberra, 1979

Horner, D.M., *High Command, Australia and Allied Strategy, 1939–1945*, Allen & Unwin, Sydney, 1982

—— *Australian Higher Command in the Vietnam War*, SDSC, Canberra, 1986

Horner, David, *Making the Australian Defence Force*, Oxford University Press, Melbourne, 2001

Horton, Dick, *Ring of Fire, Australian Guerrilla Operations Against the Japanese in World War II*, Macmillan, Melbourne, 1983

James, Harold and Shiel-Small, Denis, *The Undeclared War, The Story of the Indonesian Confrontation, 1962–1966*, Leo Cooper, London, 1971

Jeapes, Tony, *SAS: Operation Oman*, Kimber, London, 1980

Johnson, L.D., *The History of 6 RAR-NZ (ANZAC) Battalion, Volume Two, 1967–1970*, published by the battalion, Singapore, 1972

Kelly, F.J., *US Army Special Forces 1961–1971*, Department of the Army, Washington, 1973

Kruzel, Joseph (ed.), *American Defense Annual, 1988–1989*, Lexington Books, Lexington, Mass., 1988

Ladd, James D., *SAS Operations*, Robert Hale, London, 1986

Long, Gavin, *The Final Campaigns*, Australian War Memorial, Canberra, 1963

Lord, Walter, *Lonely Vigil, Coastwatchers of the Solomons*, Viking, New York, 1977

Malone, M.J. (ed.), *SAS: A Pictorial History of the Australian Special Air Service 1957–1997*, Imprimatur Books, Claremont, WA, 1998

Mark, Sir Robert, *Report to the Minister for Administrative Services on the Organization of Police Resources in the Commonwealth Area and Other Related Matters*, Australian Government Publishing Service, Canberra, 1978

McAulay, Lex, *The Battle of Long Tan*, Hutchinson, Melbourne, 1986

—— *The Battle of Coral*, Hutchinson, Melbourne, 1988

McCarthy, Dudley, *South-West Pacific Area—First Year, Kokoda to Wau*, Australian War Memorial, Canberra, 1959

McNeill, I.G., *The Team, Australian Army Advisers in Vietnam, 1962–1972*, Australian War Memorial, Canberra, 1984

Newman, K.E. (ed.), *The Anzac Battalion in South Vietnam 1967–1968*, Vol 1, published by the 2nd Battalion, The Royal Australian Regiment, Brisbane, 1968

Noonan, William, *Lost Legion, Mission 204 and the Reluctant Dragon*, Allen & Unwin, Sydney, 1987

Odgers, George, *Mission Vietnam, Royal Australian Air Force Operations 1964–1972*, Australian Government Publishing Service, Canberra, 1974

O'Leary, Shawn, *To the Green Fields Beyond*, Wilke, Brisbane, 1975

O'Neill, Robert, *Australia in the Korean War 1950–53, Volume II, Combat Operations*, Australian War Memorial and Australian Government Publishing Service, Canberra, 1985

Bibliography

—— *Vietnam Task, The 5th Battalion, The Royal Australian Regiment*, 1966/67, Cassell, Melbourne, 1968

Pemberton, Gregory, *All The Way, Australia's Road to Vietnam*, Allen & Unwin, Sydney, 1987

Petersen, Barry with Cribbin, John, *Tiger Men, An Australian Soldiers' Secret War in Vietnam*, Macmillan, Melbourne, 1988

Pimlott, Ben (ed.), *The Second World War Diary of Hugh Dalton, 1940–1945*, Jonathan Cape, London, 1986

Pocock, Tom, *Fighting General, The Public and Private Campaigns of General Sir Walter Walker*, Collins, London, 1973

Rennie, Frank, *Regular Soldier, A Life in the New Zealand Army*, Endeavour, Auckland, 1986

Richelson, Jeffrey T. and Ball, Desmond, *The Ties That Bind, Intelligence Cooperation Between the UKSA Countries*, Allen & Unwin, Sydney, 1985

Ryan, Alan, *'Primary Responsibilities and Primary Risks', Australian Defence Force Participation in the International Force East Timor*, Land Warfare Studies Centre, Study Paper no 304, Canberra, 2001

Sayce, R.L. and O'Neill, M.D. (eds), *The Fighting Fourth, A Pictorial Record of the Second Tour in South Vietnam by 4 RAR/NZ (ANZAC) Battalion, 1971–1972*, published by the battalion, Sydney, 1972

Seymour, William, *British Special Forces*, Sidgwick & Jackson, London, 1985

Simpson, Charles M., *Inside the Green Berets, A History of the US Army Special Forces*, Presidio, Novato, CA., 1983

Smith, E.D., *East of Katmandu, The Story of the 7th Duke of Edinburgh's Own Gurkha Rifles, Volume II, 1948–1973*, Leo Cooper, London, 1976

Somalia: UNOSOM II, Australian Services Contingent IV, 19 May–22 November 1994, privately published by *Australian and NZ Defender Magazine*, Brisbane, 1995

Stevens, W.G., *Official History of New Zealand in the Second World War 1939–1946, Problems of NZEF*, Department of Internal Affairs, Wellington, 1958

Strawson, John, *A History of the SAS Regiment*, Secker & Warburg, London, 1984

Stuart, R.F., *3 RAR in South Vietnam; 1967–1968*, published by the battalion, Sydney, 1969

Warner, Philip, *The Special Air Service*, William Kimber, London, 1972

Webb, J.R. (ed.), *Mission in Vietnam*, published by 4RAR/NZ, Singapore, 1969

Welsh, Nick (comp.), *A History of the Sixth Battalion The Royal Australian Regiment 1965–1985*, published by the battalion, 1986

West, F.J., *The Strike Teams: Tactical Performance and Strategic Potential*, The Rand Corporation, Santa Monica, CA., 1969

White Paper, *The Defence of Australia 1987*, presented to Parliament by the Minister for Defence, the Honourable Kim C. Beazley, MP, Australian Government Publishing Service, Canberra, 1987

Williams, I.M., *Vietnam, A Pictorial History of the Sixth Battalion, The Royal Australian Regiment 1966–1967*, published by the battalion, 1967

Willoughby, C.A. (comp.), *Operations of the Allied Intelligence Bureau, GHQ, SWPA*, GHQ, Far East Command, Tokyo, 1948

Wray, C.C.H., *Timor 1942, Australian Commandos at War with the Japanese*, Hutchinson, Melbourne, 1987

Journal Articles, etc.

Anonymous, 'Australian SAS Regiment', *Mars and Minerva*, Vol 2, No 5, December 1966

—— 'The Third's First Step', *Excalibur, The Journal of The Special Air Service Regiment, Australia*, Vol 1, No 1, December 1973

Bennett, Major P.H., 'The Case for Special Forces', *Australian Army Journal*, October 1964

Brass, Ken, 'Inside the SAS', The *Weekend Australian*, 3–4 November 1984

Clark, Major L.G., 'SAS Recondo Training', *Australian Army Journal*, August 1961

Cornford, Philip, 'The SAS, Our Ultimate Weapon', The *Weekend Australian*, 6–7 May 1978

Donnelly, C.N., 'Soviet Tactics for Operations in the Enemy Rear', *International Defense Review*, 9/1983

Eccleston, Roy, 'Warriors Without a War', The *Weekend Australian: Australian Magazine*, 21 October 1995

Eshel, David, 'Spetznaz, Soviet Special Forces . . . or Something More?', *Defence Update/85*

Freemantle, Capt. A.W., 'Patrol Lessons in Vietnam', *Army Journal*, February, 1974

Jeffery, Colonel P.M., 'Initial Thoughts on an Australian Land Surveillance Force', *Defence Force Journal*, March/April 1980

Hale, Alan, 'Paramilitarism and Paranoia', The *West Australian*, 20 March 1981

Higgins, Colonel Don, 'Foreword', in *Rendezvous: Journal of the Australian Special Air Service Association*, Vol 14, December 1997

Kelly Ross, 'Special Operations Forces in Low-Intensity Conflicts', in Charles B. Perkins (ed.), *Strategy '86, Proceedings of the 1986 Conference*, D&FA Conference Inc, Washington, 1986

Kingston, Gary, 'J Troop, 3 SAS Sqn, Somalia, April–November 1994', in M.J. Malone (ed.), *SAS: A Pictorial History of the Australian Special Air Service 1957–1997*, Imprimatur Books, Claremont, WA, 1998

Koch, Noel, 'Special Operations Forces: Tidying Up the Lines', *Armed Forces Journal International*, October 1988

Lambert, Peter, 'Heroic Sacrifices made the Everest Assault a Success', *Canberra Times*, 2 July 1988

Lindsay, General James S., 'U.S. Special Operations Command, The Professionals', *Asia–Pacific Defense Forum*, Summer 1988

Mason-Jones, David, 'Fighting the Unforeseen War', *Pacific Defence Reporter*, July 1975

McCullogh, Major S.F., 'The Role of Guerilla Warfare in Defence of Continental Australia', *Army Journal*, August 1976

Bibliography

McPhedran, Ian, 'Last of Black Hawk Tragedy Charges is Dropped', *Canberra Times*, 2 December 1997, p. 3

Millet, Major Lewis, 'Recondo Patrol of Opportunity', *Australian Army Journal*, October 1960

Murdoch, Lindsay, 'The Peacemaker: Our Diplomatic Coup in Cambodia', *Sydney Morning Herald*, 2 October 1993

Odorizzi, Charles D. and Schemer, Benjamin F., 'US and Soviet Special Operations', *Armed Forces Journal International*, February 1987

Price, Matt, 'Farewell with Prayers and Dread', *Australian*, 18 February 1998

Scott, Leisa, 'Heroes Emerge in the Darkest Hours', The *Weekend Australian*, 15–16 June 1996

Scott, Warrant Officer 2 Rod, 'Rwanda—An African Disaster: An Overview of Peacekeeping and SASR Involvement', in M.J. Malone (ed.), *SAS: A Pictorial History of the Australian Special Air Service 1957–1997*, Imprimatur Books, Claremont, WA, 1998

Smith, E.D., 'The Confrontation in Borneo, Part II', *Army Quarterly*, Vol 106, No 1, January 1976

Suvorov, Viktor, 'Spetznaz, the Soviet Union's Special Forces', *International Defense Review*, 9/1983

Taipan, 'The Men Behind the Wire', *Rendezvous*, Vol 16, December 1998

'Thermistocles', ' "The Law of War is not Cricket": Ask the Special Air Service Commandos!', *Pacific Defence Reporter*, March 1981

Thompson, Mark, 'White Knights in Terror Fight', Melbourne *Sun*, 7 February 1987

Tonna, Captain E.L.A., 'The Face of Reality—War Within Our Means', *Defence Force Journal*, November/December 1977

Vane, Captain A., 'Defence of Continental Australia—1942', *Army Journal*, November 1975

—— 'Surveillance of Northern Australia—its History, The Story of Stanner's Bush Commando, 1942', *Defence Force Journal*, January/February 1979

Wolfe, Captain Kevan, 'SAS—The Quiet Professionals', *Triad*, No 9, 1977

Yang Pei-shu, 'The Australian and New Zealand Special Air Services', *Pacific Defense Forum*, Summer 1988

Private Papers

Papers of Lieutenant Colonel G.R.S. Brathwaite
Papers of Mr H.R. Marten
Papers of Major General D. Vincent

Interviews

Captain L.C. Alver, 4 July 1988
F.J. Ayling, 5 June 1987
Colonel R.P. Beesley, AM, 22 July, 14 December 1988
Corporal P. Beresford, 9 November 2000
Colonel R.B. Bishop, 5 August 1988

537

A.M. Blacker, 2 July 1988
Captain R. Clifford, 7 November 2000
Lieutenant Colonel G.R.S. Brathwaite, 21 June 1987
Colonel D.P. Burnett, 8 February 1988
A.J. Callaghan, 5 November 1987
F.C. Cashmore, 17 June 1987
Colonel L.G. Clark, MC, 17 June 1987
Sergeant A. Cook, 9 November 2000
Captain S. Davies, 18 December 2000
Major General de la Billière, CBE, DSO, MC, 17 September 1987
Colonel L.A. Eyles, 22 November 1988
Captain A. Foxley, MBE, 19 June 1987
Major L.E. Fraser, MBE, 7 November 1987
Brigadier A.B. Garland, AM, 7 July 1987
Captain H.J. Haley, BEM, 22 June 1987
W.S. Harris, 28 October 1988
Wing Commander M.J. Haxell, DFC, 17 November 1988
Colonel D. Higgins, 15 December 2000
Staff Sergeant G. Hillman, 9 November 2000
Lieutenant Colonel W.F. Hindson, MC, 8 March 1988
Major General J.C. Hughes, AO, DSO, MC, 8 March 1988
Major General P.M. Jeffery, AO, MC, 1 November 1988
Colonel John E. Jessup, 17 August 1988
Captain C. Johns, 9 November 2000
WO1 G. Kingston, 8 December 2000
Major General K.H. Kirkland, AO, MBE, 17 December 1988
Lieutenant General Sir George Lea, KCB, DSO, MBE, 3 April 1987
Brigadier D. Lewis, 7 December 2000
General Sir Arthur MacDonald, KBE, CB, 15 November 1988
H.R. Marten, 23 July 1988
Major G. Mawkes, MBE, 2 July 1988
Lieutenant Colonel R.A. McBride, 28 October 1988
WO2 R. McCabe, 8 November 2000
Sergeant B. McCosker, 9 November 2000
Major D.H. McDaniel, 1 July 1988
Lieutenant Colonel I.D. McFarlane, 13 December 1988
Major J. McMahon, 8 November 2000
Brigadier P. McNamara, 21 February 2001
Lieutenant Colonel T. McOwan, 8 November 2000
Corporal C. Millen, 9 November 2000
J.J. Mooney, 6 November 1987
Major R.C. Moor, 29 June 1988
Colonel J.M. Murphy, MC, 30 November 1987
Major C.E. (Dare) Newell, OBE, 1 April 1987
Lieutenant Colonel T.J. Nolan, 21 June 1987
Sergeant S. Oddy, 9 November 2000

Bibliography

WO1 T. O'Farrell, OAM, 6 November 1987, 4 July 1988
Air Vice Marshal J.A. Paule, AO, DSO, AFC, 12 December 1988
Major D.S. Procopis, 1 July 1988
Sergeant A. Reid, 8 November 2000
Colonel C.A.M. Roberts, AM, 11 November 1988
J.M. Robinson, DCM, 18 June 1987
Brigadier M.J. Silverstone, 1 June 1987, 6 December 2000
Lieutenant Colonel G.C. Skardon, 17, 28 August 1987
Colonel, the Viscount Slim, OBE, 26 March 1987
Major General N.R. Smethurst, AO, MBE, 22 July 1988
Lieutenant Colonel I. Smith, 18 March 1987
Captain L.F. Smith, 5 July 1988
T.R. Swallow, 7 November 1987
Lieutenant Colonel R.J. Swarbrick, 11 November 1988
Lieutenant Colonel I. Teague, 8 August 1988
Trooper A. Thompson, 9 November 2000
K.D. Tonkin, 4 November 1987
Major E.L.A. Tonna, MM, 30 June 1988
Major J. Truscott, 8 November 2000
General Sir Harry Tuzo, GCB, OBE, MC, 26 March 1987
Sergeant K. Tyndale, 1 June 1987
Major A.G. Urquhart, 5 November 1987
Captain J. Van Der Klooster, 8 November 2000
M.J. Van Droffelaar, MM, 4 July 1988
Major C. Vaughan-Evans, 7 November 2000
Brigadier B. Wade, AM, 21 July 1987
Lieutenant Colonel J.J.A. Wallace, AM, 28 June 1988
R.L. Weir, BEM, 18 June 1987
WO M. Williams, 8 November 2000
WO2 D.H. Wright, 29 June 1988

Letters

Colonel R.P. Beesley, AM, 25, 28 October 1988
Colonel D.P. Burnett, 21 April 1988
F.C. Cashmore, 21 May 1988
G.E. Chipman, 4 December 1988
Colonel L.G. Clark, MC, June, December 1988
J.H. Coleman, 29 September 1987
Brigadier R.G. Curtis, AM, MC, 16 December 1988
Lieutenant Colonel W.N.N. Forbes, 25 January 1989
N.S. De Grussa, 26 May 1988
Brigadier A.B. Garland, AM, 15 October 1987
R.R. Gloede, 27 July 1988
B.E. Glover, April 1988
Lieutenant Colonel I.J. Gollings, 2 October 1987

Major General S.C. Graham, AO, DSO, OBE, MC, 1 April 1988
Major General W.G. Henderson, AO, DSO, OBE, 5 July 1988
Major D.G. Higgins, 9 January 1989
Lieutenant Colonel D.R. Hill, 12 July 1988
Major General R.L. Hughes, DSO, 11 March 1988
Major General J.M. Hughes, AO, DSO, MC, 1 November 1987
Brigadier O.D. Jackson, DSO, 28 March 1988
Major D.E. Lewis, 14 December 1988
B.W. Littler, 2 October 1987
G.H. Lobb, MM, May 1988
Captain M.J. Malone, OAM, 5 January 1989
H.R. Marten, 4 April 1988
J.W. Matten, 14 December 1987
K.W. McAlear, 25 April 1988
Lieutenant Colonel R.A. McBride, 27 October 1988
Major General B.A. McDonald, AO, DSO, OBE, MC, 1 March 1989
Colonel P.M. McDougall, AM, 15 December 1987
D.M. Mitchell, 23 June 1988
Colonel J.M. Murphy, MC, 5 April 1988
Lieutenant Colonel T.J. Nolan, 1 December 1988
Major General C.M.I. Pearson, AO, DSO, OBE, MC, 16 June 1988
Colonel C.A.M. Roberts, AM, 13, 27 January 1989
F.J. Roberts, DCM, 31 July 1988
Major P.A. Robottom, 11 January 1989
Major T.W. Roderick, 6 January, 17 March 1988
M.J. Ruffin, OAM, 9 August 1988
Lieutenant Colonel P.J. Schuman, MC, 10 December 1987, 14 March 1988
D.W. Scheele, DCM, 21 April 1988
Major G.L. Simpson, MC, 21 June 1988
Major General N.R. Smethurst, AO, MBE, 16 November 1988
WO2 F. Sykes, November 1987, 14 March 1988
Lieutenant Colonel I. Teague, 5 October 1988
Brigadier B. Wade, AM, 21 July, 20 October 1988
Lieutenant Colonel J.A.A. Wallace, AM, 18 December 1988
Brigadier S.P. Weir, DSO, MC, 23 August 1988
WO2 D.H. Wright, 19 April 1988
Lieutenant Colonel J.N. Woodhouse, MBE, MC, 20 June 1987
Captain B.L. Young, April 1988

Appendixes

Appendix A: Awards and decorations to members on active service with the SAS Regiment in Borneo and South Vietnam

	Date of action or period for which award was made
Military Cross (MC)	
Second Lieutenant W. F. Hindson	11 August 1967
Major J. Murphy	June 1966–February 1967
Second Lieutenant P. J. Schuman	17 July 1966
Lieutenant G. L. Simpson	7, 15 June 1968
Distinguished Conduct Medal (DCM)	
Sergeant F. J. Roberts	1 November 1969
Sergeant J. M. Robinson	15 May, October, December 1969
Military Medal (MM)	
Sergeant J. Gebhardt	17 March–9 November 1970
Corporal G. H. Lobb	17 June, 14 July 1968
Sergeant E. L. A. Tonna	June 1966–February 1967
Sergeant M. J. Van Droffelaar	26 September 1969
Mention in Despatches	
Private F. J. Ayling	21 March 1966
Major R. P. Beesley	February 1969–January 1970
Major D. P. Burnett	February 1967–February 1968
Sergeant F. C. Cashmore	18 March 1968
——	February–October 1971
Lance Corporal D. J. Cullen	February 1968–February 1969
Private N. S. De Grussa	11 August 1967
Sergeant L. E. Fraser	May, November 1967
——	February 1970–February 1971
Lance Corporal K. S. Pember	17 January 1969
Corporal J. J. Phillips	13 January 1970
Lieutenant Z. A. M. Simon	25 March, 30 October 1970
Sergeant A. F. Stewart	10–28 January 1969
Sergeant T. R. Swallow	February–October 1971
Sergeant J. L. B. Thorburn	17–18 December 1966

	Date of action or period for which award was made
Mention in Despatches	
Sergeant A. G. Urquhart	27 July 1986
Major B. Wade	February 1968–February 1969
Sergeant B. L. Young	23 May 1966

Appendix B: Active service nominal roll, Borneo and Vietnam

This roll is in two parts, covering service in Borneo and Vietnam respectively. In both parts the remarks column indicates further active service in areas where a campaign medal was awarded to the member. The asterisk in the Borneo remarks column indicates that the member also served in Vietnam with the SAS, and any additional information about that member will therefore be found in the Vietnam section. Members listed as having served with 152 Signal Squadron in Vietnam actually served with a detachment of that squadron operating under command of the SAS Squadron in Vietnam. Decorations listed are those awarded for service in units other than the Australian SAS. If a member saw service in Vietnam with a unit other than the SAS then that service is also listed. This roll was compiled by Corporal Steve Danaher, Curator of the SAS Regiment Historical Collection, with the assistance of the Central Army Records Office, Melbourne. Every effort has been made to record the detail accurately, but if there are errors or omissions he apologises to the member concerned or his family. Discrepancies should be notified to the Curator of the SAS Regiment Historical Collection, Campbell Barracks, Swanbourne, WA.

AUSTRALIAN SPECIAL AIR SERVICE REGIMENT BORNEO

214463	PTE ABSOLON Bruce Howard	2 Sqn	*
214352	PTE ADAMSON Charles Stuart	1 & 2 Sqn	*
54962	CPL AGNEW John James	2 Sqn	*
42386	LCPL AITKIN William Maxwell	1 Sqn	*
15481	LCPL ALBRECHT Ross Emil	2 Sqn	*
42886	CPL AYLES Jeffery Edward	2 Sqn	SVN HQ AFV 2/12/70–21/1/71, WO2
54219	PTE AYLING Frank John	2 Sqn	*
54840	SIG BAINBRIDGE Normal Keith	2 Sqn	*
214230	CPL BAINES George Terence	1 Sqn	* Malay Peninsula
52616	SSGT BALDWIN Ivan John	2 Sqn	*

542

53631	SGT BANNIGAN William	2 Sqn	SVN 104 Sigs 20/5/68—19/12/69, WO2 110 Sigs 18/11/70—24/5/71, WO2 Japan and Korea RA Sigs
38570	SIG BARDSLEY Stewart William	2 Sqn	*
53470	LCPL BARRETT Kevin Joseph	1 Sqn	Malaya 101 Fd Bty, GNR
213846	PTE BEEL Robert Mervyn	2 Sqn	
37343	LCPL BERGE Rodney Kennith	2 Sqn	
15656	CPL BIRCH Douglas John	1 Sqn	
16767	LCPL BLOOMFIELD Stephen Geoffery	1 Sqn	*
53865	CPL BOAG Kenneth John	1 Sqn	*
5410943	LCPL BRADLEY Terry	2 Sqn	SVN 104 Sigs 3/5/67—12/2/68, CPL
213622	PTE BRANIFF John Edward	1 Sqn	*
214287	LCPL BROWN Raymond Walter	2 Sqn	*
54581	PTE BURGESS Daniel Charles	1 Sqn	*
53817	CPL BUTCHART James Thomson	2 Sqn	*
54765	SIG BUTCHER Leonard Alexander	1 Sqn	*
42515	PTE BUTTERY Graeme Lesley	1 Sqn	SVN 6 RAR 8/5/69—16/5/70, SGT WIA
17798	PTE CARNES John Thomas	2 Sqn	SVN 1 AUST CAN UNIT 9/6/67—25/10/67, CPL Medevac from Borneo April 1966
311515	PTE CARROLL Clive Peter	2 Sqn	*
54237	PTE CASHMORE Frank Carr	2 Sqn	*
42530	LCPL CHENOWETH Edwin Ruben	1 Sqn	
311231	SGT COLEMAN John Henry	2 Sqn	*
1410935	PTE CONAGHAN Ian Frederick	2 Sqn	*
213346	LCPL CONSIDINE Joseph Bernard	2 Sqn	*
235250	LT DANILENKO Anatoly	2 Sqn	*
54752	PTE DAVIES Mark	1 Sqn	*
215118	PTE DAVIES Reginald Elliot	2 Sqn	*
15843	CPL DAVIES Robert Neville	1 Sqn	*
210788	SGT DAVIS Marsden Stephen	1 Sqn	*
5411172	PTE DELGADO Bernard John	2 Sqn	* Malay Peninsula
5411165	PTE DELGADO Vernon	2 Sqn	* Malay Peninsula
5411122	LCPL DEMAMIEL Wayne Geoffery	2 Sqn	*
37562	LCPL DENEHEY Paul Harold	1 Sqn	Killed on active service 6/6/65
54047	PTE DEVINE Harold Joseph	1 Sqn	*
42357	PTE DUNCAN John McKenzie	1 Sqn	* Malay Peninsula
36852	PTE DWYER Clement James	1 Sqn	*
53836	PTE EASTHOPE Alan	2 Sqn	*
214017	PTE ENGLISH Bruce Patrick	1 Sqn	*
16441	PTE ESTELLA Graham Phillip	2 Sqn	
214372	CPL EVELEIGH Denis Lawrence	2 Sqn	
15646	CPL FARLEY Neville Gary	1 Sqn	*
213954	PTE FARRAR William Frederick	1 Sqn	SVN 2 RAR 19/5/68, CPL SVN 2 RAR 16/5/70—26/5/71, SGT
53870	LCPL FERGUSON Norman Stanley	1 Sqn	*
16337	PTE FISHER Donald Murphy	1 Sqn	*
29707	SGT FLANNERY Joseph	1 Sqn	*

SAS: *Phantoms of War*

53797	CPL FOLEY Michael Bryan	2 Sqn	
53235	SGT FOXLEY Arthur	1 Sqn	*
53151	SGT FRASER Lawrence Edmund	2 Sqn	*
1200257	PTE GABRIEL Wallace Bruce	2 Sqn	*
215437	SIG GALLEGOS Alan James	2 Sqn	*
54909	PTE GAMMIE Ronald Lindsay	1 Sqn	*
235054	MAJ GARLAND Alfred Barrett	1 Sqn	SVN 7 RAR (Maj, 2IC) and HQ AFV as Aust LO, Long Binh 24/10/67—15/10/68. Korea 3 RAR LT, awarded AM 1980
53732	LCPL GARVIN Arthur George	1 Sqn	*
1200111	PTE GEBHARDT John	1 Sqn	*
36226	SGT GILCHRIST Ronald John	2 Sqn	*
54811	PTE GLOEDE Henry Alexander	1 Sqn	*
61097	CAPT GOLLINGS Ian John	1 Sqn	SVN AATTV 3/8/62—30/11/63, CAPT Malaya 1 and 3 RAR, 2 Lt
54533	LCPL GOMM Collin John	2 Sqn	
15187	LCPL GOSEWISCH Frederick Jacques	1 Sqn	Malaya, 1 RAR, Pte
54223	LCPL GRAFTON John Malcolm	2 Sqn	SVN AATTV 29/4/68—7/5/69, WO2 Malaya 1 RAR, Pte. Died in a parachute accident 12/8/69
15718	SGT GRIDLEY George Arthur Edgar	1 Sqn	*
19963	SIG GRIFFITHS John Leonard	2 Sqn	*
213223	SGT GROVENOR Michael Joseph	2 Sqn	*
17795	PTE GULBRANSEN Ross Leonard	2 Sqn	*
371144	LCPL HAMMOND David Anthony	1 Sqn	SVN 110 Sigs 17/1/68—14/1/69, CPL
43052	SIG HAMMOND Donald Roy	1 & 2 Sqn	
214175	PTE HARRIS William Edward	1 Sqn	*
4410614	PTE HARVEY James Franklin	2 Sqn	*
52763	CAPT HILL Deane Leighton	2 Sqn	*
14969	LT HOFFMAN Graham David	2 Sqn	SVN 2 RAR 29/4/70—1/6/71, MAJ
52628	SGT HOFFMAN John Robert	2 Sqn	SVN AATTV 11/12/67—14/8/68, WO2, served BCOF and Korea 1954 with Aust Anc Unit and 1 RAR, PTE
14441	CAPT HOLLAND Terence Henry	1 Sqn	SVN 6 RAR 7/5/69—17/6/69 MAJ. WIA GSW R forearm 10/6/69
14906	SGT HOOLIHAN John Thomas	2 Sqn	SVN AATTV 10/6/69—11/6/70, WO2. RAN Korea
14172	LT HUDSON Kenneth Ambrose	2 Sqn	Drowned on active service 21/3/66
47001	MAJ HUGHES James Curnow	2 Sqn	SVN 4 RAR 11/5/71—18/12/71, LTCOL (CO) Awarded DSO. Korea 3 RAR as LT and awarded MC. Malaya 3 RAR as Capt. Awarded AO 1982
43378	SIG HUNT Byron Charles	2 Sqn	*
212784	PTE HUNTER William John	1 Sqn	*

53584	CPL HYDE Leslie	2 Sqn	SVN 161 Recce Sqn Avn 22/12/69—27/8/70, CPL
54987	SIG JACKSON John Liddell	1 Sqn	*
43234	SIG JANSEN Michael Redfern	2 Sqn	*
212724	SGT JARVIS Ronald Leslie	1 Sqn	SVN 9 RAR 9/11/68—26/11/69, WO2
57053	CAPT JEFFERY Phillip Michael	HQ SAS Far East	SVN 8 RAR 18/11/70, MAJ Awarded MC. Awarded AM 1981, AO 1988
2412089	CPL JENNISON Christopher Arthur T.	1 Sqn	* Malay Peninsula
13446	SGT JEWELL Ian Joseph	2 Sqn	SVN AATTV 24/6/68—9/7/69 and 30/8/71—19/3/72, WO2. Malaya 1 Fd Regt RAA
16026	PTE JEWELL John Arthur	1 Sqn	*
120009	LCPL JONES Alan Terrence	2 Sqn	*
54344	PTE KELLY Edward	1 Sqn	*
13116	SGT KENNEDY William Desmond	1 Sqn	*
18580	CPL LATIMER Phillip Anthony	2 Sqn	
13394	CPL LAWRENCE Julian Broughton	1 Sqn	Malay Peninsula. Korea and Malaya 1 RAR, PTE
17971	PTE LEERENVELDT Henry Johannes	2 Sqn	*
5411077	LCPL LENNOX Andrew	1 Sqn	* Malay Peninsula
242902	LT LEVENSPIEL David	2 Sqn	*
43114	PTE LEONAVICIUS Algimantas Stanislaus	2 Sqn	
54379	PTE LEWIS Thomas William	2 Sqn	*
17896	PTE LIPINSKI Stanley	1 Sqn	*
54289	SIG LITTLE William Allen	1 Sqn	*
54150	CPL LITTLER Bryan Wilson	1 Sqn	Malay Peninsula
16540	PTE LOGUE Raymond John	1 Sqn	* Malay Peninsula
15188	LCPL LOVERIDGE John James	2 Sqn	*
54019	CPL LOWSON David Milne	1 Sqn	*
37836	SIG LUNDBERG Eric Patrick	2 Sqn	*
53982	CPL LYON Malcolm Peter	1 Sqn	*
13680	LT MARSHALL Thomas George	1 Sqn	*
54085	CPL MARTIN Vernon Walter Thomas	2 Sqn	*
214674	PTE McDONALD Samuel James	2 Sqn	*
14407	LT McDOUGALL Peter Malcolm	1 Sqn	*
54093	LCPL McHUGH Ronald Maxwell	1 Sqn	
5410259	CPL McINTYRE Donald Colin	2 Sqn	
38023	PTE McMINN Holt Frederick Noel	2 Sqn	*
2412209	PTE MONCRIEFF Robert Charles	2 Sqn	Drowned on active service 21/3/66
16908	PTE MURPHY William Allan	1 Sqn	*
43322	PTE MURRAY Allan Rex	2 Sqn	*
54324	PTE MURRELL Leslie Thomas	1 Sqn	*
42507	PTE MURTON Lynford John	1 Sqn	*
4410613	LCPL MUTCH Robert James	1 & 2 Sqn	*
5410980	PTE NEIL Raymond James	1 Sqn	*

11448	WO2 NEVILLE Daniel John	2 Sqn	SVN AATTV 6/8/63—14/12/64 and 5/12/66—5/6/68 as WO2. Then 7 RAR 10/2/70—10/3/71, WO2 Pacific Islands during WW2. Korea with 1 RAR, SGT. Awarded DCM in SVN
16362	PTE NUCIFORA Angelo	2 Sqn	*
43518	PTE NUGENT Robert Thomas	2 Sqn	*
18429	SIG O'FARRELL Denis Roderick	1 & 2 Sqn	
36949	SGT O'KEEFE John Francis	1 Sqn	*
36547	PTE O'NEILL Allan Leslie	2 Sqn	*
1410737	PTE PARRINGTON John Thomas	1 Sqn	*
2411062	PTE PAVLENKO Paul Joseph	1 Sqn	*
13824	SGT PETTIT John Gordon	1 Sqn	SVN AATTV 2/2/66—10/8/66, 20/5/68—26/2/69 and 27/1/70—4/4/70, WO2. WIA 22/7/66 GSW L elbow and stomach. KIA on 4/4/70 and awarded Posthumous MID. 2 RAR Malaya, PTE
43384	SIG PINNINGTON Christopher Robin	2 Sqn	*
16245	PTE PLATER Stanley Edgar	1 Sqn	*
12877	SGT POPE Christian Andrew	1 Sqn	SVN 2 and 6 RAR 17/3/67—9/1/68, WO2. Korea 3 RAR and Malaya 1 RAR, CPL
214032	PTE PURCELL Noel Raymond	2 Sqn	*
16417	PTE RANDLE Denis Anthony	1 Sqn	*
213892	PTE RAY Alan John	2 Sqn	
54006	LCPL REEVES Stanley	1 Sqn	*
36734	SGT REID Thomas Douglas	1 Sqn	SVN 6 RAR 31/5/66—14/6/67, WO2. RN Palestine, Yangtze and Korea
54504	SGT RENTON Edward J.	1 Sqn	
15827	PTE ROBERTS Malcolm Thomas	1 Sqn	*
213744	CPL ROBERTSON Thomas Angus	2 Sqn	*
54159	CPL ROBINSON John Murray	1 Sqn	*
16235	2LT RODERICK Trevor William	1 Sqn	*
311268	CPL ROLLASON Richard	1 Sqn	Served with British Army, Para Regt
15340	CPL ROODS William Henry	2 Sqn	*
18156	PTE ROSER Allan William	2 Sqn	*
61309	CPL RUFFIN Michael John	2 Sqn	*
214085	PTE RYAN Joseph Michael	1 Sqn	
18594	CPL RYAN John Owen	1 Sqn	Malay Peninsula, SVN 161 Recce Sqn Avn 3/5/67—18/1/68, CPL
216082	PTE RYAN Leslie Francis	2 Sqn	SVN AATTV 12/8/68—6/8/69, CAPT WIA 25/8/68 GSW R arm
17686	2LT SAVAGE David George	2 Sqn	SVN AATTV 12/8/68—6/8/69 CAPT WIA 25/8/68 GSW R arm
54901	PTE SCHEELE David Willem	2 Sqn	*
1200072	PTE SCHOENMAKER Michael Adrian	2 Sqn	SVN 1 ARU and 7 RAR 9/67—4/68. Served as PTE M.A. O'MALLON

214124	2LT SCHUMAN Peter John	1 Sqn	*
212946	SGT SCUTTS Sydney Raymond	2 Sqn	*
213438	SGT SEXTON John Francis	1 Sqn	SVN 7 RAR 2/4/67—18/3/68, SVN 7 RAR 16/2/70—10/3/71, SGT
15530	SGT SHAW Gavin Leslie	2 Sqn	*
213298	SGT SHEEHAN John Neil	1 Sqn	*
215453	SGT SHEEHAN Peter Thomas	2 Sqn	*
311517	LCPL SHELTON Arthur Trevis	2 Sqn	SVN AATTV 24/2/69—18/2/70, WO2. Served with British Army Parachute Regiment
53962	LCPL SIMCOCK James Paul	2 Sqn	*
212824	LT SKARDON Geoffrey Charles	42 RM CDO & 22 SAS att from 1 SAS Coy 1964	SVN AATTV 16/10/64— 9/10/65
17951	PTE SLOCOMBE Terrance P.	2 Sqn	*
311458	PTE SMITH Allan E.	1 Sqn	SVN 6 RAR 4/6/66—14/6/67, PTE
214166	LCPL SMITH Allan Victor	1 Sqn	*
214892	LCPL SMITH Barry Walter	1 Sqn	*
214775	PTE SMITHERS Allan Nicol	2 Sqn	*
54515	PTE SPITZ Nigel Frederick	2 Sqn	
16058	LCPL STAFFORD Robert John	1 Sqn	
54453	CPL STANDEN Frederick Barry	1 & 2 Sqn	*
16273	CPL STEVENSON Ola Sever	1 Sqn	*
29165	SGT STEWART James Thomas	2 Sqn	* Malay Peninsula
36940	LCPL SWALLOW Thomas Raymond	1 Sqn	*
36720	CPL SYKES Frank	2 Sqn	*
23770	SGT TAYLOR Kevin Arthur	1 Sqn	SVN AFV Amen and Welfare Unit 17/1/68—11/2/69, CAPT Korea 1 RAR, PTE
54329	SGT TEAR Windsor Douglas	2 Sqn	SVN AATTV 22/10/69— 29/10/70, CAPT awarded MBE. Malaya 1 RAR
33723	WO2 THOMPSON Alexander Patrick	1 Sqn	SVN 7 RAR 8/4/67—9/1/68, RSM SVN 7 RAR 10/2/70— 10/3/71, CAPT awarded MBE. Korea 1 RAR
28662	SGT THORBURN James Lawrence	1 Sqn	*
5410945	CPL TINK Clarence James	2 Sqn	
15457	CPL TONNA Emmanuel L. Antonio	1 Sqn	*
213277	SGT TURNER Cedric Clarence	2 Sqn	SVN AATTV 20/11/67— 3/12/68, WO2
213297	SGT TURTON John Barry	2 Sqn	
21321	CPL URQUHART Ashley Graham	1 Sqn	* Malay Peninsula attached to 2 Bn, Parachute Regiment
18179	PTE VAN ELDIK Joseph Edward	2 Sqn	*
38194	PTE VAN NUS Martin Phillip	2 Sqn	*
43005	PTE WALKER Leonard Albert	1 Sqn	*
214436	PTE WALSH Leo William	1 Sqn	* Malay Peninsula

53132	SGT WATERS Malcolm Burgess	1 Sqn	* Also served with 42 CDO RM in Borneo, atached during May 1964 from 1 SAS Coy
52636	SGT WEIR Roy Leslie	1 Sqn	SVN AATTV 29/7/62—19/6/63, 25/3/66—15/2/67 and 1/4/68 —26/2/69. WIA 16/2/69 Panji stake injury. SGT on first tour and WO2 for last two tours. Served in Pacific Islands WW2, BCOF and Korea with 1 RAR. Awarded BEM for services AATTV.
16949	PTE WENITONG Leonard	2 Sqn	*
54222	PTE WHITBREAD William J.	1 Sqn	*
213390	PTE WHITE Montague Robert	1 Sqn	SVN 8 RAR 17/11/69— 29/10/70, SGT
27669	SGT WHITE Peter Francis	2 Sqn	*
36387	CPL WIGG John William	1 Sqn	*
16404	SIG WILSON Donald Hector	1 Sqn	
55362	LCPL WRIGHT Daniel Henry	1 Sqn	* Malay Peninsula
61221	SGT YOUNG Barry Leon	2 Sqn	SVN 3 RAR 16/12/67— 2/12/68, SGT SVN AATTV 7/1/69—18/2/70, WO2
36724	LCPL ZOTTI Barry Michael	2 Sqn	*

AUSTRALIAN SPECIAL AIR SERVICE REGIMENT SOUTH VIETNAM

2789856	SIG ABBOTT Douglas Robert	152 Sig Sqn	9/9/69—10/9/70 Killed in parachute accident 4/3/76
214463	CPL ABSOLON Bruce Howard	1 Sqn	2/3/67—27/2/68
214352	LCPL ADAMSON Charles Stewart	3 Sqn	15/6/66—19/3/67
54962	SGT AGNEW John James	2 Sqn	5/2/68—24/2/69
218618	PTE AINSLEY Barry	1 Sqn	18/2/70—18/2/71
42386	SGT AITKEN William Maxwell	3 Sqn	16/6/66—18/3/67
15481	SGT ALBRECHT Ross Emil	2 Sqn	5/2/68—24/2/69
15481	SGT ALBRECHT Ross Emil	2 Sqn	20/1/71—11/10/71
215874	PTE ALLEN Robert Bruce	3 Sqn	15/6/66—25/3/67
215874	CPL ALLEN Robert Bruce	3 Sqn	21/2/69—20/2/70
548794	LCPL ALLISON Reginald Arthur	2 Sqn	11/8/71—11/10/71
548794	PTE ALLISON Reginald Arthur	4 RAR	28/10/68—19/5/69
5411474	PTE ALVER Leigh Clinton	2 Sqn	26/2/68—24/2/69 WIA 22/6/68 Shrapnel in leg
5411474	PTE ALVER Leigh Clinton	2 Sqn	17/2/71—6/10/71
1201288	SIG ANGEL Leslie Edward	152 Sig Sqn	21/2/69—18/2/70
3410763	SGT ANSET Alan Clive	1 Sqn	26/5/67—27/2/68
55783	PTE ANTONOVICH Graeme Mathew	1 Sqn	18/2/70—18/2/71
43382	SGT ARNOLD Peter John	1 Sqn	18/2/70—18/2/71
43382	PTE ARNOLD Peter John	1 RAR	5/6/65—3/6/66
215376	PTE ARNOLD Roger Ashley	3 Sqn	15/6/66—25/3/67
215376	LCPL ARNOLD Roger Ashley	3 Sqn	21/2/69—20/2/70

Appendixes

54219	PTE AYLING Frank John	1 Sqn	2/3/67—18/11/67 WIA 8/6/67 ankle injury (helo accident)
55171	PTE BACSKAI Arpad Laslo	1 Sqn	2/3/67—25/10/67 1 RAR Malaya, PTE
55171	CPL BACSKAI Arpad Laslo	1 Sqn	27/1/70—18/2/71
1411193	PTE BAILLIE Terrence Patrick	2 Sqn	26/2/68—3/12/68
36179	CPL BAIN Norman Patrick	3 Sqn	15/6/66—19/3/67 1 RAR Malaya, PTE
36179	WO2 BAIN Norman Patrick	AATTV	19/7/72—20/12/72 1 RAR Malaya, PTE
217552	PTE BAIN Ray Herbert	3 Sqn	21/2/69—4/11/69
54840	LCPL BAINBRIDGE Norman Keith	152 Sig Sqn	12/9/67—17/9/68
54840	CPL BAINBRIDGE Norman Keith	152 Sig Sqn	18/2/70—18/2/71
214230	SGT BAINES George Terence	1 Sqn	2/3/67—13/2/68 Accidentally killed on active service 13/2/68
52616	WO2 BALDWIN Ivan John	1 Sqn	28/1/70—25/3/71
44818	PTE BALL Gregory Francis	2 Sqn	17/2/71—7/10/71
38570	SIG BARDSLEY Stuart William	152 Sig Sqn	2/3/67—27/2/68
217585	PTE BARNBY Donald Richard	2 Sqn	17/2/71—7/10/71
38192	PTE BARNES Derek Clifford	1 Sqn	2/3/67—27/2/68 WIA 29/1/68 Shrapnel in back
218933	SIG BARNETT Eric John	152 Sig Sqn	9/9/70—9/9/71
28831	CPL BASSETT Thomas Ernest	3 Sqn	15/6/66—19/3/67 Malaya 2 RAR, PTE Borneo and Malay Peninsula 3 RAR
28831	SGT BASSETT Thomas Ernest	3 Sqn	21/2/69—20/2/70
4721793	PTE BAYLEY Robert James	2 Sqn	11/8/71—11/10/71
1201375	PTE BEARD Raymond John Keith	3 Sqn	21/2/69—18/2/70
235165	MAJ BEESLEY Reginald Patrick	3 Sqn	3/2/69—21/1/70 Borneo HQ 28 Comm Inf Bde, CAPT. Awarded AM 1982
3791202	2LT BEHM Lloyd Charles	1 Sqn	5/2/70—18/2/71
213120	PTE BELL Francis Kevin	3 Sqn	15/6/66—19/3/67
19988	PTE BELL Graham Alexander	3 Sqn	15/6/66—25/3/67
5715805	PTE BENNETT Lindsay William	2 Sqn	17/2/71—11/10/71
5715805	PTE BENNETT Lindsay William	4 RAR	25/11/68—28/5/69
3796953	PTE BEPLATE Graham John	2 Sqn	17/2/71—19/8/71
216232	CPL BERCENE Patrick Garth	2 Sqn	26/2/68—21/2/69
5411542	PTE BERRY James	2 Sqn	14/2/68—24/2/69
3411805	PTE BILBROUGH Stanley	3 Sqn	21/2/69—18/2/70
1735913	PTE BIRD Denis Alfred	2 Sqn	17/2/71—19/8/71
17096	CAPT BISHOP Ross Blake	3 Sqn	21/2/69—18/2/70
1200895	LCPL BLACKER Adrian Mark	2 Sqn	26/2/68—21/2/69
1200895	CPL BLACKER Adrian Mark	2 Sqn	17/2/71—11/10/71
6410211	PTE BLAKE Brian Harold	2 Sqn	26/2/68—21/2/69
6410211	SGT BLAKE Brian Harold	2 Sqn	17/2/71—6/10/71
16767	CPL BLOOMFIELD Stephen Geoffrey	1 Sqn	2/3/67—6/2/68
53865	SGT BOAG Kenneth John	3 Sqn	15/6/66—19/3/67 WIA 17/8/68 Shrapnel in leg
215531	PTE BONHAM Frank Robert	3 Sqn	15/6/66—19/3/67

1201963	PTE BOOTH David Nicholas	1 Sqn	18/2/70—21/1/71
55805	PTE BOVILL Kevin Albert	3 and 1 Sqn	22/12/69—10/12/70
55805	LCPL BOVILL Kevin Albert	2 Sqn	12/8/71—11/10/71
3411809	PTE BOWDEN Anthony Peter	2 Sqn	26/2/68—21/2/69
39100	CPL BOWEN Kenneth Frank	2 Sqn	17/6/68—24/2/69
39100	CPL BOWEN Kenneth Frank	2 Sqn	17/2/71—11/10/71
1732243	PTE BOWSER William Frank	2 and 3 Sqn	25/11/68—5/11/69
38387	PTE BRADSHAW Thomas Ronald	3 Sqn	15/6/66—19/3/67
38387	LCPL BRADSHAW Thomas Ronald	3 Sqn	3/2/69—20/2/70
44077	SIG BRAMLEY Creagh MacDonald	152 Sig Sqn	26/2/68—21/2/69
44077	SIG BRAMLEY Creagh MacDonald	152 Sig Sqn	17/2/71—11/10/71
1411242	PTE BRAMMER Graham John	2 Sqn	26/2/68—21/2/69
1411242	SGT BRAMMER Graham John	2 Sqn	17/2/71—10/10/71
213622	SGT BRANIFF John Edward	3 Sqn	3/2/69—16/7/69
2793957	PTE BRASSIL Graham John	2 Sqn	17/2/71—26/8/71
55433	PTE BRESSER Anthony	2 Sqn	17/6/68—24/2/69
217210	LT BRETT Robert Anthony	2 Sqn	11/5/71—22/7/71
217210	2LT BRETT Robert Anthony	5 RAR	27/1/69—10/3/70 Awarded MC and MID with 5 RAR
44951	PTE BRIGGS James Winston	1 and 2 Sqn	12/8/70—12/8/71
2795174	PTE BRIGHT James Allen	2 Sqn	11/8/71—16/10/71
3411970	LCPL BROADHURST Robert Thomas	3 Sqn	21/2/69—3/6/70
3411970	SGT BROADHURST Robert Thomas	AATTV	
2793410	PTE BROWN Colin Kennedy	1 and 2 Sqn	7/10/70—26/8/71
397815	PTE BROWN Dennis Reginald	3 Sqn	21/2/69—18/2/70
214278	LCPL BROWN Raymond Walter	1 Sqn	2/3/67—20/7/67
214278	CPL BROWN Raymond Walter	3 Sqn	21/2/69—20/2/70 WIA 2/11/69 FW side of neck
1411300	PTE BUCKLEY Patrick Joseph	2 Sqn	18/2/70—7/10/71
1411300	PTE BUCKLEY Patrick Joseph	7 RAR	7/6/67—26/4/68
3411898	PTE BULLOCK Ian Anthony	1 Sqn	2/3/67—26/2/68
3411898	CPL BULLOCK Ian Anthony	1 Sqn	18/2/70—18/2/71
1200583	LCPL BUNNEY Daryl Edward	152 Sig Sqn	26/2/68—21/2/69
1200583	CPL BUNNEY Daryl Edward	152 Sig Sqn	9/9/69—10/9/70
5716963	PTE BUNNEY Milford Douglas	2 Sqn	17/2/71—26/8/71
610420	SGT BURDON Phillip John	152 Sig Sqn	2/4/71—7/10/71
54581	CPL BURGESS Daniel Charles	3 Sqn	15/6/66—7/12/66
2794686	PTE BURKE Anthony John	2 Sqn	4/8/71—11/10/71
55203	PTE BURLEY Keith Alexander	3 Sqn	15/6/66—19/3/67
212539	SGT BURLING John Allan	3 Sqn	15/6/66—25/3/67 1 RAR Malaya, PTE
38893	LCPL BURTON Robert William	2 Sqn	26/2/68—21/2/69
53817	SGT BUTCHART James Thomson	2 Sqn	5/2/68—24/2/69
54765	CPL BUTCHER Leonard Alexander	3 Sqn	15/6/66—19/3/67

Appendixes

54765	CPL BUTCHER Leonard Alexander	152 Sig Sqn	26/2/68—24/2/69
61541	PTE BUTTON Wayne John	2 Sqn	26/2/68—24/2/69
38386	PTE BYE Peter Francis	3 Sqn	15/6/66—25/3/67
38386	CPL BYE Peter Francis	3 Sqn	21/2/69—18/2/70
5411565	PTE BYRNE James Francis	3 Sqn	10/6/69—20/2/70
5411565	PTE BYRNE James Francis	6 RAR	4/6/66—14/6/67
5411453	PTE BYRNE-KING Terence Kenneth Keith	3 Sqn	15/6/66—19/3/67
5411646	PTE BYROM Kenneth	3 and 1 Sqn	7/10/68—1/10/70 and PTE HQ 1 ALSG
5714828	SIG CALLAGHAN Allen John	152 Sig Sqn	23/9/68—13/9/69
5714828	CPL CALLAGHAN Allen John	2 Sqn and 152 Sig Sqn	9/9/70—10/9/71
215935	LCPL CALLAGHAN Paul Francis	3 Sqn	21/2/69—20/2/70
215935	PTE CALLAGHAN Paul Francis	2 RAR	9/5/67—9/1/68
1201612	PTE CAMPBELL Kerry John	3 Sqn	21/2/69—18/2/70
217573	PTE CAMPBELL William	3 Sqn	21/2/69—18/2/70
311515	PTE CARROLL Clive Peter	1 Sqn	2/3/67—27/2/68 Ex Brit Army, 3 Regt RHA
216802	PTE CARTER Wayne William	2 Sqn	26/2/68—11/9/68 Accidentally injured 5/9/68 RTA
216802	LCPL CARTER Wayne William	1 Sqn	18/2/70—18/2/71
55124	LCPL CARTMELL Kim Stuart	152 Sig Sqn	26/2/68—24/2/69
55124	LCPL CARTMELL Kim Stuart	152 Sig Sqn	17/4/69—13/9/69
55124	CPL CARTMELL Kim Stuart	152 Sig Sqn	17/2/71—11/10/71
54237	SGT CASHMORE Frank Carr	2 Sqn	5/2/68—22/2/69
54237	SGT CASHMORE Frank Carr	2 Sqn	17/2/71—11/10/71
379897	PTE CASWELL Rodney James	2 Sqn	11/8/71—10/10/71
5411566	CPL CAUSTON Reginald Alan	3 Sqn	15/6/66—14/3/67
61141	CAPT CHIPMAN Geoffrey Edward	3 Sqn	15/6/66—25/3/67 WIA 17/8/66 Shrapnel in leg
61141	MAJ CHIPMAN Geoffrey Edward	1 and 2 Sqn	25/11/70—10/10/71
3179807	PTE CHIPPENDALL Garry Wallace	1 and 2 Sqn	16/9/70—25/8/71
42567	CPL CHRISTENSEN David Leslie	3 Sqn	15/6/66—19/3/67
42567	CPL CHRISTENSEN David Leslie	5 RAR	3/2/69—10/3/70 also 1 Aust Fd Hosp
1200461	CPL CHRISTIE David Alexander	3 Sqn	15/6/66—25/3/67
218850	SIG CLARK Phillip Arthur	152 Sig Sqn	18/2/70—18/2/71
2412382	PTE CLARK Robert Francis Walter	1 Sqn	20/4/67—27/2/68
55472	CPL CLARKE Ken Ralph	1 Sqn	3/2/70—18/6/70
216927	SIG CLARKE Kerry Francis	152 Sig Sqn	26/2/68—21/2/69
216927	CPL CLARKE Kerry Francis	2 Sqn	17/2/71—11/10/71
5411622	PTE COCHRANE Daryl Raymond	1 Sqn	20/4/67—27/2/68

551

54917	PTE COFFEY Alexander Michael	3 and 1 Sqn	13/1/70—18/2/71
311251	SGT COLEMAN John Henry	2 Sqn	5/2/68—3/9/68 Royal Fusiliers in Middle East and Paras in Cyprus and Suez (2 Bn). Awarded BEM 1962
311251	WO2 COLEMAN John Henry	AATTV	10/6/64—13/4/65
55184	CPL COLLINS Terrance James	2 Sqn	5/2/68—24/2/69
1201601	CPL COMBES Murray Allan	1 Sqn	18/2/70—18/2/71
1410935	CPL CONAGHAN Ian Frederick	1 Sqn	2/3/67—26/2/68
1410935	SGT CONAGHAN Ian Frederick	1 Sqn	3/2/70—18/2/71
213546	SGT CONSIDINE Joseph Bernard	2 Sqn	26/2/68—4/2/69
215989	PTE COPEMAN Russel James	3 Sqn	15/6/66—28/2/67 WIA 18/1/67 GSW Abd, R hip and R arm. Died of wounds 10/4/67
216285	PTE COULTON Brian Colin	3 Sqn	15/6/66—19/3/67
2793732	PTE CRANE Michael John	2 Sqn	17/2/71—2/9/71 WIA 28/5/71 Shrapnel R shoulder and L ankle
3179727	PTE CROMPTON David Neil	1 and 3 Sqn	4/11/69—5/11/70
42455	SGT CROUCHER Daryl Anthony	3 Sqn	15/6/66—19/3/67
42455	WO2 CROUCHER Daryl Anthony	110 Sig Sqn	3/3/70—4/3/71
1200896	LCPL CULLEN Dennis James	2 Sqn	26/6/68—21/2/69 WIA 14/7/68 Shrapnel R butt and R arm
216987	PTE CULLEN Leslie Phillip	1 and 3 Sqn	13/1/70—14/1/71
216987	PTE CULLEN Leslie Phillip	5 and 7 RAR	10/3/67—5/3/68
217594	PTE CUZENS John Raymond	2 and 3 Sqn	25/11/68—28/11/69
235250	CAPT DANILENKO Anatoly	SAS LO HQ AFV	13/11/67—25/4/68
235250	CAPT DANILENKO Anatoly	AATTV	13/11/67—25/4/68 KIA awarded posthumous MID
54752	PTE DAVIES Mark	1 Sqn	2/3/67—27/2/68
215118	CPL DAVIES Reginald Elliot	2 Sqn	26/2/68—24/2/69
215118	SGT DAVIES Reginald Elliot	1 and 2 Sqn	13/1/71—11/10/71
15843	CPL DAVIES Robert Neville	3 Sqn	15/6/66—31/8/66 WIA 17/8/66 Shrapnel head, RTA
38268	PTE DAVIES Terence Francis	2 Sqn	5/2/68—24/2/69
210778	WO2 DAVIS Marsden Stephen	3 and 1 Sqn	14/2/67—27/2/68 British Paras WW2 and Palestine, PTE
42342	PTE DAVIDSON Thomas Walter	3 and 1 Sqn	13/1/70—14/1/71 Malaya 1 RAR, PTE, Borneo 4 RAR, PTE
217308	CPL DAW Garry	3 Sqn	21/2/69—20/2/70 Borneo 9 Ind Para Sqn, RE SPR
3797235	PTE DAZKIW Michael	2 Sqn	17/2/71—2/9/71
2244240	PTE DEATH Frederick Thomas	2 Sqn	4/8/71—7/10/71
5411465	PTE DE GRUSSA Noel Stewart	1 Sqn	2/3/67—30/8/67 WIA 11/8/67 GSW L Thigh, RTA
5411465	PTE DE GRUSSA Noel Stewart	3 Sqn	22/2/69—24/9/69

5411171	LCPL DELGADO Bernard John	1 Sqn	2/3/67—27/2/68
5411171	CPL DELGADO Bernard John	1 Sqn	18/2/70—18/2/71
541165	SGT DELGADO Vernon	2 Sqn	26/2/68—23/2/69
518679	PTE DEMBOWSKI Stefan	2 Sqn	18/2/71—11/10/71
5411122	LCPL DEMAMIEL Wayne Geoffery	1 Sqn	11/8/67—31/12/67
3411503	PTE DEMARCHI Dino	3 and 1 Sqn	12/12/66—13/9/67
38621	2LT DEMPSEY Ronald George	2 Sqn	5/2/68—21/2/69
54047	LCPL DEVINE Harold Joseph	3 and 1 Sqn	14/2/67—27/2/68 Malaya 1 RAR, PTE
42522	CPL DIDSMAN Luke Olaf	152 Sig Sqn	28/1/67—6/2/68
42522	SGT DIDSMAN Luke Olaf	152 Sig Sqn	4/2/71—11/10/71
55662	CPL DIMMACK Raymond William	3 and 1 Sqn	3/2/69—13/3/70
56047	SIG DODD John Charles	152 Sig Sqn	17/2/71—11/10/71
39380	PTE DODD John Charles	3 Sqn	3/2/69—20/2/70 Awarded BEM 1979
213154	SGT DONNELLY Ronald Knapp	1 Sqn	2/3/67—27/2/68
213154	SGT DONNELLY Ronald Knapp	1 Sqn	18/2/70—18/2/71
15325	CPL DUFFY John James Patrick	1 Sqn	2/3/67—27/2/68 Malaya 1 RAR, PTE
15325	SGT DUFFY John James Patrick	1 Sqn	3/2/70—18/2/71 Malaya 1 RAR, PTE
54477	PTE DUFFY Paul Michael	3 Sqn	25/6/66—25/11/66 Malay Peninsula, 3 RAR, PTE
54477	CPL DUFFY Paul Michael	2 Sqn	5/2/68—24/2/69
42357	SGT DUNCAN John McKenzie	1 Sqn	2/3/67—27/2/68
42357	SGT DUNCAN John McKenzie	1 Sqn	3/2/70—18/2/71
36852	CPL DWYER Clement James	3 Sqn	15/6/66—18/3/67
36852	SGT DWYER Clement James	3 Sqn	21/2/69—18/2/70
1200843	LCPL EASLEA John Michael	2 Sqn	26/2/68—21/2/69
1200843	SGT EASLEA John Michael	2 Sqn	17/2/71—16/6/71
53838	SGT EASTHOPE Alan	2 Sqn	5/2/68—24/2/69
1732019	LT EDDY Michael Ross	3 Sqn	30/9/69—1/10/70
1732019	2LT EDDY Michael Ross	2 RAR	19/5/67—5/6/68 UN Forces—Middle East
3797191	PTE ELLIS John Murdie	1 and 2 Sqn	4/11/70—9/9/71
1731438	PTE ELLIOTT David	2 Sqn	26/2/68—27/3/68 CASEVAC
1731438	SGT ELLIOTT David	3 RAR	25/2/71—8/9/71 MEDEVAC
214017	CPL ENGLISH Bruce Patrick	3 Sqn	15/6/66—18/3/67
3798466	PTE ESCOTT Geoffrey John	2 Sqn	11/8/71—7/10/71
214801	PTE EWART Malcolm	1 Sqn	2/3/67—26/2/68
61793	PTE FAGAN Michael Bernard	2 and 3 Sqn	7/1/69—18/2/70
15646	SGT FARLEY Neville Gary	3 and 1 Sqn	14/2/67—9/1/68
54793	SIG FARMER Eric Edward	152 Sig Sqn	5/2/68—24/2/69

55850	PTE FARMER Michael Desmond	2 and 1 Sqn	9/9/70—9/9/71 WIA 1/8/71 shrapnel in L shoulder Malaya 3 RAR, PTE
53870	SGT FERGUSON Norman Stanley	3 Sqn	22/9/66—26/3/67
53870	WO2 FERGUSON Norman Stanley	3 RAR	15/2/71—22/1/72
4411047	LCPL FINCH Leslie Robert	2 Sqn	26/2/68—28/2/69
4411047	CPL FINCH Leslie Robert	1 RAR	26/2/68—28/2/69
2412356	LCPL FISH Jeffery James	2 Sqn	26/2/68—8/10/68
2787344	PTE FISHER David John Elkington	2 and 3 Sqn	16/12/68—27/9/69 Killed on active service, 27/9/69
16337	CPL FISHER Donald Murphy	1 Sqn	2/3/67—27/2/68
16337	SGT FISHER Donald Murphy	1 Sqn	3/2/70—23/7/70
6708471	LCPL FISHER Norman Henry George	1 and 2 Sqn	12/9/67—7/10/68
55150	2LT FITZPATRICK Peter John	152 Sig Sqn	2/9/68—3/9/69 Awarded AM 1984
297097	LT FLANNERY Joseph	2 Sqn	13/1/71—16/10/71 Korea 2 RAR, LCPL. Malaya 2 RAR, CPL. WIA 27/12/67 shrapnel both arms. Awarded MBE 1969
297097	WO2 FLANNERY Joseph	AATTV	17/9/62—12/8/63
297097	WO2 FLANNERY Joseph	AATTV	30/6/67—12/2/68
44041	CPL FLEER Johannes Cornelis	2 Sqn	17/2/71—19/10/71
44041	CPL FLEER Johannes Cornelis	4 RAR	21/5/68—30/5/69
44041	CPL FLEER Johannes Cornelis	6 RAR	22/7/69—12/5/70 Awarded DCM with 6 RAR
217589	SIG FLEMING Gary	152 Sig Sqn	10/9/70—9/9/71
1200885	CPL FLEMMING Neale Alfred	2 Sqn	5/2/68—21/2/69
235094	MAJ FLETCHER John	SAS LO HQ AFV	3/5/67—28/11/67 Awarded GM 1961
313691	SIG FLIGHT Michael James	152 Sig Sqn	30/8/71—11/10/71
213823	CPL FOLKARD Michael Lester	3 Sqn	15/6/66—19/3/67
54887	PTE FORWARD Graham John	3 Sqn	15/6/66—19/3/67
53235	SGT FOXLEY Arthur	3 Sqn	3/2/69—20/2/70 Malaya 1 and 3 RAR, PTE. Awarded MBE 1979
6410234	PTE FRANCE Peter Aidan	1 Sqn	2/3/67—27/2/68
216039	PTE FRANKLIN George William	3 Sqn	21/2/69—18/2/70
215468	LCPL FRANKLIN Ian Frederick	2 Sqn	5/2/68—28/2/69 WIA 27/5/68
215468	LCPL FRANKLIN Ian Frederick	1 RAR	Shrapnel Throat
43716	PTE FRANZI Shea Richard	1 Sqn	2/3/67—27/2/68
43716	CPL FRANZI Shea Richard	1 Sqn	18/2/70—10/11/71 D Sqn Rhodesian SAS 1974–75
53151	SGT FRASER Lawrence Edmund	1 Sqn	2/3/67—26/2/68 Malaya, 1 RAR, PTE
53151	SGT FRASER Lawrence Edmund	1 Sqn	18/2/70—18/2/71 Awarded MBE 1982
2243903	PTE FRAZER John Allan	3 and 1 Sqn	23/9/69—24/9/70
44767	PTE FREEMAN Kenneth Wayne	2 Sqn	17/2/71—11/10/71

Appendixes

311611	LT FREEMANTLE Andrew Wayne	1 and 2 Sqn	23/9/70—10/8/71 Royal Hampshire Regt Borneo and Cyprus, re-commissioned in Royal Hamps, Northern Ireland awarded MBE and MID Brigadier 1988
1200257	PTE GABRIEL Wallace Bruce	1 Sqn	2/3/67—6/2/68
1200257	CPL GABRIEL Wallace Bruce	1 Sqn	8/7/70—16/9/70
5411541	PTE GALLAGHER Patrick Anthony	2 Sqn	5/2/68—28/2/69
215437	SIG GALLEGOS Allan James	152 Sig Sqn	2/3/67—26/2/68
215437	LCPL GALLEGOS Allan James	1 Sqn	13/2/70—18/2/71
54909	CPL GAMMIE Ronald Lindsay	3 Sqn	15/6/66—7/2/67
4410659	PTE GARAY Stephen	3 Sqn	22/9/66—25/3/67
4410659	CPL GARAY Stephen	2 AOD	6/1/70—14/1/71
53732	CPL GARVIN Arthur George	1 Sqn	2/3/67—26/2/68 Malaya 1 RAR, PTE
53732	SGT GARVIN Arthur George	1 Sqn	4/2/70—18/2/71
3166056	PTE GAVAN Keith Bernard	1 and 2 Sqn	23/9/70—16/6/71
235289	LT GAY Ian Maxwell	3 and 1 Sqn, 7 RAR	12/12/66—28/12/67 WIA 10/11/67 Shrapnel R Thigh
1200111	SGT GEBHARDT John	1 Sqn	2/3/67—26/2/68 WIA 27/12/67 shrapnel Head
1200111	SGT GEBHARDT John	1 Sqn	3/2/70—18/2/71
311332	SGT GELDHART Michael	3 Sqn	15/6/66—18/3/67 Cyprus, RAF Regt, LAC 1955–57
55682	PTE GIBLETT Eric Mervyn	1 Sqn	18/2/70—6/5/70 RTA illness
215537	PTE GIBSON John McMichael	3 Sqn	15/6/66—19/3/67
2412502	PTE GIBSON William Fulton	2 Sqn	26/2/68—21/2/69
36226	SGT GILCHRIST Ronald John	1 Sqn	2/3/67—12/12/67
36226	SGT GILCHRIST Ronald John	1 Sqn	18/2/70—18/2/71
54811	LCPL GLOEDE Henry Alexander	3 Sqn	15/6/66—19/3/67
54811	CPL GLOEDE Henry Alexander	3 Sqn	21/2/69—16/7/69
44066	2LT GLOEDE Richard Roy	1 Sqn	2/3/67—26/3/68 WIA injured ankle (helo crash) 8/6/67
213439	SGT GLOVER Barry Edward	1 Sqn	2/3/67—6/2/68
55117	SIG GOLDSWORTHY Andrew John	152 Sig Sqn	23/9/69—24/9/70
55117	SIG GOLDSWORTHY Andrew John	104 Sig Sqn	25/4/67—9/4/68
55241	PTE GOOCH Edward John	1 Sqn	18/2/70—18/2/71
55241	PTE GOOCH Edward John	7 RAR	8/6/67—26/4/68
4411081	LCPL GOODWIN Gilbert May	1 Sqn	2/3/67—6/2/68
52871	SGT GRATWICK Arthur Barrington	1 Sqn	2/3/67—27/2/68 Malaya HQ 28 Com Inf Bde, CPL, Awarded MID for Malaya
2173710	PTE GRAY Malcolm Colin	3 Sqn	21/2/69—20/2/70
217030	SIG GRAY Richard Francis	152 Sig Sqn	3/2/69—18/2/70
2793096	SIG GRAY William Robert	152 Sig Sqn	23/9/70—26/8/71
215119	PTE GRAYDON John Robert	3 Sqn	15/6/66—19/3/67

215119	LCPL GRAYDON John Robert	3 Sqn	21/2/69—21/2/70
15988	SGT GREEN Harry	3 Sqn	21/2/69—20/2/70 Killed on exercise 4/3/76
1200500	LCPL GREER Kevin Michael	1 Sqn	18/2/70—18/2/71
1200500	LCPL GREER Kevin Michael	7 RAR	17/4/67—26/4/68
15718	SGT GRIDLEY George Arthur	1 Sqn	2/3/67—26/2/68
19963	CPL GRIFFITHS John Leonard	152 Sig Sqn	2/3/67—27/2/68
19963	SGT GRIFFITHS John Leonard	152 Sig Sqn	18/2/70—18/2/71
38101	PTE GRINDAL Robert Rex	1 Sqn	2/3/67—27/2/68
55858	SIG GRONOW Alan Dudley	152 Sig Sqn	18/2/70—8/12/70
213223	WO2 GROVENOR Michael Joseph	2 Sqn	5/2/68—24/2/69
213223	WO2 GROVENOR Michael Joseph	2 Sqn	10/2/71—11/10/71
17795	LCPL GULDBRANSEN Ross Leonard	1 Sqn	2/3/67—27/2/68
1202831	SIG GUNNING Trevor Howard	152 Sig Sqn	12/8/71—11/10/71
39012	2LT GURNEY Richard Peter	152 Sig Sqn	18/8/71—16/10/71
4410943	PTE GUY John David	1 and 3 Sqn	8/2/67—6/12/67
43622	SIG HAHN David Roger	3 Sqn	15/6/66—18/3/67
43622	CPL HAHN David Roger	152 Sqn	26/2/68—21/2/69
310649	LT HALEY Herbert John A.	2 and 3 Sqn	30/9/68—15/10/69 Awarded BEM 1961 1st Guards Para Bn Palestine, Japan BCOF, Korea 3 RAR, PTE
212807	WO2 HAMMOND Horrace William	1 Sqn	2/3/67—27/2/68
212807	WO2 HAMMOND Horrace William	AATTV	1/7/64—11/6/65
217542	PTE HANCOCK Terence Stanley	3 Sqn	21/2/69—10/9/69 Medevac
411599	PTE HANNAFORD Bradley John	2 Sqn	11/8/71—7/10/71
411599	SPR HANNAFORD Bradley John	1 FD Sqn	10/6/69—11/6/70
3166224	LCPL HANNAFORD Geoffery Edward	1 Sqn	18/2/70—18/2/71
61571	2LT HANSCH Lindsay David	2 Sqn	19/8/71—7/10/71
61571	LCPL HANSCH Lindsay David	5 RAR	17/5/66—17/5/67
2788724	PTE HARDING John Anthony	3 Sqn	15/7/69—16/5/70
2788724	PTE HARDING John Anthony	6 RAR	As above
54798	PTE HARLEY Norman Edgar	1 Sqn	11/8/67—27/2/68
54798	LCPL HARLEY Norman Edgar	1 Sqn	3/2/70—18/2/71
61732	PTE HARPER John Wesley	2 Sqn	26/2/68—26/2/69
55430	SGT HARRIS Peter Gerard	2 Sqn	18/2/71—11/10/71
55430	SGT HARRIS Peter Gerard	26 Coy RAASC	11/1/69—18/1/70
5410968	CPL HARRIS Ronald Arthur	2 Sqn	17/6/68—17/1/69 Borneo A Fd Bty RAA, BDR. Killed on active service 17/1/69
55158	SGT HARRIS Roy Walter	2 Sqn	10/2/71—11/10/71
55158	CPL HARRIS Roy Walter	1 RAR	22/7/68—23/7/69
214175	LCPL HARRIS William Edward	3 Sqn	15/6/66—19/3/67
214175	SGT HARRIS William Edward	3 Sqn	3/3/69—20/2/70

4410614	PTE HARVEY James Franklin	1 Sqn	2/3/67—27/2/68 WIA 24/7/67 Shrap face
4719368	PTE HAY James Callaghan	3 Sqn	21/2/69—18/2/70
39861	PTE HAYES Christopher John	1 and 3 Sqn	21/10/69—26/11/70
5717009	PTE HAYES Francis John	2 Sqn	7/7/71—11/10/71
5411157	CPL HAYNES John Kenneth	3 Sqn	15/6/66—12/10/66 RTA illness. Malay Peninsula 2 RAR, PTE
3797343	PTE HAYWARD Stephen Charles	2 Sqn	17/2/71—26/8/71
15374	SGT HEALY Peter Lawrence	3 Sqn	15/6/66—28/1/67
15374	SGT HEALY Peter Lawrence	8 RAR	17/11/61—28/2/70 Malaya 4 RAR, PTE WIA 18/2/70 Shrap and burns face
1203261	PTE HEMMENS John Charles	2 Sqn	4/8/71—7/10/71
5714316	PTE HENDERSON William	3, 1, 2 Sqn	24/9/69—18/2/71
5714316	PTE HENDERSON William	7 RAR	8/4/67—18/3/68
37664	SGT HENNESSY John Joseph	2 Sqn	17/2/71—11/10/71 Malaya 102 Fd Bty RAA, GNR
55272	CPL HERBERT Clifford Howard	2 Sqn	17/2/71—11/10/71
5411594	PTE HERBERT Gordon Patrick	1 Sqn	20/4/67—27/2/68
216842	PTE HETHERINGTON Mervyn Patrick	1 Sqn 7 RAR	20/4/67—19/12/67
216842	CPL HETHERINGTON Mervyn Patrick	1 Div Int, AFV PRO Unit, HQ AFV	9/9/69—16/12/71
38519	SIG HICKINBOTHAM Thomas James	152 Sig Sqn	2/3/67—5/2/68
53734	CPL HIGGINS Barrie Joseph	3 Sqn	15/6/66—16/3/67 WIA 17/8/66 Shrap head
29948	SGT HIGGINS Clement	1 Sqn	3/2/70—18/2/71
29948	CPL HIGGINS Clement	3 RAR	28/12/67—28/11/68
52763	MAJ HILL Deane Leighton	SAS LO, HQ AFV	29/5/68—26/1/69
54897	SIG HILL John Roger	152 Sig Sqn	22/2/69—19/2/70
54897	SIG HILL John Roger	104 Sig Sqn	26/4/67—21/3/68
37479	LT HINDSON William Francis	1 Sqn	2/3/67—27/2/68
37479	2LT HINDSON William Francis	1 RAR	26/5/65—7/6/66
1200490	PTE HINGST John Francis	1 Sqn	2/3/67—27/2/68
1200490	CPL HINGST John Francis	1 Sqn	18/2/70—18/2/71
28757	SGT HOGG Thomas Dennis	3 Sqn	15/6/66—26/1/67 Malaya 2 RAR, PTE WIA 29/8/66 GSW left leg
218876	SIG HOLLAWAY Malcolm	152 Sig Sqn	18/2/70—18/2/71
216825	PTE HOLLIT Barry George	1 Sqn	20/4/67—27/2/68
216825	CPL HOLLIT Barry George	1 Sqn	18/2/70—18/2/71
216826	PTE HOLUSA Gordon Harry	3 Sqn	29/7/69—20/2/70
3797326	PTE HOMANN Darren Lindsay	2 Sqn	17/2/71—19/8/71

557

215951	LCPL HONINGER Michael Alexander	2 Sqn	26/2/68—21/2/69 WIA 1/1/69 shrap head
1201368	PTE HORNE John Charles	1 Sqn	18/2/70—18/2/71
39144	2LT HOWLETT Harry Nicholas	3 Sqn	3/2/69—17/12/69
43378	SIG HUNT Bryan Charles	152 Sig Sqn	2/3/67—6/2/68
55550	PTE HUNTER Allen Lindsay	1 Sqn	24/6/70—18/2/71
212784	CPL HUNTER William John	3 Sqn	15/6/66—18/3/67
212784	SGT HUNTER William John	3 Sqn	3/2/69—19/9/69 WIA 3/8/69 GSW R side neck, R arm, shrap R leg
3798540	PTE HURFORD Robert Burce	2 Sqn	11/8/71—11/10/71
1200278	SIG INALL Allan David	3 Sqn	15/6/66—19/3/67
1200278	CPL INALL Allan David	152 Sig Sqn	2/9/68—3/9/69
173010	PTE ING Alan William	3 Sqn	21/2/69—18/2/70
55231	PTE INGRAM Allan Alfred	3 Sqn	21/2/69—18/2/70
54729	2LT INGRAM Peter Merryl	3 Sqn	15/6/66—19/3/67 Awarded QC as CAPT AVN, 15/1/70
216173	LT ISON John Richard	3 Sqn	12/8/69—18/2/70
2412358	PTE ISRAEL Gerard Maxwell	1 Sqn	2/3/67—26/2/68
214778	LT IVEY Robert James	3 and 1 Sqn	3/12/66—14/11/67 Borneo and Malay Peninsula 3 RAR, 2LT
1200161	PTE JACKSON Colin James	1 Sqn	2/3/67—27/2/68
1200161	LCPL JACKSON Colin James	3 Sqn	21/2/69—18/2/70
54987	CPL JACKSON John Liddell	152 Sig Sqn	2/3/67—26/2/68
54987	SGT JACKSON John Liddell	152 Sig Sqn	3/2/70—18/2/71
1200579	LCPL JACKSON Keith Lex	3 Sqn	21/2/69—18/2/70 Borneo 4 RAR, PTE
1200579	PTE JACKSON Keith Lex	2 RAR	19/5/67—9/1/68
5715865	SIG JACQUES Alvin	152 Sig Sqn	18/2/70—18/2/71
43234	CPL JANSEN Michael Redfern	152 Sig Sqn	2/3/67—27/2/68 Commonwealth Monitoring Force Rhodesia 1980
43234	CPL JANSEN Michael Redfern	152 Sig Sqn	21/2/69—18/2/70
2412089	CPL JENNISON Christopher Arthur	3 Sqn	15/6/66—19/3/67
2412089	SGT JENNISON Christopher Arthur	3 Sqn	14/2/69—20/2/70
3411814	PTE JENSEN Donald Henry	1 Sqn	20/4/67—6/2/68
3411814	CPL JENSEN Donald Henry	1 Sqn	18/2/70—18/2/71
1201250	PTE JESSER John Rawson	3 Sqn	21/2/69—18/2/70
16029	SGT JEWELL John Arthur	3 Sqn	15/6/66—19/3/67
16029	SGT JEWELL John Arthur	3 Sqn	3/2/69—20/2/70
255351	PTE JOHNS Maxwell Roy	2 Sqn	4/8/71—11/10/71
38894	LCPL JOHNSON Christopher Edward	3 and 1 Sqn	7/10/69—8/10/70
215112	PTE JOHNSON Ian Karl	2 Sqn	26/2/68—21/1/69
215112	PTE JOHNSON Ian Karl	1 RAR	27/5/65—11/6/66
3411764	PTE JOHNSTON Kenneth Norman	2 Sqn	26/2/68—7/1/69
3411764	CPL JOHNSTON Kenneth Norman	2 RAR	16/5/70—30/12/70 Awarded MM and WIA 15/12/70 shrap

217770	PTE JONES Adrian Francis	3 Sqn	21/2/69—18/2/70
1200009	CPL JONES Alan Terrence	152 Sig Sqn	26/2/68—24/2/69
3797374	PTE JONES Brian Raymond	2 Sqn	17/2/71—2/8/71
55780	2LT JONES Brian Richard Alan	2 Sqn	17/2/71—10/4/71 Killed on active service 10/4/71
6410236	PTE JONES David Bruce	3 and 2 Sqn	17/6/68—18/6/69
4718390	PTE KACZMAREK Wladyslaw	3 and 1 Sqn	9/12/69—26/11/70
4718390	PTE KACZMAREK Wladyslaw	2 RAR	22/3/67—25/3/68
38835	LCPL KELLY Brian Patrick	2 Sqn	5/2/68—6/11/68 WIA 23/10/68 Fractured Pelvis (BC)
38835	CPL KELLY Brian Patrick	1 Sqn	27/5/70—18/2/71
54344	LCPL KELLY Edward	3 Sqn	15/6/66—16/2/67
54344	SGT KELLY Edward	3 Sqn	3/2/69—2/7/69
4721188	PTE KELLY Grant Hamilton	2 Sqn	17/2/71—19/8/71
1200290	PTE KENNEDY Brian Patrick	2 Sqn	5/2/68—21/2/69
1200290	PTE KENNEDY Brian Patrick	2 Sqn	17/2/71—11/10/71
13116	SGT KENNEDY William Desmond	3 Sqn	15/6/66—25/3/67 Korea 3 RAR CPL, awarded MM with 2 RAR Malaya, CPL, awarded MBE 1971
13116	WO2 KENNEDY William Desmond	3 Sqn	3/2/69—18/2/70
13116	SGT KENNEDY William Desmond	AATTV	29/7/62—19/11/63
43727	PTE KERKEZ Milutin	3 Sqn	15/6/66—18/3/67
43727	LCPL KERKEZ Milutin	3 Sqn	21/2/69—18/2/70
43935	LCPL KERKEZ Ratomir	2 Sqn	5/2/68—7/1/69 WIA 26/4/68 shrap L arm
1200673	SIG KEYS Allan Leslie	152 Sig Sqn	13/8/67—3/9/68
1200673	SIG KEYS Allan Leslie	152 Sig Sqn	9/9/69—10/9/70
216980	CPL KIDNER Jeffery	2 Sqn	26/2/68—21/2/69
216980	SGT KIDNER Jeffery	2 Sqn	17/2/71—22/7/71
424321	PTE KILSBY Robert Graeme	2 Sqn	17/2/71—7/10/71
55201	LCPL KIRK Milton Geoffery	152 Sig Sqn	26/2/68—24/2/69
53498	SGT KIRWAN Alan John	3 Sqn	15/6/66—26/1/67 Malaya 1 RAR, PTE
53498	CAPT KIRWAN Alan John	AATTV	17/2/71—26/10/71
3103493	PTE KLUCZNIAK Richard	3 Sqn	21/2/69—20/2/70
216491	2LT KNOX James	1 Sqn	2/3/67—27/2/68
424136	LCPL KOVALEFF Fedor	2 Sqn	17/2/71—7/10/71
216666	PTE KRASSOVSKY Michael Robert	1 Sqn	2/3/67—27/2/68
216666	LCPL KRASSOVSKY Michael Robert	1 Sqn	18/2/70—18/2/71
43291	LCPL LAMPARD Kerry Charles	2 Sqn	5/2/68—28/2/69
43291	LCPL LAMPARD Kerry Charles	1 RAR	5/2/68—28/2/69
43291	PTE LAMPARD Kerry Charles	1 RAR	26/5/65—6/5/66
1200610	PTE LANSDOWN Barry John	1 Sqn	13/1/70—4/2/71
1200610	PTE LANSDOWN Barry John	1 and 6 RAR	29/4/66—14/6/67
55840	PTE LAURENSON George	1 Sqn	18/2/70—18/2/71

55287	PTE LAVERY Ronald Stanley	2 Sqn	26/2/68—24/2/69 WIA 29/3/68 Shrap L side
1736013	PTE LAWRENCE Ian Raymond	1 and 2 Sqn	28/10/70—19/8/71
1200793	2LT LAWSON-BAKER Christopher	152 Sig Sqn	19/8/70—26/8/71
3797401	PTE LEE Graham	2 Sqn	17/2/71—19/8/71
17971	CPL LEERENTVELD Henry Johannes	2 Sqn	26/2/68—7/1/69
215340	CAPT LEGGETT Craig Earl	1 Sqn	18/2/70—18/2/71
215340	LT LEGGETT Craig Earl	1 RAR	7/6/65—8/12/65 30/6/65 GSW leg
2784114	PTE LEHMAN Anthony Walter	2 and 3 Sqn	30/11/68—28/11/69
44116	SIG LEITH Jeffery Keith	152 Sig Sqn	27/1/70—18/2/71
55658	PTE LENNON Frank	1 Sqn	18/2/70—11/2/71
55658	PTE LENNON Frank	HQ 1 ALSG	
5411077	CPL LENNOX Andrew	3 Sqn	15/6/66—11/12/66
311600	CAPT LETTS Robin David	2 Sqn	10/2/71—10/10/71 1 Bn Royal Green Jackets Borneo as LT, awarded MID. 22 SAS Borneo CAPT, awarded MC. South Arabia CAPT 22 SAS. Awarded AM 1988
242902	CAPT LEVENSPIEL David	1 Sqn	2/3/67—5/3/68
2782588	CPL LEWIS David Robert	3 Sqn	2/2/69—22/2/70 UN Observer Mid East 83–84
5717036	SIG LEWIS Gwyn Lawson	152 Sig Sqn	17/2/71—26/8/71
5411575	CPL LEWIS Lyndon James	1 Sqn	18/2/70—18/2/71
54379	CPL LEWIS Thomas William	2 Sqn	26/2/68—24/2/69
519487	CPL LIDDINGTON Leslie Alphonso	3 and 1 Sqn	2/6/69—3/6/70
3411471	PTE LIEFFTING Simon	3 and 1 Sqn	12/12/66—23/12/67
3411471	CPL LIEFFTING Simon	3 Sqn	21/2/69—18/2/70
17896	PTE LIPINSKI Stanley	3 Sqn	15/6/66—25/3/67
54289	SIG LITTLE William Allen	3 Sqn	15/6/66—19/3/67
13734	WO2 LIVOCK Edward Cordwell	1 Sqn	3/2/70—18/2/71 Malaya 2 RAR
3411804	CPL LOBB Gary Harold	2 Sqn	5/2/68—21/2/69 WIA 14/7/68 Shrap R arm, L abd
5713881	CPL LOGAN Kevan Bryson	3 Sqn	21/2/69—21/2/70
16540	LCPL LOGUE Raymond John	3 Sqn	15/6/66—19/10/66 Medevac RTA
1200201	PTE LORIMER Robert John	3 Sqn	15/6/66—25/3/67
1200201	CPL LORIMER Robert John	3 Sqn	21/2/69—17/6/69 WIA 7/6/69 Shrap chest RTA
15188	CPL LOVERIDGE John James	3 Sqn	4/11/66—26/3/67
15188	SGT LOVERIDGE John James	2 RAR	12/5/70—3/12/70
54019	SGT LOWSON David Milne	152 Sig Sqn	5/2/68—23/2/69

37836	CPL LUNDBERG Eric Patrick	152 Sig Sqn	2/3/67—27/2/68
19968	LT LUNNY Kevin William	2 and 1 Sqn	13/11/67—3/6/68
19968	2LT LUNNY Kevin William	1 RAR	6/11/65—3/6/66
4410957	SIG LYNCH Robert James	152 Sig Sqn	18/12/67—13/8/68
53982	SGT LYON Malcolm Peter	152 Sig Sqn	2/3/67—25/2/68
52982	SGT LYONS Phillip	3 and 1 Sqn	15/6/66—14/6/67
52982	SGT LYONS Phillip	1 Sqn	3/2/70—18/2/71
313257	SIG MACAULLAY Douglas Robert	152 Sig Sqn	18/2/70—18/2/71
313257	SIG MACAULLAY Douglas Robert	104 Sig Sqn	18/2/70—18/2/71
3411783	PTE MacDONALD Alick James	1 Sqn	20/4/67—27/2/68
54964	PTE MacDONALD Victor Roy	1 Sqn	18/2/70—18/2/71
54964	PTE MacDONALD Victor Roy	1 RAR	26/5/65—9/6/66
43594	PTE MACH George	3 Sqn	20/10/69—21/8/70
43594	PTE MACH George	6 RAR	31/5/66—14/6/67
62059	PTE MAHER Stephen Ralph	2 Sqn	17/2/71—7/10/71
3790704	LCPL MALCOLM John	2 and 3 Sqn	10/12/68—28/11/69
217115	PTE MALONE Michael John	3 Sqn	21/2/69—18/2/70 Awarded OAM 1989
13680	LT MARSHALL Thomas George	3 Sqn	15/6/66—21/2/67 Malaya 2 RAR, CPL
311619	CPL MASON Clive	1 and 2 Sqn	28/11/70—11/10/71 Royal Marines Borneo. Rhodesia D Sqn SAS and Selous Scouts. KIA Mozambique 1977
448363	SIG MASON Wayne John	152 Sig Sqn	17/2/71—7/10/71
55696	PTE MATHEWS Jeffery Norman	2 Sqn	17/2/71—28/8/71
18302	PTE MATTEN John Wesley	3 Sqn	16/6/66—25/3/67
18302	CPL MATTEN John Wesley	3 Sqn	21/2/69—20/8/69
43705	PTE MATTHEWS Gregory Allan	3 Sqn	15/6/66—25/3/67
43705	PTE MATTHEWS Gregory Allan	3 Sqn	21/2/69—11/7/69
5411577	CPL MAVRICK Edwin Andrew	2 Sqn	26/2/68—23/7/69
5411577	CPL MAVRICK Edwin Andrew	1 Sqn	18/2/70—18/2/71
38571	PTE MAWKES Gregory	1 Sqn	20/4/67—27/2/68 Awarded MBE, 1982
38571	CPL MAWKES Gregory	1 Sqn	18/2/70—18/2/71
54085	SGT MARTIN Vernon Walter Thomas	2 Sqn	8/2/68—24/2/69
55191	PTE McALEAR Kimberley William	2 Sqn	26/2/68—24/2/69
55191	CPL McALEAR Kimberley William	2 Sqn	17/2/71—9/9/71
39520	2 LT McBRIDE Robin Alan	1 Sqn	3/2/70—18/2/71
216247	PTE McCALLUM Ross Adair	3 Sqn	15/5/66—25/3/67
216247	CPL McCALLUM Ross Adair	3 Sqn	21/2/69—18/2/70
173152	PTE McCALMAN Gregor David	2 Sqn	17/2/71—11/10/71
38318	SIG McCANN Leonard	3 Sqn	15/6/66—25/3/67
38318	CPL McCANN Leonard	152 Sig Sqn	21/2/69—19/2/70

55477	PTE McCARTHY Dennis Edward	3 Sqn	21/2/69—18/2/70
55477	LCPL McCARTHY Dennis Edward	2 Sqn	17/2/71—7/10/71
3790692	LCPL McCORMACK William James	3 Sqn	21/2/69—18/2/70
2269809	PTE McCRONE Michael William	3 Sqn	21/2/69—18/2/70
16599	PTE McCULLOUGH Norman Herbert	1 Sqn	18/2/70—18/2/71 Borneo and Malay Peninsula 4 RAR, PTE
16599	PTE McCULLOUGH Norman Herbert	2 RAR	19/5/67—12/12/67
515942	SIG McDONALD Garry William	152 Sig Sqn	29/1/71—11/10/71
214674	LCPL McDONALD Samuel James	1 Sqn	2/3/67—8/11/67
14407	CAPT McDOUGALL Peter Malcolm	SAS LO HQ AFV	6/9/66—9/3/67 Awarded AM 1989
6420	WO2 McFADZEAN James Hughes	2 Sqn	26/2/68—24/2/69 RAN WW2 Pacific, Korea 3 RAR, WIA 21/1/52, awarded MID, CPL
6420	WO2 McFADZEAN James Hughes	AATTV	21/4/65—23/3/66
312638	PTE McGAVACHIE Phillip George	1 and 2 Sqn	21/1/71—19/5/71
312924	PTE McINNES Stephen Paul	2 Sqn	17/2/71—11/10/71
42790	LCPL McKELVIE Nicol Laurie	3 Sqn	21/2/69—20/2/70 Borneo 4 RAR, PTE
55148	LCPL McKENZIE James Henry	2 Sqn	5/2/68—24/2/69
54893	CPL McKENZIE Kevan George	152 Sig Sqn	17/2/71—7/10/71
54893	LCPL McKENZIE Kevan George	527 Sig Tp	29/7/66—13/5/67
1201320	PTE McKENZIE Trevor Ernest	3 Sqn	21/2/69—16/10/69
36923	PTE McLEOD George	3 Sqn	15/6/66—18/3/67
45254	PTE McMAHON Michael Jack	2 Sqn	17/2/71—7/10/71
38023	CPL McMINN Holt Frederick Noel	1 Sqn	27/3/67—25/10/67
1200466	SIG McMULLEN Ronald Graeme	3 Sqn	15/6/66—18/3/67 WIA 17/8/66 Shrap head and R shoulder
1200466	LCPL McMULLEN Ronald Graeme	152 Sig Sqn	15/3/68—24/2/69
217583	PTE MEALIN Lawrence Archibald	3 Sqn	18/8/69—13/8/70 Killed on exercise 22/6/76
1731527	PTE MEISENHELTER Richard Frederick	2 Sqn	26/2/68—7/1/69
3792562	SIG MELLINGTON Michael Wayne	152 Sig Sqn	21/2/69—28/11/69
53416	SGT METTAM Raymond Walter	2 Sqn	26/2/68—24/2/69
1201616	LCPL MICKELBERG Raymond John	3 Sqn	21/2/69—20/2/70 Accidentally WIA 2/11/69 shrap head
1201931	PTE MILLER Donald Tillotson	1 Sqn	18/2/70—14/1/71
2244126	PTE MILLER Ewan Mosman	2 Sqn	4/8/71—11/10/71 Killed on exercise 26/2/81
2787502	PTE MITCHELL Dennis Michael John	2 and 3 Sqn	1/12/68—12/9/69
216281	PTE MOLDRE Urmas	2 Sqn	26/2/68—28/8/68
216281	CPL MOLDRE Urmas	AATTV	22/12/70—2/12/71
313002	LCPL MONTGOMERY John	2 Sqn	17/2/71—7/10/71
13968	SSGT MOONEY John Joseph	3 Sqn	25/5/66—19/3/67
44062	PTE MOORE Bernard Charles	2 Sqn	26/2/68—21/2/69

44062	LCPL MOORE Bernard Charles	2 Sqn	17/2/71—7/10/71
216584	PTE MOORE Ian Charles	3 Sqn	21/2/69—20/2/70
214677	PTE MOORE Reginald Thomas	3 and 1 Sqn	18/1/67—27/2/68
218407	PTE MOORE Wayne	3 and 1 Sqn	14/1/70—14/1/71
217902	PTE MORGAN Brian James	3 Sqn	2/9/69—14/1/70 Accidentally injured 24/12/69 loss sight in L eye
53902	SGT MOSSMAN Robert William	3 Sqn	15/6/66—12/5/67
53902	SGT MOSSMAN Robert William	5 RAR	15/6/66—12/5/67
217238	SIG MOUNSEY John Gordon	152 Sig Sqn	21/2/69—18/2/70
217238	LCPL MOUNSEY John Gordon	152 Sig Sqn	11/8/71—11/10/71
218726	SIG MOUNSEY Robert Michael	152 Sig Sqn	18/2/70—18/2/71
37015	SGT MULBY John Thomas	1 Sqn	3/2/70—18/2/71
37015	CPL MULBY John Thomas	5 RAR	12/5/66—12/5/67
55356	SIG MULHALL Allen Joseph	152 Sig Sqn	21/2/69—21/1/70
4410936	LCPL MULHOLLAND Robert John	2 Sqn	5/2/68—23/7/68
335048	MAJ MURPHY John Matthew	3 Sqn	16/5/66—25/3/67 Korea 1 RAR, Malaya as Aust Staff, CAPT Medevac
335048	MAJ MURPHY John Matthew	AATTV	27/8/63—20/9/64
16908	CPL MURPHY William Allan	1 Sqn	2/3/67—8/11/67
16908	CPL MURPHY William Allan	1 Sqn	2/3/67—8/11/67
43322	CPL MURRAY Allan Rex	2 Sqn	26/2/68—24/2/69
54324	CPL MURRELL Leslie Thomas	1 Sqn	2/3/67—2/8/67
54324	WO2 MURRELL Leslie Thomas	AATTV	29/7/70—29/7/71
42507	CPL MURTON Lynford John	3Sqn	15/6/66—25/3/67
4410613	CPL MUTCH Robert James	1 Sqn	2/3/67—27/2/68
5411659	CPL NAGLE Eric James	2 Sqn	26/2/68—21/2/69 WIA 28/6/68 GSW R elbow RTA but returned to finish tour
38359	PTE NAGLE William Lawrance	3 Sqn	15/6/66—18/3/67 Author of *Odd Angry Shot*
5410890	CPL NEIL Raymond James	3 and 1 Sqn	12/12/66—6/6/67 WIA 6/6/67 inj R knee RTA
5410890	SGT NEIL Raymond James	3 Sqn	12/2/69—20/2/70
36135	WO2 NEILSON Frederick John	2 and 1 Sqn	13/1/71—16/10/71 Borneo and Malay Peninsula 3 RAR CPL
5715722	LCPL NELSON John Francis .	2 Sqn	17/2/71—30/6/71
5715722	PTE NELSON John Francis	4 RAR	30/9/68—19/5/69 and HQ AFV
518790	PTE NISBETT William Francis	2 Sqn	17/2/71—11/10/71
14419	SGT NOLAN George Anthony	3 Sqn	15/6/66—22/3/67 Borneo and Malay Peninsula 4 RAR, SGT. Awarded MBE 1981
216494	2LT NOLAN Terrence John	2 and 3 Sqn	12/8/68—20/8/69 Awarded AM 1989
16362	LCPL NUCIFORA Angelo	Sqn	2/3/67—13/9/67 RTA illness
43518	CPL NUGENT Robert Thomas B.	2 Sqn	8/2/68—18/6/68
3179740	PTE O'BRIEN John Desmond	3 and 1 Sqn	7/10/69—8/10/70

2412371	CPL O'FARRELL Terrence	2 Sqn	26/2/68—24/2/69 WIA 1/6/68 shrap L hand WIA 17/7/68 shrap R leg awarded OAM 1988
2412371	SGT O'FARRELL Terrence	2 Sqn	17/2/71—11/10/71
36949	SGT O'KEEFE John Francis	1 Sqn	2/3/67—27/2/68
36949	SGT O'KEEFE John Francis	1 Sqn	18/2/70—18/2/71
36547	CPL O'NEILL Allan Leslie	2 Sqn	26/2/68—11/9/68 RTA illness Malaya 3 RAR, PTE
54435	PTE OPIE Raymond Maitland	3 Sqn	25/5/66—19/3/67
3411506	PTE O'REILLY John Anthony	3 Sqn	15/6/66—19/3/67 Eire Army with UN Forces in Congo 1961—62
218723	PTE ORR Donald Campbell	3 and 1 Sqn	25/11/69—8/10/70
18490	PTE O'SHEA Geoffrey	1 Sqn	2/3/67—10/10/67 Died of illness on active service 10/10/67
2184296	PTE OWEN David Lloyd	3 and 1 Sqn	13/1/70—7/10/70
55768	PTE PAGE Michael Alexander	1 and 2 Sqn	2/12/70—11/10/71
5717060	PTE PALM Barry Michael	2 Sqn	17/2/71—26/8/71
4721238	PTE PARKER Geoffrey Thomas	2 Sqn	4/8/71—11/10/71
61527	PTE PARKER John Vernon	1 Sqn	2/3/67—26/2/68
520130	SGT PARKER Anthony Norman	152 Sig Sqn	12/8/71—11/10/71
1410737	SGT PARRINGTON John Thomas	1 Sqn	2/3/67—7/7/67
1410737	SGT PARRINGTON John Thomas	3 Sqn	3/2/69—20/2/70
54402	CPL PAULIN Edward James	3 Sqn	15/6/66—17/10/67
44937	SIG PAULL Kym Malcolm	152 Sig Sqn	23/9/70—16/9/71
2411062	LCPL PAVLENKO Paul Joseph	3 Sqn	15/6/66—19/3/67 Malaya 1 RAR, PTE
2411062	WO2 PAVLENKO Paul Joseph	AATTV	27/5/70—26/5/71
3799281	PTE PEACE Gary William	2 Sqn	11/8/71—11/10/71
219441	LCPL PEACOCK Rhett	2 Sqn	17/2/71—11/10/71
55187	LCPL PEMBER Kim Stanley	2 Sqn	26/2/68—24/2/69
55187	SGT PEMBER Kim Stanley	2 Sqn	18/2/71—11/10/71
5411230	CPL PERCY Ross Thornton	1 Sqn	2/3/67—27/2/68
5411230	SGT PERCY Ross Thornton	1 Sqn	4/2/70—12/9/70
43825	CPL PERRY Reginald Robert	2 Sqn	10/2/71—1/10/71
43825	PTE PERRY Reginald Robert	8 Fd Amb	23/4/67—30/1/68
44434	PTE PETRIE John Denis	3 Sqn	21/2/69—10/12/69
43848	CPL PHILLIPS Joseph John	3 Sqn	21/2/69—18/2/70
1732372	PTE PIERCE Stanley Norbert	2 Sqn	17/2/71—11/10/71
1732372	PTE PIERCE Stanley Norbert	7 RAR	18/7/67—26/4/68
217359	PTE PILE John Daniel	3 Sqn	19/8/69—20/8/70
311554	SIG PILKINGTON Roy	152 Sig Sqn	9/9/68—3/9/69 Royal Marines 1959–65
43384	LCPL PINNINGTON Christopher Robin	152 Sig Sqn	2/3/67—27/2/68
43384	CPL PINNINGTON Christopher Robin	152 Sig Sqn	21/2/69—19/2/70
16245	PTE PLATER Stanley Edgar	3 Sqn	15/6/66—25/3/67 WIA 17/8/66 shrap back

Appendixes

213728	LT PROCOPIS David Sanford	2 Sqn	5/2/68—21/2/69
56268	PTE PROSSER Robert Francis	2 Sqn	11/8/71—16/10/71
216334	SIG PUCKERIDGE Henry Samuel John	3 Sqn	15/6/66—25/3/67
1200532	PTE PULLIN Gregory William	2 and 3 Sqn	14/10/68—15/10/69
214032	PTE PURCELL Noel Raymond Arthur	1 Sqn	2/3/67—6/2/68
55166	PTE PURSELL Gordon Alfred	1 and 2 Sqn	11/8/67—3/9/68
55166	CPL PURSELL Gordon Alfred	1 Sqn	18/2/70—18/2/71
54961	CPL PYE Dean	1 and 2 Sqn	13/1/71—11/10/71
55078	SIG QUINN Darryl Norman	152 Sig Sqn	21/2/69—19/2/70
55078	SIG QUINN Darryl Norman	103 Sig Sqn	14/8/66—6/5/67
218005	PTE RADWELL Michael John	2 Sqn	4/8/71—7/10/71
218005	PTE RADWELL Michael John	5 RAR	3/2/69—10/3/70
1201737	PTE RAITT James Gray	1 Sqn	2/9/69—3/9/70
55232	CPL RALPH Brian Edward	3 Sqn	21/2/69—20/2/70
54864	CPL RAMSAY Ian Ronald	2 Sqn	26/2/68—23/2/69 Borneo 3 RAR, PTE
54864	WO2 RAMSAY Ian Ronald	AATTV	30/6/71—5/7/72
16417	PTE RANDLE Denis Anthony	3 Sqn	15/6/66—19/3/67
16417	CPL RANDLE Denis Anthony	2 Sqn	22/11/68—7/1/69
216722	LCPL RANKIN Brian Thomas	1 Sqn	18/2/70—11/2/71
216722	SPR RANKIN Brian Thomas	1 Fd Sqn	2/3/67—9/1/68
1200797	LCPL RASMUSSEN Ian Theodore	3 Sqn	21/2/69—18/2/70
1200797	CPL RASMUSSEN Ian Theodore	2 Sqn	17/2/71—7/10/71
1200797	PTE RASMUSSEN Ian Theodore	1 and 6 RAR	11/3/66—7/3/67
56139	PTE RATCLIFFE John Allan	2 Sqn	11/8/71—11/10/71
55689	PTE REECE Dorian Thomas	2 Sqn	22/6/71—16/10/71
54006	LCPL REEVES Leslie Stanley	3 Sqn	15/6/66—9/3/67
215478	PTE REID Anthony Leon	3 Sqn	15/6/66—19/3/67 Awarded OAM 1983
215478	CPL REID Anthony Leon	1 Sqn	18/2/70—6/5/70 Medevac RTA
55380	PTE REID Denis Richard	2 Sqn	17/6/68—24/2/69
55380	CPL REID Denis Richard	2 Sqn	17/2/71—11/10/71
43939	PTE REMYNSE John	1 Sqn	18/2/70—6/8/71
6410243	PTE RICE Joseph Bernard	2 and 3 Sqn	25/11/68—6/8/69
55669	PTE RICHARDS Felix Henry Garnet	2 Sqn	17/2/71—11/10/71
3165730	SGT RICHARDS Paul Malcolm	1 and 2 Sqn	20/1/71—11/10/71
3165730	SGT RICHARDS Paul Malcolm	2 Sqn HQ 1ALSG	6/5/68—7/5/69 also D and E Pl Sgt
3179834	PTE ROBB William E.	1 and 2 Sig Sqn	23/9/70—16/2/71
217048	PTE ROBERTS Anthony Trevethan	2 Sqn	26/2/68—21/2/69
217048	CPL ROBERTS Anthony Trevethan	AFV PRO unit	27/1/70—14/1/71

217048	CPL ROBERTS Anthony Trevethan	AATTV	25/2/71—26/8/71
57085	LT ROBERTS Christopher Andrew MacDonald	3 Sqn	3/2/69—19/2/70 Awarded AM 1985
215875	PTE ROBERTS Frederick John	3 Sqn	15/6/66—25/3/67
215875	SGT ROBERTS Frederick John	3 Sqn	3/2/69—18/2/70
15827	SGT ROBERTS Malcolm Thomas	1 Sqn	2/3/67—27/2/68
15827	WO2 ROBERTS Malcolm Thomas	AATTV	5/8/70—30/6/71 WIA 4/6/71 shrap face, L arm, result amputation finger and thumb
54801	PTE ROBERTSON Richard Thomas	1 Sqn	2/3/67—6/2/68
213744	SGT ROBERTSON Thomas Angus	2 Sqn	5/2/68—24/2/69
54159	SGT ROBINSON John Murray	3 Sqn	15/6/66—23/3/67
54159	SGT ROBINSON John Murray	3 Sqn	21/2/69—20/2/70
16235	2LT RODERICK Trevor William	3 Sqn	15/6/66—16/3/67
16235	2LT RODERICK Trevor William	SAS att 1 RAR	13/8/65—2/10/65
16235	CAPT RODERICK Trevor William	4 RAR	1/5/71—18/12/71
44175	PTE RODGERS Stephen Woodrow	1 Sqn	18/2/70—18/2/71
15340	SGT ROODS William Henry	1 Sqn	2/3/67—27/2/68
38806	PTE ROSEBOROUGH John	1 Sqn	27/5/70—18/2/71
38806	PTE ROSEBOROUGH John	2 Sqn	7/7/71—7/10/71
38806	PTE ROSEBOROUGH John	3 RAR	9/5/67—14/5/68
216059	PTE ROSE John David	3 Sqn	25/5/66—18/3/67
18156	LCPL ROSER Allan William	2 Sqn	2/3/67—2/3/68
18156	SGT ROSER Allan William	1 Sqn	3/2/70—18/2/71
216452	PTE ROSS Terry John	3 Sqn	18/1/67—6/2/68
39874	PTE ROSSITER Neville Harold	1 Sqn	18/2/70—22/4/70 WIA GSW L thigh and R groin
61309	SGT RUFFIN Michael John	2 Sqn	5/2/68—24/2/69 Awarded OAM, 1985
1201137	PTE RUSKA Edward	3 Sqn	21/2/69—17/12/69
2790626	2LT RUSSELL Brian William	2 Sqn	17/2/71—7/10/71
216082	PTE RYAN Leslie Francis	2 Sqn	26/2/68—24/2/69
1201313	PTE SAMS Laurence Charles	1 Sqn	13/5/70—18/2/71
311578	PTE SAXTON Paul Richard	3 Sqn	29/7/69—20/2/70 1 Bn Gloucestershire Regt, Cyprus
54901	SGT SCHEELE David Willem	2 Sqn	26/2/68—24/2/69 WIA 17/6/68 GSW R knee and shrap L thigh, awarded DCM with AATTV
54901	SGT SCHEELE David Willem	2 and 1 Sqn	20/1/71—11/10/71
54901	SGT SCHEELE David Willem	AATTV	26/3/69—24/9/69
44292	PTE SCHNEIDER Graeme Colin	2 Sqn	24/6/68—24/2/69
214124	2LT SCHUMAN Peter John	3 Sqn	25/5/66—11/12/66
214124	CAPT SCHUMAN Peter John	4 RAR	1/5/71—9/3/71
43791	2LT SCHWARZ Brian Rex	152 Sig	26/8/69—27/8/70
43791	LCPL SCHWARZ Brian Rex	552SigTp	22/4/66—6/5/67
2784645	SGT SCHWARZE Peter Brian	2 and 3 Sqn	3/2/68—29/9/69
2784645	SGT SCHWARZE Peter Brian	3 Sqn	5/10/69—20/2/70
214621	SIG SCUTTS Phillip Robert	3 Sqn	15/6/66—19/3/67
212946	WO2 SCUTTS Sydney Raymond	2 Sqn	5/2/68—24/2/69 WIA 17/4/68 shrap chest and 25/9/68 shrap face

Appendixes

1203013	SIG SEMPLE Eric Graham	152 Sig Sqn	17/2/71—7/10/71
15530	SGT SHAW Gavin Leslie	1 Sqn	2/3/67—27/2/68
213628	SGT SHEEHAN John Neil	SAS att 1 RAR	13/8/65—5/10/65 Awarded OAM, 1980
213628	WO2 SHEEHAN John Neil	AATTV	2/2/66—7/3/66 WIA 12/2/66 shrap back and legs
215452	SGT SHEEHAN Peter Thomas	2 Sqn	26/2/68—23/2/69
216574	LCPL SHEPHERD Anthony Joseph	152 Sig Sqn	27/11/67—3/12/68
216574	PTE SHEPHERD Anthony Joseph	2 Sqn	27/11/67—3/12/68
5717087	PTE SHUARD Robert Terrance	1 Sqn	23/9/70—28/6/71
53962	LCPL SIMCOCK James Paul	1 and 2 Sqn	18/12/67—17/12/68
235360	LT SIMON Zoltan Arpad Marton	1 Sqn	3/2/70—18/2/71 WIA 31/7/70 shrap head
235306	LT SIMPSON Gordon Lyall	2 Sqn	5/2/68—31/7/68 WIA 14/7/68 frag both legs result amputation below knees
215356	LCPL SINGER Anthony Garry	2 Sqn	17/2/71—11/10/71
215356	GNR SINGER Anthony Garry	1 Fd Regt RAA	22/4/66—31/8/66 Medevac RTA
55189	PTE SLATTERY Edward	3 Sqn	22/2/69—20/2/70
55189	CPL SLATTERY Edward	2 Sqn	17/2/71—11/10/71
17951	CPL SLOCOMBE Terrance P.	1 Sqn	2/3/67—27/2/68
214166	LCPL SMITH Allan Victor	3 and 1 Sqn	27/1/67—6/2/68
214166	SGT SMITH Allan Victor	1 Sqn	3/2/70—18/2/71
3166210	PTE SMITH Barry Edward	1 Sqn	18/2/70—3/12/70
214892	CPL SMITH Barry Walter	3 Sqn	15/6/66—18/3/67
1201047	LCPL SMITH David John Montgomery	3 and 1 Sqn	22/12/69—24/12/70
1201047	CPL SMITH David John Montgomery	3 RAR	16/12/67—5/12/68
1411301	PTE SMITH Graham Douglas	2 Sqn	26/2/68—21/2/69
1411301	CPL SMITH Graham Douglas	1 and 2 Sqn	18/2/70—7/10/71
13990	SGT SMITH James Robert Alexander	3 Sqn	21/2/69—20/2/70
55334	PTE SMITH John Robert	1 Sqn	18/2/70—18/2/71
55334	PTE SMITH John Robert	3 RAR	16/12/67—6/11/68
38793	LCPL SMITH Keith Gilbert	2 Sqn	26/2/68—21/2/69
38793	LCPL SMITH Keith Gilbert	2 Sqn	17/2/71—28/4/71
216841	LCPL SMITH Kevin Laurance	2 Sqn	26/2/68—22/2/69 Awarded OAM 1986
3179747	PTE SMITH Lennard Francis	1 and 2 Sqn	28/11/70—7/10/71
3411626	PTE SMITH Michael John	1 Sqn	2/3/67—26/2/68
214775	LCPL SMITHERS Allan Nicol	1 Sqn	24/2/67—26/2/68
313052	PTE SMITHWICK Hartley Noel	2 Sqn	17/2/71—7/10/71
141191	PTE SPOLLEN Barry	2 Sqn	26/2/68—7/1/69
54453	CPL STANDEN Frederick Barry	152 Sig Sqn	2/3/67—6/12/67
54453	SGT STANDEN Frederick Barry	152 Sig Sqn	21/2/69—19/2/70

SAS: Phantoms of War

136592	PTE STANLEY Robert Edward	1 Sqn	16/4/70—15/10/70
16273	SGT STEVENSON Ola Sever	1 Sqn	14/2/67—27/2/68
16273	WO2 STEVENSON Ola Sever	AATTV	27/8/69—13/9/70 Awarded OAM 1984
54791	SGT STEWART Alan Frederick	2 Sqn	5/2/68—23/2/69 Borneo and Malay Peninsula 3 RAR, PTE
215300	PTE STEWART Allan William Kennedy	3 Sqn	15/6/66—18/3/67
215300	PTE STEWART Allan William Kennedy	2 and 3 Sqn	14/10/68—14/5/69
38202	PTE STEWART Charles David	1 Sqn	2/3/67—2/2/68
313542	PTE STEWART Douglas	2 Sqn	11/8/71—7/10/71
1410999	PTE STEWART James Cameron	3 Sqn	15/6/66—12/5/67
1410999	PTE STEWART James Cameron	5 RAR	15/6/66—12/5/67
29165	SGT STEWART James Thomas	5 RAR	26/2/68—30/9/68
1201710	PTE STEYGER Bernard Louis	2 and 1 Sqn	28/10/70—7/10/71
3411486	PTE STILES Ian Leonard William	3 Sqn	15/6/66—25/3/67 D Sqn Rhodesian SAS 1973-75
3411486	SGT STILES Ian Leonard William	3 Sqn	21/2/69—18/2/70
38983	PTE STONES Brian Robert	2 Sqn	26/2/68—21/2/69 WIA 10/10/68 shrap L ear, face and leg
38178	PTE STYNER Rolf Kurt	1 and 2 Sqn	24/10/67—8/12/68 Borneo and Malay Peninsula 3 RAR, PTE
313675	SGT SUTTON Peter Francis	152 Sig	11/8/71—11/10/71
39372	PTE SUTTON Robert Keith	3 Sqn	21/2/69—18/2/70
36940	SGT SWALLOW Thomas Raymond	1 and 2 Sqn	8/1/68—20/11/68 WIA 8/11/68 shrap chest RTA
36940	SGT SWALLOW Thomas Raymond	1 Sqn	17/2/71—11/10/71
4721267	PTE SWANSON Rodney Gordon	1 and 2 Sqn	2/12/70—17/9/71
36720	SGT SYKES Frank	1 Sqn	2/3/67—27/2/68
36720	WO2 SYKES Frank	AATTV	29/7/70—23/6/71
44982	PTE TAYLOR Neville Anthony	1 Sqn	18/2/70—18/2/71
44982	LCPL TAYLOR Neville Anthony	2 Sqn	11/8/71—7/10/71
1200778	SIG TAYLOR Robert Roy	152 Sig Sqn	26/2/68—21/2/69
1200778	LCPL TAYLOR Robert Roy	152 Sig Sqn	16/9/70—7/10/71
35803	MAJ TEAGUE Ian	1 Sqn	27/1/70—3/12/70 Malaya 1 RAR, LT
35803	CAPT TEAGUE Ian	AATTV	14/4/64—11/11/64
35803	CAPT TEAGUE Ian	HQ AAFV	3/2/65—11/12/65
55003	PTE TEBRINKE Peter Bernard	3 Sqn	15/6/66—19/3/67
37983	SIG THOMSON Ian David	3 Sqn	15/6/66—19/3/67
28662	SGT THORBURN James Lawrence	3 Sqn	15/6/66—19/3/67 Malaya 2 RAR, PTE
28662	WO2 THORBURN James Lawrence	AATTV	9/7/72—20/12/72
250334	PTE THURGAR John Duncan	1 Sqn	18/2/70—29/10/70 WIA 16/7/70 shrap R arm, awarded Star of Courage whilst serving with AFP in Cyprus 1980 also RAN service

568

312754	PTE TIERNAN Barry John	2 Sqn	17/2/71—7/10/71
55664	PTE TIERNEY Brian William	1 Sqn	18/2/70—14/1/70
5411644	PTE TONKIN Kevin Dennis	2 Sqn	26/2/68—24/2/69
5411644	CPL TONKIN Kevin Dennis	2 Sqn	17/2/71—11/10/71
15457	SGT TONNA Emmanuel Lorenzo Antonio	3 Sqn	15/6/66—19/3/67
15457	SGT TONNA Emmanuel Lorenzo Antonio	3 Sqn	21/2/69—20/2/70
3112331	PTE TREWIN Gordon Douglas	3 Sqn	29/7/69—18/2/70 WIA 17/9/69 shrap L heel
3411909	Pte TULIP Brian James	3 and 1 Sqn	9/12/69—10/12/70
3411909	Pte TULIP Brian James	7 RAR	17/4/67—26/4/680
55113	LCPL TWYNAM-PERKINS Allan Jones	152 Sig Sqn	21/2/69—8/10/70
55113	SIG TWYNAM-PERKINS Allan Jones	709 Sig Sqn	17/7/66—27/4/67
21321	SGT URQUHART Ashley Graham	3 Sqn	15/6/66—23/3/67
21321	WO2 URQUHART Ashley Graham	AATTV	5/2/68—29/10/69
21321	WO2 URQUHART Ashley Graham	AATTV	6/1/71—5/3/71
312574	PTE VALE Gerry David	1 Sqn	18/2/70—18/2/71 Awarded OAM 1986
3411630	PTE VAN AS Pieter John	1 Sqn	2/3/67—28/11/67
4718068	SGT VAN DROFFELAAR Michael Jan	3 Sqn	21/2/69—18/3/70 WIA 3/11/69 shrap face
4718068	LCPL VAN DROFFELAAR Michael Jan	2 RAR	9/5/67—9/1/68
18179	PTE VAN ELDIK Johannes Edward	1 Sqn	2/3/67—27/2/68
18179	CPL VAN ELDIK Johannes Edward	1 Sqn	18/2/70—18/2/71
38194	PTE VAN NUS Martin Phillip Munro	1 Sqn	2/3/67—6/2/68
38194	PTE VAN NUS Martin Phillip Munro	32 Sml Ships	10/10/69—20/12/69
2184244	PTE VARLEY Peter Andrew	2 and 1 Sqn	29/7/70—29/7/71
57044	MAJ WADE Brian	2 Sqn	5/2/68—23/2/69 Malaya, CAPT att 1/3 East Anglian Regt. Awarded AM 1988
57044	CAPT WADE Brian	AATTV	29/7/62—6/9/63
43005	LCPL WALKER Leonard Albert	1 Sqn	2/3/67—25/10/67
167295	PTE WALKER Stanley James	3 Sqn	21/2/69—18/2/70
16270	CPL WALLACE Darryl William	2 Sqn	26/2/68—19/11/69 Borneo and Malay Peninsula 3 RAR, PTE
1411308	PTE WALLIS Rodney Errol	2 and 3 Sqn	22/11/68—3/9/69
214436	CPL WALSH Leo William	3 Sqn 6 RAR	15/6/66—14/6/67 RAN Korea and Malaya
214436	SGT WALSH Leo William	6 RAR	7/5/69—16/5/70
216270	PTE WARD John Patrick Joseph Francis	3 and 1 Sqn	18/1/67—6/2/68
216270	CPL WARD John Patrick Joseph Francis	2 Sqn	18/2/70—21/1/71
216953	PTE WARING Leslie	2 Sqn	17/2/71—11/10/71
216953	LCPL WARING Leslie	103 Sig Sqn	16/4/67—8/5/68
55353	PTE WARREN Ernest Jack	2 and 3 Sqn	5/8/68—6/8/69

53132	WO2 WATERS Malcolm Burgess	3 Sqn	3/2/69—20/2/70
2412391	PTE WATSON Terance Rex	2 Sqn	17/2/71—11/10/71
2412391	PTE WATSON Terance Rex	2 RAR	16/12/67—15/11/68
29836	CPL WEARING Henry Robert	3 Sqn	3/2/69—28/11/69
29836	CPL WEARING Henry Robert	1 RAR	26/5/65—30/11/65
254931	PTE WEARNE Graham Richard	2 and 1 Sqn	7/10/70—19/8/71
16949	LCPL WENITONG Leonard	1 Sqn	2/3/67—20/12/67
5411024	CPL WHEATLEY Eric Gordon	152 Sig	2/3/67—13/12/67
54222	SGT WHITBREAD William J.	1 Sqn	2/3/67—27/9/67
27669	SGT WHITE Peter Francis	2 Sqn	26/2/68—10/9/68 Malaya 105 and 100(A) Bty, GNR
5410368	SGT WHITAKER Leslie Ronald	3 Sqn	15/6/66—19/3/67
215409	PTE WHITESIDE Harry Thomas	1 Sqn	20/4/67—20/8/68 Borneo and
215409	LCPl WHITESIDE Harry Thomas	1 and 7 RAR	Malay Peninsula 3 RAR, PTE
36387	SGT WIGG John William	3 Sqn	15/6/66—31/8/66 RN Korea and Malaya WIA 3/3/69 mine incident
36387	WO2 WIGG John William	AATTV	20/1/69—21/1/70
55410	LCPL WILD Richard Eugene	152 Sig Sqn	18/2/70—18/2/71
44040	SIG WILLIAMS Allan Stewart	3 Sqn	15/6/66—23/3/67
44040	WO2 WILLIAMS Allan Stewart	AATTV	28/10/69—29/10/70
55399	PTE WILLIAMS Barry James	3 Sqn	21/2/69—20/2/70
218751	PTE WILLIAMS Jeffery James	2 Sqn	17/2/71—7/10/71
44703	PTE WILLSHIRE James Brenton	1 Sqn	18/2/70—18/2/71 awarded OAM 1984
44703	LCPl WILLSHIRE James Brenton	2 Sqn	7/7/71—7/10/71
43838	PTE WILSON Donald Sam	3 Sqn	15/6/66—25/3/67 Malaya 1 & 3 RAR, PTE
3797676	PTE WILSON Gordon Bruce	2 Sqn	17/2/71—26/8/71
622147	PTE WINES Peter James	1 Sqn	5/2/70—6/2/71
37903	CAPT WISCHUSEN Ian Edward Kenneth	2 Sqn	26/2/68—4/3/69
37903	CAPT WISCHUSEN Ian Edward Kenneth	AATTV	15/10/70—20/9/71
216844	PTE WOOD John Hilton	2 Sqn	26/2/68—24/2/69
216844	LCPL WOOD John Hilton	2 Sqn	10/2/71—11/10/71
215806	PTE WOODHOUSE James William	3 Sqn	15/6/66—18/3/67
215806	SGT WOODHOUSE James William	1 Sqn	5/2/70—18/2/71
217734	PTE WOOLDRIDGE Archibald Edgar	3 Sqn	29/7/69—18/2/70
54045	CPL WREN Colin Cuthbert	1 and 2 Sqn	13/11/67—3/12/68
54045	PTE WREN Colin Cuthbert	1 RAR	26/5/65—3/6/66
5270	WO2 WRIGHT Allan Kenneth	3 Sqn	15/6/66—23/3/67 WW2 Pacific Islands, 2/11 Bn AIF and Z Force, Japan with BCOF. Awarded MBE, 1966
55362	CPL WRIGHT Daniel Henry	2 Sqn	26/2/68—4/3/69
55362	SGT WRIGHT Daniel Henry	2 Sqn	17/2/71—11/10/71
55362	SGT WRIGHT Daniel Henry	AATTV	29/4/69—22/7/69
55873	PTE ZONIA Segio	1 Sqn	29/7/70—18/2/71
36724	CPL ZOTTI Barry	152 Sig Sqn	2/3/67—8/11/67

Appendix C: Awards and decorations to members of the SAS Regiment for service in East Timor—September 1999– February 2000

Distinguished Service Cross (DSC)
Lieutenant Colonel T. J. McOwan, CSM

Medal for Gallantry (MG)
Sergeant S. Oddy

Distinguished Service Medal (DSM)
Major J. F. McMahon
Captain J. G. Hawkins

Medal of the Order of Australia (OAM)
Warrant Officer Class Two R. T. McCabe

Commendation for Gallantry
Trooper B. Burgess

Commendation for Distinguished Service
Warrant Officer Class Two W. P. Maher
Corporal A. W. Cameron

Chief of Defence Force Commendation
Major J. B. R. Phillips
Captain C. R. Evans
Sergeant L. Kouts
Corporal P. M. Beresford
Corporal K. G. Fennell

Commander INTERFET/DJFHQ Commendation
Captain A. G. Williams
Sergeant N. G. Coenan
Sergeant V. J. Hartley
Sergeant P. O. McKenzie
Corporal S. A. Stennet

Commander Australian Theatre Commendation
Major G. Walsh
Captain C. D. Johns
Warrant Officer Class Two P. Whitely
Corporal T. J. Ridley

Appendix D: SAS Roll of Honour

The following members died while on operations or in training while serving with the SAS Regiment.

PRIVATE ANTHONY FRANK SMITH. Drowned in the Avon River, 5 August 1960, while attempting a river crossing.

LANCE CORPORAL PAUL HAROLD DENEHEY. Gored by an elephant, Borneo, June 1965.

LIEUTENANT KENNETH AMBROSE HUDSON. Presumed drowned while attempting a river crossing, Borneo, 21 March 1966.

PRIVATE ROBERT CHARLES MONCRIEFF. Presumed drowned while attempting a river crossing, Borneo, 21 March 1966.

PRIVATE RUSSEL JAMES COPEMAN. Died of wounds in Australia, 10 April 1967, several months after evacuation from South Vietnam.

PRIVATE GEOFFREY O'SHEA. Died from encephalitis, South Vietnam, 10 October 1967.

SERGEANT GEORGE TERENCE BAINES. Killed in a grenade accident, South Vietnam, 13 February 1968.

PRIVATE TREVOR WILLIAM IRWIN. Drowned in the Collie River, 14 June 1968, while attempting a river crossing.

CORPORAL RONALD ARTHUR HARRIS. Accidentally shot, South Vietnam, 17 January 1969.

LIEUTENANT CHARLES GEORGE EILER. Fatally injured at Pearce, Western Australia, as a result of a parachute accident, 12 August 1969.

SERGEANT JOHN GRAFTON. Fatally injured at Pearce, Western Australia, as a result of a parachute accident, 12 August 1969.

PRIVATE DAVID JOHN ELKINGTON FISHER. Missing, presumed dead, after falling from an extraction rope in South Vietnam, 27 September 1969.

SECOND LIEUTENANT BRIAN RICHARD ALAN JONES. Accidentally shot, South Vietnam, 10 April 1971.

SERGEANT HARRY GREEN. Killed in a vehicle accident while on exercise at Carnarvon, Western Australia, 10 October 1972.

CORPORAL DOUGLAS ROBERT ABBOTT. Killed in a parachuting accident, Western Australia, 4 March 1976.

CORPORAL LAWRENCE ARCHIBALD MEALIN. Killed in a training accident, Fremantle, 22 June 1976.

LANCE CORPORAL PETER CLIFFORD WILLIAMSON. Accidentally shot while training, Swanbourne, 10 October 1980.

SERGEANT MURRAY TONKIN. Killed in an aircraft crash, Philippines, 26 February 1981.

SERGEANT EWAN MOSMAN MILLER. Killed in an aircraft crash, Philippines, 26 February 1981.

SIGNALLER GREGORY ROLF FRY. Killed in an aircraft crash, Philippines, 26 February 1981.

TROOPER DAVID HUGH O'CALLAGHAN. Killed while diving in Bass Strait, 16 April 1982.

LANCE CORPORAL PETER WILLIAM RAWLINGS. Killed in a parachuting accident, Perth, 24 July 1982.

CORPORAL STEPHEN J. DALEY. Killed in an aircraft accident near Nowra, New South Wales, 12 November 1991.

SERGEANT PAUL RICHARD KENCH. Killed while diving in Bass Strait, 9 December 1992.

CORPORAL GORDON RANK HOLLAND. Died of natural causes while training, Swanbourne, 21 May 1993.

SIGNALLER LACHLAN ROBERT MARTIN. Killed in a parachuting accident, Bindoon, Western Australia, 14 May 1996.

CORPORAL MIHRAN AVEDISSIAN. Killed in Black Hawk helicopter crash, 12 June 1996.

TROOPER MICHAEL JOHN BIRD. Killed in Black Hawk helicopter crash, 12 June 1996.

TROOPER GORDON ANDREW CALLOW. Killed in Black Hawk helicopter crash, 12 June 1996.

TROOPER JONATHON GALUS SANDFORD CHURCH. Killed in Black Hawk helicopter crash, 12 June 1996.

CORPORAL ANDREW CONSTANTINDIS. Killed in Black Hawk helicopter crash, 12 June 1996.

SERGEANT HUGH WILLIAM ELLIS. Killed in Black Hawk helicopter crash, 12 June 1996.

TROOPER DAVID FROST. Killed in Black Hawk helicopter crash, 12 June 1996.

TROOPER GLEN DONALD HAGAN. Killed in Black Hawk helicopter crash, 12 June 1996.

LANCE CORPORAL DAVID ANDREW JOHNSTONE. Killed in Black Hawk helicopter crash, 12 June 1996.

TROOPER TIMOTHY JOHN MCDONALD. Killed in Black Hawk helicopter crash, 12 June 1996.

LANCE CORPORAL DARREN ROBERT OLDHAM. Killed in Black Hawk helicopter crash, 12 June 1996.

SIGNALLER HENDRICK PEETERS. Killed in Black Hawk helicopter crash, 12 June 1996.

CORPORAL DARREN JOHN SMITH. Killed in Black Hawk helicopter crash, 12 June 1996.

CAPTAIN TIMOTHY JAMES STEVENS. Killed in Black Hawk helicopter crash, 12 June 1996.

CORPORAL BRETT STEPHEN TOMBS. Killed in Black Hawk helicopter crash, 12 June 1996.

Index

Please note: page numbers in *italics* are photographs

574

Index

Index

Index

Index

Quang Giao, 342, 345
Queenscliff, 139
Quodling, Capt, R.W., 433

radio communications, 103, 187–8, 225, 244–5, 269, 280, 283
Raitt, Tpr J.G., 349
Randwick (NSW), 424
Raya Ridge, 167; map 3
recondo course, 51–2, 210, 302
Regional Force Surveillance Units, 416–17
Reid, LCpl A., 459
Rendezvous (magazine), 481
Richards, Sgt P.M., 311, 384, *384*, 447, 448
Rickerby, Maj B., 475
Ridley, Tpr T., 463
Ring, R., *44*
Roberts, Col C.A.M., xiii
    commands 1 Sqn, 411
    CO SASR, 438–41
    CO Special Forces, 452
    photograph, *439*
    troop commander 3 Sqn, 300, 302
    3 Sqn patrols, 319–20, 327–30, 332
Roberts, Sgt F.J.
    photograph, *341*
    3 Sqn first tour, 205
    3 Sqn second tour, 332, 334, 339–42, 346
Robertson, Capt D.G., *68*
Robertson, Fl Lt W.N., 351
Robinson, Sgt J.M.
    ambush patrol, 132–3
    attack on Lumbis, 125
    description of, 106
    photograph, *314*
    post Vietnam, 45
    recon patrol, 106–8, 122
    3 Sqn first tour, 194, 201, 204
    3 Sqn second tour, 300, 320–3, 328, 337, 342
Robottom, Maj P.A., 436–7, 440
Rock, the, 438, 441, *442*
Roderick, Capt T.W.
    ambush patrol, 131–3
    hearts and minds patrols, 89–92
    photograph, *173*
    recon patrols Borneo, 106, 124, 129–30
    troop commander 1 Sqn, 85

3 Sqn first tour, 172, 192–3, 197, 320, 388
Rogerson, Cpt M., 511, *512*
Roods, Sgt W.H., 162, 215
Roser, Sgt A.W., *212*, 225–6, 349
Ross, LCpl T.J., *212*
Rossiter, Tpr N.H., 350
Rottnest Island, 35, 39
Rowell, Lt Gen Sir Sydney, 28–30
Royal Australian Air Force (RAAF)
    5th Aviation Regiment, 467–8, 469, 485, 515
    No 9 Squadron
        begins insertions, 189
        co-operation with SAS, 206, 220–1, 243, 389
        deployed to Vietnam, 171, 178
        early reluctance, 182
        extraction techniques, 242–3
        Fisher falls off rope, 336
        gunship attacks, 367–8
        helicopter crash, 338–9
        increased activity, 260–1
        insertion techniques, 241–2
        operational problems, 230
        photographs, *284, 330*
        regard for Simpson, 265
        rope extractions, 266–7
        tractor ambush, 2, 15, 17
        training with SAS, 211
        use restricted, 252, 255, 279, 318
        1 Sqn first tour extractions, 218, 222, 224, 226
        1 Sqn second tour extractions, 351–2, 354, 366
        2 Sqn first tour extractions, 256, 259, 261–2, 274, 288
        2 Sqn first tour insertions, 256, 280
        3 Sqn first tour extractions, 196, 202
        3 Sqn second tour, 320, 322, 324, 326, 330, 341
Royal Australian Navy
    Clearance Diving Team, 431, 434
    coastwatchers, 20–1
Royal Navy, Royal Marines
    40 Commando, 69
    42 Commando, 56, 76, 157
    SBS, *see* separate entry
RPKAD, 145, 152, 403

589

Webb, Sgt M., 411, 425
Weir, WO2 R.L.
  AATTV, 57
  ambush patrol, 132–3
  career, 110
  elephant incident, 111–21
  hearts and minds, 89, 92
  joins SAS, 42
  LRRP Wing, 267–8
  photograph, 267
  recon patrol, 106, 108–9, 124
  troop sgt 1 Sqn, 86
  survival training, 397
Weir, Brig S.P., 333–4, 343, 346, 350, 354
Wells, Lt Gen Sir Henry, 30–2
West Australian newspaper, 430
Western Command, 50, 139–41, 143, 171, 210–11, 227–8, 301, 308, 393
Western Sahara, 455, 456
Westmoreland, General William, 184–5, 237–8
Whitaker, Sgt L.R., 200
White, Tpr J., 76–7
White, Pte M.R., 124
White, Sgt P.F.
  Sarawak, 145, 152, 155, 162
  2 Squadron first tour, 253, 261, 266
White, Cpl S., 463
Whitlam government, 398
Wigg, WO2 J.W., 182–3, 318
Williams, Cpl (RAAF), 190
Williams, Pte B.J., 317, 326, 345
Williams, Capt G.E., 68
Williams, Sgt M., 469
Williamson, LCpl P.C., 432
Williamtown, 31, 34–6, 41–2, 46, 54, 210
Willis, Sgt T., 456
Wills, Brig Sir Kenneth, 27

Wilson, Pte D.S., 185–6
Wilson, Harold, 79
Wilson's Promontory, 22
Wilton, Gen Sir John, 31, 170, 172, 235
Wingate, Brig Orde, 24
Wingate Gray, Lt Col M., 83, 145
Wischusen, Capt I.E.K., 252
Wood, Sgt M., 458
Woodhouse, Lt Col J.N., 64–5, 267
Woods, Sir Colin, 430
Woods, Lt W.G., 38, 40
Woolagoodja, S., 397
Wright, WO1 A.K., 174, 175, 200, 296, 297
Wright, WO2 D.H.
  photographs, 5, 295, 378
  on SAS, 518
  tractor ambush, 1, 4–8, 11–14
  2 Sqn first tour, 256, 268
  2 Sqn second tour, 311, 373–6, 377
Wright, Lt J.R. (USN), 411

Xuyen Loc, 323; map 5
Xuyen Moc, map 5
  1 Sqn first tour, 221, 229
  1 Sqn second tour, 349–50, 353–4, 356, 359, 362, 368
  2 Sqn first tour, 253, 262, 268, 274, 280
  2 Sqn second tour, 372–3
  3 Sqn first tour, 176
  3 Sqn second tour, 343, 347

Yampi Sound, 404
Young, WO2 B.L., 154–5, 161–5, 308, 448
Young, Tpr S., 450

Z Special Unit, 26, 38